T0341464

Fortran 2018 with Parallel Programming

Fortran 2018 with Parallel Programming

Subrata Ray

CRC Press
Taylor & Francis Group
Boca Raton London New York

CRC Press is an imprint of the
Taylor & Francis Group, an **Informa** business

A CHAPMAN & HALL BOOK

CRC Press
Taylor & Francis Group
6000 Broken Sound Parkway NW, Suite 300
Boca Raton, FL 33487-2742

© 2020 by Taylor & Francis Group, LLC
CRC Press is an imprint of Taylor & Francis Group, an Informa business

No claim to original U.S. Government works

Printed on acid-free paper

International Standard Book Number-13: 978-0-367-21843-0 (Hardback)

Library of Congress Control Number: 2019946209

Visit the Taylor & Francis Web site at
http://www.taylorandfrancis.com

and the CRC Press Web site at
http://www.crcpress.com

This work is dedicated to the memory of

Professor Chanchal Kumar Majumdar

and

Professor Suprokash Mukherjee

Contents

Preface

Since the early days of machine computing, there has been a constant demand for *larger* and *faster* machines. The two terms essentially mean machines with larger memory and more speed than that of the existing available machines. During the past 70 years, there have been dramatic changes in the fields of computer hardware and software—from vacuum tubes to VLSI (very large scale integration) and from no operating system to very sophisticated, time-sharing operating systems. There are three obstacles that computer designers face while aiming to increase the speed of the computer. First, the density of the active components within a chip cannot be increased arbitrarily. Second, with the increase of the density of the active components within VLSI chips, heat dissipation becomes a severe problem. Third, the speed of a signal cannot exceed the speed of light according to the special theory of relativity proposed by Einstein. Thus, a different approach to the problem has been thought of.

Instead of having a single processor, if several processors (each may not be very fast and can be inexpensive) participate in parallel for computation, the speed of calculation can be increased considerably, and in fact, using inexpensive processors controlled by special software and hardware, the speed of a supercomputer can be achieved if hundreds of processors work together in parallel.

This book contains an introduction to parallel computing using Fortran. Fortran supports three types of parallel modes of computation: Coarray, OpenMP and Message Passing Interface (MPI). All three modes of parallel computation have been discussed in this book. In addition, the first part of the book contains a discussion on the current standard of Fortran, namely, Fortran 2018.

The first part of the book can be used to learn the modern Fortran language even if the reader has not yet been exposed to the earlier versions of Fortran. The book should be read sequentially from the beginning. However, a reader who is conversant with the earlier versions of Fortran may skip the introduction to Fortran and go directly to the new features of the language.

As Fortran is mainly used to solve problems related to science and engineering, standard numerical methods have been used as a vehicle to illustrate the application of the language. However, knowledge beyond the level of elementary calculus is not required to understand the numerical examples given in the book. The emphasis of the book is on programming language, not on sophisticated numerical methods. The programming examples given in the book are simple, and to keep the code readable, the codes are not always optimized. It is expected that a reader, after proper understanding of the language, would be able to write *much more efficient* codes than the codes given in the book.

Programming tips and style have been introduced at appropriate places. They serve simply as guidelines. It is well known that every experienced programmer has his or her own programming style.

To keep the size of the book reasonable, all available features of Fortran 2018 have not been discussed. Moreover, only the essential components of Coarray, OpenMP and MPI, which are required to write reasonably useful programs, have been discussed. It is hoped that readers, after going through this book, will refer to relevant manuals and be able to write parallel programs in Fortran to solve their numerical problems.

The book is full of examples. Most of the examples have been tested with Intel Corporation's Fortran compiler, ifort, version 19.3, GCC gfortran version 7.3.0 and the Fortran compiler version 6.2 of Numerical Algorithm Group (NAG). The Fortran part is based on the draft Fortran 2018 report published on July 6, 2017. At the time of writing, these compilers do not support all the proposed features of Fortran 2018, but Intel, Free Software Foundation, Inc., and NAG will add further support for these features over time.

Subrata Ray

Acknowledgments

The author wishes to record his deep sense of gratitude to his colleagues, friends and associates who helped him to prepare this manuscript during the various phases of this work.

Abhijit Kumar Das
Ananda Deb Mukherjee
Ankush Bhattacharjee
Ashish Dutta
Biplab Sarkar
Debasis Sengupta
Gayatri Pal
Indrajit Basu
Indrani Bose
Koushik Ray
Minakshi Ghosh
Prasanta Kumar Mukherjee
Pushan Majumdar
Ramaprasad Dey
Ranjit Roy Chowdhury
Robert Dyson
Sandip Ghosh
Sankar Chakravorti
Santosh Kumar Samaddar
Sarbani Saha
Satrajit Adhikari
Satyabrata Roy
Siddhartha Chaudhuri
Souvik Mondal
Swapan Bhattacharjee
Utpal Chattopadhyay

The Numerical Algorithms Group Ltd., Oxford, UK, provided the author with a free license to use their Fortran compiler. Intel Corporation allowed the author to use the trial version of their Fortran compiler ifort. The free GCC gfortran compiler was also used. National Council of Education, Bengal, and Institute of Business Management, NCE, Bengal, allowed the use of their computer laboratory during the preparation of this book.

Finally, the author wishes to thank his family for their encouragement during the preparation of this manuscript.

Author

Dr. Subrata Ray is a retired senior professor of the Indian Association for the Cultivation of Science, Kolkata. In his career spanning over 40 years, he has taught computer software in universities, research institutes, colleges and professional bodies across the country. As a person in charge, he has set up several computer centers, in many universities and research institutes, almost from scratch. Though his field of specialization is scientific computing, he has participated in developing many systems and commercial software.

He has an MSc, a Post MSc (Saha Institute of Nuclear Physics) and PhD from Calcutta University and has served several renowned institutes like Tata Institute of Fundamental Research, Indian Institute of Technology, Kharagpur, Regional Computer Centre, Calcutta, University of Burdwan and Indian Association for the Cultivation of Science.

He is associated with the voluntary blood donation movement of the country and is an active member of Association of Voluntary Blood Donors, West Bengal. He offers his voluntary services to Eye Care & Research Centre and National Council of Education, Bengal.

He is also an amateur photographer.

He is married to Sanghamitra, and they have a daughter, Sumitra. He lives in Kolkata with his brother Debabrata and sister Uma.

1

Preliminaries

A computer is a machine that can perform basic arithmetic operations at a very high speed. It can take logical decisions and select alternative paths depending upon the program logic. It can communicate with the external world via its input/output devices. To get a job done, the computer needs to be instructed through a computer program. A computer program is a set of instructions through which one instructs a computer to perform a specific job. The computer's processor understands only one language, called the machine language. The machine language is machine dependent and is difficult for humans to learn. To circumvent this difficulty, several artificial languages (sometimes called high-level languages) have been developed. These artificial languages are very easy to learn and are practically machine independent. However, this requires translation to the machine language of the particular machine. The translation is done by the computer itself through a system program called a compiler. The compiler, while translating, checks the grammar of the language; if the source program is free from grammatical errors, it generates the machine language version of the *source program*, called the *object program* for the machine, which is subsequently linked (using a system program called the linker) to various libraries of the system. The resultant code, called the *executable code*, is executed by the machine. As different machines use different machine languages, the compilers are naturally machine dependent. Therefore, that a particular machine can *execute* a program written in a high-level language implies that the compiler for that high-level language is available to the computer system.

Fortran—one such programming language—is the abbreviation of **For**mula **trans**lation. It is widely used in solving scientific and engineering problems that require a lot of numerical computation. In this book, Fortran stands for Fortran 2018, the current version of Fortran.

It must be mentioned at this point that no computer can directly execute any program written in Fortran or any other, so-called high-level, language like Fortran. The compiler for the corresponding language must be available to the computer so that the translated version of the program written in a high-level language may be executed by the computer. As this translation—Fortran to machine language of a particular machine—is transparent to the programmer, one may assume that the computer is executing the Fortran program.

The compiler generates an object program only when the source is free from grammatical errors. In case of any grammatical error being flagged by the compiler, the programmer has to go back to the source, make the necessary correction(s) to the source and recompile the source to get the object program. The object program will not be generated until all grammatical errors are removed from the source program.

A program, free from grammatical errors, may not give a correct result. The program must be free from *logical errors*. A logical error is an error in the program logic at the source level. For example, a particular program requires addition of two numbers, but the programmer, by mistake, has performed multiplication instead of addition. A *runtime error* may occur during the execution of the program. Suppose a program has to divide two numbers. The division process is valid so long as the second number (denominator) is not zero. Division by zero is not a valid arithmetic operation. This error will show up during the execution of the program should the denominator become zero. The program behaves normally so long as the denominator remains nonzero.

Therefore, to obtain a correct result from a program, the following three conditions must be satisfied:

- The program must be free from grammatical errors.
- The program must be free from logical errors.
- There should not be any runtime error.

In this book, we frequently use the term processor. According to the Fortran report, a processor is a combined object, consisting of software (compiler, operating system, etc.) and hardware, that converts a Fortran program into its machine language equivalent and executes the same.

1.1 Character Set

The programming language and its syntax are described by a set of characters. The character set that is available to a Fortran programmer consists of (a) all letters of the English alphabet, both uppercase (A–Z) and lowercase (a–z); (b) underscore character (_); (c) all digits (0–9); (d) several special characters, such as brackets, colon and full stop; and (e) several unprintable characters, such as tab, linefeed and newline characters.

Some characters may appear only within comments, character constants, input/output records and edit descriptors. The English letters, numerals and the underscore character are collectively called alphanumeric characters.

Normally, Fortran is case insensitive; that is, it does not distinguish between the upper-case and lowercase letters. There are, however, exceptions (character strings and input/output). In addition, specifiers like file names in open and inquire statements (Chapter 9) may make it necessary to distinguish between lowercase and uppercase letters. This is processor dependent and will be discussed at appropriate places in the text. Table 1.1 shows the list of special characters.

In this book, we consider only ASCII set of characters. ASCII, abbreviation of American Standard Code for Information Interchange, is an industry standard for electronic communication. The processor may also support other types of character sets.

TABLE 1.1
Lists of Special Characters

Character	Name	Character	Name
	Blank	;	Semicolon
=	Assignment (equal)	!	Exclamation sign
+	Plus	"	Quote
−	Minus	%	Percent
*	Asterisk	&	Ampersand
/	Slash	~	Tilde
\	Back slash	<	Less than
(	Left parenthesis	>	Greater than
)	Right parenthesis	?	Question mark
[	Left square bracket	'	Apostrophe
]	Right square bracket	`	Grave accent
{	Left curly bracket	^	Circumflex accent
}	Right curly bracket	|	Vertical line
,	Comma	$	Currency sign
.	Decimal point	#	Number sign
:	Colon	@	Commercial at

1.2 Identifiers

Identifiers are used to specify various objects permitted by the language. An identifier (a) must start with a letter of the English alphabet, (b) may contain other digits and letters or the underscore character, (c) must not contain any special characters or blanks and (d) must have a length not exceeding 63 characters. It is obvious that the identifier must comprise at least one character, and in that case, it must be a letter of the English alphabet only. The first character cannot be an underscore character. However, the last character can be an underscore.

The following are valid identifiers:

```
xmax, counter, basic_pay, i, volt, name__of_a_person
```

(Note two successive underscore characters after name.)

The following are invalid identifiers:

```
1xy (starts with a digit)
x)y (contains a special character)
x min (contains a blank character)
```

The identifiers ABC, Abc, aBC and abC, or any combination of uppercase and lowercase A, B and C, are equivalent, as Fortran normally does not distinguish between the uppercase and the lowercase letters.

Usually, identifiers are so chosen that they have some relation with the actual objects they refer to. For example, the identifier volt is a natural choice for denoting the voltage of an electrical circuit. One can as well choose z43p8 to represent voltage; however, it appears to be a poor choice because the readability of the program is diminished once such a choice is made. Needless to say, the length of the identifier must be of reasonable size. Though the language permits 63 characters to represent an identifier, rarely more than 8 or 10 characters are used to represent an identifier. Unnecessarily long identifiers will invite typing errors and perhaps reduce the readability of the program.

Several characters like (2 and Z), (1 and I) and (O and 0) look similar. Therefore, care should be taken while using similar characters within the same identifier. For example, identifiers like O0 (oh zero) should be avoided. One may invite further trouble if one chooses another identifier OO (oh oh) in the same program unit. It must be understood that, for the Fortran compiler, both O0 (oh zero) and OO (oh oh) are valid but different identifiers—it is the human programmer who may mix up these two similar-looking identifiers.

1.3 Intrinsic Data Types

There are five types of intrinsic data in Fortran. They are integer, real, complex, logical and character. The integer, real and complex types are used for numeric computation; logical and character are nonnumeric data types. In addition, there are extended precision real data types, known as double precision and quadruple precision (not a standard Fortran—Intel's extension) and double precision complex. All intrinsic data types are associated with a kind parameter. The kind parameter may be explicit, or if the kind parameter is not present, the intrinsic data assume the default kind parameter. The intrinsic type character is usually associated with a len parameter, which determines the length of the character string. The kind parameter is discussed in Chapter 5.

1.4 Constants and Variables

In any programming language, we normally use two types of objects—constants and variables.

A quantity whose value remains fixed during the execution of a program is called a constant. The compiler identifies the constant—its type and value—from its appearance. Constants do not have any name associated with them. In other words, a constant conveys both its type and its value to the compiler. On the other hand, a variable may change its value during the lifetime of a program. A variable must have a name attached to it. It may also remain undefined during the lifetime of a program.

As there are five types of intrinsic data (plus two extended types) in Fortran, there are five types (plus two extended types) of intrinsic constants and variables in Fortran.

1.5 Integer Constants

An integer constant is a whole number; that is, it does not contain any decimal point. It contains only digits and a leading sign, if necessary. An integer constant may be positive, negative or zero. Zero is neither a positive nor a negative number. In addition, the numerical values of 0, +0 and -0 are same. Negative constants are prefixed by a minus sign, and positive constants may optionally be prefixed by a plus sign. Unsigned integer constants are assumed to be positive. For instance, constants 10 and +10 are equivalent. Normally, a leading positive sign is not used, as it is optional. The integer constant, say, 127, tells the compiler its type; in this case it is an integer, and its magnitude is 127. The maximum and minimum values that an integer constant may assume are processor dependent. The typical values are 2147483647 (maximum: 2 to the power 31 minus 1) and –2147483648 (minimum: minus 2 to the power 31). By default, integer constants are treated as decimal numbers (base 10). Valid integer constants are 2, 35, –7432, 12345, 0 and –4321.

Invalid integer constants are as follows:

```
2.0 (contains decimal point)
37- (negative sign is not prefixed)
12345678901234 (most probably exceeds the capacity of the processor but
                might not at some point in future)
2a3 (contains a non-digit character)
```

Leading zeros of an integer constant are ignored. For example, 01, 001, 00001 and 1 are all equivalent.

1.6 Real Constants

A real constant is a real number containing one and only one decimal point. The decimal point may be explicit or implicit (possible in scientific notation—to be discussed shortly). A real constant may be positive, negative or zero. A real negative constant is prefixed by a negative sign. An unsigned real constant is assumed to be positive. The numerical values of +0.0 and –0.0 are same even if the processor can make distinction between +0.0 and –0.0. Like for an integer constant, a leading plus sign for a positive real constant is optional. A real constant, in the standard form, contains digits, one decimal point and is prefixed by a plus or minus sign, if necessary. Valid real constants are 3.1415926, –25.3456, 12345.678 and 0.0.

Invalid real constants are as follows:

```
36 (contains no decimal point)
35.3.2 (contains more than one decimal point)
3a3563.25 (contains a letter 'a')
45.2(2) (contains special characters)
5 6.0 (contains a blank)
```

If there is no digit before or after the decimal point, zero is assumed. For example, 2. is treated as 2.0. Similarly, .25 is a valid real constant, which is same as 0.25. The use of real constants like 2. and .25 is strongly discouraged. This reduces the readability of the program. These numbers should be written as 2.0 and 0.25, respectively. It is needless to mention that just a decimal point does not represent any real constant, that is, not 0.0. The maximum and minimum values and the precision of the real number are processor dependent. A real number may be written with more digits after the decimal point than the number of digits the processor can support. The excess digits are not considered by the system. Real constants may also be represented by powers of 10, known as scientific form. This makes it very convenient to express very small or very large numbers. In this form, a real constant consists of two parts: an integer or real number followed by an exponent. The exponent is denoted by letter 'E' or 'e'. The number 1.24e4 is actually 1.24×10^4, as 'e4' stands for 10 to the power 4. There is no space between 'e' and the real or the integer part (also called fractional part). The exponent must be an integer and may be signed. The sign is placed after the exponent symbol. Unsigned exponents are assumed to be positive. Valid real constants in scientific notation are 1.24e4, -111.90e10, 77.345e-3 and -123.345e-5. The values of the above real constants are, respectively, 1.24×10^4, -111.90×10^{10}, 77.345×10^{-3} and -123.345×10^{-5}.

Invalid real constants are as follows:

```
1.24e4- (wrong position of the minus sign)
3.25 e5 (space between the fraction and the exponent symbol)
777.24e2.5 (exponent must be an integer)
6.935e500 (probably exceeds the capacity of the processor)
```

As mentioned earlier, the decimal point in a real constant may be implicit. For example, 123e4 is a real constant, though it does not contain any decimal point. The default decimal point is assumed between '3' and 'e'. It is, thus, equivalent to 123.0×10^4. Leading zeros of the fractions are ignored. For example, 0123.24e4, 00123.24e4 and 123.24e4 are all equivalent. Similarly, leading zeros in the exponent are also ignored. The real constants 2.345e+2, 2.345e+02, 2.345e02 and 2.345e2 are equivalent.

1.7 Double Precision Constants

Double precision quantities are real numbers but are more precise than their single precision counterparts. It is well known that the computer internally uses binary numbers, and all decimal numbers do not have an exact binary representation. For example, when 0.1 is converted to binary, the binary number may not be exactly 0.1—it is probably 0.09999.... A computer is a finite bit (binary digit) machine, and the number of bits used determines how close the binary number is to the decimal counterpart. For infinite precision arithmetic, the two numbers would have been identical. By increasing the number of bits to store a real number, the binary counterpart can be made closer to the decimal value. Double precision numbers take more memory than do the single precision numbers and consume more central processing unit (CPU) time to perform any arithmetic operations compared with the corresponding single precision numbers.

Double precision constants are more precise than the corresponding single precision values. Such a constant is expressed in scientific notation. To indicate ten to the power for double precision constants, 'D' or 'd' is used; π, correct up to 15 decimal places, is written as 3.141592653589793d0.

The rules discussed in connection with the single precision real numbers (in scientific notation) are equally valid for the double precision quantities.

1.8 Complex Constants

Complex numbers are widely used in science and engineering. A complex number consists of two parts—real and imaginary. The Fortran compiler can handle complex numbers according to the rules of the complex algebra. The real and the imaginary parts may separately be positive, negative or zero.

Both the real and the imaginary parts are integers or real numbers. The real and the imaginary pair are enclosed in brackets.

Valid complex constants are as follows:

```
(2.0, 5.0), (-3.373, 5.397), (-4467.23, 891.45)
(3.124E2, -4.935E-7), (33, 25), (-15.72, -21)
```

Since the real and the imaginary parts are either integers or real numbers, the discussions regarding real and integer numbers are equally valid for the real and the imaginary parts, respectively.

1.9 Double Precision Complex Constants

Double precision complex constants have double precision quantities as the real and the imaginary parts. Valid double precision complex constants are as follows:

```
(1.23456789d0, 9.87654321d0), (-3.14159265d0, 1.987665432d0)
```

1.10 Quadruple (Quad) Precision Constants

Intel's Fortran compiler, ifort, supports quadruple precision real numbers. These numbers are more precise than the corresponding double precision numbers. This is not a standard Fortran feature. Quadruple precision constants have a precision of 33 places after the decimal. To indicate quadruple precision, 'q' (or 'Q') is used in scientific notation in place of 'e' or 'd'. Valid quadruple precision constants are as follows:

```
3.1415926535897932384626433383279502q0,-7.98765432198765432198765432112345q1
```

1.11 Logical Constants

Integer and real constants can assume innumerable values. There are only two logical constants: .true. and .false.. These constants are bound by two periods (uppercase letters may be used).

1.12 Character Constants

Character constants are zero or more characters enclosed in quotes or apostrophes. Examples of character constants are 'A', 'ABC' and 'West Bengal'.

The delimiters (apostrophe or quote) are not part of the string. Note that a blank is also a character (blank between West and Bengal). The characters 'A' and 'a' are not same. It is case sensitive. In addition, character '1' and number 1 are different, and they are stored in different ways inside the machine. Conventional arithmetic operations are not permitted with character constants like '1'. To represent the apostrophe as a character constant, either two successive apostrophes are used or it is enclosed in quotes: 'don''t', "don't". Similarly, to represent the quote as a character constant, either two successive quotes are used or it is enclosed in apostrophes: """", '"'. Null is represented as two successive apostrophes (or quotes) with nothing in between. Any graphic character supported by the processor can also be part of a character constant.

Although Fortran supports other types of character sets, in this book only the ASCII character set is considered. All these character sets have one thing in common: all of them have one blank character.

1.13 Literal Constants

The constants mentioned earlier are also known as literal constants. Literal constants do not have names attached to them.

1.14 Variables

An object whose value may vary during the execution of a program is called a variable. A variable can store only one value at a time, which may change during the execution of the program. A variable must have a name attached to it. A variable is identified by its name, type and value. If no value is assigned to it, it remains unassigned or undefined. Note the word *may* in the definition. The variable may or may not change its value during the execution of the program and may even remain undefined during the lifetime of the program.

By default, if the variable name starts with i, j, k, l, m or n (or uppercase letters), it is an integer variable. All other variables that start with letters other than i-n are real variables. An integer variable can store only an integer quantity, and similarly, a real variable can store only a real quantity. In spite of the preceding default rule, it is a good programming practice to define each variable explicitly. The modern programming practice is to switch off this default feature, that is, i-n rule, by appropriate declaration, and in that case, it is mandatory to declare each variable.

1.15 Variable Declarations

An integer variable can store only an integer quantity. It is declared as follows:

```
integer :: a
integer :: b
integer :: c, d
```

In the preceding declarations, a, b, c and d are declared as integer variables, and therefore, these variables can store only integer quantities. It is apparent from these declarations that more than one variable may be declared by a single declaration—in that case, the variables are separated by a comma. The first two declarations may be combined as follows:

```
integer :: a, b
```

Blanks between integer and ':: ' and between ':: ' and the variable name can be introduced to increase the readability.

An identifier (variable name) may be treated like a box. The name of the box is the name of the variable. The content of the box is undefined. When a value is assigned to a variable, the content of the box is the value of the variable.

In an identical manner, other intrinsic types are defined.

```
real :: x
real :: p, q
double precision :: d1
double precision :: d3, d4
complex :: c
complex :: z, y
double complex :: dc1
double complex :: dc2, dc3
character :: ch
character :: e, f
logical :: 13, 14
```

By default, the number of characters that a character variable can store is 1; that is, the length is 1. A character variable may store more than one character if it is declared in an appropriate manner.

```
character (len=10) :: ch
```

The variable ch can store 10 characters. The length can also be specified as character *10 ch. Also, character(4) :: ch and character(len=4) :: ch are equivalent. In this case, just an integer without len= is assumed to be the length of the string.

1.16 Meaning of a Declaration

A declaration is a placeholder for a variable; it merely reserves location(s) for a variable and defines the type of the variable. No value is assigned to the variable. A suitable Fortran statement must be used to assign a value to a variable. A variable can store only one value at a time. The same variable cannot be declared more than once in a program unit. For example, the declarations

```
integer :: x
real :: x
```

within the same program unit will give rise to Fortran error because the variable 'x' cannot be an integer and a real variable at the same time. Unassigned variables should not be used, as the result of such computation is unpredictable.

1.17 Assignment Operator

The assignment operator (=) is used to assign a value to a variable. For example, if a variable `first` is declared as integer, the variable `first` is assigned to a value in the following manner:

```
integer :: first
...
first = 10
```

Subsequently, if 20 is assigned to `first`, the old value 10 will be lost and now `first` will contain 20. The instruction `first = 20` assigns 20 to `first`; the old value 10 is lost.
    We will discuss this assignment operator in detail in Chapter 2.

1.18 Named Constants

A symbolic name may be attached to a constant. The symbolic name becomes an alias for the constant. The alias behaves just like a literal constant, and it cannot be modified during the execution of the program. A true constant, say, π, may thus be used in this manner.

```
real, parameter :: pi=3.1415926
```

In the preceding declaration, pi is the symbolic name of 3.1415926 because of the presence of the attribute 'parameter' with the real declaration; a comma separates real and parameter. In the preceding declaration, pi is not a real variable—it is just another name of 3.1415926. During compilation, 3.1415926 will replace each occurrence of pi. Since pi is alias of 3.1415926, pi cannot be assigned to a different value; pi=4.25 is not allowed as named constants by definition cannot be modified. The reason is not difficult to guess. During compilation, 3.1415926 will replace pi, so the statement pi=4.25 will become

```
3.1415926 = 4.25
```

which is clearly not a valid Fortran statement.
    Moreover, the program unit cannot have any variable named pi as pi has already been declared as an alias for 3.1415926. The following will generate compilation error:

```
real, parameter :: pi = 3.1415926
integer :: pi
```

It is a good programming practice to assign a symbolic name to a true constant like π so that even by mistake the constant cannot be modified during the execution of the program. An alternative way to represent a named constant is through the parameter statement:

```
parameter (named constant=value,...)
```

Example: `parameter (pi=3.1415926)`

Either the type of the named constant, declared by the parameter statement, is declared or it follows the default i-n rule. For example, in the case of `parameter (ip=2.3)`, each occurrence of `ip` is substituted by 2 and not `2.3` since, without any declaration, `ip`, being an integer can store only an integer quantity. Therefore, truncation will take place. A single parameter statement may define more than one named constant: `parameter (pi=3.1415926, e=2.303, lpt=6)`.

Named constants are assigned values at the time of compilation. Therefore, it cannot contain anything whose value is not known during compilation. In an identical manner, other intrinsic type constants can be attached to a symbolic name.

```
integer, parameter :: limit=100
double precision, parameter :: dpi=3.1415926589793d0
complex, parameter :: zpar=(10.0,30.0)
logical, parameter :: l3=.true.
character, parameter :: start= 'a'
```

For a character named constant, an asterisk may be used as the length of the named constant; the compiler from the declaration can find out the length of the named constant (allocates locations to store the constant):

```
character (len=*), parameter :: city= 'kolkata'
```

From the declaration, the compiler can ascertain that the named constant `city` should have a length 7 to accommodate the string `'kolkata'` and allocates locations accordingly.

The preceding declaration is same as `character (len=7), parameter :: city= 'kolkata'`.

1.19 Keywords

Fortran contains several keywords like `integer` and `real`. However, the keywords are not reserved words and may be used as identifiers. This is strongly discouraged. For example, 'do' is a Fortran statement and also a Fortran keyword. It is permitted to have an identifier named do. The compiler will identify the do statement from its appearance; it will also correctly treat the do identifier. However, for the sake of readability, this should be avoided. These types of keywords are called statement keywords. Normally, a keyword cannot have embedded space in free form. The keyword, say, 'read' cannot be written as 're ad'. However, this is allowed in fixed form (not discussed in this book). If a name follows a keyword, the keyword and the name must be separated by a blank. Blank is optional for some single keywords that consist of two

TABLE 1.2

Adjacent Keywords

Blanks Are Optional		
block data	end file	end team
double complex	end forall	end type
double precision	end function	end where
else if	end if	error stop
else where	end interface	go to
end associate	end module	in out
end block	end procedure	select case
end block data	end program	select rank
end critical	end select	select type
end do	end submodule	
end enum	end subroutine	

Blanks Are Mandatory		
case default	interface assignment	recursive subroutine
do while	interface operator	recursive *type-spec*
implicit *type-spec*	module procedure	*type-spec* function
implicit none	recursive function	*type-spec* recursive

keywords, such as 'end do'. In this case, enddo and end do are the same. However, blank is mandatory for keywords like do while and implicit none. Table 1.2 lists such adjacent keywords where blanks are optional and mandatory.

Argument keywords are discussed along with subprograms. Keywords are also used to identify an item within a list. They are used as keyword=value so that the position of the keyword within the list is not important.

1.20 Lexical Tokens

If a Fortran statement is broken into basic language elements, the smallest meaningful objects are called lexical tokens. The lexical tokens consist of names, operators, literals, keywords, labels, assignment signs, commas, etc. In other words, a Fortran statement is a combination of lexical tokens. Names, constants and labels are usually separated from adjacent lexical tokens by one or more blanks or an end of line. For example, a+b consists of three lexical tokens: a, b and +.

1.21 Delimiters

A delimiter consists of a pair of symbols, which determines a part of a Fortran statement. Examples of delimiters are

```
/ . . . . . . /
( . . . . . . )
[ . . . . . . ]
(/ . . . . . . /)
```

These are discussed at appropriate places in the text.

1.22 Source Form

A Fortran source program consists of one or more lines. A line may contain zero or more characters. Fortran statements may be written in two different forms: (a) fixed form and (b) free form. The current trend is to write programs in free form; therefore, fixed form is not discussed in this book.

1.23 Free Form

In free form, a Fortran statement can be extended to 132 characters per line if characters of default kind are used. However, if the line contains characters other than the default kind, the number of characters that a line can accommodate is processor dependent. The statement may start anywhere within this field. A line is usually divided into several fields: (a) statement number field, (b) statement field, (c) comment field and (d) continuation field. A line may not contain all fields; even a line may be totally empty. If the last non-blank character of a particular line is ' & ', the next line (if it is not a comment line) is considered as the continuation of the previous line. A total number of 255 continuation lines are allowed (per statement). Note that ' & ' character is not a part of the statement when used as a continuation character.

Figure 1.1 is equivalent to X=Y+Z. No line can contain a single ' & ' as the only non-blank character. In addition, no line can contain one ' & ' character followed by ' ! ' character. A statement number, if any, should be placed at the beginning of the line. There must be a blank or a tab character after the statement number. There may be any number of blanks before the statement number. A statement may have a statement number between 1 and 99999. Leading zeros of the statement numbers are ignored. No two statements in a program unit can have the same statement number (one exception is discussed in Chapter 12). It is not required to assign a statement number to every statement. However, there are statements that must have a statement number. The statement number is used when a statement is required to be referred by other statement(s). If a line is continued, only the first line can have a statement number. There cannot be any blank within the statement number (Figures 1.2 through 1.4).

If the character ' ! ' is typed anywhere in a line, the rest of the line, except within a character string, is treated as a comment (Figure 1.5). Comments are used for documentation. The compiler does not try to translate the comment. Comments can be

1	2	3	4	5	6	7	8	9	10	11	12
						X	=	Y	+		&
						Z					

1	2	3	4	5	6	7	8	9	10	11	12
1	4	7				X	=	Y	+	Z	

1	2	3	4	5	6	7	8	9	10	11	12
			1	0	5		X	=	Y	+	z

1	2	3	4	5	6	7	8	9	10	11	12
1		2	5			X	=	Y	!	e	r	r

1	2	3	4	5	6	7	8	9	10	11	12
5		A	=	2	!	c	m	t			

FIGURE 1.1 through 1.5
(see text).

placed anywhere within the program unit; it may be placed before the first statement. It may also be placed after the last statement. It may be placed between two continuation lines. Comments placed between continuation lines do not contribute to the calculation of continuation lines.

A comment line cannot have a statement number (gfortran gives a warning):

```
100 ! This is a comment
```

This is not a valid statement. Furthermore, a comment line cannot be continued.

A blank line is treated as a comment. In fact, judicious use of blank lines increases the readability of the program. If the continuation symbol '&' is typed after the comment character '!', the character '&' becomes a part of the comment and is not considered a continuation character (Figure 1.6). This will generate compilation error; '&' is not considered as the continuation character in this case, so the next line is not treated as continuation of the previous line. A comment may appear after the line continuation character:

1	2	3	4	5	6	7	8	9	10	11	12
							X	=	Y	+	!	&
							Z				

1	2	3	4	5	6	7	8	9	10	11	12
							X	=	Y	+	&	
	Z										

```
    a=b+ & ! This is a comment
    c
```

1	2	3	4	5	6	7	8	9	10	11	12
							X	=	Y	+	&	
&	Z										

Normally, continuation starts from the first character of the next non-commented line. However, if it is necessary to start the con-

FIGURE 1.6 through 1.8 (see text).

tinuation from a particular position of the next line, then the '&' character must be typed just before the desired character of the next line.

The statement where the continuation line starts with a blank (Figure 1.7) will be treated as follows:

```
    X=Y+ Z (one blank between '+' and Z)
```

On the other hand, where the continuation line starts with '&', the statement will be treated as follows (Figure 1.8):

```
    X=Y+Z (no blank between '+' and 'Z')
```

In these situations, both mean the same, as normally Fortran ignores blanks. This effect will be felt when character strings are used where the presence or absence of a blank within a character string may result in a different meaning.

In free form, there cannot be any imbedded blanks within a lexical token. Blanks are used to separate various items, such as names, constants and labels, from the keywords; read 20, a, b, c cannot be written as read20, a, b, c.

However, multiple blanks may be used between tokens. This may improve the readability of the source code. Most of the times, multiple blanks between tokens are treated as a single blank.

In free form, a normal statement (a statement without continuation) terminates when either `'!'` character or the end of line is reached. Each compiler has a mechanism to identify the free and the fixed form. This is done in two ways. The first method is to use an appropriate compiler directive, say, –free or –fixed. The second method is to use the file extension. A file having extension .f (say, a.f) is considered in fixed form, and a file having extension .f90 (say, a.f90) is considered in free form.

1.24 Continuation of Character Strings

Normally, Fortran ignores blanks. A blank is considered as a character within a character string. Therefore, special consideration is needed for continuing a character string to the next line. For this purpose, an ampersand character (`'&'`) must be the last non-blank character of the first line of the character string, and each continuation line must have an ampersand character. Continuation begins from the character following the ampersand of the continuation line. Consider the following (Figure 1.9).

c	h	a	r	a	c	t	e	r	(	l	e	n	=	8	0	)	:	:	c	h	=	&					
'	A	S	S	O	C	I	A	T	I	O	N		O	F		V	O	L	U	N	T	A	R	Y	&		
&						B	L	O	O	D		D	O	N	O	R	S	'									

FIGURE 1.9
Continuation of Character String.

The variable ch is initialized along with its declaration. Lines 2 and 3 are continuation of line 1. We now consider lines 2 and 3. As the continuation starts from the first character of a line (in this case line 3), six blanks will be added before the string `'BLOOD DONORS'`. The character variable ch will be initialized to
"ASSOCIATION OF VOLUNTARY   BLOOD DONORS" (6 blank characters), and the system will add the required number of trailing blanks. However, if the intention of the programmer is to initialize the variable to "ASSOCIATION OF VOLUNTARY BLOOD DONORS", that is, only one blank between `'VOLUNTARY'` and `'BLOOD'`, the third line should have an ampersand character as shown next (Figure 1.10).

c	h	a	r	a	c	t	e	r	(	l	e	n	=	8	0	)	:	:	c	h	=	&					
c	h	a	r	a	c	t	e	r	(	l	e	n	=	8	0	)	:	:	c	h	=	&					
'	A	S	S	O	C	I	A	T	I	O	N		O	F		V	O	L	U	N	T	A	R	Y	&		
					&		B	L	O	O	D		D	O	N	O	R	S	'								

FIGURE 1.10
Continuation of Character String.

The ampersand in the third line ensures that continuation starts from the character following the ampersand, which is just a blank in this case.

If a keyword or other attributes of the language (technically called token) are split across the line for which no embedded blank is allowed, there should not be any space between the ampersand and the rest of the token in the continued line as shown in the following:

```
re&
&ad *, x
```

The preceding code is treated as read *, x. Note that the position of the ampersand in this case ensures that there is no space between 're' and 'ad'. A character constant may contain an ampersand. The last one is considered the continuation character, as shown next. If it contains more than one ampersand, the other ampersands become the part of the character string. For

```
print *, 'M/s Roy & Chatterjee & &
    &Gupta'
end
```

the output is M/s Roy & Chatterjee & Gupta. The ampersand shown in bold is considered as the continuation character. The ampersand before the continuation character is a part of the character string.

1.25 Structure of a Program

A program unit contains one or more lines, which comprises the Fortran declaration, statement, comment and included line. A line may contain zero or more characters. The end statement terminates a program unit. A program may contain more than one program unit. An executable unit must have one and only one program unit called the main program. Execution always begins from the first executable statement of the main program. In addition to the main program, an executable unit may have external subprograms, internal subprograms, modules, submodules and block data (now declared as obsolete). These topics are discussed at appropriate places. For the time being, we shall consider only one program unit; that is, the executable unit would contain only the main program.

1.26 IMPLICIT NONE

It was mentioned earlier that if the variables are not declared explicitly, Fortran applies certain default rule (i-n rule) in selecting the variable type. This default rule can be switched off by placing implicit none at the beginning of the program unit. In this case, all variables are to be declared explicitly. If implicit none is present in a program unit, the unit cannot have any other implicit statement. For example, if a variable i1 (i and 1) is declared as an integer and if it is typed as ii (i and i) in the body of the program

(typing error), implicit none will force the compiler to generate a Fortran error (undefined variable). If implicit none is absent, it will be treated as another integer variable
following the default i-n rule, and since it is undefined (no value is possibly assigned),
the result is unpredictable. The modern trend of programming is to use implicit none
in every program unit so the programmer is forced to declare all variables explicitly. Any
typing error, similar to that shown above, will be flagged as an error at the compilation
stage.

1.27 IMPLICIT

This statement can be used to treat variables that start with certain letter to be of a
particular type.

```
implicit integer (a)                ! (1)
implicit integer (b, c)             ! (2)
implicit integer (d-f)              ! (3)
implicit integer (g-i, x-z)         ! (4)
```

Statement (1) directs the compiler to treat all variables that start with 'a' as integers. That
is, the compiler will treat am, ax1, ap, etc., as integers. Statement (2) tells the compiler
to assume the variables that start with 'b' or 'c' to be integers. Statement (3) contains
d-f, which is equivalent to implicit integer (d, e, f). Statement (4) states that all
variables that start with g, h, i, x, y and z are integers. The dash sign (minus sign)
indicates a set of contiguous letters; the first and the last letter in the set are placed on the
left and the right of the dash sign. The discussion of this section is equally applicable to
other types of variables. So it will not be repeated again.

```
implicit real (a)
implicit real (x-z)
implicit real (a-h, o-z)
implicit double precision (d)
implicit double precision (e, f)
implicit complex (p-q)
implicit logical (m-n)
implicit double complex (x-z)
implicit character (r, s)
```

For a character variable, if the length parameter is not present along with the implicit
declaration, the length is assumed to be equal to 1. The length parameter can be specified
along with the implicit declaration:

```
implicit character (len=4) (u-v)
```

This declaration needs some explanation. It states that variables that start with u or v are
character variables of length 4; that is, they can store four characters.

1.28 Rules of IMPLICIT

1. If a program unit contains `implicit none`, it must not contain any other `implicit` statement.

2. A program unit cannot contain two implicit statements with the same letter:

   ```
   implicit integer (c)
   implicit real (c)
   ```
   or
   ```
   implicit integer (c-f)   ! this includes d
   implicit real (d)
   ```

   are not valid.

3. An explicit declaration overrides an `implicit` declaration.

   ```
   implicit integer (i)
   real :: i
   ```

   Here, 'i' will be treated as a real variable because of the explicit real declaration.

1.29 Type Declarations

Fortran allows declaring intrinsic variables through `type` declarations.

```
      integer :: a
and   type(integer) :: a
```

are equivalent.

   Similarly, real, complex, double precision, double complex, logical and character variables may be declared through `type` declarations as shown in the following:

```
      type(real) :: b
      type(double precision) :: c
      type(complex) :: d
      type (double complex) :: e      ! allowed in ifort, not allowed in NAG
      type (logical) :: l
      type (character) :: ch
      implicit type(integer) (a-h, o-z)! allowed in NAG, not in ifort
and   implicit integer (a-h, o-z)
```

are equivalent.

1.30 Comments on IMPLICIT Statement

Readers must have noticed that the discussion related to `implicit none` goes against the declaration `implicit integer / real / double precision / complex / character / logical / double complex`. A guideline may be formulated. `Implicit none` is certainly very safe; it isolates all undefined variables and helps to eliminate most of the typing errors related to variable names. On the other hand, `implicit integer` saves a lot of typing, especially if the program unit contains many integer variables. Some programmers use the first letter of variables to indicate the type of variables. For example, one may choose 'c' as the first letter for all complex variables and 'd' as the first letter for all double precision variables. In such a situation, `implicit double precision (d)` and `implicit complex (c)` are convenient. Therefore, it is a matter of choice. The present author prefers `implicit none`, and he feels that if more time is spent during the development phase of the program (that is, coding and typing), then debugging time is substantially reduced and the problems mentioned related to the undefined variables never crop up.

1.31 PROGRAM Statement

The optional `program` statement is the first statement of a program. It supplies the name of the program:

```
program my_first_program
```

where `my_first_program` is the name of the program.

1.32 END Statement

The end statement signifies the end of a program unit. It may contain a program and the name of the program. However, this is optional.

A typical Fortran program is of the following form:

```
program my_first_program
.
end program my_first_program
```

Once the 'end' statement is reached, the compiler starts compiling that particular unit. The end statement terminates a program unit. If the end statement contains the program name, the corresponding program name must be present with the `program` statement. In fact, it is always better to use the name of the subroutine, module and function along with the end statement. Each program unit, module subprogram and internal subprogram can have only one end statement. Such end statements are executable statement, and it is possible to jump to the statement by suitable statements (goto, if—Chapter 3). The execution of the end statement in the main program terminates the job. The execution of the end statement within a function or subprogram or subroutine subprogram or separate module subprogram is equivalent to a return statement (Chapter 12). The end statement of a module, submodule and block data is a non-executable statement.

1.33 Initialization

A variable may be initialized to a value along with its declaration. In this case, when the execution begins, the corresponding variable is not undefined; it has an initial value.

```
integer :: a=10
real :: x=1.34
integer :: b=10, c=20
```

In the first case, not only 'a' is declared as an integer but it is also initialized to 10. Similarly, x, b and c are also initialized to 1.34, 10 and 20, respectively. In case more than one variable is declared by a single declaration, all variables are to be initialized individually. For example,

```
integer :: d, e=200
```

will initialize e to 200, but d will remain uninitialized. If it is necessary to initialize both variables, it is to be done separately:

```
integer :: d=200, e=200
```

1.34 Number System

Strictly speaking, the digits are just symbols; the positions of a digit within a number determine its value. For example, when the base of the number system is 10, the number 123 is actually $1 \times 10^2 + 2 \times 10^1 + 3 \times 10^0$. Therefore, if a digit, say, 3, appears at the unit position, its value is 3. On the other hand, if the same digit appears at the position of 10, its value is 30. A number may be represented in terms of a base other than 10. The most popular bases are binary (base of 2), octal (base of 8) and hexadecimal or hex (base of 16). However, the numerical value of a particular number is independent of the number system (base)—the value of a particular number is same in all systems. In the next few sections, we indicate a base other than base 10 by means of a subscript—$(1110)_2$ stands for a binary number.

1.35 Binary Numbers

A bit is the abbreviation of **binary** digit. In the binary system, that is, when the base is 2, the available digits are 0 and 1. For example, a number $(1101)_2$ is equal to 13 in the decimal system: $1 \times 2^3 + 1 \times 2^2 + 0 \times 2^1 + 1 \times 2^0$.

1.36 Octal Numbers

Octal numbers have a base of 8. The available digits are 0 through 7. Three binary digits constitute one octal digit (the highest value is 7, that is, $(111)_2$). An octal number $(101)_8$ is equal to 65 in the decimal system: $1 \times 8^2 + 0 \times 8^1 + 1 \times 8^0$.

1.37 Hexadecimal Numbers

Hexadecimal numbers—popularly known as hex numbers—have a base of 16. The available digits are 0 through 9 and a, b, c, d, e and f (capital letters may also be used). The last six symbols are equivalent to decimal 10, 11, 12, 13, 14 and 15, respectively. Four binary digits constitute one hex digit. The highest value of a hex digit is 'f', that is, 15 in the decimal system. The hex number $(101)_{16}$ is equal to 257 in the decimal system: $1 \times 16^2 + 0 \times 16^1 + 1 \times 16^0$.

1.38 Initialization Using DATA Statement

Data statements are used to initialize variables. They are usually placed at the beginning of the program unit along with other specification statements. It is normally placed after the declaration of the variables. However, it can be placed anywhere within the program unit, but this should not be done. Two integer variables i and j may be initialized to 10 and 20, respectively, by a data statement as follows:

```
data i /10/
data j /20/
```
or
```
data i /10/, j /20/
```
or
```
data i, j /10, 20/
```

The last two declarations are equivalent. Note that either the variables are declared by appropriate declaration or the default i-n rule is followed. The following are the data statements to initialize variables l, d and c:

```
logical :: l
double precision :: d
complex :: c
data l /.true./
data d /3.1415926589d0/
data c /(2.0, 3.0)/
```

For the complex variable c, the first constant corresponds to the real part, and the second constant corresponds to the imaginary part of the variable. In the case of a complex variable, the real and the imaginary parts are enclosed in parentheses.

Character variables are also initialized in a similar manner:

```
character (len=4) :: ch
data ch / 'iacs'/
```

If the number of characters is less than the size of the variable, blanks are added at the end. If the number of characters is more than the size of the variable, the constant is truncated from the right:

```
character (len=8) :: ch
data ch / 'iacs calcutta'/
```

The variable ch is initialized to 'iacs cal' as it can store a maximum of 8 characters.

1.39 BOZ Numbers

The binary, octal or hex numbers (also known as boz numbers) are represented by the respective digits enclosed in apostrophes or quotes and prefixed by b, o or z, respectively. The following are binary, octal and hex numbers:

```
b'1001'  (decimal 9)
o'127'   (decimal 87)
z'1b7'   (decimal 439)
```

The uppercase letters B, O and Z may be substituted for their corresponding lowercase counterparts – b, o and z, respectively.

1.40 Integer Variables and BOZ Numbers

An integer variable may be initialized by a binary, octal or hex number:

```
integer :: a=b'111'
integer :: b=o'171'
integer :: c=z'1a'
```

A BOZ constant can be used to initialize an integer variable by a `data` statement:

```
integer :: p, q,r
data p/b'1111'/
data q/o'247'/
data r/z'12a'/
```

1.41 Executable and Non-Executable Statements

Fortran statements are basically of two types: executable and non-executable statements. The first one means some action. For example, x=10 is an executable statement, where the variable x is assigned to a value 10. The statement `integer :: y` is a non-executable statement. It is a declaration and merely passes information to the compiler to reserve locations for an integer variable y. The non-executable statements configure the programming environment where executions of executable statements are performed. Non-executable statements cannot be the targets of any branch statement (Chapter 3).

1.42 INCLUDE Directive

Strictly speaking `include` is not a Fortran statement; it is a directive to the compiler. The syntax of `include` is

```
include 'char-constant'
```

where the `character-constant` is usually the name of a file. The compiler replaces the `include` statement by the content of the file. The `include` statement cannot be labeled; it may be nested; that is, it may contain another `include` statement (nested `include`). The maximum number of nested `include` is processor dependent. The `include` cannot 'include' itself directly or indirectly—include 'a' may contain include 'b', but then the file 'b' cannot contain include 'a' (a recursive "call"). The `include` statement must be typed on a separate line, and it may have a trailing comment:

```
include 'myfile.f90'
```

The content of 'myfile.f90' is included at the point of inclusion. The `include` statement cannot be continued. The first included line should not be a continuation line; the last included line cannot be continued. The following program segment is unacceptable:

```
a=b+ &
include 'myfile.f90'
...
```

The file `myfile.f90` cannot have its first line as

```
&c
```

Similarly,

```
include 'myfile.f90'
&c
```

with the last line of the file `'myfile.f90'` as

```
a=b+&
```

is also not acceptable.

1.43 Statement Ordering

In a program unit, statements are ordered as shown in Appendix B. Usually, the declarations come at the beginning of the program unit.

1.44 Processor Dependencies

If the source line is created with characters other than the default type, the number of characters that a source line can have is processor dependent. There is no guideline how the compiler will identify the free-form and fixed-form sources. Though, normally, ASCII is the default character set, there is no specific guideline in the Fortran report in this matter. The interpretation of the char-literal-constant used with the `include` compiler directive is processor dependent. Also, the maximum number of `include` directives that may be nested is not specified in the Fortran report.

The maximum and minimum values (numeric) are processor dependent.

1.45 Compilation and Execution of Fortran Programs

The programs in this book have been tested with three compilers: ifort, nagfor and gfortran.

To compile a Fortran program using the `ifort` compiler, the instruction is (name of the source file—x.f90)

```
ifort x.f90
```

This will create an executable x.exe, which can be executed.

A Fortran program using Numerical Algorithm Group's `nagfor` can be compiled as follows:

```
nagfor -o x.exe x.f90 [x.exe is the executable file]
```

A Fortran program using the GCC gfortran compiler can be compiled as follows:

```
gfortran  -o x.exe x.f90 [x.exe is the executable file]
```

2

Arithmetic, Relational and Logical Operators and Expressions

As the computer is a machine that can perform basic arithmetic operations at a high speed, naturally Fortran is provided with arithmetic operators to perform these operations on arithmetic expressions. Alongside, relational operators can test a relation. They return either a true or false value. For example, if a question is asked, "Is x greater than y?" The answer is either yes (true) or no (false). In addition to this, five logical operators are also available that return either a true or false value. We first consider arithmetic operators.

2.1 Arithmetic Operators

Two types of arithmetic operators are available to a Fortran programmer – binary and unary operators. Binary operators require two operands. The binary arithmetic operators are shown in Table 2.1.

Examples of binary operators are as follows:

```
a + b   (add a to b)
a * b   (multiply a by b)
a - b   (subtract b from a)
a / b   (divide a by b)
a ** b  (a to the power of b)
```

Unary operators require a single operand. Table 2.2 shows the unary arithmetic operators. As unsigned integers, real constants or variables are treated as positive numbers, unary plus is rarely used; +5 is same as 5. The unary minus changes the sign of a variable or a constant. An example of the unary minus is -5, where the sign of 5 is changed. Similarly, the magnitude of -x is the value of x with its sign reversed.

TABLE 2.1

Arithmetic Operators (Binary)

Symbol	Meaning
**	Exponentiation (to the power)
/	Division
*	Multiplication
+	Addition
-	Subtraction

TABLE 2.2

Unary Operators

Symbol	Meaning
+	Unary plus
-	Unary minus

It may be noted that the same symbols '+' and '-' are used to indicate both the unary and binary operations. The compiler can determine from the context the meaning of the operators—whether it is a binary or a unary operator.

2.2 Arithmetic Expressions

Arithmetic expressions are formed using constants, variables and other objects as permitted by the language and arithmetic operators discussed in the previous section. Examples of arithmetic expressions are as follows:

```
x + y + z
5 * j + k
p + q - c**3 / f + 27.35
```

2.3 Assignment Sign

The symbol '=' is used to assign a value to a variable. The general form of an assignment statement is as follows:

```
variable = expression
```

The expression on the right-hand side of the assignment sign is evaluated, and the value thus obtained is stored in the variable. As a variable can store only one value at a time, the current value of the variable is lost, and a new value is stored in its place. Examples of assignments are as follows:

```
i = 2
area = length * width
s = u * t + 0.5 * f * t**2
```

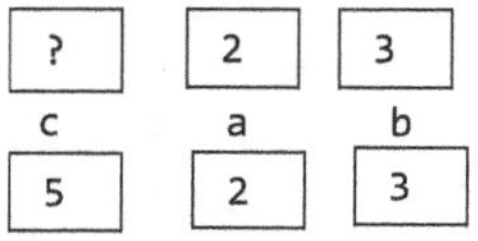

FIGURE 2.1
Arithmetic operation.

Consider the following expression:

```
c = a + b
```

Let us assume that the values of a and b are 2 and 3, respectively. The value of c is not our concern at this moment. Before the expression is evaluated, the contents of a, b and c are as shown in the upper panel of Figure 2.1.

When a is added to b, the result is 5, and it is stored in location c. However, a and b will retain their old values. Therefore, at the end of the operation, the values are as shown in the lower panel of Figure 2.1.

The symbol for the assignment sign, that is, '=', must not be confused with the equal sign used in algebra.

For example, i = i + 1 is a valid Fortran statement, which is not an algebraic equation. Had it been so, canceling i from both sides would give us, 0 = 1, which is clearly not acceptable. The proper meaning of this statement is to increment i by 1. To be more specific, in this case the current value of i is taken, 1 is added to it and the result is stored in the same location, i. If, for example, the value of i is 10 before the execution of the statement, it is 11 at the end of the operation, and the result is stored in location i.

One can write similar statements:

```
i = i - 1
i = i * j
```

In an assignment operation, unless the same variable appears on both sides of the assignment sign, the variables appearing on the right-hand side of the assignment sign are not modified – they retain their old values, and the variable on the left-hand side of the assignment sign gets a new value.

2.4 Rules for Arithmetic Expressions

The following rules must be followed while writing arithmetic expressions:

Rule 1: Arithmetic operations are not allowed on the left-hand side of the assignment sign. For example, a + b = c is not a valid arithmetic expression. The left-hand side of the assignment sign must be a variable. Arithmetic operation on the left-hand side of the assignment sign is allowed only to calculate the address of a variable and the like. This is discussed in Chapter 6, when we discuss array.

Rule 2: No arithmetic operation is assumed like algebra: (a+b) (a-b) is not taken as (a+b) * (a-b). The multiplication operator in this case must be specified explicitly.

Rule 3: No two arithmetic operators may appear side by side: c = a * -b is not a valid Fortran statement. Should such situation arise, it must be enclosed in parentheses: c = a * (-b). However, some compilers do not flag this as error. This rule appears to have been violated in case of exponentiation operator '**'. However, it must be remembered that the exponentiation operator is a single entry – it is not two successive multiplication operators.

Rule 4: Arithmetic expressions may contain parentheses and also nested parentheses (i.e., parentheses within parentheses). In case of nested parentheses, the nearest left and right parentheses form a pair. If an expression contains parentheses, the number of left parentheses must be equal to the number of right parentheses. In case of nested parentheses, computation proceeds from the inner to the outer parentheses.

The Fortran statement f = a + (b + c * (d + e) will be rejected by the compiler because of unmatched parenthesis. It should have been, f = a + (b + c * (d + e)).

Rule 5: If a and b are real numbers, a ** b can be evaluated only if a is a positive quantity. This is because when both a and b are real numbers, a ** b is calculated as $e^{b \ln(a)}$, where ln is logarithm to the base e. If a is negative, ln (a) is not defined.

2.5 Precedence of the Arithmetic Operators

An arithmetic expression may contain different kinds of operators. Therefore, it is necessary to specify a rule regarding how the expressions like d = a / b * c are going to be evaluated. If the division is performed before the multiplication, the expression becomes algebraically d = (a / b) × c. However, if the multiplication is performed before the division, the expression becomes d = a / (b × c). Needless to say, the results of the two sets of calculations are different. This may be verified by assuming the values of a, b and c as 6, 3 and 2, respectively. In the first case, d = (6 / 3) × 2 = 4, and in the second case, d = 6 / (3 × 2) = 1.

Arithmetic operators are assigned different priorities. The priority of the exponentiation operator is the highest and that of the assignment operator is the lowest. The priority of the multiplication and the division operators is same and less than that of the exponentiation operator. The priority of the addition and the subtraction operators is same and lower than that of the multiplication and the division operators. The priority of unary plus and minus operators is in between the multiplication/division and the addition/subtraction operators. In any arithmetic expression, the high-priority operators are evaluated before the low-priority operators. For example, in case of the expression e = a + b * d, multiplication, b * d, is performed first and then a is added to get the result. If an arithmetic expression contains operators having same priority, computation proceeds from left to right. In case of d = a + b - c, the addition will be performed before the subtraction (Table 2.3).

There is one exception to the preceding rule. For exponentiation, the evaluation proceeds from right to left; for a ** b ** c, b ** c is performed first and then the result is used as the power of a. If brackets are used to indicate the order of evaluation, then (a ** b) ** c and a ** (b ** c) are not equivalent. This may be verified by assuming a = 2, b = 3 and c = 4. Substituting these values, (a ** b) ** c becomes 2^{12} and a ** (b ** c) = 2^{81}. When a complex number c1 is raised to the power of another complex number c2, the result is the principle value of $c1^{c2}$.

TABLE 2.3

Precedence of Arithmetic Operators

High	Exponentiation	**
↓	Multiplication and division	/, *
	Unary minus and plus	-, +
	Addition and subtraction	+, -
Low	Assignment	=

The parentheses have the highest priority. Inside the parentheses, the preceding rules are followed. It may be noted that -2**2 is -4 but (-2)**2 is 4. Also, a/(b*c) and a/b*c are not the same. In the first case, b*c is evaluated first and then a is divided by the result. In the second case, without the parentheses, computation proceeds according to the default priority rules. The priority of the division and the multiplication being equal, computation proceeds from the left to the right and the division is performed before the multiplication. The result is multiplied by c.

In the case of nested parentheses, computation starts from the innermost one.

Consider the expression $a + (b * (c * (d + e / f)))$. The innermost parentheses containing expressions involving the division and the addition are evaluated first, the default priority rules being used (division before addition). The result is then multiplied by c, which is then multiplied by b and the result is added to a. In Figure 2.2, the numbers indicate the order of evaluation. The first is indicated by 1 and the second by 2 and so on. The rule of thumb is that in case of any doubt, parentheses may be used to indicate the intention. Extra balanced parentheses do not cause any harm. The expression $d = a / b * c$ is same as $d = (a / b) * c$.

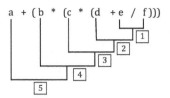

FIGURE 2.2
Evaluation of arithmetic expression.

Sometimes compilers are smart enough to change the order of the evaluation of the arithmetic expressions to make it more efficient. In Table 2.4, expressions and allowable alternative forms used by the compiler are shown. In these expressions, x, y and z are any type of numeric operands and a, b and c represent any arbitrary real or complex variables. Table 2.5 shows expressions that the compiler will never convert to the alternative forms. In this case, i and j are integers.

TABLE 2.4

Allowable Alternative

Expression	Alternative Form
x + y	y + x
x * y	y * x
-x + y	y - x
x + y + z	x + (y + z)
x - y + z	x - (y - z)
x * a / z	x * (a / z)
x * y - x * z	x * (y - z)
a / b / c	a * (b * c)
a / 5.0	0.2 * a

TABLE 2.5

Non-allowable Alternative

Expression	Non-allowable Alternative Form
i / 2	i * 0.5
x * i / j	x * (i / j)
i / j / a	i / (j * a)
(x + y) + z	x + (y + z)
(x * y) - (x * z)	x * (y - z)
x * (y - z)	x * y - x * z

2.6 Multiple Statements

Two or more Fortran statements, separated by a semicolon, may be placed in a line:

```
c = a + b; d = 10
```

This is same as follows:

```
c = a + b
d = 10
```

In this case, only the first statement may have a statement number. The semicolon is not a part of the Fortran statement. Two or more successive semicolons separated by zero or more blanks constitute a single semicolon; a=2 ; ; b=3 is same as a=2 ; b=3. If a line containing a complete Fortran statement is terminated by a semicolon, it is ignored. It is treated as a statement separator; a=10; is the same as a=10. Some compilers give compilation error if the first non-blank character in a line is a semicolon.

As multiple statements decrease the program readability, this is not recommended. However, in this book we use this semicolon to save some spaces in the book.

2.7 Mixed-Mode Operations

For a numeric expression involving variables or constants of different types, conversion takes place before the expression is evaluated. First, we consider expressions involving reals and integers.

In an expression involving a real and an integer constant or variable on the two sides of a binary arithmetic operator, the integer is converted into a real number before the calculation takes place. For example, the expression a+2 will be calculated as (a is a real variable): integer 2 will be converted to real 2.0 by the processor and 2.0 will be added to a. The result of (a+2) will be real.

Similarly, during the assignment operation, if the type of the variable (or the result) on the right-hand side of the assignment is different from the type of the variable on the left-hand side, an automatic type conversion takes place. An integer is converted into a real number, keeping the magnitude same – integer 2 is converted to real 2.0. However, a real number is converted into an integer by truncating the fractional part—real 4.56 is converted to integer 4.

One important point should be noted, the operands determine the type of the operation and accordingly type conversion takes place. Consider the expression i = j + a * 2, where i and j are integers and a is a real number (Figure 2.3). The steps required to perform this computation are as follows: (a) integer 2 is converted to a real number and stored in a temporary location, (b) priority of multiplication operator is more than that of addition, (c) 2.0 is multiplied by a—the result of the computation is real and is stored in a temporary location within the system, (d) j is converted to a real number because the result of the computation a * 2 is real and is stored in a temporary location, (e) the addition is performed in the

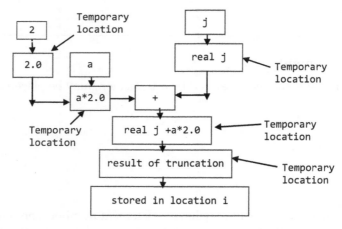

FIGURE 2.3
Mixed-mode arithmetic. i = j + a * 2

real mode—the result of addition is stored in a temporary location and (f) as the left-hand side of the assignment operator is an integer, the result of the computation is converted to an integer and is stored in location i.

Operations involving an integer or a real quantity with a double precision variable or constant are performed by first converting the integer or real quantity into double precision. However, it must be noted that the real number, thus converted, does not become more precise. The digits after the normal precision are meaningless. If single precision π (3.1415927) is converted to a double precision number, then the digits after the last digit on the right (say, 6) are not correct. Similarly, when a single precision expression is equated to a double precision variable, the single precision number is converted into a double precision number before it is stored. Again, extra digits so added to make it a double precision number do not have any significance. Thus, by equating a single precision number to a double precision variable, the resulting double precision number does not become more precise compared to the single precision number. The precision may be illustrated by considering the following program:

```
program testdbl
implicit none
double precision:: d1,d2
d1 = 2.0/3.0
d2 = 2.0d0/3.0d0
print *, d1,d2
end program testdbl
```

It is known that the result of the arithmetic operation is 0.666666666666.... The outputs of the program (using NAG Fortran compiler) are 0.6666666865348816 and 0.66666666666666, respectively. Note that the displayed value of d1 is correct up to 7 significant figures. Consider the following program:

```
program testdbl2
implicit none
double precision:: d1,d2
d1 = 3.1415926535897932
d2 = 3.1415926535897932d0
print *, d1,d2
end program testdbl2
```

In absence of 'd0', 3.1415926535897932 is treated as a single precision constant and additional digits after, say, 6 significant digits are removed to make it a single precision constant. Again, this single precision constant, when equated to a double precision variable d1, is converted into a double precision constant, but the digits so added do not have any significance. The outputs of the program are, respectively, 3.1415927410125732 and 3.1415926535897931.

We further illustrate mixed-mode arithmetic with double precision quantities in the following example:

```
double precision:: d1, d2, d3, d4
d1 = 1.1/3.1; d2 = 1.1d0/3.1; d3 = 1.1/3.1d0; d4 = 1.1d0/3.1d0
print *,"d1=",d1; print *,"d2=",d2; print *,"d3=",d3; print *,"d4=",d4
end
```

The results are as follows (shown in the same line):

```
d1 = 0.3548387289047241, d2 = 0.3548387205935669,
d3 = 0.3548387173683413, d4 = 0.3548387096774194
```

The results need explanation. The right-hand side of the expression involving d1 is in single precision. Therefore, the calculation is performed in single precision mode, and subsequently, the result is converted into double precision. The result is correct up to 7 places of decimal. One of the quantities on the right-hand side of the expression involving d2 is double precision. Therefore, the other single precision number 3.1 is converted into a double precision number. However, 3.1 when converted into a double precision number is less accurate than 3.1d0. Similar logic holds for d3, where 3.1 is a double precision number but 1.1 is not. Thus, in the case of d4, where both 1.1 and 3.1 are double precision numbers, the result is the most accurate and it is correct up to, say, 14 places of decimal.

In mixed-mode operations involving an integer or a real number with a complex quantity, the integer or the real number is converted into a complex number with the imaginary part set to 0. Thus, c*2.0, where c is a complex variable calculated as c * cmplx(2.0, 0.0). In a similar way, when a real number or an integer is equated to a complex variable, the real number or the integer becomes the real part of the complex variable with 0 as the imaginary part.

In arithmetic operations involving a double precision real and a complex variable, the double precision variables are first converted into a double complex number (both the real and the imaginary parts are double precision quantities). Subsequently, the complex variable is converted into a double precision complex number. The result is a double complex number.

This may be verified with the help of the following program:

```
double precision:: d=2.1234567891234d0
complex:: c=(2.0, 3.0)
print *, d+c
end
```

2.8 Integer Division

Integer division deserves special attention. Improper use of integer division may invite serious problems. If i and j are integers, i divided by j is calculated in the integer mode and the result of the computation is an integer (whole number). Following this logic, 3 / 2 is 1 and 2 / 3 is 0. Even if a is a real variable, the expression a = 5 / 2 is calculated in the integer mode. The result is 2 and because a is real, the result of the computation is converted to 2.0 and 2.0 is stored in a. The computer itself does the conversion. Now, consider the expressions i = 5.0 / 2.0 and j = 5 / 2. In the first case, 5.0 / 2.0 is a real number and it is 2.5. As i is an integer, 2.5 is truncated to 2 before it is stored in i. In the second case, 2 is stored in j and no type conversion takes place as all the operands and the variable on the left-hand side of the assignment sign are of the same type. It may be noted that the result of the computation in the preceding two cases are same; however, the way they are evaluated is different.

Consider the expression x = y * 10 ** (-2). The value of x is 0, irrespective of the value of y. This is because 10 ** (-2) is evaluated as 1 / 100. As both 1 and 100 are integers, computation is performed in the integer mode and the result is 0. One should be extremely careful while translating algebraic expressions like

```
a = 4 / 3 x π x r³
s = u x t + 1 / 2 x f x t²
```

into Fortran; 4 / 3 is 1 and 1 / 2 is 0. It is necessary to write at least one integer constant as a real number or better both as real numbers:

```
a = 4.0 / 3.0 * 3.1415926 * r ** 3
s = u * t + 1.0 / 2.0 * f * t ** 2 ! 1.0 / 2.0 may be written as 0.5
```

Often the result of computation unexpectedly turns out to be 0. In such a situation, one should look for an integer division similar to that shown earlier. It is advised to avoid mixed-mode operations as far as practicable.

2.9 List-Directed Input/Output Statement

We now introduce the free-formatted input/output statements. These are also called list-directed input/output statements.

An input statement reads a value of a variable from an external device, namely, the keyboard. Similarly, an output statement displays the value of a variable, the result of a computation or some message on an output device, namely, screen.

Various kinds of input/output devices are available. Some devices are used only for input and some devices are used only for output, while some devices are used for both input and output. Free-formatted input/output statements are

```
read *, list
print *, list
```

respectively, where list is a list of items to be read or written. If the list contains more than one element, the elements are separated by comma:

```
read *, a, b, c
print *, a, b, c
```

Therefore, read *, x will read data from the keyboard and store in location x erasing the existing value of x. Similarly, print *, y will display the current value of y on the screen. When a read * statement is encountered, the computer waits until the required input is supplied through the keyboard.

Normally, for numbers, one or more blanks are used as delimiters; read *, i, j, k will have the corresponding data from the keyboard as 10 20 30 so that 10, 20 and 30 will be stored in locations i, j and k, respectively. Usually, while supplying inputs, list items are separated by one or more blanks. However, characters like comma, tab or carriage return may also be used as delimiters. For example, in the earlier case, data may be entered in the following manner also:

```
     10, 20, 30 <enter>
or,  10 <enter>
     20 <enter>
     30 <enter>
or,  10, 20 <enter>
     30 <enter>
```

or various such combinations where `<enter>` indicates the enter key of the keyboard.

For real numbers, normally the decimal point is typed. If the decimal point is absent, it is assumed just before the delimiter. If x, y and z are declared as real, and the data corresponding to the read statement is entered as 10 20 30, it is taken as 10.0, 20.0 and 30.0 for x, y and z, respectively.

Real numbers may be entered as input in scientific notation also for the earlier case: 1.4e2 1.245e-4 3.25. In this case, x, y and z are assigned to $1.4 \times 10^2, 1.245 \times 10^{-4}$ and 3.25, respectively. To display a message on the screen, it needs to be enclosed in apostrophes (or quotes).

```
     print *, 'The result is = ', r
```

If the value of r is, say, 2.5, this print statement will display The result is = 2.5 on the screen. Within apostrophes, a blank is also treated as a character and the number of blanks between '=' and 2.5 on the screen depends on the number of blanks between '=' and the closing apostrophe within the print statement. There must be a comma between the message and the list element.

The read and print statements are equally valid for double precision variables.

```
     double precision:: d1
     read *, d1; print *, d1
```

The read and print statements can be used for complex variables also. As the complex number consists of two parts, two numbers are to be supplied during the input operation for each complex variable. Similarly, two numbers are displayed during the output operation. The first and the second numbers correspond to the real and the imaginary parts, respectively. The data, in response to the read statement, is supplied as (r, i), where r is the real part and i is the imaginary part.

```
     complex:: c1
     read *, c1
```

If the data supplied from the terminal is (2.0, 3.0), 2.0+i 3.0 will be assigned to c1. Similarly, print *, c1 will display the real and the imaginary parts within brackets.

To read a logical variable from the keyboard,

```
     read *, l  ! l is a logical variable
```

is to be used. The data must be of the following type: .TRUE. or TRUE for the true value and .FALSE. or FALSE (lowercase letters are also allowed) for the false value. In addition, if the first non-blank character is T or F (uppercase or lowercase) or a period followed by T or F (uppercase or lowercase), `true` or `false` value is, respectively, read in. Note that if TREU (intentional spelling mistake) is typed in place of TRUE, `true` value is assumed because the first non-blank character is T.

To print a logical variable, `print *, l` is used, where l is a logical quantity or expression. The `print` statement displays either T or F depending on whether l is `true` or `false`.

For characters, it is necessary to enclose the data in apostrophes (or quotes) when the data contains leading or trailing or embedded blanks. If the data does not contain leading, trailing or embedded blanks, apostrophes are optional. It is always better to enclose the character data in apostrophes (or quotes):

```
character (len=20):: ch
read *, ch; print *, ch
```

If the data is `Indian Association`, `'Indian'` is stored in ch because the blank between `'Indian'` and `'Association'` prevents reading the data beyond `'Indian'`. Here, the blank acts as a separator. Therefore, the `print` statement will print `'Indian'` (without the apostrophe). If the data is enclosed in apostrophes, ch becomes `'Indian Association'`. If the data corresponding to `read` statement is bbbIndian (b stands for a blank; 3 blanks in front), the variable ch will be assigned to blank as the second blank acts a separator. However, if it is desired that ch should be assigned exactly like the data, the data must be enclosed in apostrophes.

If the list item is a pointer (Chapter 16) or an allocatable variable (Chapter 15), the pointer should point to a target and the allocatable variable must have its status allocated.

List-directed input/output statements are very convenient. However, the programmer has practically no control over its appearance on the screen. For example, while using the `print` statement, the programmer has little control where the value will appear on the screen and how many digits will be displayed after the decimal point for a real number.

2.10 Variable Assignment—Comparative Study

We have just seen that a variable may be assigned in three different ways—through initialization, through assignment and through the `read` statement. A guideline may be prescribed as follows:

- True constants like π (3.1415926) and e (2.303) should be declared as named constants.
- If the initial value is required for a variable, then it should be initialized along with the declaration.
- If the same program is to be executed for different sets of values, then the corresponding variables should be read from outside; `read` statements should be used.

2.11 Library Functions

Several commonly used functions are available in the system as library functions to calculate square root, absolute value, trigonometric functions, etc. A library function is called (invoked) by its name and correct number and type of arguments are supplied within parentheses. For example, the library function sqrt calculates square root of a real (double precision, complex, etc.) quantity and takes one argument. The function nint takes one real number as its argument and returns an integer nearest to its argument; nint(3.7) returns 4 and nint (2.3) returns 2. The function floor takes one real number as its argument and returns the greatest integer less than or equal to its argument; floor(9.8) returns 9.

2.12 Memory Requirement of Intrinsic Data Types

Memory requirements of intrinsic data types (in binary digits, Chapter 10) can be obtained using the library function storage_size. This function takes one intrinsic data type as its argument. It returns the size of the data type in memory (binary digits).

```
integer:: a
real::b
double precision::c
complex::d
character::e
double complex:: f
print *, 'Integer, Size =            ',storage_size(a)
print *, 'Real, Size =               ',storage_size(b)
print *, 'Double Precision, Size = ',storage_size(c)
print *, 'Complex, Size =            ',storage_size(d)
print *, 'Character, Size =          ',storage_size(e)
print *, 'Double Complex, Size =    ',storage_size(f)
end
```

The outputs are as follows:

```
Integer, Size =                  32
Real, Size =                     32
Double Precision, Size =         64
Complex, Size =                  64
Character, Size =                 8
Double Complex, Size =          128
```

2.13 Programming Examples

The following example converts miles to kilometers:

```
program miletokm
implicit none
real, parameter:: factor=1.609 ! mile to km conversion factor
integer:: mile, yard
```

```
real:: km ! Marathon distance - 26 miles 385 yards=42.185 km
mile=26; yard=385
km=factor*(mile+yard/1760.0) ! 1 mile=1760 yards
print *, mile, 'Mile and ', Yard, 'yards = ', km, 'Kilometers'
end program miletokm
```

In this program, `yard/1760.0` is very crucial; `yard` is an integer and if `1760` is written in place of `1760.0`, then the result of the division would be 0.

The next program calculates the escape velocity from the earth. The escape velocity is defined as the minimum velocity that a projectile requires to escape from the earth. It is dependent on the mass (m), radius (r) of the earth and the universal gravitational constant G. It is given by `sqrt (2.0*G*m/r)`.

```
program escape
implicit none
real::rearth = 6378.0e3, earthm = 5.98e24 ! radius in meter, mass in kg
real::ev, G=6.67300e-11 ! gravitational constant
ev=sqrt(2.0*G*earthm/rearth)
print *,'Escape Velocity - Earth: ', ev/1000.0, 'km/sec'
end program escape
```

Output:

```
Escape Velocity - Earth: 11.18623 km/sec
```

If the argument is an expression, then the expression is evaluated and the square root of the result is calculated. The final program converts seconds to hours, minutes and seconds. It uses the property of integer division—an integer divided by another integer is an integer (whole number). The library function `mod(i, j)` returns the reminder of `i/j`.

```
program convt
implicit none
integer:: hh, mm, ss, t, second=7540
t=second/60
ss=mod(second,60) ! reminder of second/60. mod is a library function
hh=t/60; mm=mod(t,60)
print *, second, 'seconds = ', hh, 'hour ', mm,' minute ', ss, 'second'
end program convt
```

The output is:

```
7540 seconds = 2 hour 5 minute 40 second
```

2.14 BLOCK Construct

A `block` construct usually contains both declarations and statements. The construct is terminated by `end block`.

```
block
.
end block
```

The `block` construct may have a label. If `end block` has a label, it should be same as the label used with `block` construct.

```
thisblock: block
              .
           end block thisblock
```

Certain statements like `common`, `equivalence`, `implicit`, `intent`, `namelist`, `optional`, `statement function` and `value` cannot be used within a `block` construct. We have not yet introduced these statements. Some of these statements will be introduced later. Consider the following block construct:

```
block
 integer:: i
 .
 do i=1,10
 .
 enddo
 .
end block
```

The variables declared within a `block` construct are lost when the `block` is exited. In case the variable declared above the `block` construct (as shown later—known as global variable) has the same name as the variable declared within the `block`, called local variable for the `block`, the local variable always prevails (visible) over the global variable having the same name within the block. The global variable is not available within the `block` when there is a name conflict. The global variable reappears when the `block` is exited.

```
program block_demo
integer:: i            ! global to the block
i=27
 block
   integer:: i         ! local to the block
   i=77; print *, i
 end block
print *, i
end program block_demo
```

Inside the `block`, the value of `i` is 77. The global `i` is not available within the `block`. When the `block` is exited, the global `i` with its value as 27 reappears.

2.15 Assignment of BOZ Numbers

Normally, `boz` numbers cannot be used directly to assign a value to a variable. Integer, real, double precision and complex variables may be assigned to binary, octal and hex constants through the library functions `int`, `real`, `dble` and `cmplx`, respectively.

This is illustrated through the program shown next. The decimal value corresponding to the binary constants are indicated through in-line comments.

```
program num_sys
implicit none
integer:: i, j,k
real:: a, b,c
complex:: d
i=int(b'11')                        ! i=3
j=int(o'16')                        ! j=14
k=int(z'a1')                        ! k=161
a=real(int(b'111'))                 ! a=7.0
b=real(int(o'73'))                  ! b=59.0
c=real(int(z'ab'))                  ! c=171.0
d=cmplx(real(int(b'11')), real(int(o'16')))  ! (3.0, 14.0)
end program num_sys
```

If a real variable x is equated to `real(z'123456789')` (total 9 hex digits), both the gfortran and nagfor compilers give compilation error. However, ifort compiler takes the rightmost 8 hex digits. This is true for other boz numbers. The number of binary, octal, or hex digits should be such that it should not exceed the capacity of the processor (32 bits or 64 bits as the case may be).

2.16 Initialization and Library Functions

A variable may be initialized with standard library functions, which can be evaluated at the compilation time.

```
real:: a=sqrt(3.0)
```

In this case, a is initialized to the square root of 3.0, that is, 1.7320508.

2.17 Relational Operators

A relational operator tests a relation. It returns either `true` or `false`. For example, if a question is asked, "Is x greater than y?" The answer is either yes (`true`) or no (`false`). The relational operators are `lt`, `le`, `gt`, `ge`, `eq` and `ne`. These operators are bound by periods. There should not be any space between the periods and the operator. The operators are, respectively, less than, less than or equal to, greater than, greater than or equal to, equal to and not equal to. Either the symbolic notations (in both uppercase and lowercase letters) or the equivalent mathematical notations may be used. The symbol and alternative symbols may be mixed freely within an expression (Table 2.6).

TABLE 2.6

Relational Operators

Symbol	Math Symbol	Meaning
.lt.	<	Less than
.le.	<=	Less than or equal to
.gt.	>	Greater than
.ge.	>=	Greater than or equal to
.eq.	==	Equal to
.ne.	/=	Not equal to

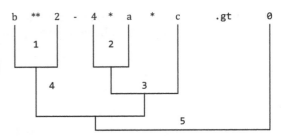

FIGURE 2.4

Precedence rules of relational operators. **(1)** b**2, **(2)** 4*a, **(3)** 4*a*c, **(4)** b**2-4*a*c, **(5)** b**2-4*a*c .gt. 0 .0

2.18 Precedence Rule of Relational Operators

The priority of all the relational operators are same, and it is lower than the priority of arithmetic operators. In an expression involving arithmetic and relational operators, the arithmetic operators are evaluated first and then the relations are tested (Figure 2.4). For b**2 - 4.0 * a * c.gt. 0, the order of evaluation is shown using serial numbers 1, 2, 3, 4 and 5.

2.19 Relational Operators and Complex Numbers

The relational operators .lt., .le., .gt. and .ge. cannot be used for comparison between two complex numbers or variables. Only .eq. (equal) and .ne. (not equal) may be used to compare two complex numbers or variables. Two complex numbers are considered to be equal if both the real and the imaginary parts are separately equal. If z1 and z2 are complex variables with z1=cmplx(2.0,3.0) and z2=cmplx(2.0,3.0), then z1 and z2 are equal but the complex variables c1 and c2, where c1=cmplx(3.0,4.0) and c2=cmplx(4.0,3.0), are not equal, as their real and imaginary parts are separately not equal.

Although the arithmetic if has been declared as obsolete feature, it may be mentioned that the arithmetic expression involving a complex number cannot be used with arithmetic if.

2.20 Logical Operators

There are five logical operators. All logical operators are bound by periods (Table 2.7). There should not be any space between the periods and the logical operators.

NOT: The logical operator .not. is a logical negation. It changes true to false and vice versa; .not. (a>b) is true if a is not greater than b, that is, a>b is false. It is false if a greater than b is true. For l = .not. (a>b), true value is assigned to l if a is not greater than b.

TABLE 2.7

Logical Operators

Operator	Meaning
.not.	Logical negation
.and.	Logical conjunction
.or.	Logical inclusive disjunction
.eqv.	Logical equivalence
.neqv.	Logical nonequivalence

AND: The logical operator .and. is a logical conjunction. If 11 and 12 are logical, then 11.and. 12 is true if both 11 and 12 are true. It is false if one of them or both are false. For 1 = a>b.and. p<q, 1 is true if a is greater than b and at the same time p is less than q. Otherwise, it is false.

OR: The operator for the logical inclusive disjunction is .or.. If 11 and 12 are logical, then 11.or. 12 is true if either 11 or 12 or both are true. It is false if both of them are false. For 1 = a<=b.or. p==q, 1 is true if either a is less than or equal to b or p is equal to q; that is, one or both the relations are true. It is false only when both are false.

EQV: The logical operator .eqv. stands for logical equivalence. If 11 and 12 are logical, then 11.eqv. 12 is true if both 11 and 12 are true or both 11 and 12 are false. For 1 = a.eq. b.eqv. p.ne.q is true if both the logical expressions are true or both the logical expressions are false.

NEQV: Finally, the logical operator .neqv. is logical nonequivalence. If both 11 and 12 are logical, then 11.neqv. 12 is true if either 11 or 12 is true. It is false if both 11 and 12 are true or both 11 and 12 are false. For 1 = a>b.neqv. p<=q is true if one of the logical expression is true. It is false otherwise.

If 11 and 12 are logical, then Table 2.8 shows the results of various logical operations. A logical expression may contain any number of logical operators.

TABLE 2.8

Truth Table for Logical Operators

11	12	.not. 12	11 .and. 12	11 .or. 12	11 .eqv. 12	11 .neqv. 12
T	T	F	T	T	T	F
T	F	T	F	T	F	T
F	T	F	F	T	F	T
F	F	T	F	F	T	F

Note: "T" stands for true, and "F" stands for false.

2.21 Precedence Rule of Logical Operators

The priority of the logical operators is lower than that of the relational operators. Among the logical operators, the priority, high to low, is shown in Table 2.9.

The priority of .eqv. and .neqv. is same. Consider the following expression:

```
l = a.gt.b .or. p.le.q .and. r.eq.s
```

The expression is evaluated as

```
l = a.gt.b .or. (p.le.q .and. r.eq.s)
```

as the priority of and is greater than that of or. Thus l is true if either a is greater than b or both the conditions p<=q and r==s are true.

TABLE 2.9

Precedence of Logical Operators

High	.not.
	.and.
	.or.
Low	.eqv.
	.neqv.

2.22 Precedence of the Operators Discussed So Far

We conclude this chapter by redrawing the priority table (Table 2.10) containing all the operators discussed so far.

TABLE 2.10

Precedence of the Operators

Priority	Symbol	Meaning
High	**	Exponentiation
	*, /	Multiplication, division
	+, -	Unary plus, unary minus
	+, -	Binary addition and subtraction
	.lt., .le., .gt., .ge., .eq., .ne.	Relational operators
	.not.	Logical negation
	.and.	Logical conjunction
	.or.	Logical inclusive disjunction
	.eqv., .neqv.	Logical equivalence, nonequivalence
Low	=	Assignment

3

Branch and Loop Statements

The computer executes instructions sequentially, that is, one after another. Normally, programming logic is not so simple, and the flow of a program is not always linear. Branch statements allow branching or jumping to a particular statement as demanded by the program logic.

There are two types of branch statements: unconditional and conditional. An unconditional branch statement, when executed, allows branching to a particular statement without any condition, whereas a conditional branch statement transfers the control to a particular statement after testing a certain condition—the branch is executed depending upon the state of the condition.

In addition, programming logic often requires executing a group of instructions repeatedly, depending on certain conditions. For example, if one tries to add the series 1+2+3+. . .+1000, it is not practicable to sum the series directly by assuming that one will not use the formula n(n+1)/2, treating the series as AP series. Fortran provides instructions to execute statements repeatedly, depending on some conditions. It is an iterative process. These instructions are collectively called loop instructions. Two instructions, (i) do and (ii) do while, are used to perform loop operation.

3.1 GO TO Statement

The goto statement is an unconditional branch statement. The syntax is goto nnn, where nnn is a statement number, which must refer to an executable statement within the same (scoping) unit as goto.

When a goto statement is encountered, the control is unconditionally passed to the statement indicated by the number mentioned in the goto statement. The statement where the control is transferred may be before or after the goto statement; for example, goto 20 will transfer the control to the statement number 20 unconditionally. There may be a space between go and to. The target of a goto statement may be the initial statement of associate, case, do, do concurrent, forall, if, select type and where; however, the target cannot be the end associate, end select, enddo, end forall, or endif statement unless goto is executed from within the corresponding block (these statements are discussed at appropriate places in the text).

Indiscriminate use of goto makes a program much unstructured and should be avoided as far as practicable. In fact, there are schools that advocate goto-less programming. There are other schools that believe it is still possible to write a structured program with restricted use of goto, especially when the jumps are very small (say, a few lines). However, the current version of Fortran provides several other statements, which facilitate writing a goto-less program with ease. In this book, we do not use the goto statement to write any serious program. There is no programming example in this book involving goto.

3.2 Block IF

The syntax of block if is

```
if (cond) then
   stmt
endif
```

In all subsequent discussions, whenever we refer to stmt or stmt-1 or the like, we mean either a single Fortran statement, called a simple statement, or a group of Fortran statements, called a compound statement or a block bound by block and end block. Statement stmt could also be an empty statement.

The block if statement tests a condition. If the condition is true, all statements as permitted by the program logic up to endif are executed. If the condition is false, the statements up to endif are skipped, and the statement following endif is executed (Figure 3.1).

```
if (a > b) then
   i=i+1      ! single statement
endif
if (p .lt. q) then
   i=i-1        ! statement 1
   print *, i   ! statement 2
endif
```

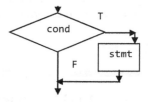

FIGURE 3.1
Block if.

In the second case, if the condition is true, both statements are executed. Block if may be labeled. However, this is optional.

```
check: if(x.eq.y) then
       c=a+b
       d=sqrt(c)
       endif check
```

There is a colon between the label and the block if. Needless to say, when the if statement has a label, the endif statement must also have the same label.

3.3 IF-THEN-ELSE

This construct allows the execution of a group of instructions (or a single instruction) if a condition is true and another group of instructions (or a single instruction) if the condition is false (Figure 3.2).

```
if (cond) then
   stmt-1
else
   stmt-2
endif
if (a>b) then
   i=i+1                    ! true path
else
   i=i-1                    ! false path
endif
```

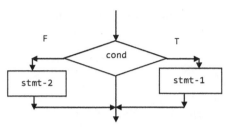

FIGURE 3.2
If-then-else.

In the preceding program segment, if the condition is true, that is, a is greater than b, i is incremented by 1, and if the condition is false, i is decremented by 1. As already mentioned, there can be any number of statements between then and else and between else and endif. The most important point to be noted is that, depending upon the condition, either the 'then' path or the 'else' path is chosen—both are never chosen simultaneously. These two paths are insulated from each other. Usually, the statement following endif is executed, provided there is no branch statement (say, goto) or the program is terminated within then . . . else or else . . . endif. Normally, such statement is not present.

The following program tests whether the input number is even or odd:

```
program even_odd
implicit none
integer :: num
print *,  'Type a positive integer '
read *, num
if(num/2*2-num .eq. 0) then ! note integer division
   print *, num, ' is an even number '
else
   print *, num, ' is an odd number '
endif
end
```

Since num is an integer, num/2 (2 is also an integer) will be calculated in integer mode. Let us consider two cases: n=4 and 3; num/2*2 - num is 0 and -1 when n=4 and n=3, respectively (4/2 is 2, and 3/2 is 1). Therefore, the property of the integer division is used to distinguish between the odd and the even number.

The condition decides the position of even and odd print statements. For example, if the condition is modified to if (num/2*2 - num .ne. 0), the same program is to be modified as shown in the following:

```
if (num/2*2-num .ne. 0) then
   print *, num, ' is an odd number '
else
   print *, num, ' is an even number '
endif
```

`If-then-else` may be labeled, provided the corresponding `if` and `endif` are labeled:

```
check: if (a>b) then
         i=i+1
       else check
         i=i-1
       endif check
```

or

```
check: if (a>b) then
         i=i+1
       else
         i=i-1
       endif check
```

Moreover, all of them must have the same label as shown in the preceding program segment.

3.4 ELSE-IF

The `else` part of `if-then-else` may comprise another `if` statement.

```
if (cond-1) then
   stmt-1
else if (cond-2) then
      stmt-2
   else if (cond-3) then
         stmt-3
                .
         else if (cond-n) then
            stmt-n
            else
            stmt
   endif
```

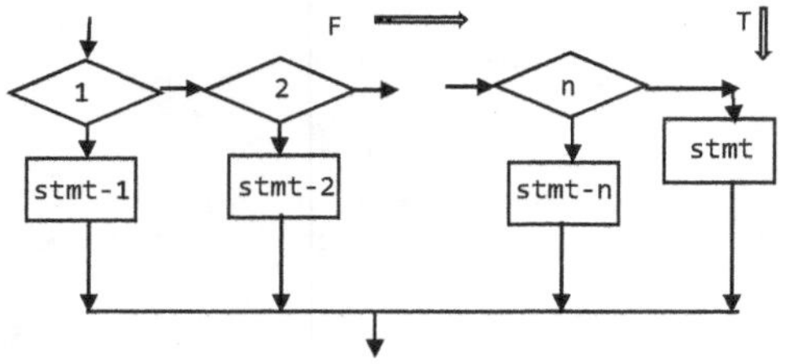

FIGURE 3.3
Else-if.

If condition `cond-1` is `true`, then statement `stmt-1` is executed, and the next statement is the statement following `endif`. If the condition `cond-1` is `false`, then condition `cond-2` is tested. If the condition is `true`, then `stmt-2` is executed, and like the previous condition, the next statement is the statement following `endif`. If all conditions are `false`, then statement `stmt` is executed. Note that there is only one `endif` in the sequence (Figure 3.3). The process is illustrated by means of a few examples:

```
program maxmin
implicit none
integer :: num1, num2 ! maximum between two numbers
print *, 'Type two integers  '
read *,num1, num2 ! assign some value to num1 and num2
if (num1 > num2) then
   print *,'num1 is greater than num2 ', num1, num2
else if (num1 < num2) then
      print *,'num1 is less than num2 ', num1, num2
```

```
      else
         print *,'num1 is equal to num2 ', num1, num2
   endif
   end program maxmin
```

The next program calculates the roots of a quadratic equation:

```
program roots
implicit none
real ::  x1,x2,a,b,c,temp,discr,twoa ! roots of quadratic equation;
print *, 'Type the values of a, b and c ' ! a*x**2+b*x+c=0
read *, a, b, c
discr = b*b - 4.0*a*c
if(discr > 0.0) then ! distinct roots
   temp = sqrt(discr)
   twoa = 1.0/(2.0*a)
   x1 = (-b+temp)*twoa
   x2 = (-b-temp)*twoa
   print *, 'Roots are : ', x1, x2
else if (discr .eq. 0) then ! such test should be avoided; chapter 10
      x1 = -b/(2.0*a) ! equal roots
      x2 = x1
      print *, 'Roots are equal ', x1, x2
   else ! not a real number
      print *, 'Roots are complex number'
endif
end program roots
```

A few important points may be noted here. The expression `1.0/(2.0*a)` is required for calculating both `x1` and `x2`. It is better to define a temporary variable for `1.0/(2.0*a)` so that the calculation is done only once. There is no need to calculate the same expression twice. Moreover, normally multiplication is a faster operation than division, and therefore, divisions may be replaced by multiplication (`1.0/2.0=0.5`). An optimizing compiler may do the same thing.

The next program converts marks of an examination to remarks like excellent, very good, etc., according to Table 3.1.

```
program marktorem
implicit none
integer :: mark
print *, 'Enter marks ... '
read *, mark ! set marks to some value
if(mark > 90) then
   print *, 'Excellent, ', 'mark = ',mark
else if (mark > 80) then
      print *, 'Very good, ','mark = ',mark
else if (mark > 70) then
      print *, 'Good, ','mark = ',mark
else if(mark > 60) then
      print *, 'Fair, ','mark = ',mark
   else
      print *,'Repeat, ','mark = ',mark
endif
end program marktorem
```

TABLE 3.1

Mark versus Remark

Mark	Remark
>90	Excellent
>80 and ≤90	Very good
>70 and ≤80	Good
>60 and ≤70	Fair
≤60	Repeat

Else-if may be labeled, provided the corresponding if and endif are also labeled:

```
lab:    if(n.eq.0) then
            n=1
        else if (n.eq.1) then lab
            n=-1
            else lab
            n=0
        endif lab
```

3.5 Nested IF

If statements may be nested; there can be an if statement within another if statement
(Figure 3.4).

```
if (cond-1) then
  if (cond-2) then
    stmt-1
  else
    stmt-2
  endif
else
  stmt-3
endif
```

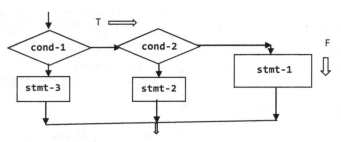

If cond-1 is true, then cond-2   **FIGURE 3.4**
is tested. If cond-2 is true, then   Nested if.
stmt-1 is executed, and if it is
false, then stmt-2 is executed. If cond-1 is false, then stmt-3 is executed. Note
that condition cond-2 is checked only when cond-1 is true. If cond-1 is false, then
stmt-3 is executed — cond-2 is not checked.

```
if(a>b) then
  if(p<q) then
    i=i+1   ! both the conditions are true
  else
    i=i-1   ! a>b is true, p<q is false
  endif
else
  i=0       ! a>b is false
endif
```

If statements can be nested to any level.

3.6 Nested IF without ELSE

As an `if` statement might not have an `else`, the number of `else` in a nested `if` is always less than or equal to the number of `then`. The number of `then` cannot be less than the number of `else`.

In the following program segment, the `else` is associated with the inner `if`—the outer `if` does not have any `else`. If condition `cond-1` is `false`, then the control is passed to the statement following the outer `endif`.

```
if (a>b) then      ! no else with this if
   if(p>q) then
      i=i+1
   else            ! associated with inner if
      i=i-1
   endif           ! associated with inner if
endif              ! associated with outer if
```

If `a>b` is `false`, then the value of `i` will not be changed, and the next statement to be executed will be the statement following the second `endif` (outer `endif`).

Consider the following program segment:

```
if (a>b) then
   if(p<q) then    ! no else for this if
      i=i+1
   endif           ! associated with inner if
else               ! associated with outer if
   i=i-1
endif              ! associated with outer if
```

If `a>b` is true, but `p<q` is false, then the value of `i` will not be changed because there is no `else` associated with the inner `if`. If `a>b` is `false`, then `i` will be decremented by 1.

3.7 Rules of Block IF

1. It is possible to come out of the block `if` by means of a branch statement:

```
   if (a>b) then
      ...
      goto 10 !allowed
      ...
   endif
10    ...
```

2. However, it is not permitted to enter into block if from outside (without touching the if statement):

```
      go to 100 ! not allowed
      . . .
      if(p<q) then
100   . . .
      endif
      goto 20  ! not allowed
      if(a>b) then
         .
      else
         .
20    . . . .
      endif
```

Similarly, it is not permitted to jump into the 'else' path from the 'then' path, and vice versa:

```
      if (a<b) then
         .
       goto 30 ! not permitted
         .
      else
         .
30    . . . . . .
      endif
```

3. It is legal if such transfer is caused within the 'then' or 'else' block:

```
      if (a>b) then
         .
       goto 10  ! allowed
         .
10    . . . . . .
      else
         .
       goto 30  ! allowed
         .
   30 . . . .
      endif
```

4. If labels are used for nested if, the inner label must terminate before the outer label:

```
outer:   if (cond-1) then
            .
inner:      if (cond-2) then
            .
            endif inner ! ok
            .
         endif outer
```

5. The `else-if` statement cannot appear after the `else` statement.

6. One can jump onto the `endif` statement from either the `'then'` path or the `'else'` path.

3.8 CASE Statement

Case statements allow multiway branching. The syntax of the `case` statement is as follows:

```
select case (expr)
  case(low-1:high-1)
    stmt-1
  case(low-2:high-2)
    stmt-2
      .
  end select
```

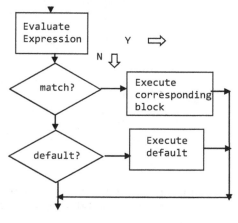

FIGURE 3.5
Case statement.

The `low` and `high` are the lower and upper bounds of the selected expression, respectively. Both the `high` and the `low` or one of them must be present. When the expression `expr` is evaluated, it should return an integer (logical and character data types will be considered later). A particular path is chosen, depending on the value of the expression. If the value of the expression is between (and including) `low-1` and `high-1`, then `stmt-1` is executed, and the next statement is the statement following `end select`. Similarly, if the value of the expression lies between `low-2` and `high-2`, then `stmt-2` is executed.

A particular path is automatically insulated from other paths—`case` does not fall through. If the value of the expression does not match with any of the `case` statements, then the `case` statement is ignored (`case default` will be discussed shortly) and the executable statement following the `end select` is the next statement executed. If `stmt-1`, `stmt-2`, etc. do not contain any branch statement (say, `goto`) or the program terminates, the next executable statement following the `end select` is always executed.

The `case` statement may have a label. However, this is optional.

```
lab:   select case (expr)
          .
       end select lab
```

There are several types of `case` statements.

1. **Type I:** When only one integer (or expression) is present, the `case` is selected for that particular value.

```
select case (index)
  case(1)
    n=n+1
  case(2)
    n=n-1
```

```
   case(3)
    n=n+2
   case(4)
    n=n-2
  end select
```

When the value of index is 1, n is incremented by 1, and the next statement is the statement following end select. Similarly, for index=2, 3 and 4, the respective path is chosen. Note that, after choosing a particular path and exhausting the statements in the path, the control is passed to the statement following end select.

2. **Type II:** If the upper and lower bounds are specified, a particular path is chosen when the value of the expression falls within (including) the range.

```
  select case (index)
   case(1:3)    ! index=1 or 2 or 3
    n=n+1
   case(4:6)    ! index=4 or 5 or 6
    n=n-1
   case(7:14)   ! index=7 or 8 or 9... or 14
    n=n+2
   case(15:20)  ! index=15 or 16 or 17 ... 20
    n=n-2
  end select
```

In the preceding example, when the value of the index is 1, 2 or 3, n is incremented by 1; when the index is 4, 5 or 6, n is decremented by 1, and similarly for other paths.

3. **Type III:** When either the upper or the lower bound is missing, but the colon is present, the upper or lower bound is, respectively, assumed to be all permissible values. For example, (20:) indicates all values greater than or equal to 20. In a similar way, (:1) indicates 1, 0 and all negative integers (all values less than or equal to 1).

```
  select case (index)
   case( :0)    ! index less than or equal to 0
    n=n+1
   case(1:3)
    n=n-1
   case(4:10)
    n=n+2
   case(11: )   ! index greater than or equal to 11
    n=n-2
  end select
```

In the case of case(:0), the lower bound is missing; for all values of the index less than or equal to 0, n is incremented by 1. Similarly, for all values of the index greater than or equal to 11, n is decremented by 2.

4. **Type IV:** It is possible to specify discrete as well as continuous values for the expressions, provided they do not overlap.

```
select case(index)
 case(1,3,5,9:12,14)
 stmt-1
end select
```

Statement `stmt-1` will be executed if the value of the control variable (in this case, `index`) is 1 or 3 or 5 or any value between (and including) 9 and 12 or 14. Similarly, for

```
select case(index)
 case(1:3, 5: )
 stmt-2
end select
```

Statement `stmt-2` will be executed if the value of the control variable is 1 or 2 or 3 or any value greater than or equal to 5. Also, for

```
select case(index)
 case( :0,2,3,7: )
 stmt-3
end select
```

Statement `stmt-3` will be executed when the control variable has a value less than or equal to 0 or 2 or 3 or any value greater than or equal to 7.

3.9 CASE DEFAULT

If the value of an expression does not match with any `case`, then the `select case` statement is ignored. However, if it is necessary to choose a default path — should there be no match—case `default` may be used (Figure 3.5).

```
select case (index)
 case(1,2)           ! 1 and 2
    n=n+1
 case(3:6)           ! 3, 4, 5, 6
   n=n-1
 case(7:10)          ! 7, 8, 9, 10
   n=n+2
 case(11:20)         ! 11 to 20
   n=n-2
 case default        ! none of the above
   n=0
end select
```

The case `default` is chosen in the preceding program segment when the `index` is less than 1 or greater than 20 and n is set to 0. The case `default` does not have to be the last item of the list of `case` statements, and there must be a space between `case` and `default`. There cannot be more than one case `default` within a `select case` construct.

3.10 CASE and LOGICAL

The selection criterion of a case statement may be a logical constant, logical variable or
logical expression.

```
logical :: l=.false.
integer::i=0
select case (l)
  case (.true.)         ! follow this path if l is true
    i=i+1
  case (.false.)        ! follow this path if l is false
    i=i-1
end select
```

3.11 Nested CASE

The case can be nested; that is, there can be one case within another.

```
select case(i)
  case(1)
  select case(j)
    case (1)            ! i=1, j=1
      .
    case(2:3)           ! i=1, j=2 or 3
      .
    case(4: )           ! i=1, j=4,5,6,....
      .
  end select
  case(2)               ! i=2
    .
end select
```

The values of i and j for different paths are shown through inline comments.

3.12 EXIT Statement and CASE

If an exit statement is executed within a case statement, the control is passed to the
statement end select. In other words, the case statement is exited.

```
      integer:: index
        .
11:   select case (index)
        case (1)
          .
```

```
    case (2)
       if(cond) then    ! some condition
          exit 11
       endif
         .
    case (3)
         .
    case default
         .
end select 11
```

If the condition (cond) is true when the value of the index is 2, the exit statement is executed, and the control is passed to the statement following the statement end select.

3.13 Rules of CASE

1. There should not be any overlap among the various paths while specifying the case. The following is not acceptable:

```
select case (index)
   case(1:2)
      n=n+1
   case(2:3)                ! overlap, not allowed
      n=n-1
end select
```

2. A statement within a particular case may have a statement number. It is possible to jump to a statement by goto or if, within the same path. However, such jump is not allowed from one path to another. In addition, a jump is not allowed from outside the case to inside the case.

```
        goto 10              ! not allowed
         .
        select case(index)
           case(1:2)
              .
10            i=i+1
              .
        end select
        select case(index)   ! allowed
           case(1:2)
              n=n+1
              if(n>20) then
                 goto 20      ! this is allowed
              endif
20            .....
           case(3:4)
              .
        end select
```

```
      select case(index)
        case(1:2)
          n=n+1
            .
          if(n>20) then
            goto 30    ! not allowed
          endif
            .
        case(3:4)
          n=n-1
            .
30          .
      end select
```

3. A `case` statement may be labeled. In the case of nested statements, the inner one must terminate before the outer one.

4. A block `if` may contain a `case`. However, the terminal point of the `case` must be before the block `if`.

```
      outer:    if(cond-1) then
                    .
      inner:      select case(expr)
                    .
                  end select inner
                    .
                endif outer
```

5. A `case` statement may contain a block `if` statement. In this case, the block `if` must terminate before the `end select` statement.

6. It is possible to jump to the `end select` statement from within the `case` construct.

7. For a labeled `case` statement, if an `exit` statement with the label is executed, the control is passed to the corresponding `end select` statement with the same label. For a nested `case`, the inner `case` may transfer control to the statement following the `end select` of the outer `case` using the label of the outer `case`.

8. The path cannot be empty, like `case ( )`.

3.14 Programming Example

The following program converts marks to grades, using `case`:

```
program marktogr
implicit none
integer :: mark !    read mark
print *,'type a positive integer between 0 and 100'
read *, mark
select case (mark)
```

```
      case(91:100)
        print *, 'Excellent ', mark
      case(81:90)
        print *, 'Very good ', mark
      case(71:80)
        print *, 'Good ', mark
      case(61:70)
        print *, 'Fair ', mark
      case(0:60)
        print *, 'Repeat ', mark
      case default
        print *, 'Bad data ', mark
    end select
    end program marktogr
```

3.15 DO Statement

This statement is used to execute one or more instructions repeatedly, depending upon a loop count. There are several forms of do statements. The most popular form is as follows:

```
    do int_var = m1, m2, m3
    .
    end do
```

where int_var is an integer variable (also called a control or loop variable) and m1, m2 and m3 are integer constants (scalar) or integer expressions. If m1, m2 and m3 are expressions, they are evaluated before the do loop is entered. A space between end and do is optional:

```
    do i=1, 10, 1
    .
    enddo
```

There may be an optional comma between do and the integer variable:

```
    do, i=1, 10, 1
    .
    enddo
```

The preceding do statement works as follows.

The number of times the loop will be executed (t) is calculated by evaluating the following expression:

```
    t=max((m2-m1)/m3+1,0)
```

The library function max returns the maximum value of the two arguments. It is obvious that the minimum value of t is 0. If t is greater than 0, then the loop is entered and int_var is initialized to m1. The statements following do, up to enddo, are executed depending on the programming logic. When the enddo statement is reached, t is decremented by 1 and int_var is incremented by m3 (in each iteration). This process continues until t becomes 0, and that is when the do loop is exited. If the increment is 1, then the increment (m3) may be omitted.

```
do i=1,10        ! increment 1
.
enddo
```

Although a do statement may have a terminal statement other than enddo, in this book we always terminate the do statement using enddo. This procedure will help the programmer to avoid a number of possible mistakes that might creep up because of improper use of the terminal statement (Figure 3.6).

A do statement may have a label. However, it is optional; do and enddo must have the same label:

```
loop: do i= 1, 10
.
enddo loop
```

The next two examples are invalid do statements:

```
11:     do i= 1, 10
.
enddo 12 ! labels do not match
do j= 1, 100
.
enddo 12 ! do does not have label
```

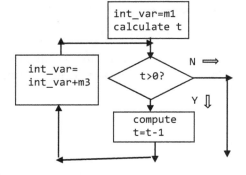

FIGURE 3.6
Do statement.

The use of do statements is illustrated in the following examples:

```
program add
implicit none
integer, parameter :: limit=100      !  s=1+2+3+... +100
integer :: sum=0, i                  !  the counter is set to zero
do i=1, limit
  sum=sum+i ! 1+2+3 . . +100
enddo
print *, sum
end program add

program factorial
implicit none
integer, parameter :: limit=6        !  p=1x2x3...x6
integer :: prod=1, i
do i=1, limit                        ! for product the counter is set to 1
  prod=prod*i
enddo
print *, prod
end  program factorial
```

The next problem is to find all numbers between 100 and 999 such that the number is equal to the sum of the cubes of the digits, for example, $153 = 1^3 + 5^3 + 3^3$. The algorithm to solve the problem is to extract the individual digits from all numbers between 100 and 999 and check whether they satisfy the aforementioned conditions.

```
program numcube
implicit none
integer :: num, i1, i2, i3, it
print *, 'The numbers are:'
do num=100, 999        ! extract digits
   i1=num/100          ! integer division gives the left most digit
   it=mod(num,100)     ! mod(i,j) is the remainder of i/j
   i2=it/10            ! integer division, extract the second digit
   i3=mod(it,10)       ! right most digit
   if(i1**3+i2**3+i3**3 .eq. num) then
      print *, num
   endif
enddo
end program numcube
```

It may be verified that there are four such numbers between 100 and 999 that satisfy the condition. They are 153, 370, 371 and 407—known as Armstrong numbers.

3.16 Negative Increment

The increment of a do loop (m3) may be negative. However, in that case, the initial value of the control variable (m1) should be greater than or equal to the final value (m2) so that t (loop count) becomes greater than 0.

```
sum=0 ! s=100+99+98+... +1
do i=100, 1, -1
   sum=sum+i
enddo
```

3.17 Infinite Loop

Another form of do is possible when no control variable is required. It forms an infinite loop, and unless there is some mechanism to come out of the loop, the loop will never terminate.

```
do
.
enddo
```

The statements between do and enddo are repeatedly executed.

3.18 EXIT Statement

When the `exit` statement is executed within a do loop, the control is passed to the statement following the enddo. The following program segment adds 1 to 100:

```
sum=0
i=1
do ! infinite loop
  sum=sum+i
  i=i+1
  if(i > 100) then     ! comes out of the loop when i exceeds 100
    exit
  endif
enddo
print *, sum
```

When i exceeds 100, the `exit` statement is executed, and the control is passed to the statement following the enddo, that is, the `print` statement.

3.19 CYCLE Statement

When a `cycle` statement is executed within a do loop, all instructions following the `cycle` statement up to enddo (terminal statement) are skipped, and a fresh cycle for the do loop begins. Obviously, necessary modifications of different variables take place before the new cycle begins. The difference between `exit` and `cycle` statements may be noted at this stage. The former, when executed, transfers the control outside the current loop, whereas, the latter initiates, after skipping all instructions up to the terminal statement of the loop, a new cycle from that point. The following program segment, though not an elegant program, shows how to add all even numbers between 1 and 100 using `cycle`:

```
sum=0
do i=1, 100
  if(mod(i,2) /=0) then
    cycle   ! skip all odd numbers
  endif
  sum=sum+i
enddo
```

This program demonstrates the `cycle` statement. An elegant way to do the same job is to use a do statement with 2 as an increment:

```
sum=0
do i=2, 100, 2
  sum=sum+i
enddo
```

3.20 DO WHILE

The syntax of do while is as follows:

```
do while (cond)
   .
enddo
```

The space between do and while is mandatory. There may be an optional comma between do and while. The condition cond is tested; if it is true, all statements permitted by the program logic following do while up to enddo are executed. The control is then passed to the top of the loop, and a fresh cycle begins. The condition is tested at the beginning of each cycle. The process continues until the condition becomes false, and in that case, the control is passed to the statement following enddo.

Any positive number can be reduced to 1 by arithmetic operations. If the number is odd, it is multiplied by 3, and then 1 is added; if the number is even, it is divided by 2. This is continued until the number reduces to 1.

```
program reduce
implicit none
integer :: num, count= 0
print *, 'Type one positive integer '
read *, num ! read the number
do while (num /= 1)
   if (mod(num,2) /= 0) then
      num = 3*num + 1          ! num is odd
   else
      num = num/2             ! num is even
   endif
   count=count+1
enddo
print *, 'Convergence achieved in ', count,' Cycles:', ' Num = ', num
end program reduce
```

The next program calculates the GCD between two numbers.

```
program gcd
implicit none
integer :: first, second, num1, num2, temp
print *, 'type two positive integers'
read *, first, second  ! say gcd of 81 and 45 required
num1=first
num2=second
temp=mod(num1,num2)
if (temp .eq. 0) then
   print *, 'The gcd of ',first, 'and ', second, 'is = ', num2
   stop                 ! gcd obtained so stop
endif
do while(temp .ne. 0)
```

```
   num1=num2    ! second number becomes new first number
   num2=temp    ! new second number becomes reminder of num1 and num2
   temp=mod(num1,num2) ! continues till the reminder becomes zero
enddo
print *, 'The gcd of ',first, 'and ', second, 'is = ', num2
end program gcd
```

The statement `stop` terminates the job at that point. The `stop` statement may contain a number (integer) or a character string. This is displayed when the `stop` statement is executed. This helps to identify the `stop` statement if there are more than one `stop` statements within the program.

```
stop 10
stop 'End of data'
```

In this case, the `stop` statement is necessary if the GCD is obtained before entering the do `while` loop (say, when `first=12`, `second=4`).

The trapezoidal rule can be used to integrate a function numerically:

$$\int f(x) \ dx = h[(f(a)+ f(b))/2+ f(a+h)+ f(a+2h)+...+ f(a+nh)]$$

where a and b are the limits of integration and `h= (b-a) /n`, with n an even number. We now integrate a function `f (x) =x**3*exp (-x**3)`, and the limits of integration are 1 and 4; `exp` is a library function that calculates the e to the power its argument (real).

```
program trap
implicit none
integer :: i, n=100
real :: a=1.0, b=4.0
real :: sum=0.0, h, x
h=(b-a)/n                  ! length of the interval
sum=sum+0.5*(a**3*exp(-a**3) + b**3*exp(-b**3)) ! first and last point
x=a+h
do i=2,n
   sum=sum+(x**3*exp(-x**3))   ! f(a+h) + f(a+2h) …
   x=x+h
enddo
sum=sum*h
print *, 'The result of integration is ', sum
end program trap
```

A better method to integrate a function numerically is Simpson's one-third rule. In this method,

$$\int f(x) \ dx = h/3 [(y_1+y_{n+1})+ 4(y_2+y_4+y_6+...+ y_n)+ 2(y_3+y_5+y_7+...+y_{n-1})]$$

where `h = (b-a) /n`, n is an even number, and a and b are the limits of integration.

```
program simpson
implicit none
integer :: i, n=100
```

```
real :: a=1.0, b=4.0, h, sum=0.0, x
h=(b-a)/n
sum=sum+(a**3*exp(-a**3)+ b**3*exp(-b**3))+ & ! terminal points
   4.0*(b-h)**3*exp(-(b-h)**3)
x=a+h
do i=2,n-2,2                            ! note the increment
   sum=sum+4.0*(x**3*exp(-x**3))        ! 4y_2+ 4y_4...
   x=x+h
   sum=sum+2.0*(x**3*exp(-x**3))        ! 2y_1+2y_3
   x=x+h
enddo
sum=h/3.0*sum
print *, 'The result of integration ', sum
end program simpson
```

It is interesting to observe the changes in the result by varying the value of n, that is, the number of strips between a and b.

3.21 Nested DO

The do statements may be nested; there can be a do within another do. These are, respectively, called the inner and outer loops.

```
do i=1, 10
 .
   do j=1, 100
    .
   enddo
 .
enddo
```

During the execution, for each value of the control variable of the outer loop, all values of the control variable of the inner loop are taken. This means that the control variables of the inner loop vary more rapidly than that of the outer loop. This is illustrated by the following program segment:

```
do i=1,2
   do j=1,2
      print *, 'i = ', i, 'j = ', j
   enddo
enddo
```

The outputs of the program are as follows (shown in one line):

```
i = 1 j = 1, i = 1 j = 2, i = 2 j = 1, i = 2 j = 2
```

Following the property of the do loop, the whole process may be summarized. When the nested loop is entered from the top, i is initialized to 1. The number of times the i loop is to be executed, t_i, is calculated, which in this case is 2. With this value of i

(i.e., 1), the j loop is entered, and j is initialized to 1. The number of times the j loop
is to be executed, t_j, is calculated. Again, it is 2. For i=1, the j loop is executed for
j=1 and 2. When the j loop is exited, i is set to 2, and the j loop is entered again.
The same procedure is followed; that is, j is set to 1, and t_j is again calculated. Again,
it is 2. For i=2, the j loop is executed for j=1 and 2. Having completed the loop for
i=2, the i loop is exited, and the control is passed to the next statement following the
enddo of the outer loop.

Figure 3.7 shows that, for a given value of the outer index, in this case i, all permissible
values of the inner loop index, in this case j, are taken. The same concept may be extended
to a nested loop with a depth greater than 2 (Figure 3.8).

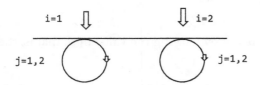

FIGURE 3.7
Nested do.

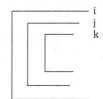

FIGURE 3.8
Nested do (depth greater than 2).

The readers may verify that the outputs of the program segment are as shown after the
source code.

```
do i=1,2
   do j=1,2
      do k=1,2
         print *, i,j,k ! Fig 3.8
      enddo
   enddo
enddo
```

The outputs are as follows (shown in one line):

1 1 1, 1 1 2, 1 2 1, 1 2 2, 2 1 1, 2 1 2, 2 2 1, 2 2 2.

The digit program (i.e., the number is equal to the sum of the cubes of the digits) shown
earlier with a single do statement may also be written with the help of a nested do.

```
program digit
implicit none
integer :: i,j,k,ic,jc,ih,jt
   do i=1,9                ! first digit
      ic=i**3              ! i cube
      ih=100*i             ! 100i
      do j=0,9             ! second digit
         jc=j**3           ! j cube
         jt=10*j           ! 10j
         do k=0,9          ! third digit
            if(ic+jc+k**3 .eq. ih+jt+k) then ! condition
               print *,i,j,k
```

```
          endif
        enddo
      enddo
    enddo
  end program digit
```

`ic+jc+k**3` is the sum of the cube of the digits, and `ih+jt+k` is the number, where the digits from the left are, respectively, `i`, `j` and `k`. It may be noted that, in the earlier case, digits were extracted from the number, and in this case, the number is constructed from the digits.

3.22 CYCLE, EXIT and the Nested Loop

When a `cycle` or an `exit` is executed within a loop, it acts on that particular loop only. For example, if the `exit` statement is executed within an inner loop, the control is transferred to the statement following the corresponding `enddo`.

```
do i=1, 50
 .
  do j=1, 100
  .
  if(i+j .lt. 100) then      ! some condition
    exit
  endif
  enddo
 .
enddo
```

Similarly, if `cycle` is executed within a loop, a fresh `cycle` for that loop starts.

```
do i=1, 50
 .
  if(a+b > c+d) then    ! some condition
    cycle
  endif
  do j=1, 100
  .
  enddo
 .
enddo
```

If loops are labeled, then it is possible for an inner loop to transfer the control to the outside of the outer loop.

```
outer:   do i=1, 25
          .

          .
inner:     do j=1, 125
          .
```

```
            if(i+j .eq. 70) then
               exit outer
            endif
                .
            enddo inner
                .
      enddo outer
```
←——

Similarly, a `cycle` statement executed within an inner loop may begin a fresh cycle for the outer loop.

```
      outer: do i=1, 25
                .
      inner: do j=1, 125
                .
            if(i+j .eq. 70) then
               cycle outer
            endif
                .
            enddo inner
                .
      enddo outer        ←————————
```

3.23 Termination of DO Loop

A do loop is exited under the following circumstances:

- When the loop is satisfied; that is, the iteration count becomes 0.
- For a do-while loop, the condition becomes false.
- An exit statement within the loop is executed.
- A goto or if statement is executed to bring the control outside the loop.
- A stop or return (Chapter 12) statement is executed within the loop.
- A cycle statement is executed within an inner loop to bring the control to the tail of the outer loop.

3.24 Rules of DO Statement

Some of the rules of a do loop were introduced in earlier sections. Here, all relevant rules of the do loop are summarized. It may be noted that the compiler may not complain if some of the rules are violated, but such violations of rules may lead to undesirable results.

1. The loop index should not be modified inside the loop from the maintenance point of view of the program. As the number of times the do loop is to be executed is calculated when the loop is entered for the first time, this modification does not change the number of iterations.

```
num=20
do i=1,num
 .
 num=num-1
enddo
print *, i, num
```

The values of i and num just outside the loop are 21 and 0, respectively. The loop index should never become undefined (deallocated, see Chapter 15) within the loop.

2. m1, m2 and m3 are integer constants or integer expressions (earlier versions of Fortran allowed real numbers).

3. If the increment m3 is absent, it is assumed to be equal to 1. If it is present, it should be nonzero.

```
do i=1, 10, 1 and do i=1,10 are equivalent.
```

4. The number of times the do loop is executed is determined by the following expression:

```
t=max((m2-m1)/m3 + 1,0)
```

The minimum value of the expression is 0. If the value of t is 0, the loop is skipped. The following do loop is skipped, as t is 0.

```
do i=10, 1
 .
enddo
```

5. If m1, m2 and m3 are (all or any one of them) variables or expressions involving variables, the variables may be modified inside the do loop. This modification does not change the number of iterations as this is calculated when the loop is entered for the first time, using the formula mentioned in rule 4.

6. The increment may be negative; in that case, m1 should be greater than or equal to m2 to make t greater than 0. The iteration count is 0 when m1>m2 and m3>0 or m1<m2 and m3<0.

7. It is prohibited to jump into a do loop without touching the do statement.

```
goto 100    ! not allowed
 .
do i=1, 80
 .
100 continue  ! this is a dummy statement, it does not do anything
 .
enddo
```

The reason being that unless the do statement is touched, proper initialization of the control variables is not performed.

8. Jumping into the top of the do loop from within the loop without touching the enddo is prohibited.

```
200  do j=1, 100
        .
     if(sum.gt.100) then
        goto 200    ! not allowed
     endif
        .
     enddo
```

The control variable is modified when enddo is reached, which is not done in the preceding case.

9. However, it is possible to jump to the top of the loop from outside the loop.

```
        goto 90   ! allowed
        .
90      do k=10, 20
        .
        enddo
```

10. It is prohibited to jump to the terminal statement of do from outside the do loop.

```
        goto 501   ! not allowed
        .
        do k=1, 25
        .
501     enddo
```

11. It is permissible to jump to the terminal statement of the do loop from inside the do loop. In that case, enddo must have a statement number, but it is not necessary as the same thing can be done with the `cycle` statement.

12. Premature exit, that is, exit from the do loop before exhausting all cycles, is possible.

```
        do k=1, 75
        .
        if (a>b) then ! some condition
            goto 20
        .
        endif
        enddo
20      continue
```

On premature exit from the loop, the loop index retains its last value. For example, if in the preceding program segment the value of k is, say, 37 when a>b becomes true, the value of k will retain this value outside the do loop. However, on normal exit from the do loop, the loop index has a value for which the counter has become 0. It is advisable not to assume this value of the loop index outside the loop. It is better to redefine this variable before it is used again.

```
do i=1,20,4
.
enddo
```

On normal exit, the value of the loop index i is 21 just outside the loop.

13. The nested loop must have a different loop index; otherwise, it will violate rule 1 (Figure 3.9).

14. For nested loops, the inner loop must terminate before the outer loop.

```
do 70 k=1, 45
   .
   do 75 l=1, 40
   .
70   continue        ! not allowed
75   continue        ! not allowed
```

FIGURE 3.9
Nested loop (not allowed).

15. For a numbered do loop, the inner and outer loops may terminate at the same point (Figure 3.10).

```
do 10 i=1,4
  do 10 j=1,10
  .
10   continue
```

Each loop should have a separate terminal statement. The following program may cause trouble:

```
do 10 i=1, 10
.
if(i.eq.5) then
   go to 10
endif
.
do 10 j=1, 10
.
10   continue
```

FIGURE 3.10
Nested do (same terminal point).

It is not clear whether the goto statement will transfer the control to the terminal statement of the inner or the outer loop—such statement should be avoided. Some of the present day compilers do not allow this type of branch statement. In fact, in this book, all do statements will be terminated by an unnumbered enddo statement, and therefore, this situation will never arise.

16. If a do loop contains a case statement, the case statement must end before the enddo. In addition, if a case statement contains a do, the enddo must come before the end select.

17. If the do loop contains an if statement, the if statement must terminate before the enddo. Similarly, the do statement inside an if statement must terminate before the endif.

18. It is possible to jump from the inner loop to the outer loop and jump back to the inner loop. However, it is strongly discouraged and is totally against the modern concept of structured programming.

19. Finally, there can be a do statement within an `if` or a `case` block. The general rule is that the inner one must terminate before the outer one. This is true for other two statements – `case` and `if`.

3.25 Remark about Loop Statements

A rule of thumb may be suggested regarding the usage of do and do while statements. If the number of times the loop is to be executed is known beforehand, do enddo is used. For example, to sum the series s=1+2+3+...+100, it is preciously known that the loop is to be executed 100 times, and therefore, do enddo is preferred. However, when it is not known how many times the loop is to be executed beforehand, do while is a better choice than do enddo. The program logic can always be modified such that do while may be replaced by do, and vice versa. Moreover, it is always better to avoid numbered do statements. It may be mentioned that nesting among do while is allowed like the normal do. In addition, nesting between do and do while is also permitted.

4

Handling of Characters

Character variables and constants were introduced in Chapter 1. Since the operations involving characters and behavior of the intrinsics associated with the characters are different from other intrinsic variables of the Fortran, a separate chapter is provided for characters.

4.1 Assignment

The general rules for assignment are also applicable to characters. A character variable may be assigned to a character constant or expression:

```
character (len=6) :: ch
ch = 'ABCDEF' ! Figure 4.1
```

A	B	C	D	E	F
A	B	C	D		
P	Q	R	S	T	U

FIGURE 4.1
Assignment of character variable.

If the number of characters on the right-hand side of the assignment sign is less than the number of characters that the variable can accommodate, blanks are added to the right.

```
ch = 'ABCD' ! system will add 2 blanks, Figure 4.1
```

If there are more characters on the right-hand side of the assignment sign than the variable can store, then it is truncated from the right.

```
ch = 'PQRSTUVW' ! VW will not be stored, Figure 4.1
```

It is stored as 'PQRSTU' as the variable can store only six characters.

4.2 Concatenation

Two or more character variables or constants or their combinations may be concatenated (joined) by the concatenation operator, which is denoted by two successive division signs (//); 'abc' // 'def' will become 'abcdef'.

The operand must be of the same kind (Chapter 5). The priority of the concatenation operator is lower than that of the addition or subtraction (binary) operator but is higher than that of the relational operator.

```fortran
character (len=2) :: ch= 'pq'
if ('a' // 'b' .eq. ch) then
  .
endif
```

Since the priority of the concatenation operator is higher than the priority of the relational operator, 'a' and 'b' are joined to form 'ab' and subsequently compared with the character variable ch. The following program calculates the approximate blood volume of a person (male or female) and the amount of blood he or she can donate. The program is explained through comments.

```fortran
program blood_vol
implicit none
real, parameter :: male=76.0, female=66.0, donate=8.0
real :: bv, weight, togive ! male 76ml,female 66ml per kg of body weight
character :: sex           ! can be "m" or "f"
print *, "Type m (male) or f (female)"
read *, sex
print *, "Type the body weight in kg"
read *, weight             ! no check on weight
if (sex=="m") then
  bv=male*weight
  togive=donate*weight     ! Can donate 8 ml per kg of body weight
  print *, "Male having weight= ", weight, "kg has ", bv, " ml blood"
  print *, "He can safely donate ", togive, " ml of blood"
else if (sex=="f") then
      bv=female*weight
      togive=donate*weight
      print *, "Female having weight= ", weight, "kg has ", &
        bv, " ml blood"
      print *, "She can safely donate ", togive, " ml of blood"
    else
      print *, "Data error - sex can be m or f"
endif
end program blood_vol
```

4.3 Collating Sequence

The question whether a particular character is greater than or less than another character is determined by the collating sequence. In the ASCII character set, the sequence number starts with zero and goes up to 127. A character that has a sequence number 65 is less than a character having sequence number 66. The sequence number is actually the internal representation of the character in integer. The character 'A' is less than the character 'B'. If the characters are arranged in ascending order, the character 'A' will come before the character 'B'. Similarly, character '0' is less than character '1'. The following sequence is always maintained:

```
'A' < 'B' < 'C'    ....        < 'Z'
'a' < 'b' < 'c'    ....        < 'z'
'0' < '1' < '2'    ....        < '9'
'^' < '0'          ....        < '9'< 'A'
'^' < 'A'          ....        < 'Z'
'^' < '0'          ....        < '9'< 'a'
'^' < 'a'          ....        < 'z'
```

where '^' indicates a blank character. Characters having the collating sequence number greater than 127 are processor dependent. Fortran supports other types of characters (Unicode). If the program depends on the absolute value of a character in the collating sequence, the program may not behave properly in a machine that uses a character set other than ASCII.

4.4 Character Comparison

Characters can be compared by means of relational operators:

```
character :: c1, c2
.
c1 = 'x'
c2 = 'a'
if (c1 > c2) then
.
else
.
endif
```

In the preceding example, since 'x' appears later than 'a' in the collating sequence, c1 is greater than c2, and 'then' path is chosen. Relational operators cannot be used to compare a numeric type with a character string.

4.5 Comparison of Character Strings

Two character strings may be lexically compared. If they are not of equal length, blanks are added at the end of the shorter string to make the length of the two strings equal. Two strings are said to be equal if each of the characters of the first string is exactly identical to the corresponding character of the second string. For example, strings 'ABCD' and 'ABCD' as well as 'pq ' (two blanks at the end) and 'pq' are equal. In the second case, the second string is shorter than the first string, and therefore, it is extended by adding two blanks at the tail. The first unmatched character of the two strings determines whether one string is greater or less than the other string. The string having the first unmatched character less in the collating sequence is less than the other string.

For example, 'ABCD' is less than 'ADBC' as the first unmatched character (second character) of the first string is ('B'), which is less than the corresponding character ('D') of the second string (Figure 4.2).

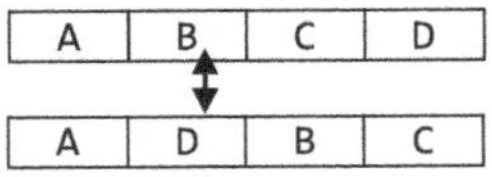

FIGURE 4.2
Character comparison.

Note that less or greater is determined by the first unmatched character. Characters beyond the first unmatched character are not considered. If two character strings having a length equal to zero are compared, they are considered equal.

4.6 Lexical Comparison Functions

Fortran provides four lexical comparison functions. All these functions take two strings as arguments. They are llt, lle, lgt and lge. The comparison always takes place according to the ASCII collating sequence even if the processor uses some other character set (say, EBCDIC—Extended Binary Coded Decimal Interchange Code). Note the difference between these functions and relational operators <, <=, > and >= introduced earlier. The former always uses the ASCII collating sequence, and the later returns true or false, depending upon the character set being used by the processor. However, if the processor uses the ASCII character set, then both are same. These lexical comparison functions are preferred over the relational operators for string comparison as they are machine independent (always use the ASCII character set). In subsequent discussions of this section, str-1 and str-2 refer to two different strings.

llt(str-1, str-2) returns true if str-1 is less than str-2; otherwise it returns false.

lle(str-1, str-2) returns true if str-1 is less than or equal to str-2; otherwise it returns false.

lgt(str-1, str-2) returns true if str-1 is greater than str-2; otherwise it returns false.

lge(str-1, str-2) returns true if str-1 is greater than or equal to str-2; otherwise it returns false.

The if statement may be used along with the lexical comparison function.

```
logical :: l1
character (len=6) :: c1, c2
.
if (llt(c1, c2)) then
.
else
.
endif
```

4.7 Length of a String

The library function `len` takes one string as its argument and returns an integer as the length of the string.

```
integer :: length
length = len ('abcd')              ! length is 4
length = len ('uvwx    ')          ! 4 blanks after x, length is 8
```

Len can return the length of a character variable even if it is not defined. In that case, the length is the number of characters the variable can store.

```
integer :: i
character (len=10) :: ch
i=len(ch) ! i=10
```

Even before the character variable is initialized, the `len` function returns `10` as the length of the string because by definition the variable `ch` can hold `10` characters.

The library function `len_trim` takes one character string or a character variable/expression as its argument and returns an integer as the length of the string without the trailing blanks.

```
integer :: i
i = len_trim ('abcd')        ! i=4
i = len_trim ('abcde   ')    ! i=5 (3 blanks at the end)
```

In the second example, `len_trim` returns 5 as the length of the string, since trailing blanks are not considered while calculating the length of the string. However, it is obvious that embedded blanks are considered while calculating the length of the string using `len_trim`.

```
i = len_trim ('abcd  ef    ')! i=8 (2 blanks after 'd', 4 blanks
                             ! at the end)
```

In this case, `len_trim` returns 8—the embedded blanks are included for the calculation of the length of the string. Both the intrinsics, `len` and `len_trim`, may take an additional argument `kind` (Chapter 5).

4.8 Trimming and Adjusting a String

Three library functions are available to trim and adjust a string. They are `trim`, `adjustl` and `adjustr`.

TRIM: Trim takes a string or character variable as its argument and returns a string without the trailing blanks. This function is especially useful when it is necessary to join two strings after removing trailing blanks (Figure 4.3).

S	a	n	k	a	r								
C	h	a	k	r	a	v	o	r	t	i			

FIGURE 4.3

(a, b) Trim example.

```
character (len=15) :: name, title
character (len=20) :: fname
name = 'Sankar'
title = 'Chakravorti'
fname = trim (name) // ' ' // trim (title)
```

fname:

| S | a | n | k | a | r | | C | h | a | k | r | a | v | o | r | t | i | | |

The `trim` function will remove all the trailing blanks after 'Sankar'. This is then concatenated with a blank and then with the string 'Chakravorti'. The variable fname will have two blanks at the end because fname can accommodate 20 characters and the total length of the string is 18.

ADJUSTL: The library function `adjustl` takes an identical argument as in the preceding case. It removes the leading blanks and adds these blanks at the end (Figure 4.4).

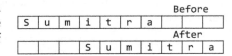

FIGURE 4.4
Adjustl example.

```
character (len=10) :: name
name = '    Sankar' ! 4 blanks before Sankar
name =  adjustl (name)
```

This function, therefore, makes a string left adjusted, keeping the length of the string same.

ADJUSTR: The function `adjustr` makes a string right adjusted; that is, it removes the trailing blanks and adds the same in front of the string (Figure 4.5).

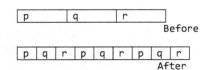

FIGURE 4.5
Adjustr example.

```
name = 'Sumitra    ' ! 3 blanks at the end
name = adjustr (name)
```

The variable name now contains '   Sumitra'.

4.9 REPEAT

The library function `repeat` takes two arguments, a string or character variable and an integer count. It returns a string with the input string concatenated repeat count times (Figure 4.6).

| p | | q | | r | |

Before

| p | q | r | p | q | r | p | q | r |

After

```
repeat ('pqr', 3), returns
 'pqrpqrpqr'
```

FIGURE 4.6
Repeat example.

The repeat count must be positive. Repeat with zero repeat count returns a null string. Thus,

```
print *, len(repeat('pqr',0)), returns zero
```

4.10 Character–Integer Conversion

In character–integer conversion, an integer may be converted into a character according to its position in the collating sequence. Similarly, a character may be converted into an integer, where its position in the collating sequence is returned as an integer. There are two sets of these functions: one for the ASCII character set and the other for the processor-dependent character set.

The library functions achar and iachar are used for the ASCII character set. The function achar takes one integer between 0 and 127 as its argument and returns the corresponding character (len=1) in the collating sequence.

```
character :: ch
integer :: ic=99
ch=achar(ic)
```

The value returned is 'c'. Similarly, if ch= 'A', ic = iachar(ch), will return 65.

The pair char and ichar is similar to the pair achar and iachar except that the former is used for the processor-dependent character set. Both the intrinsics, iachar and ichar, may take an optional argument kind, which is discussed in Chapter 5. If the system is using the ASCII character set, these two pairs are same.

4.11 Character Substring

A single or a group of contiguous characters may be referenced from a character string using character substring representation by providing the start and end positions of the substring within the string. The positions are indicated by integers or integer expressions. A colon separates the start and end positions. Character position 1 starts from the beginning of the string and goes up to n, where n is the length of the string. The lower bound must be greater than or equal to 1, and the upper bound must be less than or equal to the length of the character string (Figure 4.7).

```
character (len=15) :: c1,c2,c3
character(len=49):: c= 'Indian Association for the Cultivation of Science'
```

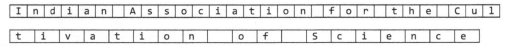

FIGURE 4.7
Character substring example.

```
c1 = c(1:6)      ! c1 = 'Indian'
c2 = c(8:18)     ! c2 ='Association'
c3 = c(20:30)    ! c3 = 'for the cul'
```

If the lower bound is not supplied, it is assumed to be 1. The expressions c1 = c (: 6) and c1 = c (1: 6) are equivalent.

If the upper bound is not supplied, it is assumed to be the end of the string.

```
c3 = c (40:) ! c3 ='of Science'
```

Note that c(:) refers to the entire string. Also, c (i: i) refers to the ith character of the string. If the starting value is greater than the end value, a null string is returned. The expression i = 10; j = 8; c1 = c (i: j) will return a null string. The substring may also appear on the left-hand side of the assignment sign.

```
character(len=13) :: bld = 'Blood Volume '! one blank at the end
bld(7:13) = 'Storage'
```

The character positions 7 to 13 of bld have been replaced. Note that the character positions 1 to 6 remain unchanged (Figure 4.8). We conclude this section with another example.

| B | l | o | o | d | | V | o | l | u | m | e | |
| B | l | o | o | d | | S | t | o | r | a | g | e |

FIGURE 4.8
Replacement of substring.

```
character (len=4) :: c
c='1234567890' (4:7); print *, c
end
```

The output is 4567; positions 4 to 7 of the string are stored in variable c. A character variable is said to be defined if all its elements are defined. For example,

```
character (len=6):: ch
ch(1:4)='abcd'
```

defines the first four characters of ch. However, as the next two characters remain undefined, variable ch is considered as undefined. Thus, ch(5:5) and ch(6:6) should not be used till they are properly defined. Consider the statement ch='abcd'. As the right-hand side contains 4 characters, the system will add two blank characters at the end, and ch is assigned to 'abcd   ' (2 blanks at the end). In this case ch(5:5) and ch(6:6), that is the fifth and sixth characters, contain blank character as such ch is said to be defined.

A program to invert a character string using character substring is as follows:

```
program invert
implicit none
integer :: pt,i,ln
character(len=*),parameter :: buf="computer program"
character(len=len(buf)) :: kuf
ln=len(buf); print *, buf
do i=1,ln
  pt=ln-i+1 ! last character of buf is stored as the
  kuf(i:i)=buf(pt:pt) ! first character of kuf.
end do ! Similarly for other characters.
print *, kuf
end program invert
```

Output:
```
computer program
margorp retupmoc
```

4.12 Programming Examples

The following program reads a character string and converts all the uppercase letters to the corresponding lowercase letters. The programming trick is that we know

```
iachar('A') < iachar('B') < iachar('C').......... < iachar('Z')
iachar('a') < iachar('b') < iachar('c').......... < iachar('z')
```

and the relative position of any uppercase letter with respect to 'A' in the collating sequence is the same as the relative position of the corresponding lowercase letter with respect to 'a' in the collating sequence. Therefore, to convert an uppercase letter, say, 'D' to 'd', the following statement is utilized:

```
achar(iachar('D') - iachar('A') + iachar('a'))
```

The expression iachar('D') - iachar('A') will give the relative position of 'D' with respect to 'A', which is 3. When this is added to iachar('a') and converted to character type by achar, it gives character 'd'. Neither the absolute value of 'D' nor 'A' and 'a' in the collating sequence nor the numerical value of the difference between 'A' and 'a' in the collating sequence (actually 32) is required (Figure 4.9).

1	2	3	4
A	B	C	D
a	b	c	d

FIGURE 4.9
Conversion—uppercase to lowercase.

```
program up2low
implicit none
integer :: asmall= iachar('a'), acap = iachar('A')
character (len=80) :: buf
character :: ct
integer :: siz, i
read *, buf
siz = len_trim (buf)
print *,buf(1:siz)
do i=1, siz
   ct=buf(i:i) ! convert if it is an uppercase letter
   if (lge(ct, 'A') .and. lle(ct, 'Z')) then
     buf(i:i)=achar(iachar(ct) - acap + asmall)
   endif
enddo
print *,buf(1:siz)
end program up2low
```

If the input stream contains characters other than uppercase letters, these characters must be allowed to pass through, and only uppercase letters are to be changed. All the uppercase characters are greater than or equal to 'A' and less than or equal to 'Z'. The if statement in the preceding program converts only uppercase letters to the corresponding lowercase letters.

The next program determines whether a string is a palindrome or not. A palindrome is a string of characters that when written in the reverse order remains the same as the original string. For example, 'MADAM' or 'madam' is a palindrome ('Madam' is not a palindrome as the lowercase letters are not same as the uppercase letters within a string). The program will check the first character with the last character, second with the last but one and so on. Any mismatch indicates that the string is not a palindrome.

```
program palin
implicit none
integer :: it, siz, i
character (len=80) :: buf
logical :: l=.true.
read *, buf ! input will be Oct 8, 2018 which is 8102018
siz=len_trim(buf)
do i=1, siz/2    ! mid point obtained by integer division
  it=siz-i+1     ! compare first with the last character
  if(buf(i:i) .ne. buf(it:it)) then ! compare 2nd with the last but one
    print *, 'Not a palindrome : ',buf
    l=.false.
    exit
  endif
enddo
if (l) then   ! l remains true if there is no mismatch
  print *, 'The string is a palindrome : ', buf(1:siz)
endif
end program palin
```

The next program locates a substring within a string using brute-force method. The program logic is simple and is explained through comments.

```
program search
implicit none
character(len=80) :: str= "the quick brown fox jumps over the lazy dog"
character(len=80) :: sch="e"                ! search for a every occurrence
integer :: i,j,l,m,k,t                      ! of substring within a string
logical :: index,global=.false.
l=len_trim(str)
m=len_trim(sch)
print *,"String: ",trim(str); print *,"Substring: ",trim(sch); print *
do i=1,l-m+1
  index=.true.                              ! becomes false when match fails
  t=i ! remember the index of the outer loop
  do j=1,m
  k=i+j-1
  if(str(k:k) .ne. sch(j:j)) then
    index=.false.
    exit                                    ! exit from inner loop
  endif
enddo
if(index) then                              ! print the matched string
  print *,"Found at position: ",t,"--   ", str(t:t+m-1)
  global=.true.     ! global remains false if there is no match
endif
enddo
```

```
      if (.not. global) then
         print *, "String not found"
      endif
      end program search
```
Outputs:
```
   String: the quick brown fox jumps over the lazy dog
   Substring: e
   Found at position:        3 --  e
   Found at position:       29 --  e
   Found at position:       34 --  e
```

The logical variable global is set to true when a match is found, and it remains false if no match is found.

4.13 Library Functions INDEX, SCAN and VERIFY

Fortran provides a few more useful library functions to handle strings, such as index, scan and verify. All these intrinsics take an argument (fourth) kind, which is the kind of the returned integer by the intrinsics. This is discussed in Chapter 5.

INDEX: The index function takes four arguments. The first three are a string, a substring and an optional logical parameter. The logical parameter, if not present, is assumed to be false.

If the logical parameter is either absent or false, index returns the starting position (as an integer value) of the substring within the string.

   If the substring is not present, it returns zero.

```
      index ('Subrata', 'a') ! Figure 4.10
  or, index ('Subrata', 'a', .false.), returns 5.
```

If the logical parameter is true, index returns the starting position of the substring from the end.

S	u	b	r	a	t	a

S	u	b	r	a	t	a

```
      index ('Subrata', 'a', .true.),
      returns 7 (Figure 4.10)
```

FIGURE 4.10
Index example.

If the substring is the null string and the third parameter is either absent or false, index returns 1. However, if the third parameter is true, it returns an integer equal to the length of the string +1.

```
      index ('Subrata', ''), returns 1
  and, index ('Subrata', '', .true.), returns 8.
```

The conversion of lowercase letters to uppercase letters may be done conveniently using the intrinsic index. The logic is to search for a lowercase letter using index and replace the character by the corresponding uppercase letter.

```
program low_to_up2
implicit none
character(len=26) :: cap= "ABCDEFGHIJKLMNOPQRSTUVWXYZ"
character(len=26) :: small= "abcdefghijklmnopqrstuvwxyz"
character(len=80) :: text
integer:: ln, i, pos
read *, text ! read a text from the keyboard; maximum 80 characters
ln=len_trim(text)
do i=1, ln
  pos=index(small, text(i:i)) ! check for small letter
  if(pos .ne. 0) then ! if there is match, convert to corresponding UC
    text(i:i)=cap(pos:pos)
  endif
end do
print *, text
end program low_to_up2
```

SCAN: This intrinsic also takes four arguments. The first three are a string, a set of characters and an optional logical parameter. If the logical parameter is absent, it is assumed to be `false`.

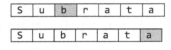

FIGURE 4.11

Scan example.

If the logical parameter is absent or `false`, the position of first occurrence of any character from the set of characters is returned. If none of the members of the set is present in the first string, zero is returned.

```
    scan ('Subrata', 'ab')
or, scan ('Subrata', 'ab', .false.) will return 3 (Figure 4.11).
```

If the logical parameter is `true`, the position of the last character from the (i.e., scanning starts from the end) string that matches with any member from the set is returned.

```
scan ('Subrata', 'ab',.true.), returns 7 (Figure 4.11)
```

The expression `scan ('Subra', 'xyz')` returns zero as no character from the second string is present in the first string.

We now use `scan` to count the number of individual characters within a string.

```
program count
implicit none
character (len=80) :: text
integer :: length, i, pos
integer, dimension(26) :: num=0
character(len=26), parameter :: capital='ABCDEFGHIJKLMNOPQRSTUVWXYZ'
character(len=26), parameter :: small='abcdefghijklmnopqrstuvwxyz'
character(len=52):: capsmall=capital//small
read *, text              ! maximum 80 character
text=trim(adjustl(text)) ! put the leading blanks, if any, at the end
length=len_trim(text)     ! length of the string without trailing blanks
print *, 'Input String: ',text
do i=1, length
```

```
      pos=scan(capsmall, text(i:i))
      if (pos .ne. 0) then          ! match found
        if (pos .gt. 26) then ! if LC, increment the corresponding UC counter
          pos=pos-26 ! for z pos will be 52, change it to 26
        endif
      num(pos)=num(pos)+1 ! increment the counter corresponding to the letter
    endif
  end do
  do i= 1, 26
    print *, capsmall(i:i), ' : ',    num(i)
  end do
  end program count
```

VERIFY: Like `scan`, `verify` also takes four arguments. The first three are a string, a set of characters and an optional logical parameter. If the logical parameter is absent, it is assumed to be `false`. If the logical parameter is absent or `false`, the function returns zero when all the characters in the string are members of the set (second argument); otherwise it returns the position of the first left-most character, which is not present in the set.

> `verify ('subrata', 'abrstu')`
>
> or, `verify ('subrata', 'abrstu', .false.)`, returns zero. However,
>
> `verify ('subrata', 'suba')`, returns 4 as `'r'` is not present in the
>   set (2nd argument).

If the logical parameter is `true`, the position of the right-most character of the string, which is not a member of the set, is returned.

The expression `verify ('subrata', 'suba', .true.)` returns 6 as `'t'` is not a member of the set. The next program counts the number of words within a string. The words are separated by a blank or punctuation mark.

```
program Word_count
implicit none
character(len=26) :: cap= "ABCDEFGHIJKLMNOPQRSTUVWXYZ"
character(len=26) :: small= "abcdefghijklmnopqrstuvwxyz"
character(len=53) :: alpha ! count number of words in a line.
character(len=80) :: text    ! first replace punctuation mark by blank
integer:: ln, pos, p, pl, siz, wdcount=0 !replace multiple blanks
alpha=cap//small // " "     ! contains all small and cap letters and blank
print *, "Input:"
read *, text ! inefficient algorithm to keep the logic simple
print *, text ! read a text from the keyboard; maximum 80 characters
text=adjustl(text)       ! send the leading blanks if any to the right
ln=len_trim(text)        ! length of the string without trailing blanks
p=1        ! trailing blanks will not be considered
pl=1       ! present character position from the left
do while (pl <= ln)
  pos=verify(text(p: ), alpha)       !scans: position p to the end
    if(pos .eq. 0) then              !end of the string or
    exit                             !no more non alpha character
    endif
  pl=pl+pos
```

```
 text(pl-1:pl-1)=" "   ! replace the non alpha character by blank
 p=pl    ! next time scan from character position p to end of the string
end do
text=adjustl(text)      ! blanks are sent to the end
print *, "Output:"
print *, text
siz=len_trim(text) ! length without the trailing blanks
do
 pos= index(text(1:siz), "  ") ! search for two successive blanks
  if (pos .eq. 0) then
   exit
  endif
  text(pos: ) =text(pos+1:) ! replace by a single blank, decrement the
  siz=siz-1                 ! length of the string by 1
enddo
text=text //" "
ln=len_trim(text)+1       ! length of the string with a trailing blanks
p=1                       ! trailing blanks will not be considered
pl=1                      ! present character position from the left
do while (pl < ln)
 pos=index(text(p:ln), " ")    ! scans: position p to the end
 if(pos .eq. 0) then           ! end of the string or
    exit                       ! no more non alpha character
 endif
 pl=pl+pos
 print *, text(p:pl-1)
 wdcount=wdcount+1
 p=pl  ! next time scan from character position p to end of the string
end do
print *, "No of Words: ", wdcount
end program word_count
```

4.14 CASE and CHARACTER

The control variable of a case statement can be characters as well. The following program counts the number of vowels from a given character string:

```
program count_v
implicit none
character (len=80) :: ch
integer :: siz, i, vc=0, oc=0
read *, ch
siz=len_trim(ch)
do i=1,siz
   select case (ch(i:i))       ! Examine each character one by one
     case('A','E','I','O','U','a','e','i','o','u')
        vc=vc+1 !   increment the counter for vowel
     case default
        oc=oc+1 !   increment the counter for other
   end select
enddo
```

```
print *, 'No of vowels : ', vc
print *, 'No of other characters: ', oc
end program count_v
```

The case statement may be of the following form: case('ME': 'YOU'). For example,

```
character(len=4) :: ch='SHE'
select case (ch)
  case('ME' : 'YOU')
     print *, ch
end select
```

will display 'SHE' as it is between 'ME' and 'YOU'. We conclude this section with another example:

```
character(len=15):: status
read *, status ! different path will be chosen depending upon status
select case (status)
 case("ready")
  print *, "ready"
 case("on your mark")
  print *, "on your mark"
 case ("get set")
  print *,"get set"
 case ("go")
  print *, "go"
 case default
  print *, "Wrong Command"
end select
```

4.15 NEW LINE

This function returns the new line character. The argument is of type character. For the default character type (ASCII), print *, iachar(new_line('a')), returns 10.

5

Precision and Range

One of the main features of Fortran is the portability of source code. When a variable or a constant is declared, the compiler allocates storage locations for the variable or the constant with the default processor-dependent range and precision. Fortran allows the programmer to specify the precision and the range of a variable through the kind attribute. This makes the program machine independent. Fortran provides library functions or intrinsics selected_int_kind and selected_real_kind to specify the range and precision of variables or constants of integer and real types, respectively. If the kind attribute is not specified, the compiler assumes the default kind, which is compiler dependent.

A word of caution: beginners are advised to use the default kind (i.e., precision and range) for constants and variables. This concept of kind looks attractive, but in reality most of the current compilers support basically three types of real variables: the so-called standard precision, double precision and extended precision. The default kind (compiler-dependent kind number, an integer) for the Intel ifort compiler may be obtained by executing a simple program, shown next, and using the kind intrinsic.

```
print *, 'default kind for ifort compiler....'
print *, 'default integer kind           ',kind(0)
print *, 'default real kind              ',kind (0.0)
print *, 'default double precision kind  ',kind (0.0d0)
print *, 'default quadruple kind         ',kind (0.0q0)
print *, 'default complex kind           ',kind(cmplx(0.0,0.0))
print *, 'default complex double precision ', &
         kind(cmplx(0.0d0,0.0d0,kind=kind(0.0d0)))
print *, 'default logical kind           ',kind(.false.)
end
```

The output is as follows:

```
default kind for ifort compiler....
default integer kind                  4
default real kind                     4
default double precision kind         8
default quadruple kind               16
default complex kind                  4
default complex double precision      8
default logical kind                  4
```

5.1 SELECTED_INT_KIND

Intrinsic `selected_int_kind` takes one integer, say, *x*, as its argument. The argument determines the range of the integer variable or constant corresponding to this `kind`; it is -10^x < n < 10^x, where n is an integer.

```
      integer, parameter :: siz=selected_int_kind(2)
      integer(siz) :: x
or,   integer(kind=siz) :: x
or,   integer(kind=selected_int_kind(2)) :: x
```

The integer variable x is of kind 'siz'. The argument of `selected_int_kind` in this case is 2, and therefore, the variable of `kind=siz` should be able to handle integers between (and including) –99 and +99. The intrinsic returns –1 if it fails to allocate an integer in the range with the specified argument.

```
      integer, parameter :: big=selected_int_kind(9)
      integer (big) :: b
```

The integer variable b may assume values between -10^9 and $+10^9$. Based on the declaration, it is the responsibility of the compiler to allocate the required number of bytes (storage) for the variable to store the integer having this range of values.

Coming back to the declaration, `selected_int_kind(2)`, one must not think that there will be integer overflow if the value of the stored number exceeds 99, say, 100. The declaration merely guarantees that the compiler will allocate enough storage location for the variable such that it can accommodate any number from –99 to +99. Memory is allocated in terms of bytes. Eight binary digits make one byte. One byte of storage can accommodate any integer from –128 to +127. Though the variable created using `selected_int_kind(2)` is supposed to accommodate any integer between –99 and +99, it, in fact, will accommodate any integer between –128 and +127. Similarly, `selected_int_kind(3)` tells the compiler to assign storage such that it can accommodate any integer from –999 to +999. Actually, the compiler will allocate 2 bytes (16 bits), and this amount of memory can actually accommodate integers between (and including) –32768 and 32767. The compiler allocates the minimum amount of storage to cover the range specified in the `selected_int_kind` intrinsic. Since the memory is allocated in terms of bytes, normally the memory allocated by the compiler would cover a range beyond the range specified in the declaration. This, however, does not go against the declaration. The declaration merely ensures the upper and lower limits of the variable. Table 5.1 illustrates this for the NAG Fortran compiler. Note that Table 5.1 is compiler dependent as different people (compiler designers) use different integers to represent `kind` values internally.

Actually, `selected_int_kind` returns a compiler-dependent integer parameter called `kind`, which is subsequently used to define a variable or a constant. From Table 5.1, it is apparent that the current compiler (NAG) supports four types of integers having the `kind` parameter as 1, 2, 3 and 4. In fact, for the NAG compiler, one can define a variable through these numbers directly:

```
      integer (kind=1) :: k1
      integer (kind=2) :: k2
      integer (kind=3) :: k3
      integer (kind=4) :: k4
```

TABLE 5.1

The Kind Parameters of the NAG Fortran Compiler (Integer)

Argument of selected_ int_kind	Range of Integer Desired	Number of Bytes Allocated	Value of the kind Parameter	Maximum Value Permitted
1	$-10^1 < n < 10^1$	1	1	127
2	$-10^2 < n < 10^2$	1	1	127
3	$-10^3 < n < 10^3$	2	2	32767
4	$-10^4 < n < 10^4$	2	2	32767
5	$-10^5 < n < 10^5$	4	3	2147483647
6	$-10^6 < n < 10^6$	4	3	2147483647
7	$-10^7 < n < 10^7$	4	3	2147483647
8	$-10^8 < n < 10^8$	4	3	2147483647
9	$-10^9 < n < 10^9$	4	3	2147483647
10	$-10^{10} < n < 10^{10}$	8	4	9223372036854775807
11	$-10^{11} < n < 10^{11}$	8	4	9223372036854775807
12	$-10^{12} < n < 10^{12}$	8	4	9223372036854775807
13	$-10^{13} < n < 10^{13}$	8	4	9223372036854775807
14	$-10^{14} < n < 10^{14}$	8	4	9223372036854775807
15	$-10^{15} < n < 10^{15}$	8	4	9223372036854775807
16	$-10^{16} < n < 10^{16}$	8	4	9223372036854775807
17	$-10^{17} < n < 10^{17}$	8	4	9223372036854775807
18	$-10^{18} < n < 10^{18}$	8	4	9223372036854775807
19	$-10^{19} < n < 10^{19}$		Not	Supported

The constants may also refer to these kind parameters directly, where the constant and the kind value are separated by an underscore (_), for example, 2_1, 3456_2, 423456_3 and 123456789123_4. However, use of these compiler-dependent kind values is discouraged because the numerical value of the kind parameters may be different for another compiler. If used, such program becomes compiler dependent. A better way is not to use the numerical value for the kind parameters directly:

```
integer, parameter :: s1=selected_int_kind(1)
integer, parameter :: s2=selected_int_kind(2)
integer, parameter :: s3=selected_int_kind(6)
integer, parameter :: s4=selected_int_kind(12)
integer(kind=s1) :: k1
integer(kind=s2) :: k2
integer(kind=s3) :: k3
integer(kind=s4) :: k4
```

The constants may also be defined in terms of s1, s2, s3 and s4 as 2_s1, 3456_s2, 423456_s3 and 123456789123_s4, respectively.

It is apparent that if the preceding procedures are followed, the program becomes compiler independent, as the program does not use numerical value of kind for a particular compiler directly. We mention again that in this example we used the properties of the NAG Fortran compiler. It may be different for other compilers. Fortran supports at least one type of integer having kind=selected_int_kind(18).

```
integer, parameter :: very_big=selected_int_kind(18)
integer(kind=very_big) :: int
```

It may appear from the definition of selected_int_kind that the compiler allocates memory in a continuous fashion to accommodate different ranges. In practice, as memory is allocated in units of bytes, and so long as a particular chunk of memory can accommodate different ranges, change of kind does not take place. For example, one byte of memory can accommodate integers from -128 to 127. Therefore, selected_int_kind(1) and selected_int_kind(2) will have same kind parameter. The former wants a range between -9 and +9, while the latter requires a range between -99 and +99. However, selected_int_kind(3) requires a range between -999 and +999, and one byte of memory cannot accommodate such range. Therefore, a new kind parameter is required that needs two bytes of memory. These two bytes cannot accommodate integers between 10^{-5} and 10^{+5}. So again, a change of kind takes place for this range. In short, several ranges may have identical kind values.

5.2 Precision and Range of Real Numbers

The precision of a real number is a measure of exactness of the number. It is specified in terms of the number of digits that are used to express the number reliably. When a real number is stored in a computer, the number of bits used for the fractional part determines the precision of the number. Though real numbers are stored as binary digits, precision is expressed both in binary as well as in decimal form. For standard precision, precision of a real number is approximately 6 (decimal places), and for double precision, it is 15 (decimal places).

The range for real numbers is a measure of the biggest number that can be processed and is the maximum exponent (power of ten) that the processor can handle. For standard precision, it is 37.

5.3 SELECTED_REAL_KIND

This intrinsic returns an integer value of the kind parameter corresponding to the precision p, range r (power of ten) and radix, where p, r and radix are integers (scalar)—the first, second and third parameters of the function, respectively (the third parameter is optional; radix is the base of the numerical model, which is 2; normally this is omitted).

```
integer, parameter :: rn=selected_real_kind (6,37)
integer, parameter :: rd=selected_real_kind (15, 307)
real (kind=rn) :: x
real (kind=rd) :: y
```

Like the intrinsic function selected_int_kind, the compiler-dependent kind values are not used directly. In this case, the variable x has a precision 6 and range 37, that is, 10 to the power 37. Similarly, the variable y has a precision 15 and range 307, that is, 10 to the power 307. All valid kind numbers are positive integers. The intrinsic returns -1 if support from the processor for precision p is not available but support for range r is available for a particular supported `radix` (normally this is 2).

```
integer, parameter :: s=selected_real_kind (40,37)! s=-1
```

Similarly, the intrinsic returns −2 if support from the processor for precision p is available but support for the range r is not available for a particular supported `radix`.

```
integer, parameter :: s=selected_real_kind (6,400)! s=-2, NAG compiler
```

The intrinsic returns −3 if support from the processor is simultaneously not available for precision p and range r for a particular supported `radix`.

```
integer, parameter :: s=selected_real_kind (40,400)! s=-3, NAG compiler
```

The intrinsic returns −4 if the processor supports the precision p and the range r individually but not simultaneously for a particular supported `radix`.

```
integer, parameter :: s=selected_real_kind (31,307)! s=-4, NAG compiler
```

The intrinsic returns −5 if the processor does not support any real type with radix equals to `radix`.

```
integer, parameter :: s=selected_real_kind (6,37,3)! s=-5, NAG compiler
```

If more than one `kind` is available for the precision p and range r, the intrinsic returns the one with the smallest decimal precision.

If both `kind` and exponent are present, the exponent letter should be `'e'` (or `'E'`). A real constant with or without the exponent letter `'e'` is treated as the default real constant. A double precision real constant is indicated by letter `'d'` (or `'D'`). Table 5.2 shows precision, range and `kind` parameters of real numbers for the NAG Fortran compiler.

TABLE 5.2

The `Kind` Parameters of the NAG Fortran Compiler (Real)

Argument of selected_ real_kind		Number of Significant Digits (Binary)	Value of kind Parameter	Precision Allocated	Highest Exponent	Maximum Permitted Value
p	r					
1	1	24	1	6	37	3.4028235E+38
6	37	24	1	6	37	3.4028235E+38
7	37	53	2	15	307	1.7976931348623157E+308
1	38	53	2	15	307	1.7976931348623157E+308
15	37	53	2	15	307	1.7976931348623157E+308
15	307	53	2	15	307	1.7976931348623157E+308
20	200	106	3	31	291	8.98846567431157953864652\ 595E+307
31	291	106	3	31	291	8.98846567431157953864652\ 595E+307

From Table 5.2, it is observed that the current NAG compiler supports three types of real variables or constants—the so-called standard precision, double precision and extended precision. The precision and the range for standard precision are 6 and 37, respectively. The precision and range for double precision are 15 and 307, respectively; for extended precision, they are 31 and 291, respectively. One interesting point may be observed from Table 5.2. When the desired precision is 1 and the range is 38, the compiler allocates locations for precision 15 and range 307 because, to accommodate a range greater than 37, the next available range for this compiler is 307 and for which the precision is 15. To be more precise, there is no exact kind corresponding to precision 1 and range 38. The nearest one available to the compiler has a precision 15 and range 307. Therefore, to accommodate range 38, the compiler chooses a kind that in this case has a range 307. This particular kind will increase not only the range to 307 but also the precision 1 to 15. The requirement was of precision 1 and range 38, but to accommodate this range and precision, the compiler is forced to choose a kind that has a precision 15 and range 307.

Note that if r is not specified and only p is specified, the system will select appropriate kind to meet the requirement for this p. For example, `selected_real_kind(p=1)`, `selected_real_kind(p=7)` and `selected_real_kind(p=16)` will select kind=1, 2 and 3, respectively, for the NAG compiler. Similarly, `selected_real_kind(r=1)`, `selected_real_kind(r=38)` and `selected_real_kind(r=100)` will select kind=1, 2 and 2, respectively, for the NAG compiler. The reason why, for r=100, kind=2 becomes clear if one examines Table 5.2 carefully. In the absence of p, it is found from Table 5.2 that kind=2 can accommodate r=100. However, if p is set to any valid value greater than 15, kind would be 3.

5.4 SELECTED_CHAR_KIND

This intrinsic takes one character string as its argument and returns the kind parameter corresponding to its argument. The character string can be one of the following:

- DEFAULT
- ASCII
- ISO_10646

```
integer, parameter :: s=selected_char_kind('ASCII')
character(kind=s) :: ch
```

The intrinsic returns −1 if the argument is not supported. While assigning an expression to a character variable, the kind type of the expression and the variable must be the same.

5.5 KIND Intrinsic

The kind intrinsic takes one real, integer or character constant or variable as its argument and returns the kind (a compiler-dependent integer) of its argument. The NAG compiler returns 1 and 2 when the kind intrinsic has argument 0.0 and 0.0d0; print *, kind(0.0) and print *, kind(0.0d0) will display 1 and 2, respectively. The output will be same if the following program is executed:

```
real :: r1
double precision :: r2
print *, kind(r1); print *, kind(r2)
end
```

Also, 2.0_2 is equivalent to 2.0d0, that is, a double precision constant.

However, print *, kind(0) will display 3 (the NAG compiler) because this is the default kind for an integer. The NAG compiler returns 1 when print *, kind('a') is executed. Consider the following program:

```
program int_kind
implicit none
integer, parameter :: nano=selected_int_kind(2)
integer, parameter :: small=selected_int_kind(3)
integer, parameter :: medium=selected_int_kind(5)
integer, parameter :: large=selected_int_kind(10)
integer (nano) :: a
integer (small) :: b
integer (medium) :: c
integer (large) :: d
print *, kind(a); print *, kind(b)
print *, kind(c); print *, kind(d)
end program int_kind
```

The outputs are 1, 2, 3 and 4, respectively. The result follows from the numbers given in Table 5.1. For gfortran and ifort compilers, they are 1, 2, 4 and 8, respectively.

Using the property of the intrinsic selected_int_kind that the intrinsic returns −1 for non-existent kind, the following program prints the available kind number of a particular compiler for integers:

```
program avail_kind
implicit none
integer :: kold=0, knew, i=0 ! NAG compiler
do while (.true.)
  i=i+1
  knew=selected_int_kind(i)
  if(knew < 0) then
    exit            ! non-existent kind, that is some negative integer
  endif
  if(knew .ne. kold) then
    kold=knew    ! change of kind
   print *, "Kind= ",knew, "  9 (",i,")"
  endif
enddo
end program avail_kind
```

The output is as follows:

```
Kind=  1 9 ( 1 )
Kind=  2 9 ( 3 )
Kind=  3 9 ( 5 )
Kind=  4 9 ( 10 )
```

This again demonstrates that the NAG compiler supports four types (kind) of integers having kind=1, 2, 3 and 4. Other compilers may or may not support four kinds of integer. Even if they support, the kind number may be different. For ifort and gfortran compilers, the kind numbers will be 1, 2, 4 and 8.

Real and double precision variables may also be defined with the help of kind.

```
real (kind=kind(0.0)) :: r5
real (kind=kind(0.0d0)) :: r6
real (kind=1) :: r7    ! compiler dependent
real (kind=2) :: r8    ! compiler dependent
real (kind=3) :: r9    ! compiler dependent
```

Some explanation is perhaps required for the declarations kind=kind(0.0) and kind=kind(0.0d0); kind (0.0) returns the kind parameter corresponding to constant 0.0 (single precision real number), which is 1 for the NAG compiler. This 'kind' is kind intrinsic. When this is equated to the kind attribute, it declares r5 as a single precision real variable. Therefore, the first kind is the attribute of the declaration, and the second kind is a 'call' to the intrinsic kind, which returns the 'kind number' corresponding to its argument. The argument is same for kind=kind(0.0d0).

5.6 KIND and COMPLEX Constants

The real and imaginary part of a complex constant may be real or integer (either or both). When both of them are real, the kind of the complex constant is the kind of the part having greater precision (unless the precision of both the parts is same). If one of them is an integer, the kind of the complex constant is the kind of the other part. If both of them are integer, kind of the complex constant is of the default real kind.

```
print *,kind((2,3))          ; print *,kind((2.0,3))
print *,kind((2, 3.0d0))     ; print *,kind((2.0,3.0))
print *,kind((2.0d0,3))      ; print *,kind((2d0,3.0))
print *,kind((2.0d0,3.0d0))
end
```

The outputs for the NAG compiler (shown in a single line) are 1 1 2 1 2 2 2.

5.7 KIND and Character Handling Intrinsics

Character handling intrinsics iachar, ichar, index, len, len_trim, scan and verify may take kind as their argument. The integer returned by these intrinsics is of type specified by the kind attribute if kind is present; otherwise the returned integer is of the default type. For example, print *, kind(len_trim('abcde ', kind=selected_int_kind(2))) displays 1 when the NAG compiler is used.

Thus, the returned integer is of type `kind=1`. If `kind=selected_int_kind(2)` is absent, the returned integer is of the default type having `kind=3` for the NAG compiler.

5.8 Quadruple (Quad) Precision

Intel's ifort compiler supports quadruple precision real numbers. These numbers are more precise than the corresponding double precision real numbers. A quad precision variable is defined as follows:

```
real (kind=kind(0.0q0)) :: q, r ! for intel ifort compiler
```

Note that `0.0q0` stands for the quadruple precision real number having magnitude `0.0`. However, this is not a standard Fortran notation. The precision of such numbers may be verified by executing the following program using the ifort compiler:

```
print *, 2.0q0/3.0q0; end
```

The result is `0.6666666666666666...6` (approximately 34 places). The NAG compiler accepts the following program:

```
integer, parameter :: big=selected_real_kind(31,291)
real (kind=big) :: a
a=2.56789812726374849439392721234512341_big; print *, a; end
```

The output from the program segment will be `2.56789812726374849439392721`.

5.9 DOUBLE COMPLEX

The complex declaration with a proper `kind` is required to declare a `double complex` variable.

```
complex (kind=kind(0.0d0)) :: c1
```

The library function `cmplx` takes a third optional argument, `kind`. The double complex variable `c1` is equated to double precision real and imaginary part with this parameter of kind.

```
c1=cmplx(1.23456789123456d0, 2.34567891234567d0, kind=kind(0.0d0))
```

Without the correct argument of `kind`, `c1` will not be double complex in the true sense, as without a proper `kind`, the real and the imaginary parts (despite the presence of `d0`) will be truncated to normal real quantities first and then converted to double precision before storing them as the real and imaginary parts of `c1`, as `c1` has been declared as double complex. Therefore digits, say, after 6 places of decimal, are meaningless.

```
program dbl_cmplx
implicit none
complex (kind=kind(0.0d0)) :: c1, c2
c1=cmplx(1.23456789123456d0, 2.34567891234567d0, kind=kind(0.0d0))
c2=cmplx(1.23456789123456d0, 2.34567891234567d0)
print *, c1 ! NAG compiler was used
print *, c2
end program dbl_cmplx
```

The outputs are as follows:

```
(1.2345678912345599,2.3456789123456701)
(1.2345678806304932,2.3456788063049316)
```

As the declaration double complex is not a part of the Fortran standard, it is better to use declaration similar to complex (kind=kind(0.0d0)) for double precision complex variables. The real and imaginary parts of a complex number may be of extended precision for the NAG compiler.

```
integer, parameter :: s=selected_real_kind(31,291)
print *, cmplx(1.23456789123456789123456789123456_s, &
          9.87654321987654321987654321123456_s, kind=s)
end
```

The Intel ifort compiler accepts declaration complex (kind=kind(0.0q0)) :: c, where both real and imaginary parts are of type quadruple precision. The following program illustrates this feature:

```
complex(kind=kind(0.0q0)) :: c1=cmplx(2.1q0,3.1q0,kind=kind(0.0q0))
print *, c1 ! ifort compiler
end
```

The output is as follows:

```
(2.10000000000000000000000000000000000,3.10000000000000000000000000000000000)
```

5.10 IMPLICIT and SELECTED KIND

The implicit statement can be combined with selected_int_kind, selected_real_kind and selected_char_kind.

```
implicit real (selected_real_kind(15,307)) (a-h)
abc=2.1115d0; print *, abc ! NAG Compiler
end
```

The implicit statement will ensure that the variable names that start with any letter between a and h are of type real having precision 15 and range 307. The print statement will display 2.1114999999999999, indicating that, indeed, abc is actually a double precision variable.

5.11 Type Parameter Inquiry

These are used to inquire about the kind of a data object (`kind`) and the length of a character variable (`len`). It always returns a scalar.

```
integer, parameter :: s=selected_real_kind(6, 37)
real(kind=s) :: a
character(len=10) :: b ! NAG compiler
print *, b%len ,a%kind, b%kind ! note the notation
end
```

This returns 10, 1 and 1, corresponding to `len(b)`, `kind(a)` and `kind(b)`.

5.12 Named Kind Constants

The named constants `int8`, `int16`, `int32`, `int64`, `real32`, `real64` and `real128` have been defined in the module (Chapter 12) `iso_fortran_env`. They, respectively, define an 8-bit integer, a 16-bit integer, 32-bit integer, 64-bit integer, 32-bit real, 64-bit real and a 128-bit real. To use these named constants, the module must be used (Chapter 12; use `iso_fortran_env`). For example, `real(kind=real64) :: x` defines a double precision variable x. The `kind` values available for integer, real, logical and character that a particular compiler supports can be obtained from the named array constants `integer_kinds`, `real_kinds`, `logical_kinds` and `character_kinds` defined in `iso_fortran_env`.

```
use iso_fortran_env          ! NAG compiler
print *, "Integer_kinds = ",  integer_kinds
print *, "Real_kinds = ",     real_kinds
print *, "Logical_kinds = ",  logical_kinds
print *, "Character_kinds = ",character_kinds
end
```

The output is as follows:

```
Integer_kinds = 1 2 3 4
Real_kinds = 1 2 3
Logical_kinds = 1 2 3 4
Character_kinds = 1 2 3 4
```

6

Array and Array-Handling Intrinsics

Fortran programs often require handling of vectors and matrices. It is practically impossible to name each and every element of vectors or matrices. For example, a 100 by 100 matrix has 10000 elements. To access each element, 10000 different variable names would be required in such a naming scheme, which is clearly impractical.

An array is a built-in data structure of Fortran where all elements are of the same type. An array is referred to by its name and usually accessed by subscripts. An array may be accessed in four different ways: (1) the array as a whole, (2) element by element, (3) a portion of the array and (4) by means of vector subscripts.

An array may be of single dimension, or it may have more than one dimension. A single dimensional array is called a vector, and a two-dimensional array is called a matrix. Fortran allows an array to have a maximum of 15 dimensions (minus corank, discussed in Chapter 20).

6.1 Array Declaration

An integer single-dimensional array of size 10 is declared as follows:

```
      integer, dimension (10) :: a
or    integer :: a
      dimension a(10)
```

The keyword dimension may be omitted if the upper bound is given within parentheses:

```
      integer :: a(10)
```

If the i-n rule is followed, type declaration may be omitted.

```
      dimension x(10)  ! real array x of dimension 10
```

Array a is referred to by its name, and its elements are accessed as a(1), a(2), ..., a(10) or in general a(i), where i is usually an integer constant or expression. Depending upon the current value of i, a particular element of an array is accessed. The lower bound of the array is, by default, 1, and the upper bound is specified along with the array declaration; in this case it is 10. Therefore, the subscript of the array, in this case, must have a value between (and including) 1 and 10. For an integer array, the individual element of the array can store only integer quantity. The dimension of the array is an integer constant or could be a symbolic name associated with a constant (allocatable arrays are discussed in Chapter 15) (Figure 6.1).

```
integer, parameter :: s=10
real, dimension(s) :: a
```

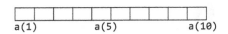

a(1)   a(5)   a(10)

FIGURE 6.1
Array a.

In the declaration, the first item is the type of the array, that is, the type of its content—like `integer`, `real`, `complex`, `double precision`, `logical`, `character`, `double complex` or a user-defined type (Chapter 7). Next, the dimension of the array followed by the name of the array is specified. The array declaration merely reserves locations for the variable—no value is assigned to the array elements.

```
        real, dimension (100) :: c, d      ! c & d are real arrays of
                                           ! dimension 100
        complex, dimension (50) :: z1, z2  ! complex arrays of
                                           ! dimension 50
Again,  integer, dimension (100) :: x, y(40)
```

declares array `x` with dimension 100 and `y` with dimension 40. If `y(40)` is replaced by just `y`, both `x` and `y` will have an upper bound of 100.

A character array may also be declared in an identical manner. However, if the individual elements are expected to store more than one character, the size of each element must be specified. Obviously, the size of all array elements must be the same.

```
        character(len=20),dimension (200) :: ch
```

This defines a character array `ch` of dimension 200, having a length equal to 20 characters per element. Note the difference between `character, dimension(50) :: ch1` and `character(len=50) :: ch2`. The first one defines an array of size 50, and each element can store one character. The second one defines a scalar, which can store 50 characters.

It is also possible to specify the lower bound of the array.

```
        real, dimension(-5: 5) :: x ! lower bound -5 upper bound 5
```

A colon separates the upper and lower bounds. In the preceding case, the lower bound of `x` is -5, and the upper bound is 5. The permissible subscripts for the array are -5, -4, -3, -2, -1, 0, 1, 2, 3, 4, 5. The upper bound must be greater than or equal to the lower bound. If the upper bound is less than the lower bound, a zero-sized array is created.

Just like scalars, the precision and range of the array elements may be specified. It is obvious that, in such a case, all elements of the array must be of the same precision and range as specified in the declaration. In the absence of the precision and range, the array will assume the default precision and range.

```
        integer, parameter :: s=selected_int_kind(9)
        integer(kind=s), dimension (100) :: a
or      integer(kind=selected_int_kind(9)), dimension (100) :: a
        integer, parameter :: p=selected_real_kind(15, 307)
        real(kind=p),dimension(200) :: b
or      real(kind=selected_real_kind(15,307)),dimension(200) :: b
```

6.2 Multidimensional Array

An array can have more than one dimension, where the upper bounds (or the lower–upper combination) of the two dimensions are separated by a comma (Table 6.1).

```
real, dimension(3,3) :: x
real, dimension(-1:1, -2:0) :: p
```

TABLE 6.1

x and p Array Elements

Array Elements					
x			p		
1, 1	1, 2	1, 3	-1, -2	-1, -1	-1, 0
2, 1	2, 2	2, 3	0, -2	0, -1	0, 0
3, 1	3, 2	3, 3	1, -2	1, -1	1, 0

Again, by default, the lower bound in each dimension is 1. In this case, x is a two-dimensional array having 3 rows and 3 columns. As in the case of single dimension, the lower bounds may be other than 1. To access a single element, two subscripts are required for a two-dimensional array. The upper and lower bounds are separated by a colon, and each dimension is separated by a comma. The array elements of p array are (column-wise) p(-1,-2), p(0,-2), p(1,-2), p(-1,-1), p(0,-1), p(1,-1), p(-1, 0), p(0, 0) and p(1, 0).

Similarly, arrays having more than two dimensions may be declared:

```
integer, dimension(3, 4, 5) :: r       ! 3-d array
real, dimension(1 : 2, -1 : 5, 3) :: t ! lower bound of 3rd dimension
                                       ! is 1
logical, dimension(3, 3, 3) :: l
character, dimension(4, 4, 3) :: ch
```

6.3 Storage Arrangement of Two Dimensional Array

All multidimensional arrays are mapped into the single dimension of memory addresses. A two-dimensional array is stored column-wise; array a of 2 by 2 is mapped into a single dimension as shown in Table 6.2.

TABLE 6.2

Elements and Locations of x and b Arrays

Element	Location
a(1, 1)	1
a(2, 1)	2
a(1, 2)	3
a(2, 2)	4

Element	Location	Element	Location
b(1,1,1)	1	b(1,1,2)	5
b(2,1,1)	2	b(2,1,2)	6
b(1,2,1)	3	b(1,2,2)	7
b(2,2,1)	4	b(2,2,2)	8

Similarly, for a three-dimensional array b of dimension 2 × 2 × 2, relations between the array elements and the locations are indicated in Figure 6.2. The same concept may be extended for arrays of higher dimensions. Fortran compilers arrange storage so that successive array elements can be accessed by the very fast machine code increment operator.

We now discuss how a two-dimensional array d (3 × 3) is mapped into a single dimension (Table 6.3).

Symbolically, if the dimension of the array is (I, J) (in this case it is 3 × 3), the (i, j)th element can be computed by the following formula:

TABLE 6.3

d Array

1, 1	1, 2	1, 3
2, 1	2, 2	2, 3
3, 1	3, 2	3, 3

```
(i, j)th element = i + (j - 1) * I
```

This may be verified for d(1, 2). Substituting, we get 1 + (2 -1) * 3 = 4.

As the elements are stored column-wise, the (1, 2)th element is stored in the fourth location, and that is what we get from the formula.

For a three-dimensional array of dimension (I, J, K), the (i, j, k)th element is mapped as follows:

```
(i, j, k)th element = i + (j - 1) * I + (k - 1) * I * J
```

All multidimensional arrays are mapped into a single dimension in the memory. To access an element of a 15-dimensional array, the processor has to perform a calculation similar to the preceding one that involves several additions, subtractions and multiplications (Appendix I). This will reduce the execution speed. The rule of thumb is that unless it is absolutely essential, higher dimensional arrays should be avoided.

6.4 Characteristics of Array

An array is characterized by three attributes: rank, size and shape.

RANK: The rank of an array is defined as the number of dimensions of the array—a single-dimensional array has rank 1, a two-dimensional array has rank 2 and so on. Fortran supports arrays up to rank 15. Some compilers support arrays up to rank 31 (not a standard Fortran). An ordinary variable, also called a scalar, has rank 0. The library function rank is used to find the rank of an array.

```
integer, dimension(4,3) :: a
print *, rank(a)
```

This code will return 2.

SIZE: The size of an array is defined as the total number of elements of an array or the number of elements in a particular direction. First we consider a two-dimensional array. The total size of the array as well as the size of the row or column separately can be determined by the size intrinsic (Table 6.4).

```
integer, dimension(4,3) :: a
print *, size(a)          ! total size is 4*3 =12
print *, size(a, dim=1) ! total no of rows 4
print *, size(a, dim=2) ! total no of columns 3
```

TABLE 6.4

a Array

1,1	1,2	1,3
2,1	2,2	2,3
3,1	3,2	3,3
4,1	4,2	4,3

dim=2 ⟶　　dim=1 ⇓

Expression size(a) returns 12, the total number of elements of a=(4×3).

Expression size(a,dim=1) returns 4, the number of rows, and size(a,dim=2) returns 3, the number of columns. Readers may verify that, for an array x having dimension 4×5×6,

```
print *, size (x,dim=1); print *, size (x,dim=2);
print *, size (x,dim=3)
```

will display 4, 5 and 6.

SHAPE: The extent of an array is defined as the number of elements in a particular dimension. The shape of an array is the set of all extents of the array. The shape intrinsic returns a single-dimensional array of size equal to the rank of the input array to the intrinsic (shape).

```
integer, dimension (4, 5, 6) :: x
```

The extents of the array x are 4, 5 and 6, and the shape of the array is [4 5 6]. Now consider the following program:

```
integer, dimension (-10:10,20) :: x
integer, dimension(2) :: sz      ! the size of sz must be at least 2
sz=shape (x)                     ! as the rank of x is 2
print *, sz
end
```

Since x is a rank 2 array (2-d array), the size of rank 1 array sz must be at least 2. The print statement will display two numbers, 21 and 20, which are the number of elements (extents) in the first and second dimensions, respectively.

Two arrays are said to be conformable if the shapes of the two arrays are identical.

```
integer, dimension(3,3) :: a, b
```

Arrays a and b are conformable.

```
integer, dimension(3,3) :: c
integer, dimension(0:2, 2:4) :: d
```

In this case also, the shape of the arrays c and d are the same—they are conformable. This may be verified by executing the following statements:

```
print *, shape (c)
print *, shape (d)
```

Both the print statements will print [3 3] (without brackets). Two zero-sized arrays may have the same rank, but shapes may be different. One of the arrays may have shape (0 5) and the other (5 0) or (0 4). These types of arrays are not conformable. However, a scalar is always conformable with any array, and thus, `zero-sized-array=scalar` is a valid statement.

A zero-sized array is always considered as defined. The shape of a scalar returns a zero-sized rank 1 array.

```
integer :: x
integer, dimension(0:2) :: a=[10,20,30] ! array constant is described
a(size(shape(x)))=100                    ! in the next section
print *, a
end
```

The output is as follows:

```
100 20 30
```

This demonstrates that `size(shape(x))` returns 0 as x is a scalar.

6.5 Array Constants

An array constant of rank 1 can be constructed by enclosing the constants between "(/" and "/)" or "[" and "]." For example, (/ 1, 2, 3, 4 /) or [1, 2, 3, 4] defines an array constant of size 4 with individual elements as 1, 2, 3 and 4. This is also called an array constructor. An implied do loop can be used to define the aforementioned array constant.

```
        (/ (i, i=1, 4) /)
or      [(i, i=1, 4)]
```

The value of i starts from 1 and goes up to 4 with a step of 1 (default step, if the step is not explicitly mentioned). The step can be other than 1.

```
        (/ (i, i=2, 10, 2) /)
or      [(i, i=2, 10, 2)]
```

is equivalent to

```
        (/ 2, 4, 6, 8, 10 /)
or      [2, 4, 6, 8, 10]
```

In this book, we use the square bracket notation to indicate array constants. Needless to say, array constants may be of data type other than integers.

```
[10.0, 20.0, 30.0, 40.0, 50.0]                ! real
[.true., .false., .false., .true.]            ! logical
[1.2e2, 9.23e-1, -25.73e4, 33.75]             ! real
[(1.0, 2.0),(3.0, 4.0),(5.0, 6.0),(7.0, 8.0)] ! complex
```

The last one is a complex array constant. A few other array constants using the implied do are as follows: [(i*10,i=1,5)] results in an array constant [10, 20, 30, 40, 50]. Similarly, [(i*10, i=1, 10, 2)] is the same as [10, 30, 50, 70, 90].

6.6 Initialization

The dimension statement merely reserves locations for variables—no value is assigned to them: they are undefined. The dimensioned variable may be assigned to a value along with the declaration.

```
integer, dimension(4) :: inp = 100
```

100	100	100	100
inp(1)			inp(4)

FIGURE 6.2
Array inp.

All elements of inp are initialized to 100.

```
real, dimension (3,3) :: x = 10.0
```

All elements of x are set to 10.0.

```
integer, dimension (4) :: b = [10, 20, 30, 40]
```

Element b(1) is initialized to 10, b(2) to 20, b(3) to 30 and b(4) to 40.
An array may also be initialized by an implied do.

```
integer, dimension(10) :: a = [(i, i=1,10)] ! a(1)=1, a(2)=2, …, a(10)=10
```

The implied do loop may be of the following type:

```
integer, dimension (5) :: z = [(3*i, i=1,5)]
```

In this case z(1) to z(5) are initialized to 3, 6, 9, 12 and 15. The implied do loop, like the ordinary do loop, can have an increment not equal to 1.

```
integer, dimension (5) :: w=[(3*i, i=2, 10, 2)]
```

This initialization results in setting w(1) to 6, w(2) to 12, ..., w(5) to 30.

6.7 Initialization with DATA Statement

An array can also be initialized by a data statement. A number of options are available.

```
real, dimension (4) :: a
data a /1.0, 2.0, 3.0, 4.0/
```

The array without any subscripts corresponds to the whole array. Therefore, a(1), a(2), a(3) and a(4) are initialized to 1.0, 2.0, 3.0 and 4.0, respectively. For a two-dimensional array, the elements are stored column-wise. The next declaration

```
integer, dimension (2, 2) :: b
data b /10, 20, 30, 40/
```

will initialize b(1, 1) to 10, b(2, 1) to 20, b(1, 2) to 30 and b(2, 2) to 40. A complex dimensioned variable requires two constants for each location: one for the real part and the other for the imaginary part.

```
complex, dimension (2) :: c
data c /(2.0, 3.0), (4.0, 5.0)/
```

The element c(1) is initialized to (2.0, 3.0), and c(2) is initialized to (4.0, 5.0). A logical variable can have a value that is either .true. or .false..

```
logical, dimension (4) :: l
data l /.true.,.false.,.false.,.true./
```

Character arrays are also initialized in a similar way:

```
character (len=4), dimension (3) :: ch
data ch / 'SINP', 'TIFR', 'IACS' /
```

The same concept can be extended to multidimensional arrays. Dimensioned quantities may also be initialized element by element.

```
integer, dimension (4) :: a
data a(1), a(2), a(3), a(4) / 100, 200, 300, 400/
```

The number of variables present in the data statement must be the same as the number of constants given as data. The compiler will flag an error if this rule is violated.

```
integer, dimension (4) :: d
data d / 1, 2, 3, 4, 5/    ! error
```

In case of a dimensioned quantity, the dimension statement must appear before the data statement.

```
data ia /1, 2, 3, 4, 5/    ! not allowed
integer, dimension(5) :: ia
```

The correct procedure is to interchange these two statements.

6.8 Repeat Factor and Initialization

Several consecutive elements of an array can be initialized by a constant through a single instruction.

```
real, dimension (10) :: x
data x / 10 * 100.0/
```

Expression `10 * 100.0` indicates that all `10` elements of array x are initialized to `100`. Here, `10` is the repeat factor. This can be mixed with the usual initialization methods. For example, in the preceding case, if it is necessary to initialize the first four elements to `1.0`, `2.0`, `3.0` and `4.0`, the next four elements to `100.0` and the last two elements to `0.0`, it can also be achieved through the `data` statement.

```
data x / 1.0, 2.0, 3.0, 4.0, 4 * 100.0, 2 * 0.0/
```

Only a few contiguous elements can be initialized:

```
data x(1:5) /5 * 10/
```

Here, only x (1) to x (5) are initialized.
The repeat factor must be a positive number, normally greater than zero. It may be zero.

```
integer, dimension(5) :: ia
data ia /0 * 10, 2, 3, 4, 5, 6/
print *, ia
end
```

The output is as follows: 2 3 4 5 6. The quantity with repeat factor 0 is ignored.

6.9 DATA Statement and Implied DO Loop

The `data` statement with the implied `do` is a very powerful tool to initialize a dimensioned quantity in a particular order. In the following program segment, b represents a rank 1 array of size 10 and x represents a rank 2 array of shape (2, 2):

```
integer :: i, j
integer, dimension (10) :: b
real, dimension (2, 2) :: x
data (b (i), i=1, 10, 2) / 5 * 100/
```

The preceding statement will initialize b(1), b(3), b(5), b(7) and b(9) to 100. The elements b(2), b(4), b(6) and b(10) are not affected (initialized). Similarly,

```
data ((x(i, j), j=1, 2), i=1, 2) / 1.0, 2.0, 3.0, 4.0 /
```

will set x(1, 1), x(1, 2), x(2, 1) and x(2, 2) to 1.0, 2.0, 3.0 and 4.0, respectively. Note that

```
data x /1.0, 2.0, 3.0, 4.0/
```
or
```
data ((x(i, j), i=1, 2), j=1, 2) / 1.0, 2.0, 3.0, 4.0/ ! i inner loop
```

will set x(1, 1), x(2, 1), x(1, 2) and x(2, 2) to 1.0, 2.0, 3.0 and 4.0, respectively.

6.10 Named Array Constant

A named array constant is a rank 1 array constant having a name attached to it. A named array constant is declared with the `parameter` attribute and cannot be modified during the execution of the program.

```
integer, parameter, dimension(4) :: ia=[25, 50, 75, 100]
```

The elements of `ia` may be accessed in the usual manner through subscripts. However, `ia` (being a named constant) cannot be modified and therefore cannot appear on the left-hand side of the assignment sign.

It is possible to replace the upper bound of the named constant by an asterisk (implied shape array). If the lower bound is absent, it is assumed to be 1. The upper bound is determined from the number of constants present within the initialization string. It is actually the lower bound plus the number of initialization constants with the initialization string minus 1.

```
integer, parameter, dimension(0:*) :: ia=[10, 20, 30, 40, 50]
```

Here, the lower bound of `ia` is 0 and the upper bound is 4 (0 + 5-1), as there are five integers within the initialization string.

6.11 Character Variable and Array Constructors

When the length of the different elements of a character array are not same, it is supposed to give a Fortran error (ifort does not flag this as error; however, NAG compiler and gfortran flag this as error).

```
character(len=20),dimension(4) :: c
c=["abc", "abcdef", "p", "asdfghj"]
print *,c
end
```

To circumvent this programming error, the current version of Fortran allows the type specification with an array constructor:

```
[type spec :: array-construction-list]
```

The preceding program may be modified as follows:

```
c=[character(len=20) :: "abc", "abcdef", "p", "asdfghj"]
```

This declaration ensures that the length of each element will be 20; the compiler will add the required number of blanks, if required.

6.12 Array Elements

One of the methods of accessing array elements is through subscripts. The subscript must be an integer or an integer expression. The expression is evaluated before accessing the array element. The number of subscripts must be the same as the rank of the array.

Rank 1 array: It is accessed by a single subscript: `a(1)`, `a(i)`, `a(3*i)`, `a(3*i+3)`. The current value of i determines the actual array element.

Rank 2 array: It is accessed by two subscripts separated by a comma: `b(1,2)`, `b(i,3)`, `b(i,j)`, `b(2,j)`, `b(2*i,3*j+2)`. Again, the current values of i and j determine the actual array element.

For an array of rank 3, three subscripts are required. An array can have a maximum of 15 subscripts, as the rank of an array cannot exceed 15. The subscripts must be the within the bounds of the array. If the subscript is out of bounds, the result is unpredictable.

6.13 Array Assignment and Array Arithmetic

An array element may be used like a scalar variable.

```
a(1)   = 0            ! a, b, c are 1-d array; i, j, k are defined
b(2)   = a(2) * c(3)
b(i)   = a(j) - b(k)
b(i+2) = a(3*j) * b(3)
p(1, 2) = 110.0       ! p, q, r are 2-d array
p(i, j) = q(i, k) * r(k, j)
```

An array may be assigned as a whole, that is, the same value to all its elements.

```
integer, dimension(10) :: a, b
a = 100
b = 200
```

The statement a = 100 sets all elements of a to 100, and similarly b = 200 sets all elements of b to 200.

A rank 1 array may be equated to an array constructor.

```
integer, dimension (4) :: ia
ia = [50, 60, 70, 80]
```

This assigns ia(1), ia(2), ia(3) and ia(4), respectively, to 50, 60, 70 and 80.

Arithmetic operations are allowed involving the whole array. A scalar, when operating on an array, modifies each of the array elements identically according to the arithmetic operator; ia = ia*10 is equivalent to   ia(1) = 10 * ia(1); ia(2) = 10 * ia(2); ia(3) = 10 * ia(3); ia(4) = 10 * ia(4).

Now, consider the following program segment where the arrays ia and ib are of identical shape (Figure 6.3).

```
integer, dimension (5) :: ia, ib
ib = ib + ia
```

The preceding statement is equivalent to

```
ib(1) = ib(1) + ia(1)
ib(2) = ib(2) + ia(2)
   .
ib(5) = ib(5) + ia(5)
```

The following program segment demonstrates arithmetic operations involving the array as a whole:

```
integer, dimension (4) :: a, b, t
a = [10, 20, 30, 40]
b = [100, 200, 300, 400]
t= a + b; t= a - b
t= a * b; t= b / a
end
```

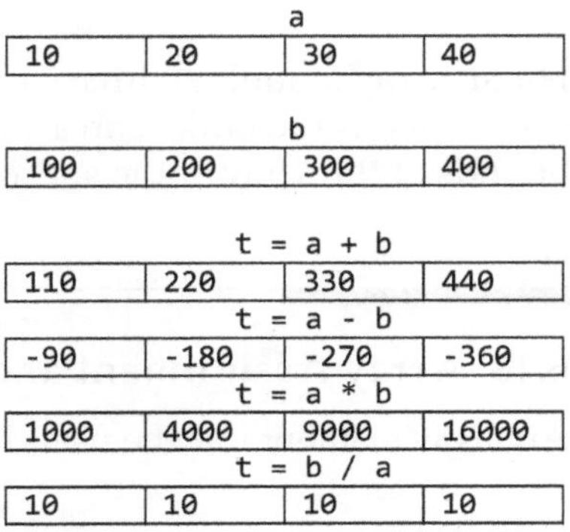

a			
10	20	30	40

b			
100	200	300	400

t = a + b			
110	220	330	440

t = a - b			
-90	-180	-270	-360

t = a * b			
1000	4000	9000	16000

t = b / a			
10	10	10	10

FIGURE 6.3
Arithmetic operation involving whole array.

We discuss the addition operation. The rest follows the same logic.

```
t = a + b
```

is equivalent to

```
t(1) = a(1) + b(1)
t(2) = a(2) + b(2)
t(3) = a(3) + b(3)
t(4) = a(4) + b(4)
```

Now, consider the following program segment:

```
integer, dimension(9) :: a
integer, dimension(-4:4) :: b
integer, dimension(21:29) :: c
   .
c=a+b
end
```

As the arrays a, b and c are conformable, the statement c = a + b is equivalent to

```
c(21) = a(1) + b(-4)
c(22) = a(2) + b(-3)
c(23) = a(3) + b(-2)
   .
c(29) = a(9) + b(4)
```

The next program swaps the contents of two arrays:

1	3
2	4

c

100	300
200	400

d

100	900
400	1600

x

```
program swap
integer, dimension (4) :: a=[10, 20, 30, 40]
integer, dimension (4) :: b=[100, 200, 300, 400]
integer, dimension (4) :: t
t = a; a = b; b = t
print *, a; print *, b
end
```

FIGURE 6.4

Arrays c, d and x.

Now, suppose c and d are 2 × 2 matrices as shown in Figure 6.4.

If x is a 2 × 2 matrix, x = c * d will return matrix x as the product of the corresponding matrix elements of c and d. Note that this type of operation is possible, provided arrays are conformable.

6.14 Array Section

A portion of the array can be accessed. Subscript triplets are often used to access a section of an array. We first consider a single-dimensional array.

An array without any subscript represents the whole array. An array name with just a colon in place of the subscript for each dimension, like a(:) for a one-dimensional array, also represents the whole array. An array name with a subscript triplet like a(i:j:k) represents array elements a(i), a(i+k) and a(i+2k) until the subscript exceeds j. For example, for a one-dimensional array a of size 10 with lower bound 1, a(1:10:2) represents a(1), a(3), a(5), a(7) and a(9). However, a(2:10:2) represents a(2), a(4), a(6), a(8) and a(10). The subscript triplet is similar to the do index, and if the third one, called the stride, is absent, it is assumed to be 1. Similarly, a(5:10) represents a(5), a(6), a(7), a(8), a(9) and a(10). The stride cannot be zero. If it is negative, the starting value must be greater than or equal to the end value. Therefore, a(10:1:-3) represents a(10), a(7), a(4) and a(1).

```
integer, dimension(10) :: a=100
a(10:1:-3)=[10,20,30,40]
print *, a
end
```

The output is as follows:

40 100 100 **30** 100 100 **20** 100 100 **10**

The array elements a(2), a(3), a(5), a(6), a(8) and a(9) are not affected. If the starting value is greater than the end value and the stride is positive, the array section is a zero-sized array. If the starting index is omitted and replaced by a colon, it is assumed to be the lower bound of the array. Similarly, if the end index is omitted and replaced by a colon, it is taken as the upper bound.

Now consider a three-dimensional array b of dimension (4, 5, 6). The array section b(2:4, 4, 3:4) is a rank 2 array, having shape [3, 2] and size 6 with elements b(2,4,3), b(3,4,3), b(4,4,3), b(2,4,4), b(3,4,4) and b(4,4,4).

It is not required that the subscript value should be within the bounds of the array so long as the array elements selected are within the bounds of the array. Array a(4:12:3) is acceptable as the selected elements are a(4), a(7) and a(10) in spite of the fact that 12 is greater than the upper bound of the array.

The following program segment defines the 'a' array for this section (Table 6.5).

```
integer, dimension (4, 4) :: a
integer, dimension (2, 2) :: b
integer, dimension (4) :: c, d
integer :: i, j
do i = 1, 4
  do j = 1, 4
    a(i, j) = 100*i+j
  enddo
enddo
```

TABLE 6.5

a Array

101	102	103	104
201	202	203	204
301	302	303	304
401	402	403	404

Now the statement c(:) = a(1,:) will assign c(1), c(2), c(3) and c(4) to the respective elements of the first row of a, that is, 101, 102, 103 and 104, respectively.

Thus, the preceding statement is equivalent to

```
do i = 1, 4
  c(i) = a(1, i)
enddo
```

Similarly, d(:)=a(:,2) will copy the second column of a to d. The elements of d are now d(1) = 102, d(2) = 202, d(3) = 302 and d(4) = 402.

Using the same notation, a portion of contiguous elements of an array may be assigned to a value without affecting other array elements not present in the statement.

```
c(1:3) = [10,20,30]
```

The element c(4), not present in the statement, is not affected. Here, c(1:3) refers to c(1), c(2) and c(3). The array index could be a triplet consisting of a starting location, an end location and an increment (stride). Somewhat similar to the implied do statement, one can specify the increment (default is 1).

Hence, (1:10:3) refers to the elements 1, 4, 7 and 10. The index triplet may be used for higher dimensions also. The following program illustrates all aforementioned cases:

```
integer :: i
integer, dimension(10) :: a=[(i,i=1,10)]
print *, a                ! whole array
print *, a(:)             ! whole array
print *, a(2:5)           ! a(2),a(3),a(4),a(5)
print *, a(2:10:2)        ! a(2),a(4),a(6),a(8),a(10)
print *, a(5:)            ! a(5),a(6),a(7),a(8),a(9),a(10)
print *, a(:5)            ! a(1),a(2),a(3),a(4),a(5)
print *, a(::2)           ! a(1),a(3),a(5),a(7),a(9)
end
```

One may verify that the outputs of the preceding program are as follows:

```
1 2 3 4 5 6 7 8 9 10
1 2 3 4 5 6 7 8 9 10
2 3 4 5
2 4 6 8 10
5 6 7 8 9 10
1 2 3 4 5
1 3 5 7 9
```

A portion of the array can be extracted by using an array section:

```
b(:,:) = a (2:3,2:3)
```

The 2 × 2 b array will have the elements b(1,1)=202, b(1,2)=203, b(2,1)=302 and b(2,2)=303.

Both the sides of the assignment statement may contain array sections from the same array. In fact, there can even be overlapping elements. However, the right-hand side always uses the unmodified (existing) array elements for the overlapping portions (Figure 6.5).

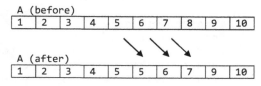

FIGURE 6.5 Array assignment.

```
integer, dimension (10) :: a=[(i, i=1,10)]
a(6:8) = a (5:7)
print *, a
```

This instruction when expanded results

```
a(6) = a(5)
a(7) = a(6)
a(8) = a(7)
```

with the assumption that the unmodified values of a(5), a(6) and a(7) are used on the right-hand side of the assignment sign. Although a(6)=a(5) will modify a(6) to 5, yet this value of a(6) will not be used in the next statement, a(7)=a(6); the old unmodified value of a(6), which is 6, will be used instead.

The print statement will display the a array as: 1 2 3 4 5 5 6 7 9 10.

In general,

- x(i,:) corresponds to the entire ith row.
- x(:,j) corresponds to the entire jth column.
- x(i, 1:j:k) corresponds to the 1st, (1 + k)th, (1 + 2k)th, ..., elements of the ith row.
- x(1:i:k, j) corresponds to the 1st, (1 + k)th, (1 + 2k)th, ..., elements of the jth column.

Finally, for an array of characters, it is possible to access a substring within a particular element of the character array.

```
character (len=20), dimension (100) :: ch
ch (4)(3:6) = 'SINP'
```

will store `'SINP'` in the character positions 3 to 6 of the fourth element of the character array ch.

6.15 Array Input

The read statement is used to input either the individual elements or the array as a whole or the selected elements using the implied do loop.

```
integer, dimension(10) :: a
read *, a(1), a(2), a(4)    ! only a(1), a(2) and a(4) are read
read *, a                   ! read all elements of a
read *, (a(k), k=1, 8)      ! read a(1) to a(8)
read *, (a(k), k=1, 10, 2)  ! read a(1),a(3),a(5),a(7),a(9)
read *, a(1:3)              ! read a(1), a(2) and a(3)
read *, a(1:10:2)          ! read a(1),a(3),a(5),a(7),a(9)
```

Similarly, for two-dimensional arrays, either the individual elements or the whole array or some elements using the implied do may be read.

```
real, dimension (2,2) :: b
read *, b(1,1), b(2,2)      ! only two elements are read
read *, b                   ! read column-wise b(1,1), b(2,1),
                            ! b(1,2), b(2,2)
read *, ((b(i, j), j=1, 2), i=1, 2)   ! implied do read row-wise
```

The last two read statements need some clarification. As the array is stored column-wise, read *, b will assign data in the following order: the first item will be stored in b(1,1), the second item will be stored in b(2,1) and so on. If the supplied data are 10, 20, 30 and 40, read *, b will create the matrix b as follows: b(1,1)=10, b(2,1)=20, b(1,2)=30 and b(2,2)=40. However, the last statement uses the implied do loop, and because the inner index j takes all values before a change in the outer index i takes place, this read statement will assign data in the following order: b(1,1), b(1,2), b(2,1) and b(2,2).

6.16 Array Output

The print statement can display a particular element, the whole array or a selected portion of the array using the implied do statement.

```
print *, a(1), a(3) ! only a(1), a(3) using a's of the
                    ! previous section
print *, a            ! prints the whole array
print *, (a(i), i=1,j)  ! prints a(1), a(2), ... a(j)
print *, a(1:3)        ! prints a(1), a(2), a(3)
print *, a(1:10:2)      ! prints a(1), a(3), a(5), a(7) and a(9)
print *, b(1,1), b(2,2) ! prints only b(1,1) and b(2,2)
print *, b             ! prints the complete array column-wise
print ((b(i,j), j=1, 2), i=1, 2) ! prints the complete array row-wise
```

6.17 Programming Examples

The next few programs illustrate the use of arrays. The following program displays the maximum value and the corresponding location of an array:

```
program maxval
implicit none
integer, parameter :: sz=10
real, dimension(sz) :: a
real :: am
integer :: mloc, i       ! to find the maximum value and its location
a = [-40.0, 23.97, 0.0, 37.25, 11.92, -17.1, &
    100.0, -123.25, 36.25, 22.12] ! to find the largest value from
                                ! an array
mloc=1 ! let maximum=first element, compare and redefine
        ! if necessary
am=a(1)
do i= 2, sz
 if(am .lt. a(i)) then
   am=a(i) ! am the maximum of a. mloc is the corresponding array
           ! location
   mloc=i
 endif
enddo
print *, 'location ', mloc, ' contains ', am, ' (highest value)'
end program maxval
```

By changing the `if` statement, it is easy to find the minimum of a, and in this case, `mloc` is the corresponding array location.

```
am=a(1)
do i= 2, sz
 if(am .gt. a(i)) then
   am=a(i)
   mloc=i
 endif
enddo
```

The next program shows how to multiply two matrices. Two matrices a and b may be multiplied if the number of rows of matrix a is the same as the number of columns of matrix b. The c(i, j)th elements of the product is given by $c_{ij} = \sum a_{ik} * b_{kj}$ (sum over all k).

```fortran
program matrix_mul
implicit none
integer, parameter :: sz =3
real, dimension(sz,sz) :: a, b, c   ! square matrix for simplicity
integer :: i, j, k
real :: sum, x
do i =1, sz
  do j= 1, sz
    call random_number(x)         ! initialize the matrices using
                                  ! random number
    a(i,j)=x*100.0                ! between 0 and 1 using library function
    call random_number(x)
    b(i,j)=x*10.0
  enddo
enddo
print *, a
print *, b
do i=1, sz
  do j=1, sz
    sum=0.0
      do k=1, sz
        sum=sum+a(i,k)*b(k,j)   ! c_ij = ∑a_ik * b_kj [sum over all k]
      enddo
    c(i,j)=sum
  enddo
enddo
print *, c
end program matrix_mul
```

We now use a very inefficient algorithm to sort an array in ascending order.

```fortran
program sort
implicit none
integer, parameter :: sz=10
real, dimension(sz) :: a
integer :: i,j
real :: x, temp                  ! a array is initialized with some
                                 ! random value
  do i=1,sz
   call random_number(x)
   a(i)=x*100.0

  enddo
print *, a
do i=1, sz-1
  do j=i, sz
      if (a(i) > a(j)) then  ! interchange if not in order
        temp=a(i)            ! after the first j loop a(i) contains
        a(i)=a(j)            ! smallest number
        a(j)=temp
      endif
```

```
      enddo
   enddo
   print *, a
   end program sort
```

The preceding program compares the first element of the array with the rest (second to sz); at the end of the j loop, a(1) contains the smallest value among all array elements. Next, the second element of the array is compared with the third to sz elements. At the end of the j loop, a(2) would contain the second smallest value among all array elements. This process continues, and at the end of i-loop, array 'a' becomes a sorted array.

The following program merges two sorted arrays (ascending order) into a single array:

```
program merge_list
implicit none
integer, dimension(10) :: a=[7,20,40,51,79,105,127,190,207,315]
integer, dimension(8) :: b=[-1,36,42,81,112,136,167,328]
integer :: i,j,k,sa,sb
integer, dimension(50) :: c ! the size of c array must be
                            ! at least sa+sb
sa=size(a); sb=size(b)
print *,a; print *,b
i=1; j=1; k=1
do while(i<=sa.and.j<=sb) ! exit if one of the array (or both)
                          ! is exhausted
  if(a(i)<b(j)) then
   c(k)=a(i)
   i=i+1
   k=k+1
  else
   c(k)=b(j)
   j=j+1
   k=k+1
  endif
 enddo
  do while(i<=sa)      !  add the elements from a array if available
   c(k)=a(i)
   i=i+1
   k=k+1
  enddo
 do while(j<=sb)       !  add the elements from b array if available
  c(k)=b(j)
  j=j+1
  k=k+1
 enddo
print *,c(1:sa+sb)
end program merge_list
```

The following program searches an element from an already sorted array using a technique known as binary search:

```
program bin_search
implicit none
integer, dimension(10) :: a=[1,33,45,96,107,109,127,145,205,300]
```

```
integer :: low, high, mid, d, loc
logical :: found=.false.
print *, 'type an integer ... '
read *, d                    ! element to be searched
low=1
high=10
mid=(low+high)/2             ! mid point
do while (low < mid .and. mid < high)
 if(a(mid) .eq. d) then
   loc=mid                   ! found, print the result
   found=.true.
   exit
 endif
 if(a(mid) .gt. d) then
   high=mid
 else
   low=mid
 endif
 mid=(low+high)/2
 enddo
if(.not. found) then       ! if already found skip
  if(a(low).eq.d) then
   loc=low
  else if(a(high).eq.d) then
        loc=high
       else
        loc=0 ! not found
  endif
endif
if (loc.eq.0) then
 print *, d, ' is not in the list'
else
 print *, d, ' is at the position ', loc
endif
end program bin_search
```

The method is to locate the mid-point (mid) and to check whether the number (d) is less than or equal to or greater than the corresponding array element (a(mid)). If d is equal to a(mid), d is present in the array at location mid. If d is not equal to a(mid), the low or high variable is adjusted, and again the same procedure is repeated until either the element is found within the array or it is not found within the array. If the number is present, the corresponding position is displayed. If it is not present, 0 is displayed. This method can be applied to a sorted array. The algorithm may be found in any standard textbook on data structures.

The next program uses the least square technique to fit a straight line through the 10 points shown on the left (Table 6.6).

The equation of a straight line is $y = a_2x + a_1$.

The values of a_2 and a_1 can be obtained by solving the following linear simultaneous equations:

$$m \, a_1 + (\textstyle\sum x_i) \, a_2 = \textstyle\sum y_i$$
$$(\textstyle\sum x_i) a_1 + (\textstyle\sum x_i{}^2) a_2 = \textstyle\sum x_i \, y_i$$

where m is the number of points, and the range of the summation is 1 to m. The derivation may be found in any standard textbook on numerical analysis. Here, $\sum$ indicates the summation. Thus,

$$a_2 = (\textstyle\sum y_i \, \textstyle\sum x_i{}^2 - \textstyle\sum x_i \, \textstyle\sum x_i \, y_i) \, / \, [m \, \textstyle\sum x_i{}^2 - (\textstyle\sum x_i)^2]$$
$$a_1 = (m \, \textstyle\sum x_i \, y_i - \textstyle\sum x_i \, \textstyle\sum y_i) \, / \, [m \, \textstyle\sum x_i{}^2 - (\textstyle\sum x_i)^2]$$

The following program uses these expressions directly:

```
program least_sq
implicit none
integer, parameter :: np=10     ! number of points
real, dimension(np) :: x,y
integer :: i
real :: sy, sxsq, sx, sxy, a1, a2,t
x=[1.0,2.0,3.0,4.0,5.0,6.0,7.0,8.0,9.0,10.0]
y=[9.81, 14.32, 15.86, 17.73, 23.41, 26.37, 29.78, &
   32.51, 33.23, 38.5]
sy=0.0 ; sxsq=0.0; sx=0.0; sxy=0.0
do i=1, np
 sy=sy+y(i)
 sxsq=sxsq+x(i)**2
 sx=sx+x(i)
 sxy=sxy+x(i)*y(i)
enddo
t=np*sxsq-sx*sx
a1=(sy*sxsq-sx*sxy)/t
a2=(np*sxy-sy*sx)/t
print *, "The eqn of the best fit line: y = ",a2, " * x + ",a1
end program least_sq
```

The equation of the best-fit line turns out to be y = 3.1 * x + 7.05.

TABLE 6.6

x vs y

x	y
1.0	09.81
2.0	14.32
3.0	15.86
4.0	17.73
5.0	23.41
6.0	26.37
7.0	29.78
8.0	32.51
9.0	33.23
10.0	38.50

6.18 Array Bounds

As mentioned earlier, each dimension of the array has a lower bound and an upper bound. Fortran provides two intrinsics to find out the lower and upper bounds of an array. It is the responsibility of a programmer not to access an array element outside its bounds. The compilers may have an option for checking the bounds of the arrays during the execution of the program. If this facility is invoked, the program will take more time compared to the case when this option is not invoked.

However, when this option is not invoked, in case the program tries to access an element outside the bounds of an array, the result is unpredictable. In this case, the program will access the adjacent location.

6.19 LBOUND

The intrinsic lbound returns an integer array of rank 1 and size equal to the rank of the input array containing the lower bounds in each dimension.

```
real, dimension(-10:10,0:35) :: a
integer, dimension(2) :: lb
.
lb=lbound (a); print *, lb
end
```

The program sets lb(1) to –10 and lb(2) to 0. They are, respectively, the lower bounds of the input array in each dimension. lbound returns a scalar, if the lower bound of an array in a particular direction is desired through the dim argument.

```
j=lbound(a, dim=1)    ! j is an integer (scalar), it is -10
k=lbound(a, dim=2)    ! k is an integer (scalar), it is 0
```

Expression dim=1 returns the lower bound of the first dimension, and dim=2 returns the lower bound of the second dimension. An interesting case arises if the input to lbound is a single-dimensional array, say, z. Both lbound(z) and lbound(z, dim=1) will return the same value, but the type of the returned values is different—the former returns a single-dimensional integer array of size 1 and the latter returns an integer (scalar). Lbound may take an optional argument kind. If kind is absent, the intrinsic returns an integer of default kind. If, however, kind is present, it returns an integer of that particular kind.

```
integer, dimension(10) :: a
print *, kind(lbound(a, kind=selected_int_kind(2)))
end
```

It displays 1 when NAG compiler is used.

6.20 UBOUND

The intrinsic ubound is identical to its sister intrinsic lbound, except that it returns the upper bound. Using the declaration of the previous section and defining integer, dimension(2) :: ub, this program segment returns the upper bounds in the respective dimensions.

```
ub=ubound (a)         ! ub(1)  = 10, ub(2)=35
j=ubound(a, dim=1)    ! j is a scalar, it is 10
k=ubound(a, dim=2)    ! k is a scalar, it is 35
```

The discussions related to the input of a single-dimensional array to lbound are equally valid for ubound.

Ubound(z), where z is a single-dimensional array, returns a rank 1 integer array of size 1, but ubound (z, dim=1) returns an integer (scalar). Both will contain the upper bound of z for the first dimension.

Ubound may take an optional argument kind. If kind is absent, the intrinsic returns an integer of default kind. If, however, kind is present, it returns an integer of that particular kind.

```
integer, dimension(10) :: a
print *, kind(lbound(a, kind=selected_int_kind(4)))
end
```

It displays 2 when NAG compiler is used.

Although the examples are given using single- and two-dimensioned arrays, the same can be extended to arrays having more dimensions as permitted by the language.

6.21 RESHAPE

Intrinsic reshape is used to construct an array of specified shape (i.e., the number of elements in each dimension) from the elements in a given array source. It takes four arguments: source, shape, pad and order. Out of these, the last two are optional. The first argument is the array with the values that the user wants, and the second argument is an integer array of rank 1. We illustrate reshape with the help of examples. In these examples, if the arguments are supplied in order, that is, source, shape, pad and order, it is not necessary to write source= array-name, shape= ... etc. Expressions source= and shape= must be used when the arguments are supplied in any order (Table 6.7).

Type I argument:

```
integer, dimension(3,3) :: b
integer, dimension(9) :: x
integer, dimension(2) :: s
x = [10, 20, 30, 40, 50, 60, 70, 80, 90]
s =[3,3]
b=reshape(source=x, shape=[3,3])
```

TABLE 6.7

b Array

10	40	70
20	50	80
30	60	90

This will construct a 3 × 3 array b whose elements are assigned column-wise:

```
b(1,1) = 10, b(2,1) = 20, b(3,1) = 30, b(1,2) = 40, b(2,2) = 50,
b(3,2) = 60, b(1,3) = 70, b(2,3) = 80, b(3,3) = 90.
```

The preceding `reshape` instruction may also be written in the following forms:

```
        b=reshape(x,  [3,  3])
or      b=reshape([10,20,30,40,50,60,70,80,90],  [3,3])
or      b=reshape(shape=[3,3],  source=x)
or      b=reshape(x,  s)
or      b=reshape(shape=s,  source=x)
or      b=reshape  (shape=[3,3],  source=[10,20,30,40,50,60,70,80,90])
```

or any combination of the these forms. All aforementioned forms are equivalent. The number of elements in source (in this case, it is 9) must be greater than or equal to the product of the elements of the shape array (in this case, it is 3 × 3=9). For example, b=reshape([1,2,3,4,5,6,7,8,9,10], [3,3]) is acceptable, but b=reshape([1,2,3,4,5,6,7,8], [3,3]) is not because the number of elements in source is less than the product of extents (in this case, 9) of the input array (unless pad is used; ifort—no error).

Type II argument:

Reshape can have a third optional array-valued parameter pad. If the number of elements in source is less than the product of the elements of the shape array (in this case, 3 × 3=9), the unfilled elements are padded with pad (Table 6.8).

```
    b = reshape(source=[1,2,3,4,5,6,7],  shape=[3,3],  pad=[99])
```

will create the b array shown on the right-hand side.

If the elements to be filled are more than the number of elements in pad, the elements of pad are used repeatedly until all elements are filled. In this case, 99 is used twice to fill in elements (2,3) and (3,3). In the case of b=reshape(source=[1,2,3,4,5,6], shape=[3,3], pad=[88,99]), 88 and 99 are used if required repeatedly to fill in all elements of the b array.

TABLE 6.8

b Array

1	4	7
2	5	99
3	6	99

1	4	88
2	5	99
3	6	88

Type III argument:

Reshape accepts a fourth optional array-valued parameter order. The order integer array is of the same shape as the shape parameter of reshape, and the elements are permutations of 1, 2, ..., n, where n is the size of shape. If the order parameter is absent, it is assumed to be 1, 2, 3, ..., n. The argument order tells the order in which the index will vary during storing. The use of order is illustrated with the help of the following examples:

```
    integer, dimension (2, 2, 2) :: b
    b=reshape(source=[1,2,3,4,5,6,7,8], shape=[2,2,2], order=[1,2,3])
```

In this case, order=[1, 2, 3] is the natural order, so that the first index will vary most rapidly followed by the second and third indices. The natural order is given as follows: b(1,1,1) = 1, b(2,1,1) = 2, b(1,2,1) = 3, b(2,2,1) = 4, b(1,1,2) = 5, b(2,1,2) = 6, b(1,2,2) = 7, b(2,2,2) = 8. The natural order is the default value of order, so without the order parameter, reshape will also generate the same b array. However,

```
b=reshape(source=[1,2,3,4,5,6,7,8], shape=[2,2,2], order=[3,2,1])
```

will force the third index to vary most rapidly followed by the second and first indices. Therefore, the elements of the b array would be as follows: b(1,1,1) = 1, b(1,1, 2) = 2, b(1,2,1) = 3, b(1,2,2) = 4, b(2,1,1) = 5, b(2,1,2) = 6, b(2,2,1) = 7, b(2,2,2) = 8.

Similarly, b=reshape(source=[1,2,3,4,5,6,7,8], shape=[2,2,2], order=[2,1,3]) will force the second index to vary most rapidly followed by the first and third indices: b(1,1,1) = 1, b(1,2,1) = 2, b(2,1,1) = 3, b(2,2,1) = 4, b(1,1,2) = 5, b(1, 2,2) = 6, b(2,1,2) = 7, b(2,2,2) = 8.

Type IV argument:
It is possible to use both the pad and order arguments simultaneously.

```
b=reshape(source=[1,2,3,4,5,6], shape=[2,2,2], pad=[88,99], &
    order=[3,2,1])
```

will set the two unfilled elements to 88 and 99, respectively. So, b(2,2,1) = 88 and b(2,2,2) = 99.

It is perhaps clear that pad may be absent but order may be present. We close this section with two more examples. The first one uses a two-dimensional array, that is, a matrix with the order argument.

```
integer, dimension (2,2) :: a
a=reshape([10,20,30,40], [2,2])
```

This will generate a matrix with elements 10, 20, 30 and 40 column-wise. Now,

```
a=reshape(source=[10,20,30,40], shape=[2,2], order=[2,1])
```

will generate the 'a' matrix with elements 10, 20, 30 and 40 row-wise because of the order parameter.

Thus, this 'a' array is actually the transpose of the previous 'a' array. Finally, consider the following program:

```
integer, dimension (2,2,2,2) :: w
integer, dimension (4,4) :: x      ! generate w array
do i=1, 2
  do j=1, 2
    do k=1, 2
      do l =1, 2
        w(i,j,k,l)=10*i+50*j+70*k+90*l
      enddo
```

```
      enddo
     enddo
    enddo
    print *, w
    x=reshape(w,[4,4],order=[1,2])
    print *, x
    x=reshape(w,[4,4],order=[2,1])
    print *, x
    end
```

The elements of w array will be as follows: w(1,1,1,1) = 220, w(2,1,1,1) = 230, w(1,2,1,1) = 270, w(2,2,1,1) = 280, w(1,1,2,1) = 290, w(2,1,2,1) = 300, w(1, 2,2,1) = 340, w(2,2,2,1) = 350, w(1,1,1,2) = 310, w(2,1,1,2) = 320, w(1,2,1, 2) = 360, w(2,2,1,2) = 370, w(1,1,2,2) = 380, w(2,1,2,2) = 390, w(1,2,2,2) = 430, w(2,2,2,2) = 440.

For natural ordering, reshape will create an x array having the elements as follows:

x(1,1) = 220, x(2,1) = 230, x(3,1) = 270, x(4,1) = 280, x(1,2) = 290, x(2,2) = 300, x(3,2) = 340, x(4,2) = 350, x(1,3) = 310, x(2,3) = 320, x(3,3) = 360, x(4,3) = 370, x(1,4) = 380, x(2,4) = 390, x(3,4) = 430, x(4,4) = 440.

When the order parameter is [2,1], reshape will create an x array having the following elements: x(1,1) = 220, x(2,1) = 290, x(3,1) = 310, x(4,1) = 380, x(1,2) = 230, x(2,2) = 300, x(3,2) = 320, x(4,2) = 390, x(1,3) = 270, x(2,3) = 340, x(3,3) = 360, x(4,3) = 430, x(1,4) = 280, x(2,4) = 350, x(3,4) = 370, x(4,4) = 440.

We close this section by mentioning a rule of reshape, which states that source cannot be empty.

```
    b= reshape(source =[], shape=[3,3], pad=[99]) is not allowed.
```

6.22 Vector Subscripts

Vector subscripts allow the elements of an array to be extracted or accessed in any order.

```
    integer, dimension(5) :: a
    .
    a([4,5,1,2,3]) = [10,20,30,40,50]
or  integer, dimension(5) :: a
    integer, dimension(5) :: d=[4,5,1,2,3]
    a(d) =[10,20,30,40,50]
```

This sets a(4) = 10, a(5) = 20, a(1) = 30, a(2) = 40 and a(3) = 50.
   Although an index cannot be repeated, most of the compilers do not flag this as error. The last value is taken (NAG compiler flags this as error).

```
integer, dimension(5) :: a=100
          .
a([2,4,3,2,1])=[1,2,3,4,5]
```

Majority of the compilers will generate code so that a(2) will be 4. Now, consider the following program segment (Table 6.9).

```
integer, dimension(3,3) :: a, b
integer, dimension(3) :: c, d
a=reshape([1,2,3,4,5,6,7,8,9], [3,3])
```

TABLE 6.9

a Array

1	4	7
2	5	8
3	6	9

The instruction

```
b=a([3,2,1], [1,3,2])
```

will perform the following assignment (Table 6.10):

```
b(1,1) = a(3,1), b(2,1) = a(2,1), b(3,1) = a(1,1), b(1,2) = a(3,3),
b(2,2) = a(2,3), b(3,2) = a(1,3), b(1,3) = a(3,2), b(2,3) = a(2,2)
b(3,3) = a(1,2).
```

The result would be the same if

```
c = [3, 2, 1]
d = [1, 3, 2]
b = a(c, d)
```

TABLE 6.10

b Array

3	9	6
2	8	5
1	7	4

and the b matrix will be as shown. Note that the indices on the right-hand side are obtained from the elements of c and d rank 1 arrays. For each element of the d array, all elements of c are taken to generate the indices of the right-hand side, for example, (3,1), (2,1), (1,1), (3,3), (2,3), (1,3), (3,2), (2,2), (1,2). It is also possible to use the same array both on the right-hand side and on the left-hand side of the assignment sign: a = a (c, d) will redefine the 'a' array. We shall analyze the instruction a = a (c, d) in detail. The preceding statement is equivalent to a(1,1) = a(3,1), a(2,1)=a(2,1), a(3,1) = a(1,1), a(1,2) = a(3,3), a(2,2) = a(2,3), a(3,2) = a(1,3), a(1,3) = a(3,2), a(2,3)=a(2,2), a(3,3) = a(1,2), but the existing values (i.e., values before the assignment) of the a array are always used on the right-hand side of the assignment sign. For example, in the statement a(3,1)=a(1,1), the existing value of a(1,1), which is 1, is used, though the equivalent statement a(1,1) has already been changed through the statement a(1,1)=a(3,1). The reader may verify that b= a([3,2,1], [1,2,3]) will generate b as shown in Table 6.11.
   We conclude this section with an example of a three-dimensional array.

```
integer, dimension(2, 2, 2) :: x, y
x= reshape([1,2,3,4,5,6,7,8], [2,2,2])
y=x([2,1], [1,2], [2,1])
```

TABLE 6.11

b Array

3	6	9
2	5	8
1	4	7

```
    print *, x
    print *, y
    end
```

The output is as follows:

```
1  2  3  4  5  6  7  8
6  5  8  7  2  1  4  3
```

6.23 WHERE Statement

This statement is used to modify the array elements selectively depending on the values of a logical array (mask). Stated in another way, the where statement is used to assign some elements of an array to another array under the control of a mask. There are three forms of where statements:

Form I:

```
where (logical-array-expression) array-variable = expression
```

Form II:

```
where (logical-array-expression)
  array-variable = expression
end where
```

Form III:

```
where (logical-array-expression)
  array-variable = expression
else where
  array-variable=expression
end where
```

The where construct can contain only the array assignment statement and the nested where construct (to be discussed shortly). All arrays used within the where construct must be of the same shape as the logical expression used along with the where statement. We first illustrate the use of where with a simple example. Suppose a real rank 1 array a of size 10 contains some numbers, and we want to take the square root of the elements that are positive.

```
real, dimension(10) :: a= &
  [7.0, 39.2, -47.5, 67.25, 100.39, 25.0, -49.0, 0.0, 1.0, -32.0]
where (a>0) a=sqrt(a)
  print *, a
```

The output is as follows:

```
2.6457512    6.2609906    -47.5000000    8.2006102 10.0194807    5.0000000
-49.0000000    0.0000000    1.0000000    -32.0000000
```

Note that while taking the square roots, negative numbers are skipped.

Using Form II type of the where statement, the same thing can be achieved.

```
where (a>=0.0)
  a=sqrt(a)
end where
```

The logical mask is evaluated. The square root of the elements of the 'a' array corresponding to the true value of the logical mask is taken and stored in the same location. The array elements of 'a' corresponding to the false value of the mask are left untouched. Therefore, the print statement would show that the third, seventh and tenth elements have not been modified; other array elements contain the square root of the original value.

The next program sets the elements of an array of size 50 to 99 if the elements are multiples of the first four prime numbers (2, 3, 5 and 7).

```
integer, dimension(50) :: a=[(i,i=1,50)]
print *, a
where (mod(a,2).eq.0 .or. mod(a,3).eq.0 .or. &
    mod(a,5).eq.0 .or. mod(a,7).eq.0)
  a=99
end where
print *, a
end
```

If it is desired to set the array elements of 'a' corresponding to the false value of the logical array to some value, else where may be used.

```
real, dimension(10) :: a= &
  [7.0, 39.2, -47.5, 67.25, 100.39, 25.0, -49.0, 0.0, 1.0, -32.0]
where (a>=0)
  a=sqrt(a)       ! square root for positive elements
else where
  a=-999.0        ! set to -999 for negative elements
end where
print *, a
end
```

All array elements of 'a' having negative values are set to -999.0. The space between end and where is optional.

There can be any number of else where with mask; however, one of the else where must be without the mask. If it is necessary to set all array elements of 'a' less than -48.0 to -99.0 and other negative elements to -77.0, an else where with the mask and another else where without the mask must be added to the preceding program.

```
real, dimension(10) :: a= &
  [7.0, 39.2, -47.5, 67.25, 100.39, 25.0, -49.0, 0.0, 1.0, -32.0]
where (a>=0)
    a=sqrt(a)
else where (a < -48.0)
    a=-99.0
```

```
else where
    a=-77.0
end where
print *, a
end
```

The output is as follows:

```
2.6457512   6.2609906  -77.0000000   8.2006102  10.0194807    5.0000000
-99.0000000   0.0000000   1.0000000  -77.0000000
```

The assignments are performed sequentially—first the 'where' block and then the 'else where' block. Note the difference between the if-then-else and where statements. In case of the if-then-else statement, either the 'then' block or the 'else' block is executed. Both are never executed simultaneously. In case of the where statement, both the 'where' and 'else where' blocks are executed sequentially.

The mask expression is evaluated only once. Any subsequent changes in the mask expression do not affect the mask.

```
real, dimension(5) :: a=[1.0,-2.0,3.0,-4.0,5.0]
real, dimension(5) :: b
where (a>0)
    b=sqrt(a)
else where
    a=-a
    b=sqrt(a)
end where
print *,b
end
```

The output is as follows:

```
1.0000000   1.4142135   1.7320508   2.0000000   2.2360680
```

The output indicates that the instruction a=-a did not affect the mask. If a non-elemental function (Chapter 11) appears within the where statement or the mask expression, it is evaluated without the control of the mask. The function sqrt is an elemental function, whereas the function sum is not an elemental function. The function sum(a) calculates the sum of all elements of a without any control of the mask.

```
real, dimension(5) :: a=[1.0,-2.0,3.0,-4.0,5.0]
real, dimension(5) :: b=999.0,c=888.0
where (a>0)
    b=a/sqrt(a)   ! sqrt is an elemental function
    c=a/sum(a)    ! sum is a non-elemental function
end where
print *,b
print *,c
end
```

The output is as follows:

```
1.0000000   9.9900000E+02   1.7320509   9.9900000E+02   2.2360680
0.3333333   8.8800000E+02   1.0000000   8.8800000E+02   1.6666666
```

We shall now analyze the output. The first line of the output shows that the square root corresponding to the positive elements of array a are calculated (first, third and fifth elements). As a(2) and a(4) are less than zero, they are not affected—they retain their initial value of 999.0. However, the function sum is not an elemental function. Therefore, this function is evaluated without the control of the mask and both the positive and the negative elements are taken. Thus, sum(a) returns 3.0 and this value of sum(a) is used to calculate a/sum(a) for the positive array elements of a (first, third and fifth elements); the second and fourth elements of a remain unaffected.

In case an array constructor is present within the where statement or along with the logical mask, the same is evaluated without the control of the mask.

The where statements may be nested, that is, it is possible to have one where inside another where. The rules for nested where are similar to the rules of a nested do loop.

```
real, dimension(5) :: a=[-10.0,20.0,30.0,-40.0,50.0]
real, dimension(5) :: b=[5.0,0.0,3.0,-2.0,4.0]
real, dimension(5) :: c=100.0
where (a>0.0)
 where(b>0.0)
   c=sqrt(a)/b
 end where
end where
print *,a
print *,b
print *,c
end
```

The output is as follows:

```
-10.0000000   20.0000000   30.0000000  -40.0000000   50.0000000
5.0000000      0.0000000    3.0000000   -2.0000000    4.0000000
1.0000000E+02   1.0000000E+02   1.8257419   1.0000000E+02   1.7677670
```

In this case, nesting of the where statements is not required. The following program gives an identical result.

```
where (a>0.0 .and. b>0.0)
 c=sqrt(a)/b
endwhere
```

6.24 DO CONCURRENT

This statement is used to facilitate parallel computing. The assumption is that the loops are independent. The header of do concurrent is similar to forall (now declared as obsolete).

```
integer, dimension(10) :: a
do concurrent (i=1:10)
 a(i)=10
enddo
```

It may be noted that a(1)=10 is nothing to do with a(2)=10. Therefore, these can be executed independently in parallel. It does not matter whether a(2) is set before a(1) or vice versa. Therefore, if more than one processor is available, the load can be shared among the processors. Do concurrent may have a label. In that case, enddo must have the same label.

```
mylabel: do concurrent (i=1:10)
           a(i)=100
         enddo mylabel
```

There may be an optional comma (,) between do and concurrent.

The index (in this case, i) must be an integer; the starting value, the end value and the step must be an integer or an integer expression. If the step is absent, it is assumed to be 1 (as the preceding case). However, it may not be 1 always.

```
do concurrent (i=1:10:2)
  a(i)=50
enddo
```

In this case, the step is 2. The starting value may be greater than the end value; however, in this case, the step must be a negative number.

```
do concurrent (i=10:1:-1)
  a(i)=25
enddo
```

There may be more than one index: the first one being the inner and the second one being the outer.

```
integer, dimension(3,3) :: a
do concurrent (i=1:3, j=1:3)
  a(i, j)=80
enddo
```

For each value of j, all values of i are taken. The values of i and j will be (1,1), (2,1), (3,1), (1,2), (2,2), (3,2), (1,3), (2,3) and (3,3). Variables i and j are local to the 'do' block. They are undefined outside the 'do' block. This is according to the gfortran compiler. Intel ifort compiler treats these variables in a different way. However, if there is an integer declaration outside the 'do' block having the same name, the global variable reappears after the 'do' block.

```
integer, dimension(10) :: a
integer:: i =300
do concurrent (i=1:10)
  a(i)=10
enddo
print *, i            ! this will print 300
```

It is possible to declare an additional local variable(s) inside the 'do' block:

```
do concurrent (i=1:10) local(x)
```

but the current compilers do not support this feature.

A mask expression, which is scalar and logical, is allowed within the construct. The statements after do concurrent will be executed if the mask is true; otherwise, the statements are skipped.

```
integer, dimension(3,3) :: a=100
do concurrent (i=1:3, j=1:3, i.ne.j)
  a(i, j)=80
enddo
end
```

The preceding program will not change the values of a(1,1), a(2,2) and a(3,3) because in these cases the value of the mask is false.

The do concurrent constructs may be nested. In this case, the loop index must be different. The preceding program may be written as follows:

```
integer, dimension(3,3) :: a=100
do concurrent (i=1:3)
  do concurrent (j=1:3, i.ne.j)
   a(i,j)=80
  enddo
enddo
end
```

For labeled blocks, the inner one must terminate before the outer one.

```
      integer, dimension(3,3) :: a=100
m1: do concurrent (i=1:3)
m2:    do concurrent (j=1:3, i.ne.j)
         a(i, j)=80
       enddo m2
     enddo m1
     end
```

The following rules are applicable within a do concurrent construct.

1. There shall not be a return statement, an image control statement (Chapter 20) and an exit statement within the construct. Also, a cycle statement that brings control to the outer do concurrent statement is not allowed.

2. Procedures used within it must be pure (Chapter 11).

3. IEEE routines—ieee_get_flag, ieee_set_halting_mode and ieee_get_halting_mode (Chapter 14)—shall not appear within the construct.

4. Branch statements within the do concurrent statements to transfer the control to outside the construct is not allowed.

5. The index variables are local to the block and are integer (scalar).

6.25 FORALL Statement

The loop statements discussed in Chapter 4 (do-enddo and do-while-endo) are executed cycle by cycle; that is, once the first cycle is over, the second cycle begins. Often the cycles are independent of each other. The simplest example is to set all elements of an array to 0. In this case, not only the cycles are independent of each other but the order in which the cycles are to be executed is also arbitrary. For example, it does matter if the tenth element of the array is initialized before the first element. In such a case, parallel execution within a system having more than one processor is possible. The forall statement does precisely this job. The syntax of forall is as follows:

```
forall(v1=i1:e1:s1, v2=i2:e2:s2, ..., vn=in:en:sn,mask) &
            array variable=expression
```

or

```
forall (v1=i1:e1:s1, v2=i2:e2:s2, ..., vn=in:en:sn, mask)
            array variable=expression
          .
end forall
```

The forall statement may have the type and the kind of its index along with the header.

```
forall (integer(kind=big):: v1=i1:e1:s1, ..., vn=in:en:sn, mask)
  .
end forall
```

where big is a suitable named constant.

The variables v1, v2, ..., vn are integers (scalar). They start with initial values i1, i2, ..., in (integer and scalar) and go up to e1, e2, ..., en (integer and scalar) with an increment (stride) s1, s2, ..., sn (integer and scalar). If the increments are absent, they are assumed to be 1. The optional mask, if not present, is assumed to be true always. If it is present, a particular element is processed if the mask corresponding to the element is true; otherwise, it is not processed.

```
integer, dimension(10) :: a
integer :: i
forall (i=1:10)    ! without mask
  a(i)=0
end forall
print *, a
end
```

The next program uses a mask:

```
integer :: i, j, n=4
real, dimension (4,4) :: b, x=0.0
!    set the b array
  .
forall (i=1:n, j=1:n, b(i,j) /=0)
  x(i,j)=1.0/b(i,j)
end forall
  .
end
```

This program calculates `1.0/b(i, j)` for the nonzero elements of b and stores the value in x.

Note the difference of mask between `where` and `forall`. In case of `forall`, mask can be set for the individual elements.

```
forall (i=1:n, j=1:n, b(i, j) /=0 .and. i/=j)
.
end forall
```

In all assignment statements within `forall`, the original values of the elements are always used—even when the value of an element has been modified. The modified value is never used within the `forall`. This is justified as the cycles are executed in any order. Therefore, for the sake of consistency of the results, the old values (existing values) are always used. This is demonstrated with the help of the following example:

```
integer,dimension(4,4) :: a,b=0
integer :: i, j, n=4
a=reshape([[(i,i=1,16)],[4,4]])
forall (i=2:n-1, j=2:n-1)
 a(i,j)=a(i,j-1)+a(i,j+1)+a(i-1,j)+a(i+1,j)
 b(i,j)=a(i,j)
end forall
print *, a
print *, b
end
```

Each element of `a(2:3, 2:3)` is redefined by adding the values of its four adjacent neighbors. These a and b matrices are as follows (column-wise):

```
a=[1 2 3 4 5 24 28 8 9 40 44 12 13 14 15 16]  and
b=[0 0 0 0 0 24 28 0 0 40 44 0 0 0 0 0]
```

Incidentally,

```
b(2,2) = a(2,1)+a(2,3)+a(1,2)+a(3,2)
```

Similar calculations are done for calculating `b(2,3)`, `b(3,2)` and `b(3,3)`.

6.26 Rules for FORALL

1. All indices must be integer and scalar.
2. The indices are valid only within the `forall`. The `forall` is their scope. The indices do not have any effect on the same named variable outside the `forall`.

```
i=10
forall (i=1:100)
.
end forall
print *, i
```

The preceding program segment will still print i as `10`.

3. The initial value, final value and stride of any index cannot refer to another index of the `forall`.

```
forall (i=1:n, j=i+1:n-1)
    .
end forall
```

This is not a valid `forall` statement.

4. If the mask is absent, it is assumed to be true always.
5. If the mask is present, the assignment statement is executed corresponding to the true value of the mask.
6. If the mask contains a call to a procedure, it must be pure (Chapter 12).
7. Scalar assignment statements are not allowed within `forall`—only arrays are allowed.
8. A `forall` statement may have a label, and in that case, the `end forall` must have the same label.

```
 lab: forall(...  )
        .
     end forall lab
```

9. Within `forall`, if there is a call to any procedure, the procedure must be pure.
10. The `forall` statements may be nested. In addition, the `forall` may also contain a `where` statement within it.
11. Only an array can be used on the left-hand side of the assignment sign within the `forall`.
12. A `where` construct cannot have a `forall` statement within it.

The structure of `forall` and all its rules ensure cycle-independent, parallel execution of `forall`.

6.27 EQUIVALENCE Statement

Two or more variables may share the same location within a program unit. It is like a person having a good name and a nickname. The location is the same, but it can be referred to by two or more names. This is done through the `equivalence` statement. Since a variable can store only one value at a time, `equivalence` actually means assigning two names to a single variable. The main reason for using the `equivalence` statement is to reuse the locations, which are not required at a given point. However, assigning another name to a variable is convenient because then the new name (but the old location) will have some relation with the actual quantity. We use a hypothetical example to illustrate this point.

Suppose we are using a variable `age` to denote the age of a person. In the same program unit, suppose it is necessary to use this location to denote the price of a car when operations with the age are over and the value stored in the location age is not required. It is desirable to call this location now as price. Of course, no harm is done if the same name age is used to denote the price of a car, provided one remembers that at that point the variable age does not denote the age of a person except that the program loses some amount of readability. It may be noted that the variable can store only one value at a time, so whatever be the name age or price the variable contains the age of a person or the price of a car. Therefore, it is the responsibility of the programmer to interpret the variable in a proper manner.

Another possible reason for using the equivalence statement is when two persons have developed one program unit and perhaps due to lack of communication between them, they used different names for the same variable. In that case, instead of changing the names at several places, the equivalence statement can be used. Of course, the aforementioned situations are hypothetical, and nowadays as the computer has more memory than it had 30 years ago, the equivalence statement is rarely used. In fact, unless there is a very special reason, the use of equivalence is strongly discouraged. The equivalence statement is used to indicate that two or more variables would use the same location in the same program unit.

```
real :: a, b
equivalence (a, b)
```

Since a and b use the same location, if a is set to 10,

```
print *, b
```

will display 10.

```
      real, dimension (4) :: a, b
      equivalence (a(1), b(1))
or    equivalence (a, b)
```

will make a and b equivalent. That is, a(1) and b(1), a(2) and b(2), a(3) and b(3) and a(4) and b(4) will share the same location.

```
real, dimension (2, 2) :: a
real, dimension (4) :: b
equivalence (a(1, 1), b(1))
```

In this example, a(1, 1) and b(1), a(2, 1) and b(2), a(1, 2) and b(3) and a(2, 2) and b(4) share the same location. A two-dimensional array a is mapped into single dimension column-wise.

Now, consider the following equivalence statement:

```
real, dimension (4) :: a, b
equivalence (a(1), b(2))
```

Since a(1) and b(2), a(2) and b(3) and a(3) and b(4) are sharing the same location, b(1) and a(4) are really not sharing their locations with any variable. However, b(1) may be referred to as a(0) and a(4) as b(5).

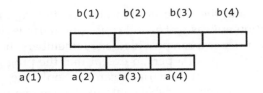

FIGURE 6.6
Array equivalence.

In a similar way, equivalence (a(2), b(1)) will make a(2) and b(1), a(3) and b(2) and a(4) and b(3) equivalent; a(1) and b(4) are not sharing their locations with any variable, but they may be treated as b(0) and a(5) (Figure 6.6).

The equivalence statement is based on the fact that the locations are assigned sequentially.

An equivalence statement between two different types of variables is not recommended. However, the following is an interesting example:

```
complex :: c=(1.0, 2.0)
real, dimension (2) :: a
equivalence (a(1), c)
print *, c
print *, a
end
```

The output will be
```
(1.0, 2.0)
1.0 2.0
```

Complex variables actually consists of two real variables—the real and imaginary parts. When the array a and the complex variable c are made equivalent, a(1) and the real part of c, and a(2) and the imaginary part of c share the same location. Therefore, a(1) is essentially the real part of c, and a(2) is the imaginary part of c. Again, this is strongly discouraged.

A single variable cannot be made equivalent to two different elements of an array.

```
real :: a
real, dimension (4) :: b
equivalence (a, b(1)), (a, b(2))
```

is not allowed.

6.28 EQUIVALENCE and Character Variables

Two character variables of similar or dissimilar length may be made equivalent:

```
character (len=4) :: a, b
equivalence (a, b)
```

This makes a(1:1) equivalent to b(1:1), a(2:2) to b(2:2) and so on.

```
character (len=4) :: a= 'iacs'
character (len=3) :: b
equivalence(a, b)
```

This will make a(1:1) and b(1:1), a(2:2) and b(2:2) and a(3:3) and b(3:3) equivalent. Thus,

```
print *, b
```

will display 'iac' since the length of the character variable b is 3.

```
equivalence (a(2: ), b(1: ))
```

will make the second character of a equivalent to the first character of b. Therefore,

```
print *, b
```

will display 'acs'.

Another interesting property of a logical variable can be seen through the equivalence statement.

```
logical :: l
integer :: a=1    ! ifort compiler
equivalence (a,l)
print *, a
print *, l
end
```

The output is
1
T

By changing a=1 to a=0, the output becomes
0
F

Another feature of the ifort Fortran compiler can be demonstrated with equivalence. If a is set to 2, the output is 2 and F. The reason is that the zeroth bit of the logical variable determines whether the value is true or false. When the variable a is 2, the zeroth bit is 0 [2 in binary $(10)_2$], and therefore, print *, l displays F (false). This may be compiler dependent and is strongly discouraged.

The diagonal element of a matrix may be extracted if a two-dimensional array a is made equivalent to a single-dimensional array e (Table 6.12):

```
integer, dimension(4,4) :: a
integer, dimension(16) :: e
integer, dimension(4) :: d
equivalence (a(1,1), e(1)) ! Table 6.12
        .
```

then, d(:) = e (1:16:5)
with   d=[101, 202, 303, 404]

TABLE 6.12

a Array

101	102	103	104
201	202	203	204
301	302	303	304
401	402	403	404

Before we conclude, we advise the readers to avoid these types of trick. One must not twist the language to extract something, which is not the intention of the designers of the language. For example, extracting the real and imaginary parts of a complex number by using the equivalence statement should be avoided; instead, real and aimag library functions are provided to perform the same task, and such programs look clean, readable and machine independent.

6.29 Programming Examples

The next program sorts an array of character strings. The technique is to sort one byte at a time from the right (from the least significant to the most significant) and place them in proper sequence with respect to this byte. When all bytes from right to left are sorted in this way, the input array becomes a sorted array. The strategy is to count the frequency of the number of occurrences of each character (characters are converted to integer according to ASCII collating sequence). Then its position in the sorted array is determined with respect to this byte, and the input array is accordingly adjusted.

```fortran
program title_sort
implicit none
integer, parameter :: ns=11, nl=20,cl=128 ! ns= number of strings
character(len=nl),dimension(ns):: str, tstr ! nl= length of each string
integer :: i,j,k,l,t ! cl = normal ascii character set has 128 characters
integer, dimension(0:cl) :: carray ! string to be sorted
str=[character(len=nl) :: "Chatterjee", "Roy", "Aich", "Dey", &
"Mukhopadhyay", "Ghosh", "Mitra", "Bag", "Bandyopadhyay", "Ray", &
"Ganguly"]
print *, "Input String:"
print *
do k=1, ns
 print *, str(k)
enddo
! take the right most byte and each cycle move one byte
! towards left that is least significant to most
! significant byte till it reaches the leftmost byte
    do l=nl, 1, -1 ! right to left byte cycle
! initialize the counter array in every "byte cycle"
       carray=0
       do i=1, ns ! string 1 to ns
        t=iachar(str(i)(1:1)) ! convert 1-th character to number
! count the frequency of occurrence of each byte
         carray(t+1)= carray(t+1)+1
        enddo
       do j=0, cl-1
! determine the actual position in the array with respect to this byte
! the exact position of the byte in the sorted array with respect to this
! byte is sum of all array elements before this byte plus the
! value of this byte
       carray(j+1)=carray(j+1)+carray(j)
       enddo
```

```
      do k=1, ns
! move to a temporary location according to sort sequence with respect
! to this byte
        t=iachar(str(k)(1:1))
        tstr(carray(t)+1)=str(k)
! increment the counter so that string having the same character goes to
! the next array element. note identical string will maintain same
! relative
! position as the input string in the sorted array
        carray(t)= carray(t)+1
      enddo
      do k=1, ns
! copy back to the original array
        str(k)= tstr(k)
      enddo
      enddo
! sort one byte at a time from the right
! when the above loop is over all bytes have been sorted.
      print *
      print *, "Sorted string:"
      print *
      do k=1, ns
       print *, str(k)
      enddo
      end program title_sort
```

The output is as follows:

```
Input String:
Chatterjee
Roy
Aich
Dey
Mukhopadhyay
Ghosh
Mitra
Bag
Bandyopadhyay
Ray
Ganguly
Sorted string:
Aich
Bag
Bandyopadhyay
Chatterjee
Dey
Ganguly
Ghosh
Mitra
Mukhopadhyay
Ray
Roy
```

The program logic is explained through comments.

We give another complete program. This numerical technique is known as the Gauss–Seidel method for solving linear simultaneous equations. We consider three linear simultaneous equations with three unknowns:

$$a_{11}x_1 + a_{12}x_2 + a_{13}x_3 = b_1 \quad (1)$$
$$a_{21}x_1 + a_{22}x_2 + a_{23}x_3 = b_2 \quad (2)$$
$$a_{31}x_1 + a_{32}x_2 + a_{33}x_3 = b_3 \quad (3)$$

Rearranging these equations, we get

$$x_1 = (1/a_{11})(b_1 - a_{12\,x2} - a_{13\,x3}) \quad (4)$$
$$x_2 = (1/a_{22})(b_2 - a_{21\,x1} - a_{23\,x3}) \quad (5)$$
$$x_3 = (1/a_{33})(b_3 - a_{31\,x1} - a_{32\,x2}) \quad (6)$$

a_{11}, a_{22} and a_{33} are not equal to 0.

We choose a guess value, say, $x_1 = x_2 = x_3 = 0$ (the guess value of x_1 is actually not required) and substitute x_2 and x_3 in equation (4) and obtain x_1. Using this value of x_1 and guess value of x_3, we obtain x_2 from equation (5). Again using new values of x_1 and x_2, we obtain a new value of x_3. This process is repeated until the difference between the sum of the old absolute values of x and the new absolute values of x is small compared to some predefined value, say, `1.0e-7`.

We shall now solve the equations:

```
4x - y + 2z   = 8
2x + 7y - z   = 13
x + 2y + 5z   = 20

program gauss_seidel
implicit none
integer, parameter :: sz=3, iter=100
integer :: count=0, i, j, k
real, dimension (sz,sz) :: a
real, dimension(sz) :: b, x
real :: sum,res, eps=1.0e-7,r
x=[0.0,0.0,0.0]     ! guess value = 0
b=[8.0,13.0,20.0]
a=reshape([4.0,2.0,1.0,-1.0,7.0,2.0, 2.0,-1.0,5.0],[3,3])
!    construct a matrix with the coefficients x, y and z (column-wise)
do k=1, iter
  count=count+1
  do i=1,3
    sum=0.0
    r=0.0
    do j=1,3 !skip when i=j
      if(i.eq.j) then
        cycle
      endif
      sum=sum+a(i,j)*x(j)
    enddo
    sum=(b(i)-sum)/a(i,i)
    res=abs(sum-x(i))
    r=r+res                ! sum of the residue
    x(i)=sum               ! redefine x's
  enddo
```

```
     if(r .le. eps) then
       exit
       endif
     enddo
     if(count < 100) then
       print *, 'Converged after ', count, ' cycle'
       print *, 'Roots are : ',x
       else
       print *, 'Convergence not achieved ', x
     endif
       end program gauss_seidel
```

The roots of these equations are 1, 2 and 3.

6.30 Array-Handling Intrinsics

Fortran contains several array-handling intrinsics. The intrinsics can be divided into several groups. Although these intrinsics may operate on any array of rank less than or equal to 15, we have chosen arrays of rank less than or equal to 2 for simplicity. However, just to get a feeling how calculations are performed for the array of rank 3 or 4, a separate section has been added and detailed calculations are shown. This can also be extended for higher dimensional arrays.

6.31 Maximum, Minimum and Finding Location

There are five intrinsics under this group: maxval, minval, maxloc, minloc and findloc. Maxval and minval, as the names imply, return the maximum and minimum values from an array. The array can be of type integer, real or character. Similarly, maxloc and minloc return the locations containing the maximum and minimum values, respectively. Both the maxval and the minval take three arguments—two of them are optional. The arguments of maxval and minval are array, dim and mask. The arguments of maxloc and minloc are array, dim, mask, kind and back. The first argument is the name of the array whose maximum and minimum values are returned (Table 6.13).

```
integer, dimension (2) :: ml
integer, dimension (3,3) :: a
integer :: ami, ama
integer, dimension(3) :: arc
a=reshape([17,32,5,11,12,33,2,8,10], [3,3])
ama=maxval(a)
ami=minval(a)
print *, ama, ami      ! ama=33, ami=2, Table 6.13
end
```

TABLE 6.13

a Array

17	11	2
32	12	8
5	33	10

The preceding program prints 33 and 2 as the highest and lowest values among the array elements.

Comparison of character strings was discussed in Chapter 4. Following the same logic,

```
print *, maxval(['ABC', 'PQR', 'XYZ', 'RST'])
```

displays XYZ as the third element is greater than all other elements of the array.

For an array greater than rank 1, say, for a two-dimensional array, a second optional argument dim may be used to find the maximum in the rows or in the columns. For a rank 2 array, dim=1 corresponds to the columns and dim=2 corresponds to the rows. When dim is used, maxval and minval return an array of rank 1 less than that of the input array and size equal to the shape of the given array in that direction (for a two-dimensional array, each row or column is a single-dimensional array). For a 3 × 3 input array, the intrinsic returns a rank 1 array of size 3 for both dim=1 and dim=2. However, for an array of shape (3, 4), dim=1 returns a rank 1 array of size 4, and dim=2 returns a rank 1 array of size 3. The elements of the return array contain the maximum value of each row or column depending on the value of the dim argument.

```
arc= maxval(a, dim=1)
```

will return the maximum value of the respective columns. Obviously, the elements of arc will be 32, 33 and 10, respectively (Table 6.14a).

TABLE 6.14 a & b

a Array

17	11	2
32	12	8
5	33	10

dim=1 ↓

b Array

17	11	2
32	12	8
5	33	10

dim=2 →

Similarly, arc= maxval(a, dim=2) will return the maximum value for the respective rows. In this case, the array elements of arc will be 17, 32 and 33 (Table 6.14b).

The third optional argument is a mask, which can be used to select an array element that passes through the test.

```
ama=maxval(a, mask=a<15)
```

It will find the maximum value from the array elements that are less than 15. Clearly, ama in this case will be 12. The array elements 17, 32 and 33 are excluded and not considered while finding the maximum value. The array elements, which are left out because of the mask, are indicated by shades (Table 6.15a).

The intrinsic minval works in an identical manner (Table 6.15b).

```
arc=minval(a, dim=1)
```

It will return 5, 11 and 2 to arc(1), arc(2) and arc(3), respectively.

TABLE 6.15a

a Array

17	11	2
32	12	8
5	33	10

TABLE 6.15b

a Array dim=1

17	11	2
32	12	8
5	33	10

Similarly,

```
arc=minval (a, dim=2)
```

It will return `2`, `8` and `5` to `arc(1)`, `arc(2)` and `arc(3)`, respectively (Table 6.15c).

```
ami=minval (a, mask=a>10)
```

It will exclude `2`, `5`, `8` and `10` as they are not greater than `10`. Naturally, the intrinsic returns `11` to `ami` (Table 6.15d).

If the `mask` is such that no element could be selected, `maxval` returns the smallest negative number (in this case, an integer) available in the system (default `kind`). Similarly, in the case of `minval`, if the `mask` is so chosen that it excludes all elements, it returns the highest positive number (in this case, an integer) available in the system (default `kind`).

Both `dim` and `mask` may be present simultaneously:

```
amc=maxval (a, dim=1, mask=a<15)
```

It will return `5`, `12` and `10` as the array elements of `ama` (Table 6.15e).

Also, `ami=minval(a, dim=2, mask=a>8)` will return an array of rank `1` with its elements as `11`, `12` and `10` (Table 6.15f).

The intrinsics `maxloc` and `minloc` determine the locations that contain the maximum and minimum of an array. It always returns an array of rank `1` and size equal to the rank of the input array if `dim` is not present. If `dim` is present, it returns an array of rank $(n-1)$, where n is the rank of the input array and shape $[d_1, d_2, \ldots, d_{dim-1}, d_{dim+1}, \ldots, d_n]$, where $[d_1, d_2, \ldots, d_n]$ is the shape of the input array.

To locate a particular location of a matrix, two subscripts are required. Therefore, to return the position of an element, `maxloc` and `minloc` require an array of rank `1` and size `2` (Table 6.16a).

```
ml=maxloc(a)   ! a(3, 2)=33; ml(1)=3, ml(2)=2
ml=minloc(a)   ! a(1, 3)=2; ml(1)=1, ml(2)=3
```

Note that, even when the input to `maxloc` or `minloc` is an array of rank `1`, the output from these intrinsics is a rank 1 array of size `1`, that is, just one element.

`Maxloc` and `minloc` always calculate the location assuming the lower bound of the array in each dimension to be equal to 1, irrespective of its declaration.

```
integer, dimension (0: 4):: a=[10, -5, 7, 14, 3]
integer, dimension (1):: loc
loc=maxloc(a)
```

TABLE 6.15c

a Array dim=2

17	11	2
32	12	8
5	33	10

TABLE 6.15d

a Array mask=a>10

17	⑪	2
32	12	8
5	33	10

TABLE 6.15e

a Array dim=1

17	11	2
32	⑫	8
⑤	33	⑩

TABLE 6.15f

17	⑪	2
32	⑫	8
5	33	⑩

a Array
dim=2 ⟶

TABLE 6.16a

a Array

17	11	2
32	12	8
5	33	10

It will return `loc(1)` as 4 and not as 3. If it is desired to have the exact location with respect to the lower bound, the last expression is to be modified as follows:

```
loc = (maxloc(a)+lbound(a)-1)
```

The same is true for multidimensional arrays also. It is also possible to use a mask to include (or exclude) certain elements while finding the location.

```
ml=maxloc(a, mask=a<16)  ! a(2,2)=12; ml(1)=2, ml(2)=2
ml=minloc(a, mask=a>20)  ! a(2,1)=32; ml(1)=2, ml(2)=1
```

If the maximum or the minimum value occurs more than once within the array, the first one is returned. For multidimensional arrays, it was mentioned earlier how they are mapped into single dimension. The first is determined after mapping the array into single dimension.

```
integer, dimension (3, 3) :: x
x=reshape([10,9,17,32,65,13,65,19,12], [3, 3])
```

TABLE 6.16b

x Array

10	32	65
9	65	19
17	13	12

It is seen that when x, a two-dimensional array, is mapped into single dimension, the fifth and seventh locations contain 65. Maxloc, in this case, returns the fifth location, which is (2, 2) (Table 6.16b).

Now, consider the following program:

```
integer, dimension (3, 3) :: x        ! Table 6.17
x=reshape([10,17,9,32,12,13,65,19,12], [3, 3])
print *,x
print *, maxloc(x, dim=1)
print *, maxloc(x, dim=2)
end
```

TABLE 6.17

x Array

10	32	65
17	12	19
9	13	12

Readers may verify that the outputs of this program are as follows:

```
10 17 9 32 12 13 65 19 12
 2 1 1
 3 3 2
```

For `dim=1`, the outputs are the values of the first dimension of the matched locations. Similarly, for `dim=2`, the outputs are the values of the second dimension of the matched locations.

Both `maxloc` and `minloc` take an optional logical argument back. If `back` is absent, it is taken as `false`. If `back=.true.`, the returned location is the last location containing the maximum/minimum value; that is, the searching takes place from the end (the first location from the end).

```
print *, maxloc([10,12,35,33,35,14], back=.false.)
```

It prints 3. On the other hand,

```
print *, maxloc([10,12,35,33,35,14], back=.true.)
```

displays 5 as the searching takes place from the rear.

Both `maxloc` and `minloc` take an optional integer parameter `kind`. If `kind` is absent, the returned integer is of default kind. If `kind` is present, the returned integer is of kind specified by the `kind` parameter.

For zero-sized arrays, `maxval` returns the largest negative value available in the system, `minval` returns the largest positive value that the system supports and both `maxloc` and `minloc` return 0.

FINDLOC: This intrinsic takes six arguments, of which four are optional. The arguments are `array`, `value`, `dim`, `mask`, `kind` and `back`. It is almost the same as `maxval`. The first argument, `array`, is an array of any standard intrinsic type. The second argument, `value`, is the same type as `array` and is a scalar. The intrinsic searches for the `value` within the elements of the array (`maxval` locates the maximum value within the array elements). The returned value is an integer array containing the location of the `array` where a match has been found, and if `dim` is not present, it is rank 1 array of the size of the rank of the input array. In case the `value` does not match with any element of the `array`, 0 is returned.

Case I

```
        print *, findloc([10,20,35,33,35],35) ! ifort used
displays [3], size is the rank of the input array.
        print *, findloc([10,20,35,33,35],35,back=.true.)
```

It displays [5]; that is, the scanning is initiated from the end of the array.

Case II

```
        integer, dimension(2,3) :: a=reshape([2,5,4,5,1,5], [2,3])
        print *, findloc(a, 5, dim=1)
        print *, findloc(a, 5, dim=2)
```

It displays [2 2 2] and [0 1]. This is independent of the declared lower bound of a. The lower bound is always taken as 1.

Case III

```
        logical, parameter :: t=.true., f=.false.
        integer, dimension(3,4) :: &
          a=reshape([3,7,13,8,13,11,13,22,25,21,13,5], [3,4])
        logical, dimension(3,4) :: &
          l= reshape([t, t, t, t, f, t, f, t, t, t, t, t], [3,4])
        print *, findloc(a, 13, dim=1, mask=l)
        print *, findloc(a, 13, dim=2, mask=l)
        end
```

The results are [3 0 0 2] and [0 4 1]. In this case, `mask` is a logical array of the same size and shape of the input array. The elements corresponding to the false value of the mask elements are not considered by the function `findloc`.

Case IV

```
        integer, dimension(3,4) :: &
        a=reshape([2,13,13,13,13,12,15,13,12,13,12,5],[3,4])
        print *, findloc(a, 13, dim=1, back=.true.)
        print *, findloc(a, 13, dim=2, back=.true.)
        end
```

The results are [3 2 2 1] and [4 3 1]. However, if back= is removed or set to false, the results will be [2 1 2 1] and [2 1 1], respectively.

6.32 SUM and PRODUCT

These two intrinsics return the sum and product of array elements, respectively. The array may be integer, real or complex. Like maxval and minval, these intrinsics also take three arguments. Two of them are optional. The first argument must be an array of numeric type (Table 6.18).

```
integer, dimension (3, 3) :: a
integer :: s, p
integer, dimension (3):: sa, pa
a=reshape ([2,5,13,8,24,35,12,1,40], [3, 3])
```

TABLE 6.18

a Array

2	8	12
5	24	1
13	35	40

Now, s=sum(a) corresponds to the sum of all elements:

```
s = 2+5+13+8+24+35+12+1+40
```

Similarly, p=product (a) corresponds to the product of all elements:

```
p = 2*5*13*8*24*35*12*1*40
```

Both these intrinsics may take an optional parameter dim (Table 6.19).

```
sa=sum (a, dim=1)
```

This returns a rank 1 array of size 3 with

```
sa(1)=2+5+13
sa(2)=8+24+35
sa(3)=12+1+40
```

In an identical manner,

```
sa=sum (a, dim=2)
```
returns,
```
sa(1)= 2+8+12
sa(2)= 5+24+1
sa(3)= 13+35+40
```

The following program will sum the series 1+2+3+...+100 without using the do loop:

```
integer :: i
print *, sum([(i, i=1,100)])
end
```

The implied do loop will generate an array constant [1, 2, 3, ..., 100], and sum will add these elements to return the sum of the series.

It is not difficult to guess the result of the product intrinsic.

TABLE 6.19

a Array

returns,
```
pa = product (a, dim=1)
pa(1)=2*5*13
pa(2)=8*24*35
pa(3)=12*1*40
```
and,
returns,
```
pa=product (a, dim=2)
pa(1)=2*8*12
pa(2)=5*24*1
pa(3)=13*35*40
```

Another example of product is to evaluate factorial n, for, say, n=7.

```
print *, product([(i, i=2,7)])
```

Both sum and product may take a second optional argument, mask. This was discussed in detail in an earlier section.

```
sa=sum(a, dim=1, mask=a<20)
```

It gives (elements passed through the mask are shown as bold in Table 6.20)

TABLE 6.20

a Array mask<10

2	8	12
5	24	1
13	35	40

gives,
Finally,
gives,
```
sa(1)=2+5+13
sa(2)=8
sa(3)=12+1
p=product(a, mask=a<10)
p=2*5*8*1
pa=product(a, dim=2, mask=a<10)
pa(1)=2*8  ! Shaded cells are excluded
pa(2)=5*1
pa(3)=1
```

6.33 Handling of Arrays of More than Two Dimensions

All preceding examples used single- or two-dimensional arrays. In this section, detailed calculations will show how three-dimensional arrays are handled by sum with the dim argument. The same logic is applicable to other intrinsics that use dim as argument. Having understood three-dimensional arrays, it is expected that the readers should be able to extend the logic for higher dimensional arrays.

It was already mentioned that when the `dim` argument is used, `sum` (or similar intrinsics) returns an array of rank 1 less than that of the input array. The following program will be used throughout this section:

```
program array_3d
implicit none
integer :: i
integer, dimension (3, 3, 3) :: a
integer, dimension (3, 3) :: b
a=reshape([(i, i=1, 27)], [3, 3, 3])
b=sum(a, dim=1); print *, b
b=sum(a, dim=2); print *, b
b=sum(a, dim=3); print *, b
end
```

TABLE 6.21

a Array

1,1,1	1	1,1,2	10	1,1,3	19
2,1,1	2	2,1,2	11	2,1,3	20
3,1,1	3	3,1,2	12	3,1,3	21
1,2,1	4	1,2,2	13	1,2,3	22
2,2,1	5	2,2,2	14	2,2,3	23
3,2,1	6	3,2,2	15	3,2,3	24
1,3,1	7	1,3,2	16	1,3,3	25
2,3,1	8	2,3,2	17	2,3,3	26
3,3,1	9	3,3,2	18	3,3,3	27

The outputs are as follows:

```
 6  15 24 33 42 51 60 69 78
12 15 18 39 42 45 66 69 72
30 33 36 39 42 45 48 51 54
```

The contents of the a array with the locations are shown in Table 6.21. When the `dim` argument is used, the intrinsic returns an array of rank 2 having a shape of (2 2), which is less than the rank of the input array. For `dim=1`, the (i, j)th of the returned array is $\sum a(:, i, j)$, where $\sum$ indicates the sum over the first index and ':' stands for all indices (in this case, 1, 2 and 3). Expanding, we get

```
b(1, 1) = a(1,1,1)+a(2,1,1)+a(3,1,1),  b(2, 1) = a(1,2,1)+a(2,2,1)+a(3,2,1),
b(3, 1) = a(1,3,1)+a(2,3,1)+a(3,3,1),  b(1, 2) = a(1,1,2)+a(2,1,2)+a(3,1,2),
b(2, 2) = a(1,2,2)+a(2,2,2)+a(3,2,2),  b(3, 2) = a(1,3,2)+a(2,3,2)+a(3,3,2),
b(1, 3) = a(1,1,3)+a(2,1,3)+a(3,1,3),  b(2, 3) = a(1,2,3)+a(2,2,3)+a(3,2,3),
b(3, 3) = a(1,3,3)+a(2,3,3)+a(3,3,3) .
```

For `dim=2`, the (i, j)th element of the returned array is $\sum A(i, :, j)$, where $\sum$ indicates the sum over the second index. Expanding, we get

```
b(1, 1) = a(1,1,1)+a(1,2,1)+a(1,3,1),  b(2, 1) = a(2,1,1)+a(2,2,1)+a(2,3,1),
b(3, 1) = a(3,1,1)+a(3,2,1)+a(3,3,1),  b(1, 2) = a(1,1,2)+a(1,2,2)+a(1,3,2),
b(2, 2) = a(2,1,2)+a(2,2,2)+a(2,3,2),  b(3, 2) = a(3,1,2)+a(3,2,2)+a(3,3,2),
b(1, 3) = a(1,1,3)+a(1,2,3)+a(1,3,3),  b(2, 3) = a(2,1,3)+a(2,2,3)+a(2,3,3),
b(3, 3) = a(3,1,3)+a(3,2,3)+a(3,3,3) .
```

Finally, for `dim=3`, the (i, j)th element of the returned array is $\sum a(i, j, :)$, where $\sum$ indicates the sum over the third index. Expanding, we get

```
b(1, 1) = a(1,1,1)+a(1,1,2)+a(1,1,3),  b(2, 1) = a(2,1,1)+a(2,1,2)+a(2,1,3),
b(3, 1) = a(3,1,1)+a(3,1,2)+a(3,1,3),  b(1, 2) = a(1,2,1)+a(1,2,2)+a(1,2,3),
b(2, 2) = a(2,2,1)+a(2,2,2)+a(2,2,3),  b(3, 2) = a(3,2,1)+a(3,2,2)+a(3,2,3),
b(1, 3) = a(1,3,1)+a(1,3,2)+a(1,3,3),  b(2, 3) = a(2,3,1)+a(2,3,2)+a(2,3,3),
b(3, 3) = a(3,3,1)+a(3,3,2)+a(3,3,3) .
```

It may be verified that the results obtained from the preceding expressions agree with the results obtained from the intrinsics.

6.34 DOT_PRODUCT

This intrinsic takes two arrays of rank 1 (also of same shape) and returns the dot product. The array may be of type integer, real, logical or complex. The dot product is defined as $\sum a_i.b_i$ [i =1,n (n=size of the array)].

```
real, dimension(4) :: a=[1.0,2.0,3.0,4.0]
real, dimension(4) :: b=[10.0,20.0,30.0,40.0]
real :: dp
 .
dp=dot_product(a, b)
```

The dot product is calculated as $1.0\times10.0+2.0\times20.0+3.0\times30.0+4.0\times40.0 = 300.0$. For logical arrays, l1 and l2, l1.and. l2 is calculated element by element. If one of the results is true, the dot_product returns true. If all results are false, the dot_product returns false.

```
logical, dimension(3) :: l1=[.true.,.true.,.false.]
logical, dimension(3) :: l2=[.false.,.true.,.true.]
print *, dot_product(l1,l2)
end
```

The result is true. For complex arrays, c1 and c2, the dot_product is calculated as $\sum$conjugate(c1)*c2.

```
complex, dimension(3) :: c1=[(1,2),(3,4),(5,6)]
complex, dimension(3) :: c2=[(5,6),(7,8),(25,30)]
print *, dot_product(c1,c2)
end
```

The print statements would display (375.0, -8.0).

6.35 Matrix Multiplication

Two matrices mat_a and mat_b can be multiplied using this intrinsic. The arrays may be integer, real, logical or complex. If the shape of mat_a is (m n) and that of mat_b is (n k), the result will have a shape of (m k). If the shape of mat_a is (m) and that of mat_b is (m k), the result will have a shape of (k). If the shape of mat_a is (n m) and that of mat_b is (m), the result will have a shape of (n):

```
real, dimension (3, 2) :: a
real, dimension (2, 3) :: b
real, dimension (3, 3) :: c
!    initialize the matrix
 .
c=matmul(a, b)
```

The multiplication is performed (for numeric array) according to the rule of matrix algebra.

```
c_ij = Σ a_ik*b_kj   [sum over k]
```

For a logical array, `matmul` multiplies the row of the first matrix by the column of the second matrix element by element using the 'and' operation, and if any one of them is true, true value is returned. This is illustrated with the help of two logical matrices (Table 6.22):

```
logical, dimension (3,3):: 11, 12, 13
11=reshape([.true.,.false.,.true.,.false.,&
 .true.,.false.,.true.,.true.,.false.],[3,3])
12=reshape([.true.,.false.,.false.,.false.,&
 .true.,.false.,.true.,.false.,.false.],[3,3])
13=matmul(11, 12)
print *, 13
end
```

TABLE 6.22

11 and 12

t	f	t
f	t	t
t	f	f

t	f	t
f	t	f
f	f	f

[t=true, f=false]

The output is as follows:

```
 T  F  T  F  T  F  T  F  T
```
`13(1,1)` is obtained as follows:
```
.true.   .and.   .true.
.false.  .and.   .false.
.true.   .and.   .false.
```

This is `true` because one of the preceding expressions is `true`. Similarly, `13(2, 1)` is `false` as none of the following

```
.false.  .and.   .true.
.true.   .and.   .false.
.true.   .and.   .false.
```
returns `true`.

6.36 TRANSPOSE of a Matrix

The intrinsic `transpose` takes a square matrix (number of rows = number of columns) as its argument and returns the transpose of the original matrix; that is, rows and columns are interchanged (Table 6.23).

```
integer, dimension (2, 2) :: a, b
a=reshape ([1, 2, 3, 4], [2, 2])
b=transpose (a)
```

The original and the transposed matrix are shown in the adjacent figure.

TABLE 6.23

a Array

1	3
2	4

Original

b Array

1	2
3	4

Transposed

6.37 Array Shift

There are two intrinsics to shift the elements of an array: eoshift (end of shift) and cshift (circular shift).

EOSHIFT: We first illustrate eoshift with an array of rank 1. In the simplest case, this intrinsic takes two arguments: an array of any type (integer, real, complex, logical or character) and a shift count. The shift count is an integer (scalar). A positive shift count corresponds to the left shift and a negative shift count corresponds to the right shift. The vacancy created by this operation is filled according to Table 6.24a. The elements shifted out are lost.

```
integer, dimension(6) :: a=[1, 2, 3, 4, 5, 6]
integer, dimension (6) :: b
b=eoshift(a, shift=2)
```

As the shift count is 2, the elements are shifted two places to the left. Similarly, b=eoshift(a, shift=-3) will shift array elements three places to the right. One can specify the boundary value by the optional boundary parameter. If the boundary parameter is present, the vacancies created by eoshift are filled with the boundary value instead of the values mentioned in Table 6.24a (Table 6.24b).

TABLE 6.24a

Type vs Value

Type	Value
integer	0
real	0.0
complex	(0.0, 0.0)
logical	false
character	blank

```
b=eoshift(a, shift=2, boundary=88)
b=eoshift(a, shift=-3, boundary=99)
```

The array may be of rank greater than 1. In that case, the shift count may be a scalar or an array of rank 1 less than the rank of the input array. If the shift count is a scalar, all elements are shifted by an equal amount.

```
integer, dimension (3, 3) :: c
integer, dimension (3, 3) :: d
c=reshape([10,20,30,40,50,60,70,80,90], [3,3])
```

TABLE 6.24b

Original, Left Shift 2 and Right Shift 3 without and with Boundary Values

1	2	3	4	5	6
3	4	5	6	0	0
0	0	0	1	2	3
3	4	5	6	88	88
99	99	99	1	2	3

Now, d=eoshift(c, shift=1) will shift all columns by 1 to the left. Similarly, when the shift is negative, like d=eoshift(c, shift=-1), all columns are shifted by 1 to the right.

The left and right shifts for a two-dimensional array may be understood by mapping the two-dimensional array into single dimension.

TABLE 6.25a & b

Shift with Boundary Value

20	50	80
30	60	90
999	999	999

Left ←   Right →

10	20	30		40	50	60		70	80	90

If the optional boundary parameter (as scalar) is specified, the vacancy created is filled in by the boundary value (Table 6.25).

```
d=eoshift(c, shift=1, boundary=999)
```

999	999	999
10	40	70
20	50	80

The preceding statement will create the matrix d as 20, 30, 999, 50, 60, 999, 80, 90, 999.

In a similar way, a negative shift count along with the boundary value

```
d=eoshift(c, shift=-1, boundary=999)
```

results in the matrix d as shown in Tables 6.25a and 6.25b.

For an array of rank greater than 1, it is possible to apply different shifts in different columns, and the vacancies created by the shifts may be filled with different values instead of a single value by using the boundary as an array of rank 1 less than the rank of the original array.

```
d=eoshift(c, shift=[1, -1, 2]) ! Table 6.26a
```

Now, the shift argument is a rank 1 array, and this shifts the first column 1 to the left, the second column 1 to the right and the third column 2 to the left.

Similarly, with the boundary value of 999, the instruction

```
d=eoshift(c, shift=[1, -1, 2], boundary=999)
```

will generate the d matrix as shown in Table 6.26b.

It is possible to specify different boundary values for different columns.

```
d=eoshift(c,shift=[1, -1, 2], &
      boundary=[99,88,77]) ! Table 6.26c
```

So, the d array becomes 20, 30, 99, 88, 40, 50, 90, 77, 77.

Eoshift may take a fourth optional argument, dim, which is a scalar. If it is absent, it is assumed to be 1. It satisfies the following inequality:

```
1 <= dim <=n
```

where n is the rank of the input array.

For the default value of dim, that is, 1, the shift takes place column-wise. The following two statements are thus equivalent:

```
      d=eoshift(c, shift=1, boundary=999)
and   d=eoshift(c, shift=1, boundary=999, dim=1)
```

In case of the aforementioned array, if dim=2, shift will take place row-wise.

```
d=eoshift(c, shift=1, boundary=888, dim=2)
```

It will generate the d matrix as 40, 50, 60, 70, 80, 90, 888, 888, 888 (Table 6.26d).

TABLE 6.26a–f

Shift under Different Conditions

20	0	90
30	40	0
0	50	0

20	999	90
30	40	999
999	50	999

20	88	90
30	40	77
99	50	77

40	70	888
50	80	888
60	90	888

40	70	777
888	20	50
60	90	999

40	70	777
888	20	50
30	60	90

Some more examples of eoshift are as follows:

```
d=eoshift(c, shift=[1,-1,1], boundary=[777,888,999], dim=2)
d=eoshift(c, shift=[1,-1,0], boundary=[777,888,999], dim=2)
```

The third row is not shifted since the shift count is 0 (Table 6.28f).

The following program is an application of eoshift on a three-dimensional array of shape (2 2 2):

```
integer, dimension (2, 2, 2) :: a, b
a=reshape([[(i, i=1, 8)], [2, 2, 2])
print *, a; print *, 'dim=1'
b=eoshift(a, 1, dim=1)
print *, b
b=eoshift(a, -1, dim=1)
print *, b; print *, 'dim=2'
b=eoshift(a, 1, dim=2)
print *, b
b=eoshift(a, -1, dim=2)
print *, b; print *, 'dim=3'
b=eoshift(a, 1, dim=3)
print *, b
b=eoshift(a, -1, dim=3)
print *, b
end
```

The outputs are as follows:

```
1  2  3  4  5  6  7  8
dim=1
2  0  4  0  6  0  8  0
0  1  0  3  0  5  0  7
dim=2
3  4  0  0  7  8  0  0
0  0  1  2  0  0  5  6
dim=3
5  6  7  8  0  0  0  0
0  0  0  0  1  2  3  4
```

Let us try to understand the output generated by eoshift. The a array created by reshape will be of the following form:

```
a(1, 1, 1) = 1, a(2, 1, 1) = 2, a(1, 2, 1) = 3, a(2, 2, 1) = 4,
a(1, 1, 2) = 5, a(2, 1, 2) = 6, a(1, 2, 2) = 7, a(2, 2, 2) = 8.
```

Consider a typical case—dim=2. The shifting will take place among the elements a(i,:,j), where ':' indicates all possible values. These are shown within parentheses: {a(1,1,1), a(1,2,1)} {a(2,1,1), a(2,2,1)} {a(1,1,2), a(1,2,2)} {a(2,1,2), a(2,2,2)}.

Now, a(1,1,1)=1 and a(1,2,1)=3. If one left shift is given by eoshift, a(1,1,1) becomes 3 and a(1,2,1) becomes 0. Similarly, one left shift to the pair {a(2,1,1), a(2,2,1)} sets a(2,1,1) to 4 and a(2,2,1) to 0. Proceeding in a similar way, it can be shown that the result is finally 3 4 0 0 7 8 0 0.

A complex variable contains two elements: real and imaginary parts.

```
complex, dimension(4) :: x, b
x=[(1.0,2.0), (3.0,4.0),(5.0,6.0),(7.0,8.0)]
b=eoshift(x, shift=1)
print *,b
b=eoshift(x, shift=-1)
print *, b
end
```

The first eoshift shifts the complex array 1 position to the left (both the real and imaginary parts), and the second one shifts the complex array 1 position to the right (both the real and imaginary parts). In the first case, the b array becomes

```
b=[(3,4), (5,6), (7,8), (0,0)]
```

and in the second case, the b array becomes

```
b=[(0,0), (1,2), (3,4), (5,6)]
```

CSHIFT: This intrinsic is used to shift the array elements in a circular fashion; that is, the elements that are shifted out are added at the other end. Like eoshift, the positive shift count indicates a left shift and the negative shift count indicates a right shift.

For an array of rank 1, the shift count is a scalar. Using the 'a' array declared (Table 6.27a) as the preceding one (eoshift),

```
b=cshift(a, shift=2) ! Table 6.27b
```

will make the b array as [3, 4, 5, 6, 1, 2]. Similarly,

```
b=cshift(a, shift=-2) ! Table 6.27c
```

will create b array as [5, 6, 1, 2, 3, 4].

In case of an array of rank 2, it is possible to shift each column or row by the same amount by using the shift count as a scalar or different amounts or directions by using an array of rank 1 less than the rank of the original array. Using the c and d arrays as declared previously,

```
d=cshift(c, shift=1) ! Table 6.28a
```

will create a d array as [20, 30, 10, 50, 60, 40, 80, 90, 70].

```
d= cshift(c, shift=-1) ! Table 6.28b
```

will create a d array as [30, 10, 20, 60, 40, 50, 90, 70, 80].

TABLE 6.27a–c

Examples of cshift

1	2	3	4	5	6

3	4	5	6	1	2

5	6	1	2	3	4

TABLE 6.28a–e

cshift with 2-d Array

20	50	80
30	60	90
10	40	70

30	60	90
10	40	70
20	50	80

In a similar way,

```
d= cshift(c, shift=[1, -1, 1])
```

40	70	10
50	80	20
60	90	30

will generate a d array as $[20, 30, 10, 60, 40, 50, 80, 90, 70]$.

The optional parameter dim, which is, by default 1, as in the previous case, may be specified; dim=1 affects the column and dim=2 affects the row. In the following, we use cshift on the d array. The outputs are shown in Table 6.28c–e.

70	10	40
80	20	50
90	30	60

```
d=cshift(c, shift=1, dim=2)          ! Table 6.28c
d=cshift(c, shift=-1, dim=2)         ! Table 6.28d
d=cshift(c, shift=[1, -1, 1], dim=2) ! Table 6.28e
```

The discussions related to the three-dimensional array in connection with eoshift are equally applicable to cshift.

40	70	10
80	20	50
60	90	30

6.38 Euclidian Norm

The intrinsic norm2 takes two arguments: a real array x and an optional integer scalar dim. It returns the L2 norm of the array. The L2 norm is defined as the square root of the sum of the squares of the array elements. The returned value is a scalar if dim is absent. If dim is present, dim must be greater than or equal to 1 and less than or equal to n, where n is the rank of the array (first argument). The returned value when dim is present is an array of rank $(n-1)$ having shapes $[d1, d_2, \ldots, d_{dim-1}, d_{dim+1}, \ldots, d_n]$, where $[d_1, d_2, \ldots, d_n]$ is the shape of x.

Case I

```
real, dimension(3) :: x=[2.0, 3.0, 4.0]
print *, norm2(x)
end
norm2 calculates sqrt(2.0**2 + 3.0**2 + 4.0**2)
```

Case II

```
real, dimension(3,3) :: y
y=reshape([1.0,2.0,3.0,4.0,5.0,6.0,7.0,8.0,9.0], [3,3])
print *, norm2(y)
print *, norm2(y, dim=1)
print *, norm2(y, dim=2)
end
```

The first print statement displays the square root of the sum of the squares of each array element. The second print statement displays an array of rank 1 of size 3 containing $\sqrt{(1**2 + 2**2 + 3**2)}$, $\sqrt{(4**2 + 5**2 + 6**2)}$ and $\sqrt{(7**2 + 8**2 + 9**2)}$ as its elements. The last print statement also displays an array of rank 1 of size 3 containing

$\sqrt{(1**2 + 4**2 + 7**2)}$, $\sqrt{(2**2 + 5**2 + 8**2)}$ and $\sqrt{(3**2 + 6**2 + 9**2)}$ as its elements.

6.39 Parity of Logical Array

This intrinsic parity takes two arguments: a logical array mask and an optional scalar integer dim, where the property of dim is the same as that of the dim discussed in section 6.38 (Table 6.29).

Case I: If dim is not present, the intrinsic returns true if the total number of true elements of mask is odd; otherwise, it returns false.

```
logical, dimension(5) :: m=[.true.,.false.,.false.,.true.,.true.]
print *, parity(m)
end
```

The preceding program returns true as the number of true elements of m is 3 which is odd.

Case II: Consider the following program:

```
logical, parameter :: t=.true.
logical, parameter :: f=.false.
logical, dimension(3,3) :: x=reshape([t,t,f, &
   f,f,t,t,t,t], [3,3])
print *, parity(x)          ! (a)
print *, parity(x, dim=1)   ! (b)
print *, parity(x, dim=2)   ! (c)
end
```

TABLE 6.29

Logical Array

t	f	t
t	f	t
f	t	t

Following the prescription related to dim in the section 6.38, the returned values from the intrinsic are (a) f (b) [f t t] (c) [f f f], where t and f stand for true and false, respectively. The readers may verify these results.

6.40 Locating and Counting Array Elements

There are three intrinsics in this category: any, all and count. The intrinsics any and all are used to locate one or more elements that meet a certain condition. The third intrinsic count counts the number of elements that satisfy a certain condition. All these instructions are illustrated by means of examples. The program segment uses a two-dimensional array containing the marks of three students. Each row indicates the marks of a particular student in different subjects. Each column indicates the marks of a particular subject for all students (Table 6.30).

```
logical :: l1
logical, dimension(3) :: l2
logical, dimension(6) :: l3
integer :: n1
integer, dimension(3) :: n2
integer, dimension(6) :: n3
integer, dimension(3,6) :: result
result=reshape(source = [60, &
   65, 70, 62, 80, 85, 75, 68, 88, 77, 92, &
   90, 55, 70, 90, 80, 93, 96], shape=[3, 6], order=[2,1])
```

TABLE 6.30

Student vs Subject-mark

60	65	70	62	80	85
75	68	88	77	92	90
55	70	90	80	93	96

⇩ Subject

Student ⟹

Therefore, the array `result` will be filled in row-wise because of the presence of the `order` argument.

ANY: This intrinsic takes two arguments, `mask` and `dim`, where `mask` is of type logical. The optional argument `dim` is an integer. In the simplest form, `mask` could be a condition:

```
mask = result > 90
```

It tests whether any student has scored more than 90 in any subject. The intrinsic returns a logical value—either `true` or `false`.

```
l1 = any (result > 90)   ! true
l1 = any (result < 50)   ! false
```

The intrinsic 'any' will return a logical array if the optional parameter `dim` (integer) is used. If `dim=1` is used, any returns a logical array of size 6. Similarly, if `dim=2`, any returns a logical array of size 3.

```
l2 = any(result > 90, dim=2)
```

It will return `l2(1)`, `l2(2)` and `l3(3)` as `false`, `true` and `true`, respectively. The statement examines whether any student scored marks greater than 90 in any subject. In a similar way,

```
l3 = any(result < 65, dim=1)
```

returns `l3(1)`, `l3(2)`, `l3(3)`, `l3(4)`, `l3(5)` and `l3(6)` as `true`, `false`, `false`, `true`, `false` and `false`, respectively. The statement is equivalent to asking a question, "Is there anyone who scored marks less than 65 in a particular subject?" Obviously, the size of the output logical array is the number of subjects. The concept may be extended to arrays of rank greater than 2.

ALL: This tests whether all elements of an array meet a certain condition. It returns `true` when the condition is met; otherwise, it returns `false`. Like any, it takes two parameters: logical and `dim` (integer), which is optional.

```
l1 = all (result > 50)   ! true
l1 = all (result > 60)   ! False
```

In the first case, 'all' tests whether every student has scored more than 50 in all subjects. The result is `true`. In the second case, since not everybody has scored more than 60 in all subjects, the result is `false`. `All` accepts a second optional parameter `dim`. For `dim=2`, it returns an array of size 3 (rows of the matrix). In this case, `all` tests whether every student has met a certain condition in each subject.

```
l2 = all (result > 65, dim=2)
```

It returns `l2(1)`, `l2(2)` and `l2(3)` as `false`, `true` and `false`, respectively. Assuming 65% is the pass mark in each subject, the preceding statement returns that only student number 2 has passed in every subject. Similarly, `l3 = all (result > 65, dim=1)` returns `l3(1)`, `l2(2)`, `l3(3)`, `l3(4)`, `l3(5)` and `l3(6)` as `false`, `false`, `true`, `false`, `true` and `true`, respectively. This tells us that all students have passed in subjects 3, 5 and 6. The input to `all` may be an array of rank more than 2.

> **COUNT:** This intrinsic counts the number of array elements that meet certain conditions. Like the previous two intrinsics, it also takes two arguments, of which the second one, `dim`, is optional. `Count` returns an integer or integer array depending on whether the optional `dim` parameter is present or not.

```
n1 = count (result >= 80)
```

If marks greater than or equal to 80 are considered as star mark, the preceding statement returns the total number of array elements that satisfy the condition, which is 9 in this case.

> Similarly, `n2 = count (result >= 80, dim=2)` returns `n2(1)`, `n2(2)` and `n3(3) as 2`, 3 and 4, respectively. This indicates that student 1 obtained star marks in 2 subjects, student 2 in 3 subjects and student 3 in 4 subjects. In a similar manner, `n3 = count (result >= 80, dim=1)` returns `n3(1)`, `n3(2)`, `n3(3)`, `n3(4)`, `n3(5)` and `n3(6)` as 0, 0, 2, 1, 3 and 3, respectively.

The array n3 now contains the number of students who received star marks in each subject. None could obtain star marks in subjects 1 and 2, whereas the number of students who received star marks in subjects 3, 4, 5 and 6 are, respectively, 2, 1, 3 and 3. Like 'any' and 'all' intrinsics, this intrinsic can be applied to arrays of rank greater than 2.

6.41 Packing and Unpacking

In this category there are three intrinsics: `pack`, `unpack` and `spread`.

> **PACK:** This intrinsic is used to generate a single-dimensional array from a multidimensional array under the control of a mask. It accepts three arguments, of which the third argument is optional. The first argument is a multidimensional array, and the second argument is a mask. Depending on the mask (`true` or `false`), the array elements of the first argument are returned as a single-dimensional array. The array elements of the multidimensional array are taken column-wise. The output array is of the same type as that of the input array.

Case I: If the `mask` is `.true.`, all elements of the multidimensional array are selected (Table 6.31).

```
integer, dimension (3,3) :: ua
integer, dimension (9) :: pa
logical, dimension (3, 3) :: m
ua=reshape ([(i, i=1, 9)], [3, 3])
pa=pack(ua,.true.) ! all true
```

and `pa(1)`, `pa(2)`, `...`, `pa(9)` are respectively 1, 2, 3, `...`, 9.

TABLE 6.31

ua Array

1	4	7
2	5	8
3	6	9

The size of `pa` is the product of extents of `ua`.

The diagonal elements of a matrix may be extracted very easily using `pack` (Table 6.32).

```
integer, dimension(16) :: il
integer, dimension(4,4) :: ib
integer, dimension(4) :: id
ib=reshape([(i, i=1, 16)], [4, 4])
il=pack(ib,.true.)
id=il(1: 16: 5)
```

TABLE 6.32

ib Array

1	5	9	13
2	6	10	14
3	7	11	15
4	8	12	16

The `reshape` will create a matrix `ib`. The intrinsic `pack` will linearize the matrix `ib` into a rank 1 array `il`. The next instruction will pick up only the first, sixth, eleventh and sixteenth elements of `il`, which are nothing but the diagonal elements of the original array.

Case II: Mask can be a logical array of the same shape and size of the input array to `pack`. The array elements of the array to `pack` corresponding to the `true` elements of `mask` are returned as an array of rank 1 (Table 6.33).

The size of the returned vector is the number of `true` elements of the array. If m is the matrix where t and f denote true and false, respectively, the returned vector will be `[1, 2, 7]` from array ua. It is perhaps clear that `print *, size (pack(ua, mask=m))` will display 3 as the size of the returned vector. If it is necessary to store the returned vector, it can be assigned to a properly dimensioned variable.

TABLE 6.33

m Array

t	f	t
t	f	f
f	f	f

```
pa = pack(ua, mask=m)
```

However, a better instruction would be

```
pa(1: count(m))=pack(ua, mask=m)
```

which will count the number of `true` elements (which is 3 in this case) of m and pack the array accordingly.

Case III: Pack can have a third argument `vector`, which is an array of rank 1 and the same type as the input array. The size of `vector` must be at least equal to the number of `true` elements of the `mask`. When the `vector` is present, `pack` returns an array of size equal to the size of the `vector`. Elements from the array vector are taken to fill in the array if sufficient number of elements are not selected by the mask If mask is a scaler and its value is true, the size of `vector` must not be less than the size of the input array.

```
pa = pack (ua, mask=m, vector=[10,20,30,40,50,60,70,80,90])
```

It will generate the pa vector containing [1, 2, 7, 40, 50, 60, 70, 80, 90] as its array elements. Note that the compiler will flag the following as an error because the size of the vector is less than the size of the pa array, and therefore, the array that will be generated on the right-hand side of the assignment sign will be of size 5. This will not be able to fill in the array pa having a size of 9.

```
pa = pack (ua, mask=m, vector=[10,20,30,40,50]) ! error
```

Therefore, the preceding statement must be replaced by

```
pa(1:count(m)) = pack (ua, mask=m, vector=[10,20,30,40,50])
```

which generates only the elements of pa corresponding to the true elements of mask and ignores the others. In this case, vector will cause no error; pa(1:count(m)) actually becomes pa(1:3).

Another way of doing the same thing is as follows:

```
integer, dimension (5) :: vec=[10, 20, 30, 40, 50]
...
pa (1: size(vec))=pack(ua, mask=m, vector=vec)
```

This will generate the array elements of pa as [1, 2, 7, 40, 50].

UNPACK: This intrinsic is the reverse of the intrinsic pack. It takes three arguments. The first is a single-dimensioned array having size of at least t, where t is the number of true elements in the masked array. The second argument, mask, is a logical array. The third argument is conformable with mask and the same as the first argument. Unpack returns a multidimensional array of the same shape and size of its second argument, which is a logical array. For the ith true element of the mask, the next available non-picked element of the first argument is picked. For the false value of the mask, the corresponding element is picked up from the array field. However, if the field is a scalar, it is used repeatedly. The elements of the output array are filled in column-wise. The third argument, field, may be a scalar or an array of the same size and shape as the mask and of the same type as that of the first argument. The elements of the returned array corresponding to the mask are discussed next (Table 6.34a).

Case I: When the third argument field is a scalar, say, 99, and the elements of pa are [1,2,3,4,5,6,7,8,9], consider the following program:

```
integer, dimension (3,3) :: ua
logical, parameter :: t=.true.
logical, parameter :: f=.false.
integer, dimension (9) :: pa
logical, dimension (3,3) :: m
pa=[1,2,3,4,5,6,7,8,9]
m=reshape([f,t,t,t,f,t,t,t,f],[3,3])
ua=unpack(pa,mask=m,field=99)
print *, ua      ! Table 6.34(b)
end
```

TABLE 6.34a

m Array

f	t	t
t	f	t
t	t	f

TABLE 6.34b

ua Array

99	3	5
1	99	6
2	4	99

The third argument `field` is substituted corresponding to the false elements of the `mask`.

Case II: If the `field` array is as shown in Table 6.34c, the returned array will have elements (column-wise): 100, 1, 2, 3, 500, 4, 5, 6, 900. Note that when `field` is an array, the elements of `field` corresponding to the false elements of `mask` are transferred to the output array (Table 6.34d).

TABLE 6.34c

field

100	400	700
200	500	800
300	600	900

TABLE 6.34d

Returned Array (see text)

100	3	5
1	500	6
2	4	900

SPREAD: This intrinsic is used to replicate a scalar to form a rank 1 array or replicate an array by adding a dimension. It accepts three arguments. The first argument, called the `source`, is a scalar or an array of rank less than 15. The second argument `dim` is a scalar such that `1<=dim<= (n+1)`, where n is the rank of the first argument. The third argument `ncopies` is an integer, and the source is replicated `ncopies` times.

TABLE 6.35a

dim=1 (see text)

1	2	3
1	2	3
1	2	3

TABLE 6.35b

dim=2

1	1	1
2	2	2
3	3	3

Case I: If the source is a scalar, the shape of the resultant array is `max (ncopies, 0)`. For example,

```
spread (source=10, dim=1, ncopies=3) results in array [10, 10, 10].
```

Case II: When the source is a vector:

```
integer, dimension(3) :: s=[1, 2, 3]
spread(source=s, dim=1, ncopies=3)
```

It results in a 3 × 3 output array having rows 1, 2 and 3 (Table 6.35a). On the other hand,
```
spread(source=s, dim=2, ncopies=3)
```
returns a 3 × 3 array of the form having columns 1, 2 and 3 (Table 6.35b).

6.42 MERGE

This intrinsic also takes three arguments, `tsource`, `msource` and `mask`. The `tsource` may be a scalar or an array, and the `msource` is of same type as `tsource`. If this is an array, it must be of same shape as `tsource`. The `mask` is logical. If `mask` is an array, it should be of same shape as `tsource` or `msource`.

Case I: When `tsource` is a scalar like `merge (20, 10, m > 15)`, `merge` returns 20 if `m` is greater than 15; otherwise, it returns 10.

Case II: In this case, `tsource`, `msource` and `mask` are the arrays as shown in Table 6.36 (`t` stands for `.true.` value, and `f` stands for

TABLE 6.36a

tsource Array

1	4	7
2	5	8
3	6	9

.false. value). When a particular array element of mask is true, the corresponding array element from the tsource is taken, and if a particular eement of mask is false, the corresponding element from the msource is taken.

```
logical, parameter :: t=.true.,f=.false.
logical, dimension(3,3):: mask
integer, dimension(3,3) :: tsource, msource, ra
mask=reshape([t,f,t,f,f,t,f,t,f],[3,3])
tsource=reshape([1,2,3,4,5,6,7,8,9],[3,3])
msource=reshape([10,20,30,40,50,60,70,80,90],[3,3])
ra=merge(tsource, msource, mask)
print *, ra
! (column-wise): 1, 20, 3, 40, 50, 6, 70, 8, 90,Table 6.37
end
```

When a particular array element of mask is true, the corresponding array element from the source is taken, and if a particular element of mask is false, the corresponding element from the msource is taken.

TABLE 6.36b

msource Array

10	40	70
20	50	80
30	60	90

TABLE 6.36c

mask Array

t	f	f
f	f	t
t	t	f

TABLE 6.37

ra Array

1	40	70
20	50	8
3	6	90

6.43 REDUCE

Array reduction using the user-defined operation is possible with this transformational intrinsic. There are two forms of reduce. In the first form, the calling parameters are array, operation, mask, identity and ordered. The last three arguments are optional. In the second form, the calling parameters are array, operation, dim, mask, identity and ordered. Again, the last three are optional. The array is of any type; operation is a pure procedure (Chapter 12) having two normal scalar arguments of the same type as the array; dim is an scalar integer that satisfies 1<=dim<=n, where n is the rank of the array; mask is logical and conformable with the array; identity is a scalar having the same type as the array; and ordered is a logical scalar. The returned value from reduce is obtained by combining the adjacent values of the array using the function operation. We consider a simple case where the integer array is of the form [10, 20, 30] and the operation is just multiplication. The output of reduce is 10 * 20 * 30 = 6000. The Fortran 2018 report may be consulted for the description of other parameters. However, it may be mentioned at this point that a similar procedure co_reduce has been discussed in Chapter 20.

7

User Defined Data Type

A user-defined data type (derived type) is a data structure created by the user consisting of elementary elements of standard types such as integer, real, logical, complex and character variables. The structure can be accessed as a whole, or the individual elements may be treated separately. This structure is quite flexible and allows mixing various kinds of variables and arrays in any proportion. For example, to prepare an employee's company employment record, it is perhaps necessary to create arrays for, say, serial number, name and address, and the synchronization among the different elements becomes the responsibility of the programmer. The programmer has to keep track of the relations among the different variables. The variables declared as user-defined data type can be accessed as a whole (all the elementary items together), or the elementary items may be manipulated separately.

7.1 Derived Type

The derived type is illustrated with an example. Consider a record of a person consisting of (1) name: 30 characters, (2) address: 50 characters, (3) age: integers, (4) blood group: 2 characters and (5) Rh factor: 1 character. The prototype of such a derived type is defined as follows:

```
type person
   character (len=30) :: name
   character (len=50) :: address
   integer :: age
   character (len=2) :: blood_group
   character (len=1) :: rh_factor
end type person
```

The type name (in this case person) cannot be the same as the intrinsic type names—integer, real, complex, logical, character, double precision and double complex. Variable donor of type person may be defined as type (person) :: donor (Figure 7.1).

Variable donor contains five elementary items of standard types—character and integer. Variable donor as a whole refers to all elementary items (name, address, age, blood_group and rh_factor) of donor. The individual elements may be referred to as

donor%name, donor%address, donor%age, donor%blood_group, donor%rh_factor

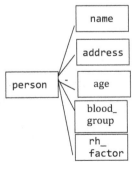

FIGURE 7.1
Personal record.

where the symbol '%' is used to fetch an elementary item from the variable name. Note that `name`, `address`, `age`, `blood_group` and `rh_factor` exist with reference to a variable—in this case, the variable name is `donor`. We will see shortly that an element of a derived type may also have a structure.

Another variable of type `person` can be defined in an identical manner:

```
     type (person) :: donor1
or   type (person) :: donor, donor1
```

Here, `donor1%name` refers to the name corresponding to `donor1` similar to the name corresponding to `donor (donor%name)`.

7.2 Assignment

A derived-type variable can be assigned to a value in many ways. Following the definition in Section 7.1, we can assign values to variable `donor`:

```
     donor=person('S RAY', 'CC, IACS', 72, 'O ', '+')
```

The assignment statement needs some explanation. First, the type of variable `person` must be specified in the assignment. Next, the data corresponding to each of its elements are given in the same order as was defined earlier.

An element is just like an ordinary variable and can be assigned in the usual manner.

```
     donor%name = 'S RAY'
     donor%address = "CC, IACS"
     donor%age = 72
     donor%blood_group ="O "
     donor%rh_factor = "+"
```

If another variable `donor1` of type `person` is equated to `donor`,

```
     donor1=donor
```

the elementary items of variable `donor1` become the same as the elementary items of the corresponding variable `donor`.

All arithmetic, relational and logical operations as permitted by the language can be used with the elementary items. For example, the following statements and the likes are valid:

```
     donor1%age=donor%age+10
     if (donor%rh_factor .eq. '+') then
        ...
     endif
```

7.3 Initialization

Initialization of user-defined data type variables can be done in two different ways. One is to initialize at the prototype level and the other at the variable level.

In the prototype level, the prototype is initialized, and any variable defined based on the prototype will have the corresponding variable initialized. This can be overridden by explicit initialization.

```
type person
   character (len=30) :: name='S RAY'
   character (len=50) :: address
   integer :: age
   character (len=2) :: blood_group
   character (len=1) :: rh_factor='+'
 end type person
type (person) :: blood_donor
```

Variable `blood_donor` will have its name and `rh_factor` initialized by 'S RAY' and '+', respectively. Other elementary items will not be initialized.

When the initialization takes place at the variable level, it can be done in two different ways.

```
type (person) :: blood_donor=person('S RAY', 'CC, IACS', 72, &
                 'O ', '+')
```

or

```
type (person) :: blood_donor
data blood_donor/person ('S RAY', 'CC, IACS', 72, 'O ', '+')/
```

The point to be noted is the presence of the data type (in this case `person` in the initialization statement).

7.4 Named Constant and Derived Type

Following Chapter 1, it is possible to define a user-defined data type named constant.

```
type mytype
   integer :: a
   real :: r
end type mytype
type(mytype),parameter :: nc=mytype(2,3.5)
type(mytype) :: x
x=nc
print *,x%a, x%r
end
```

The elementary items of the named constants of nc are initialized to 2 and 3.5, respectively. Neither can these be modified within the program nor can their components appear on the left-hand side of the assignment sign.

7.5 Keywords and Derived Types

Although Fortran keywords are not reserved words, the name of the prototype of a user-defined data type cannot be the name of the intrinsic data type—integer, real, character, logical, double precision, complex and double complex.

```
type integer ! not allowed
   integer :: a
   real :: b
end type integer
```

This may be treated as an exception. In general, Fortran keywords are not reserved words.

7.6 IMPLICIT and Derived Types

Implicit statements can be used to associate a variable that starts with a particular character to be of a particular derived type.

```
implicit type (donor) (d-e)
type donor
   character(len=30) :: name
   character (len=2) :: blood_group
   character (len=1) :: rh_factor
end type donor
d%name= 'Subrata Ray'
d%blood_group= 'O '
```

The implicit declaration ensures that any variable that starts with either d or e in this program unit is of user-defined data type donor.

7.7 Input and Output

The derived type may be read or displayed as a whole or by an individual element. Statement read *, blood_donor will expect that the data for each element be supplied sequentially during the execution as 'SR' 'CC, IACS' 71 'O' '+'. However, individual elements may also be read separately.

```
read *, blood_donor%name, blood_donor%address, blood_donor%age, &
   blood_donor%blood_group, blood_donor%rh_factor
```

The input list cannot contain any pointer (Chapter 16) or allocatable variable (Chapter 15). In a similar manner print *, blood_donor will display all the elements sequentially on the screen. Individual elements may also be displayed separately.

```
print *, blood_donor%name, blood_donor%address, blood_donor%age, &
    blood_donor%blood_group, blood_donor%rh_factor
```

We now demonstrate the use of derived type with an example. This program adds two English distances (miles and yards) d1 and d2. Note that this cannot be done through statement d3=d1+d2 because addition operation is not valid in this particular case as d1, d2 and d3 are not elementary items (real, integer or complex).

```
program distance
type dist !  add "English" distance
 integer :: mile
 integer :: yds
end type dist
type(dist) :: d1, d2, d3
read *, d1%mile, d1%yds, d2%mile, d2%yds ! data validation not done
print *," d1.miles = ", d1%mile, " d1.yds = ",d1%yds
            ! yds less than 1760
print *," d2.miles = ", d2%mile, " d2.yds = ",d2%yds
d3%mile=0
d3%yds=d1%yds+d2%yds
if (d3%yds >=1760) then  ! 1760 yds = 1 mile
d3%mile=d3%mile+1; d3%yds=d3%yds-1760
endif
d3%mile=d3%mile+d1%mile+d2%mile
print *," d3.miles = ", d3%mile, " d3.yds = ",d3%yds
end program distance
```

7.8 Substrings

A part of the string corresponding to an element (character) can also be accessed:

```
blood_donor%name(1:7)= 'Subrata'
```

This will modify characters 1 to 7 of elementary item name corresponding to blood_ donor. It is also possible to read or write a character substring.

```
print *, blood_donor%name(1:7)
```

It will print the first seven characters from blood_donor%name.

7.9 Array and Derived Types

A derived type may be a dimensioned quantity. For example, to store the information for 100 such aforementioned donors, the corresponding declaration would be

```
type (person), dimension (100) :: blood_donors
```

Now, blood_donors is an array of rank 1, and hence, it can be accessed like other arrays through a subscript; blood_donors(1) refers to the first donor, blood_donors(2) refers to the second donor and blood_donors(100) refers to the 100th donor. In general, blood_donors(i) refers to the ith donor (Figure 7.2).

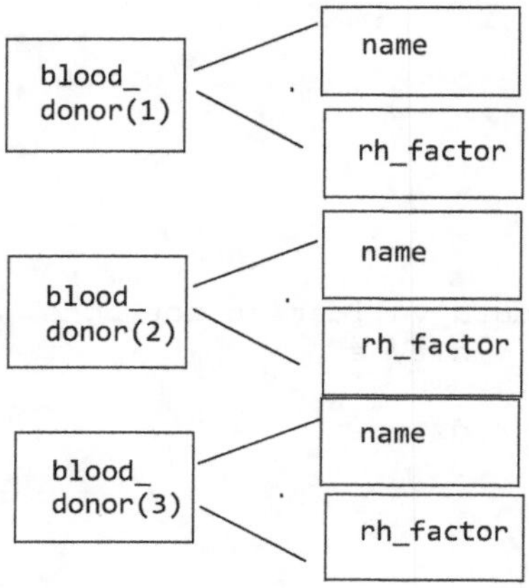

The individual elements of a particular donor can be accessed as follows:

FIGURE 7.2
Array of derived type.

```
blood_donors(1)%name='Soumya'
blood_donors(2)%blood_group='ab'
blood_donors(100)%address= &
 & '20A Fordyce Lane'
blood_donors (10)%age=35
ave = (blood_donors(5)%age+ &
 & blood_donors(9)%age) / 2.0
print *, &
 & blood_donors(3)%name(1:10)
```

The last statement prints the first 10 characters from the name of the third blood_donors. Array blood_donors can be initialized, if necessary, like the normal array.

```
type person
  character (len=30) :: name
  character (len=50) :: address
  integer :: age
  character (len=2) :: blood_group
  character (len=1) :: rh_factor
end type person
type (person), dimension(2) :: blood_donors= &
 [person('SR', 'cc, iacs', 72, 'o ', '+'), &
  person('PR', 'times now', 35, 'o ', '+')]
print *, blood_donors(1)
print *, blood_donors(2)
end
```

A data statement can also be used to initialize the array.

```
data blood_donors/&
person('SR', 'cc, iacs', 69, 'o ', '+'), &
person('PR', 'times now', 34, 'o ', '+')/
```

This will initialize blood_donors(1)%name to 'SR' and blood_donors(2)%name to 'PR'. Other elements are also initialized in the same manner.

7.10 Nested Derived Types

An element of a derived type need not be an elementary item. It may itself be another derived type. If the birthday of the donors is to be included, the `type` declaration may be modified accordingly (Figure 7.3).

```
type bday
   integer :: dd, mm, yy
end type bday
type person
   character (len=30) :: name
   character (len=50) :: address
   integer :: age
   character (len=2) :: blood_group
   character (len=1) :: rh_factor
   type(bday) :: birth_day
end type person
type (person) :: blood_donor
type (person), dimension (10) :: blood_donors
```

In this example, `blood_donor%birth_day` refers to the birthday of a `blood_donor` as a whole. To access the day (`dd`), month (`mm`) and year (`yy`), another level of reference is required.

```
blood_donor%birth_day%dd, blood_donor%birth_day%mm,
blood_donor%birth_day%yy
```

Similarly, to access an element of array `blood_donors`, an explicit reference is necessary:

```
blood_donors (1)%birth_day%dd, blood_donors (1)%birth_day%mm,
blood_donors (1)%birth_day%yy, blood_donors (2)%birth_day%dd,
blood_donors (2)%birth_day%mm, blood_donors (2)%birth_day%yy.
```

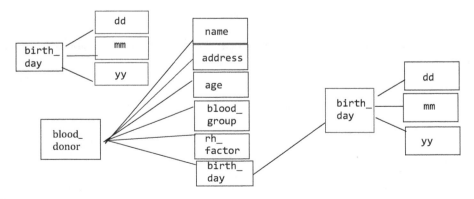

FIGURE 7.3
Nested derived type.

Such variables may be initialized as follows:

```
type (person) :: blood_donor= &
 person('AC','FORDYCE LANE', 78, 'B ','+',BDAY(20,7,1940))
type (person), dimension (2) :: bld_donors
data bld_donors/&
 person('SR', 'CC, IACS', 72, 'O ', '+',BDAY(19,6,1927)), &
 person('PR', 'MKP ROAD', 35, 'O ', '+',BDAY(16,7,1980))/
 .
end
```

Note that the data corresponding to `birth_day`, which consists of three elementary items, are supplied as `dd, mm, yy`—the way they appear in the derived types.

7.11 Arrays as Elementary Items

An elementary item may be a dimensioned quantity. For example, if we want to include telephone numbers (say, a maximum of three), the prototype for the derived type must be modified:

```
type bday
integer :: dd, mm, yy
end type bday
type person
   character (len=30) :: name
   .
   character (len=15), dimension(3) :: phone_no
end type person
type (person) :: donor_list
```

The first, second and third telephone numbers may be accessed using appropriate subscripts.

```
donor_list%phone_no (1)="03324734484"
donor_list%phone_no (2)="03324166109"
donor_list%phone_no (3)="03324245028"
```

Like any other variable, telephone numbers can be read from the keyboard and can be displayed on the screen with `read` and `print` statements.

```
read *, donor_list%phone_no (1)
print *, donor_list%phone_no (3)
```

If the list of donors is a dimensioned variable:

```
type (person), dimension(100) :: list_of_donor
```

then the telephone number of the individual donor may be accessed as follows:

```
list_of_donor(4)%phone_no(3)="22271022" !3nd ph no 4th donor
```

TABLE 7.1

Terms Associated with Derived Type

Name	Meaning
`list_of_donor`	List of all the 100 donors
`list_of_donor(i)`	All the elements of the ith donor
`list_of_donor(i)%name`	Name of the ith donor
`list_of_donor(i)%name(1:10)`	The first 10 characters of the name of the ith donor
`list_of_donor(i)%phone_no`	All telephone numbers of the ith donor
`list_of_donor(i)%phone_no(j)`	The jth phone number of the ith donor

In general, `list_of_donor(i)%phone_no(j)="24006625"` is the jth telephone number of the ith `list_of_donor`. Of course, integers i and j should have values within the boundaries specified by the declaration. The meaning of the various items associated with `list_of_donor` is summarized in Table 7.1.

7.12 SEQUENCE

The processor may not arrange the elements of a derived type in sequence. There may be gaps between elements to increase the efficiency of the program. However, in certain cases, especially when a variable of derived type is passed as an argument to a procedure (Chapter 12), it may be necessary to force the compiler to allocate the variables in consecutive locations. Statement `sequence` is used to direct the compiler to allocate consecutive locations for such variables.

```
type point3d
   sequence
   real :: x, y, z
end type point3d
type(point3d) :: a
```

Variables `a%x`, `a%y` and `a%z` will be allocated in successive locations by the compiler because of the presence of statement `sequence`. If there is a structure within a structure, and if the outer one is sequenced, the inner one must also be sequenced.

7.13 Derived Types and EQUIVALENCE Statement

Statement `equivalence` (declared as obsolete) works on the assumption that storage is arranged in sequence. Naturally, `sequence` must be used if `equivalence` with derived types is desired.

7.14 Parameterized Derived Types

Two types of parameters are available:

- Parameters known at the time of compilation (kind type)
- Evaluation of the parameter is deferred and the parameter becomes available during the run-time

```
type twodarray(k,l,m)
  integer, kind :: k=kind(0.0)
  integer, len :: l,m
  real(k), dimension(l,m) :: x
end type twodarray
```

One must not confuse the kind attribute with the kind intrinsic.

```
type(twoarray(kind(0.0d0), 20, 20)) :: x
```

It now defines an object x of derived type twodarray with three parameters—one 'kind' type and two 'len' types. All parameters are of type integer. For example, the integer parameter k has an attribute kind and parameters l and m have an attribute len. These parameters (l and m) are used to specify the bounds of the array.

These are illustrated with the following program:

```
program udtest
type ud1(k)   ! compiled with ifort
 integer, kind :: k
 real(kind=k) :: d
end type ud1
type ud2(k,l,m)
 integer, kind :: k
 integer, len :: l,m
 real(kind=k),dimension(l,m) :: r
end type ud2
type ud3(k)
 integer, kind :: k=selected_real_kind(7,37)
 real(kind=k) :: r1
end type ud3
type(ud1(kind(0.0))) :: u1
type(ud1(kind(0.0d0))) :: u2
type(ud1(kind(0.0q0))) :: u3 ! intel's extension
type(ud2(kind(0.0d0),3,3)) :: u4
type (ud3) :: u5 ! default double precision
type (ud3(kind(0.0))) :: u6
u1%d=2.158
u2%d=2.158d0
u3%d=2.158q0
print*, " u1 =", u1," u2 = ",u2
print *, " u3 = ",u3
u4%r (1,1)=2.1d0
```

```
print *, "U4(1,1) = ",u4%r(1,1)
u5%r1=2.11d0     ! default double precision
u6%r1=2.11       ! default is replaced by single precision through
                 ! kind argument
print *, "Default double precision = ", u5%r1
print *, "Forced single precision = ", u6%r1
end
```

The outputs are as follows:

```
u1 = 2.158000   u2 = 2.15800000000000
u3 = 2.15800000000000000000000000000000
U4(1,1) =   2.10000000000000
Default double precision =   2.11000000000000
Forced single precision = 2.110000
```

The outputs need some clarifications. The three variables u1, u2 and u3 are declared, respectively, as single, double and quad precisions through 'kind'-type parameter. Variable u4 is a dimensioned double precision array of size 3 × 3 through 'kind'- and 'len'-type parameters. While declaring variable u5, no 'kind'-type parameter is specified. It takes the default 'kind' from prototype ud3. Variable u6 is forced to be of single precision type, ignoring the default double precision type mentioned in prototype ud3.

We will come back to this user-defined data structure in Chapter 19 in connection with the object-oriented programming using Fortran.

8

Format Statement

List-directed input/output statements were introduced in Chapter 1. Though the list-directed input/output statements are very convenient, users do not have much control over them. Most of the time, users prefer to have output in user-defined style, and design of the output format is one of the important aspects of programming. There is a saying, "Input should be user friendly and output should be self-explanatory." The `format` statement provides adequate facilities to design both the input and the output. The editing specification, associated with the `format` statement, during the input operation converts the external form (say ASCII) into the internal form (binary). It is just the reverse for the output operation.

8.1 Edit Descriptors

The editing specification may be supplied in three different ways: (a) in-line form, (b) in a `format` statement and (c) a character variable containing the editing specification.

In-line form uses the editing specification in-line with the list elements.

```
read '(edit descriptor)', list
print '(edit descriptor)', list
```

It is apparent that if the same editing specification is to be used with another input/output statement, it is to be repeated along with the `read` or `print` statement; it cannot be reused. A `format` statement may be used many times, and in fact, the same `format` statement may be used for the input as well as for the output.

```
        read st-no, list
        print st-no, list
st-no format(edit descriptor)
```

In the preceding expression, `st-no` is a unique statement label; `format` is a non-executable statement and can be placed anywhere after the specification (declaration) statement and before the `end` statement (in case there is any in-line procedure [Chapter 12], the `format` statement should be before the `contain` statement). Usually, it is placed immediately after the declaration or just before the `end` statement. Some programmers prefer to keep the `format` statement near the corresponding `read` or `print` statement.

The `format` statement must have a unique statement number and must contain at least one pair of left and right parentheses. In case of nested parentheses, the number of left and right parentheses must be the same.

```
      read 10, a, b, c
10    format(edit descriptor)
      print 20, p, q
20    format (edit descriptor)
```

Another form of I/O statements is:

```
read(unit, st-no) list
write(unit, st-no) list
```

These forms will be discussed in detail in Chapter 9. The unit number is an integer between 1 and 99. Unit numbers 5 and 6 are usually connected to the keyboard and the screen, respectively. For example,

```
      read(5, 10) x
      write(6, 10) x
10    format(edit descriptor)
```

will read the variable x from the keyboard and display the result on the screen according to the format. In this form, the in-line edit descriptor may also be included:

```
read(5, fmt= '(edit descriptor)') list
write(6, fmt= '(edit descriptor)') list
```

The expression fmt= may be omitted.

```
read(5,'(edit descriptor)') list
write(6,'(edit descriptor)') list
```

The third form takes a character variable in place of statement number. This character variable is earlier assigned to a string of characters containing the edit descriptors.

```
character (len=20) :: for ! for is assigned to some edit descriptor
read for, a, b, c
print for, a, b, c
```

8.2 Input/Output Lists

The I/O list is the list of elements to be read or written. For the input, the list must be variable names, and for the output, the list elements may be expressions. The expression is evaluated before the list elements are displayed.

```
read(5, 10) a, b, c
write(6, 20) a+i, b-j, c
```

An output list may contain a reference to a function. The function is evaluated before write statement is executed.

```
write(6, 100) sqrt(x)
```

The I/O may contain a reference to an array. The array may be used as a whole:

```
integer, dimension (10):: a
read(5, 125) a
```

Here, during the execution of the program, data for all a's must be provided. Array elements may be referred individually: `write(6,100) a(1), a(2), a(3), a(4)`. Implied do loop may be used to read or write array elements: `write(6,200) (a(k), k=1,10)`. Also, the array elements may be referred to as `write(6, 25) a(1:3)`; this is same as `a(1), a(2)` and `a(3)`. Similarly, `write(6, 26) a(1:10:2)` will print `a(1), a(3), a(5), a(7)` and `a(9)`.

If the input/output list contains at least one item, the `format` statement must contain at least one edit descriptor. If the input/output list is empty, the `format` statement may contain just a pair of left and right parentheses.

```
     print 10
10   format ( )
```

This will print a blank line.

8.3 General Form of Format Statement

The general form of a `format` statement is `rfw.d`, where `r`, an unsigned integer, is the repeat factor. If absent, it is assumed to be 1 (exception: `'x'` edit descriptor); `r` cannot have a `kind` parameter attached to it. The repeat factor may be an asterisk (`*`). In this case, `r` is treated as a very large integer, `f` is the editing specification, `w` is the width of the field (it may be zero) and `d` is the position of the decimal point. Note that not all editing specifications may have all the aforementioned components (`r`, `w` and `d`). The editing specification or the `format` tells the computer how the data is arranged in the input device or how the data will be displayed on the output device. The exact position of the data is dependent on the relative position of the edit descriptor within the `format` statement. Consider the following two statements:

```
      read 100, j1,j2
100   format (2i5)
and   read 200, j1,j3,j2
200   format (3i5)
```

In the first case, `j2` is available between positions 6 and 10 (first 5 positions are for `j1`), and in the second case, the `j2` is available between positions 11 and 15. The edit descriptor just specifies the width of the field, but the exact position of the data is governed by other list elements, if any, and their width.

Normally, when a `read` statement is executed, data is always taken from a fresh record, and similarly when a `print` (or `write`) statement is executed, output starts from a new line.

8.4 Carriage Control

This feature, to control the movement of the carriage during printing, is no longer available in Fortran. This was used extensively till Fortran 95.

8.5 Summary of Edit Descriptors

The edit descriptors may be grouped under several categories as follows:

- integer—i
- boz—b, o, z
- real—f, e, en, es, ex, d, b, o, z
- complex—f, e, en, es, ex, d, b, o, z
- logical—l
- character—a
- general—g
- position—t, tl, tr, x
- slash—/
- blank handling—bn, bz
- string—' (apostrophe), " (quote), h
- decimal editing—dc, dp
- rounding mode—ru, rd, rz, rn, rc, rp
- colon—:
- sign—s, sp, ss
- scale—p
- user-defined derived type—dt (this will be discussed in Chapter 19)

Neither the edit descriptor nor the numbers associated with the edit descriptor may have a kind parameter. For example, format (i.5_1) or format (i_short.1) or the combination is not allowed.

8.6 Descriptor for Integer

The edit descriptor for integer is i.

```
        read 10, ind
10      format(i3)
```

The value of the variable will be read from the external device (say, keyboard). The edit descriptor i3 indicates the width the field, and the data is available between positions 1 and 3 of the input. If the data contains 107, ind will be assigned to 107. By default, Fortran ignores leading, trailing and embedded blanks and leading zeros. For a completely blank field, zero is assigned to the corresponding list element. Unsigned data is assumed to be positive. For negative numbers, minus sign is placed before the number like -47. A single read statement may read more than one quantity.

```
        read 100, j1, j2, j3
10      format (i3, i4, i6)
```

On the input device, j1 is available between positions 1 and 3, j2 is available in the next four positions between 4 and 7 and j3 is read between positions 8 and 13. The edit descriptor contains the width of the field, and the exact position of the data is relative with respect to the previous field.

```
        read 10, j1, j2
10      format (2i5)
```

In this case, the repeat factor is 2, and thus it is equivalent to i5, i5. The first five positions contain the value for j1, and the next five positions contain the value for j2. During input, 'd' part is ignored. Also, during input, w cannot be zero.

A print statement with the i descriptor can be used to display integers on the output device.

```
        print 20, j1, j2
20      format (2i5)
```

will print j1 between positions 1 and 5 and j2 between positions 5 and 10. The numbers are printed right adjusted within the field. Leading zeros are replaced by blanks. If the quantity is positive, '+' sign is not printed. For a negative quantity, a '-' sign is placed before the number. The integer format may be of the form w.d. For output, if 'd' part is present, at least 'd' number of digits are printed, if necessary, by adding leading zeros.

```
        print 41, j1
41      format(i5.5)
```

If the value of j1 is 37, the output will show 00037. If w is equal to 0, the processor will calculate minimum number of places necessary to display the number with leading sign, if necessary. Unless w is not equal to 0, d must not be greater than w. If the edit descriptor is i5.3, ^^037 is displayed, where '^' represents blanks. If the width is not sufficient to accommodate the number (when w>0), asterisks are displayed. If d=0 and data is 0, w blanks are displayed. If w is also 0 (i0.0), only one blank is displayed. If w=0, d is not equal to 0 and the data is 0, d number of zeros are displayed. The next program shows all the results. It may be noted that we have not introduced slash ('/') edit descriptor. It will be introduced at the appropriate place. At this point, it may be mentioned that the slash in the edit descriptor forces a new output record; that is, the value corresponding to the list element goes to a new line. The following program is used to show the position of the outputs relative to the left margin.

```
      integer::x=123
      print 10,x,x,x,x,x,x,x,x,x,x
10    format(i3/i4/i5/i5.4/i5.5/i2/i5.0/i0.4/i0.2/i0.0)
      end
```

The outputs are as follows (comments are added):

```
123    ! i3
 123   ! i4; one leading blank
  123  ! i5; two leading blanks
 0123  ! i5.4; leading zero to make d=4
00123  ! I5.5; two leading zeros to make d=5
**     ! i2; insufficient w, so two asterisks are displayed
  123  ! i5.0; d=0; this is same as I5
0123   ! i0.4; w=0, d=4; one leading zero to satisfy d=4
123    ! i0.2; w=0, d=2; processor calculates w
123    ! i0.0; w=0, d=0; processor calculates w
```

If the same program is executed with x=-123, the outputs will be as follows:

```
***    ! i3; insufficient field width w for -123
-123   ! i4; leading minus sign
 -123  ! i5; one leading blank and then the minus sign
-0123  ! i5.4; leading minus sign and then leading zero, d=4
*****  ! i5.5; insufficient width, w=5, d=5 - no place for sign
**     ! i2; insufficient field
 -123  ! i5.0; d=0; this is same as I5
-0123  ! i0.4; w=0, d=4
-123   ! i0.2; w=0, d=2; processor calculates w
-123   ! i0.0; w=0, d=0; processor calculates w
```

8.7 Descriptors for Real Number

There are nine edit descriptors under this category: 'f', 'e', 'd', 'en', 'es', 'ex', 'b', 'o' and 'z'. The 'f' descriptor is used for normal real numbers, and the 'e' descriptor is used when the input/output is desired in scientific notation. The 'd' edit descriptor is used for double precision quantities. The descriptors 'en' and 'es' are both used for the output. During input, both 'en' and 'es' edit descriptors are same as 'e' edit descriptor. A typical 'f' descriptor may be of the form f8.3, where 8 is the width of the field and 3 is the position of the decimal point. During input operations, if the decimal point is typed explicitly, 'd' has no effect. During input, w cannot be 0. If w is not 0, w should not be less than d.

```
      read 105, a
105   format(f8.3)
```

If the data is supplied as 25.2797, the variable a will be assigned to 25.2797 when the data is available anywhere within the field (positions 1 and 8). Leading zeros and blanks are ignored. Trailing blanks are ignored if blanks are treated as null; otherwise zero is taken.

For positive number, the plus sign is optional. If the number is negative, the minus sign is placed in front of the number. If the data does not contain a decimal point, that is, it is supplied as integer, the actual value of the variable becomes the input number multiplied by 10^{-d}, where d is the position of the decimal point. In the preceding example, if the input data is 12345678, 'a' will be assigned to 12345.678. If the field is totally blank, it is treated as zero during input.

A real number may be printed with the 'f' descriptor. The number is printed right adjusted within the field. Leading zeros are replaced by blanks, and for positive numbers the plus sign is omitted. For a negative number, the minus sign is placed before the number. If w is equal to 0, the processor will calculate minimum number of places necessary to display the number with a leading sign, if required. Unless w is not equal to 0, d must not be greater than w. If the width is not sufficient to accommodate the number (when w>0), asterisks are displayed.

```
      real :: a
      a=1.236
      print 200, a
200   format(f8.3)
```

will display 1.236, but if the format is

```
200   format (f8.2)
```

the output will be 1.24 (after rounding the last digit).

Some more examples of the 'f' edit descriptor are as follows:

```
     real:: x=123.456
     print 10, x,x,x,x,x,x,x,x,x,x
10   format(f7.3/f8.3/f9.3/f10.3/f10.4/f0.3/f0.0/f6.3/f0.4/f0.6)
     end
```

The outputs are as follows (comments are added):

```
123.456 ! f7.3; just sufficient; six places for digits, one for decimal
 123.456 ! f8.3; one leading blank
  123.456 ! f9.3; two leading blanks
   123.456 ! f10.3; three leading blanks
  123.4560 ! f10.4; two leading blanks
123.456 ! f0.3; processor calculates w=7
123.      ! f0.0; d=0 — digit after decimal, processor calculates w=4
****** ! f6.3; insufficient field width
123.4560 ! f0.4; processor calculates w=8
123.456001 ! f0.6; processor calculates w=10
```

If the same program is executed with x=-123.456, the outputs will be as follows:

```
******* ! f7.3; insufficient field width
-123.456 ! f8.3; just sufficient
 -123.456 ! f9.3; one leading blank
  -123.456 ! f10.3; two leading blanks
 -123.4560 ! f10.4; one leading blank
-123.456  ! f0.3; processor calculates w=8
```

```
-123. ! f0.0; d=0 — digit after decimal, processor calculates w=5
****** ! f6.3; insufficient field width
-123.4560 ! f0.4; processor calculates w=9
-123.456001 ! f0.6; processor calculates w=11
```

When the internal value of a variable is 0, the next program shows the output under different edit descriptor.

```
        real :: x=0.0
        print 10, x,x,x
10      format(f0.3/f0.0/f3.0)
        end
```

The outputs are as follows:

```
.000   ! f0.3; w=0
0.     ! f0.0; w=0, d=0, no digit after the decimal point
   0.  ! f3.0; w=3, one blank before zero; no digit after decimal point
```

In scientific notation, the editing symbol is 'e' like e16.6, where 16 is the width of the field and 6 is the position of the decimal point. Again during the input operation, if the decimal point is typed explicitly, 'd' part is ignored. The data consists of a string of decimal digits followed by 'e' and an integer to specify the exponent. The exponent may be signed. An unsigned exponent is assumed to be positive. For a negative exponent, a minus sign is typed between 'e' and the exponent. For example, 1.2367E4, -37.45E-7, 43.2646E-11 and -100.37E7 are valid data. The values are, respectively, 1.2367×10^4, -37.45×10^{-7}, 43.2646×10^{-11} and -100.37×10^7. If the decimal number is typed explicitly, the data may be typed anywhere within the field. It is recommended that the number be right adjusted within the field, as some old compilers may assume trailing blanks as zeros and these zeros become part of the exponent (1 becomes 10). If the decimal point is not typed explicitly, the number becomes (integer part) $\times 10^{-d} \times 10^n$, where d is the position of the decimal point and n is the exponent part. If the edit descriptor is e10.3, 0012345e4 will be read as 12.345×10^4.

The e descriptor, when used with a print statement, prints the value of the corresponding real variable right adjusted within the field. Usually, a zero is printed before the fraction, and the exponent is adjusted accordingly.

```
        real :: a
        a=12.34
        print 25, a
25      format(e11.3)
```

The output will be 0.123e+02 with 2 leading blanks. The 'd' edit descriptor is similar to the 'e' edit descriptor. It is used for double precision quantities; f, e, es, en and d are interchangeable. For example, the format statement may contain the edit descriptor f7.3, but the input may be 1.2e04.

The second form of e descriptor is rew.dex, where x is the number of digits for the exponent field during output. If the edit descriptor of the statement number 25 is replaced by e11.3e4, the output will be 0.123e+0002.

There are three other variations of the e format. They are en, es and ex. The descriptors en and es are normally used with the output statement. The en descriptor prints a variable such that the exponent part is divisible by 3 and the fractional part is between 1 and 999, except when the list element is zero.

Internal Value	Edit Descriptor	Output
0.8	en12.3	800.000E-03
1.256	en12.3	1.256E+00
1239.672	en12.3	1.240E+03

The es edit descriptor ensures that the number before the decimal is between 1 and 9 except when it is zero. The exponent is adjusted accordingly.

Internal Value	Edit Descriptor	Output
0.8	es12.3	8.000E-01
1.256	es12.3	1.256E+00
37.935	es12.3	3.794E+01

The ex descriptor is used for both input and output with hexadecimal significant number. We conclude this section with a few more examples of the edit descriptors e, en and es.

```
        real :: x=-123.456
        print 10, x,x,x,x,x,x,x,x,x,x
10      format(e15.3/e15.4/e15.4e3/e15.4e4/ &
            en15.3/en15.3e3/en15.4e4/es15.3/es15.3e3/es15.3e4)
        end
```

The outputs are as follows:

```
   -0.123e+03 ! e15.3    ; five leading blanks
  -0.1235e+03 ! e15.4    ; four leading blanks
 -0.1235e+003 ! e15.4e3 ; three leading blanks
-0.1235e+0003 ! e15.4e4 ; two leading blanks
  -123.456e+00 ! en15.3   ; three leading blanks
 -123.456e+000 ! en15.3e3; two leading blanks
-123.4560e+0000 ! en15.4e4; zero leading blank
     -1.235e+02 ! es15.3   ; five leading blanks
    -1.235e+002 ! es15.3e3; four leading blanks
   -1.235e+0002 ! es15.3e4; three leading blanks
```

8.8 Insufficient Width

If the width of the field is not sufficient to display the number, asterisks are printed. A format with edit descriptor i3 is insufficient to print a variable having a value 9999. A relation exists between the width of the field and the position of the decimal point for e edit descriptor. If the edit descriptor is ew.d, we need (a) d places for the digits after the decimal point, (b) 1 place for the sign of the exponent, (c) 2 places for the exponent (usually), (d) 1 place for the letter 'e', (e) 1 place for the decimal itself, (f) 1 place for the digit before the decimal point (usually zero) and (g) 1 place for the sign of the fraction. Therefore, w >= d+7. For the d (double precision) descriptor, where the exponent is usually a four-digit number, the relation between w and d will be w >= d+9. If the edit descriptor is of the form ew.dex, the relation between w and d will be w >= d+x+5.

8.9 Format and List Elements

We have seen that two or more variables may be read or written by a single format statement.

```
      read 15, ia, ib, ic
15    format(i4,i5,i6)
```

It may be noted that in this case, there is a one-to-one correspondence between the list elements and the edit descriptors. Two cases may arise.

Case I: If the edit descriptor contains more items than the list elements like the following:

```
      read 35, ia, ib
35    format(3i3)
```

the excess edit descriptor is ignored. If the input record contains 123456789, ia will be assigned to 123 and ib to 456. The digits following 6 are ignored, and if there is another read statement following this one, a fresh record will be read.

Case II: If the number of list elements is more than editing specification like the following:

```
      print 45, ia, ib, ic
45    format (i8)
```

the edit descriptor is used repeatedly until the list is exhausted. Each time the format is reused, a fresh record is used. To be more precise, when the format is exhausted, the nearest left bracket is traced. And between the traced left bracket and the next right bracket, edit descriptors are repeatedly used, and each time it comes to the left bracket, a fresh record is generated. In the preceding example, ia, ib and ic will be printed in three separate lines. If the format is:

```
45    format (2i8)
```

ia and ib will be printed in one line and ic will be printed in the next line. If the format contains nested parentheses:

```
      print 55, n, a, b, c
55    format(i8,(2f9.2))
```

n, a and b will be printed in the first line using the edit descriptors i8, 2f9.2. As the format has been exhausted but the list is not yet exhausted, the edit descriptor 2f9.2 will be used repeatedly until the list is exhausted, and each time this edit descriptor is reused, a fresh line will be generated. Therefore, c will be printed in the next line with the edit descriptor f9.2. Note that after printing c, the additional edit descriptor is ignored as the list has been exhausted. Though all the preceding examples were based on the print statement, the same is also true for the read statement.

8.10 Descriptors for Complex Number

Complex number consists of two parts: real and imaginary. The edit descriptors for these two parts may be same or different for both the input and the output. The first edit descriptor corresponds to the real part, and the second one corresponds to the imaginary part.

```
        complex :: c1
          .
        print 100, c1
   100  format(2e16.6)
```

Both the real and the imaginary parts are displayed using the e16.6 edit descriptor. If the format is changed to:

```
   100    format(e16.6, f12.3)
```

the real and the imaginary parts will use edit descriptors e16.6 and f12.3, respectively. The read statement requires two numbers: one for the real part and another for the imaginary part. The discussions related to real variables are equally valid for complex numbers also.

8.11 Descriptors for BOZ Numbers

Integer, real and complex variables may be assigned to a binary, octal or hex value through a read statement using b, o or z format, respectively. The same edit descriptor may be used to display a variable as binary, octal or hex number. The next program segment will assign values to the respective variables when the data is supplied through the input device (in this case the keyboard).

```
        integer :: ib, io, iz
          .
        read 10, ib
   10   format(b4)
```

If the supplied value is 1101, ib will be assigned to 13 [$(1101)_2 = 13$].

```
      read 20, io
20    format (o5)
```

If the supplied data is 00472, io will be assigned to 314 [$(00472)_8 = 314$].

```
      read 30, iz
30    format(z5)
```

If the supplied data is 002c5, iz will be assigned to 709 [$(002c5) = 709$].

Similarly, the write statement with b, o and z format will display the integer as binary, octal and hex numbers, respectively.

```
      print 40, ib; print 50, io; print 60, iz
40    format (b32)
50    format (o11)
60    format(z8)
```

Though the examples are given for integers, one may use real and complex variables as well (present NAG compiler does not allow). In addition, it is possible to use bw.d, ow.d and zw.d edit descriptors that during the output operation will ensure the presence of d digits, if necessary, by adding leading zeros.

8.12 Descriptor for Logical

The edit descriptor for a logical quantity for the input/output operation is lw, where w is the width of the field. During input, if the first non-blank character of the field is t or T or .t or .T, true value will be assigned to the corresponding list element. Similarly, if the first non-blank character of the field is f or F or .f or .F, false value is assigned to the corresponding list element. During the output operation, T or F, depending on the value of the list element, is displayed right adjusted within the field; that is, w-1 blanks followed by T or F.

```
      logical :: ll
      read 200, ll; print 200, ll
200   format(l4)
```

If the data supplied is .true., the output will be T preceded by 3 blanks. During input, the first character (t or f) determines the true or false value. In the preceding example, if the input is tabcd, true value will be assigned to ll, as the first character is t. Similarly, if the input is fxyz, false value will be assigned to ll.

8.13 Descriptor for Character

The edit descriptor corresponding to character variable is 'a'. Its form is aw, where w is width of the field. The width may be omitted, in which case the width is assumed to be the length of the variable. Three cases may arise.

Case I: The length of the variable is same as the width of the field w.

```
        character (len=4) :: x
        read 25, a; print 25, x
  25    format(a4)
```

The read statement inputs 4 characters, and the print statement prints the same 4 characters.

Case II: The length of the variable is greater than the width of the field.

```
        character (len=8) :: x
        read 35, a; print 45, x
  35    format(a4)
  45    format (a6)
```

The first four characters from the input field are read and stored in the leftmost bytes of the variable; trailing blanks are also added. If, in the preceding case, the data contains iacs, the variable x will be set to iacsbbbb, where 'b' stands for blank. The print statement outputs the first six characters of the variable x that is iacsbb.

Case III: The length of the variable is less than the width of the field.

```
        character (len=4) :: x
        read 55, x; print 56,x
  55    format(a8)
  56    format(a10)
```

The rightmost len characters (here it is 4) are stored in the variable. If the data is calcUTTA, the variable x will be set to UTTA. The print statement will display UTTA right adjusted within the field and naturally will add 6 blanks before UTTA (bbbbbbUTTA), where b stands for blank.

8.14 General Edit Descriptor

The g edit descriptor is used for generalized editing. The g descriptors gw, gw.d and gw.dee can be used with integer, real, complex, logical and character for both the input and the output operations. For integer, logical and character type data, 'd' part is ignored; w is width of the field, and for the real and complex numbers, d is the number of the decimal places and e is the number of digits in the exponent; w should be greater than 0 during input. A general rule is followed for the numeric data—integer, real and complex. Some compilers do not support gw form for integer, logical and character I/O list element. It allows only gw.d form.

For integers when w is not equal to 0, gw, gw.d and gw.dee edit descriptors are same as iw for both the input and the output. For output operation, g0 is same as i0.

For real and complex variables, during input, g descriptors behave like fw.d. For output operation, depending on the value of the list element, either e or f edit descriptor is chosen. The prescription is as follows.

First, the number is expressed in the form $0.dddd..x10^x$. If the edit descriptor is $gw.d$, f edit descriptor is chosen when x satisfies the inequality $0<=x<=d$. The width (w) and the number of digits after the decimal (d) are redefined as $w_{new} = w-4$, $d_{new} = d-x$, and 4 blanks are displayed at the end. For example, suppose a variable y has a value 123.456 and the corresponding edit descriptor is g16.4.

Now, $y = 123.456 = 0.123456 \times 10^3$. Thus, $x = 3$, $w_{new} = 16-4 = 12$ and $d_{new} = 4-3 = 1$.

Therefore, 123.456 will be displayed according to edit descriptor f12.1 with 4 blanks at the end. The number, thus, will be displayed as bbbbb123.5bbbb, where b stands for blank. The rounding is done according to the existing rounding mode. Now, consider the following program and the output from the program.

```
      real(kind=kind(0.0d0))::  y=1234.5678d0          The outputs are as follows
      print 10, y,y,y,y,y,y,y,y,y,y                    (shown side by side):
10    format('|',  g14.1,'|'/, &                       |      0.1e+04|
           '|',  g14.2,'|'/, &                         |     0.12e+04|
           '|',  g14.3,'|'/, &                         |    0.123e+04|
           '|',  g14.4,'|'/, &                         |     1235.    |
           '|',  g14.5,'|'/, &                         |     1234.6   |
           '|',  g14.6,'|'/, &                         |     1234.57  |
           '|',  g14.7,'|'/, &                         |     1234.568 |
           '|',  g14.8,'|'/, &                         |    1234.5678 |
           '|',  g14.9,'|'/, &                         |1234.56780    |
           '|',  g14.10,'|')                           |**********    |
      end
```

We shall now try to understand the output. The variable $y=1234.5678d0 = 0.12345678 \times 10^4$. Here $x=4$ and $d=1$. Therefore, y will be displayed according to the e edit descriptor e14.1 with 7 leading blanks. Now, consider the fourth line of the output. The edit descriptor, in this case, is g14.4 with $x=4$ and $d=4$. So, y will be displayed according to the edit descriptor f10.0 with 4 trailing blanks and 5 leading blanks. For line 9 of the output, the edit descriptor is g14.9 with $x=4$ and $d=9$. This time y will be displayed according to the edit descriptor 10.5 with 4 trailing blanks. The last line is interesting as the edit descriptor is g14.10. The modified edit descriptor according to the prescription given above is f10.6 and is insufficient to display the value y, and as such, asterisks are displayed.

If the edit descriptor is $gew.dee$, w_{new} and d_{new} are defined as $w_{new} = w-e-2$ and $d_{new} = d-x$ with $(e+2)$ blanks at the end. Again, if $y=123.456$ and the edit descriptor is g16.4e4, the output will be bbbbb123.5bbbbbb. The next program shows the difference between the edit descriptors g16.4e4 and g16.4.

```
      real(kind=kind(0.0d0))  :: x=123.456d0
      print 100,x,x
100   format('|',g16.4e4,'|'/'|',g16.4,'|')
      end
```

The outputs are as follows (comments are added):

```
|      123.5      |  ! five leading blanks and six trailing blanks
|        123.5    |  ! seven leading blanks and four trailing blanks
```

Readers are advised to verify the output.

For logical, gw.d and gw.dee descriptors are same as lw. During output operation, g0 is same as l1.

```
        logical :: l
        read 10, l; print 5, l
10      format(g5.2e4)
5       format(g5.0)
        print 10,l; print 20,l
20      format(l5)
        end
```

The above program will generate identical output.

For character, gw.d and gw.dee descriptors are same as aw. During the output operation, g0 is same as a descriptor without any field (w).

```
        character(len=4)::a
10      format(g4.3e3)
        read 10, a; print 10, a; print 15, a; print 20, a
15      format(g4.3)
20      format(a4)
        end
```

The preceding program will also generate identical output.

8.15 Unlimited Repeat Factor

If the repeat factor is replaced by an asterisk (*), this is treated as a very large integer.

```
        write (25, '("The result is ",*(i10))' ) ix
```

The preceding write statement will produce a single record on unit 25 with "The result is " followed by all the values of the array elements ix using i10 edit descriptor. The asterisk ensures that all the elements of ix are accommodated in a single record. In fact, if i10 is replaced by g0, the processor can choose appropriate edit descriptor with suitable field widths for various types of list elements, if present, in the list.

8.16 Scale Factor

The scale factor is used to scale numeric value during input or output operations. At the beginning of each input/output statement, the scale factor, by default, is zero. Once set, it applies to all subsequent f, e, en, es, ex, d and g edit descriptors, unless it is changed to some other value. The edit descriptor used for this purpose is kp, where k is an integer. It can also be a negative number. Comma between the edit descriptor p and its neighbor (edit descriptor) is optional. If there is no exponent in the edit descriptor, the input is multiplied by 10^{-k} while reading:

```
       read 20, a
  20   format(2pf8.2)
```

If the data is 14.2, it is stored as 0.142. If the edit descriptor is -2pf8.2, the same data is stored as 1420.0. Inputs having exponents are not affected by the scale factor. In the preceding program segment, if the input is 1.24e2, a will be assigned to 124.0. On output, for the f descriptor, the internal number is multiplied by 10^k. For example, 1.43 under the edit descriptor 2pf8.4 is displayed as 143.0000. If the edit descriptor is -2pf8.4, the displayed value is the original value multiplied by 10^{-k}; that is, the number is 0.0143. However, for the e, d and g descriptors, the mantissa is multiplied by 10^k and the exponent is multiplied by 10^{-k}. Therefore, 1.43 under 2pe12.5 and -2pe12.5 will be displayed as 14.3000e-01 and 0.00143e+03, respectively. If during output operation with g descriptor the number can be displayed using f descriptor, g descriptor behaves like f descriptor. However, if e edit descriptor is required, it behaves like e descriptor. For en and es edit descriptors, the scale factor has no effect.

8.17 Leading Signs

These edit descriptors are related to the sign of a number. For a positive number, the leading plus sign is replaced by a blank; for a negative number, the negative sign is placed before the number. There are three such edit descriptors related to sign sp, s and ss; sp or sign print forces the '+' sign of a positive number to be displayed.

```
       ia=10; print 10, ia
  10   format (sp,i4)
```

prints b+10, where b stands for blank. While sp is active, ss or sign suppress suppresses the printing of the positive sign.

```
       ia=10; ja=20; ka=30
       print 100, ia, ja, ka
 100   format (sp,i4,ss,i4,i4)
```

displays b+10bb20bb30

The third edit descriptor in this category, s, restores the default condition. If the previous format statement is replaced by:

```
 100   format (sp,i4,s,i4,i4)
```

the output will be b+10bb20bb30. The edit descriptors s, ss and sp have no effect during the input operation.

8.18 Tab Descriptors

The Tab descriptor specifies the position of the record from which the input may be obtained or the position of the output record where output of the corresponding list element would go.

Case I: The Tab edit descriptor has the form tn, where n specifies the position of the input/output record.

```
      read 110, ia, ja
110   format (t2,i2,t6,i2)
```

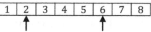

FIGURE 8.1
Tab descriptor.

will set ia to 23 and ja to 67 if the input record is 12345678 (Figure 8.1). Tab edit descriptor always counts the position from the beginning of the record. If the format statement is replaced by (Figure 8.1):

```
110   format (t6,i2,t2,i2)  ! Figure 8.2
```

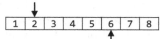

FIGURE 8.2
(see text).

ia will be assigned to 67 and ja will be assigned to 23.

```
      print 30, ia, ja
30    format (t2, i4, t10, i4)
```

will print ia from the position 2 (right adjusted in positions 2, 3, 4 and 5) and ja from the position 10 (right adjusted in positions 10, 11, 12 and 13). If the format statement is changed to:

```
30    format (t10, i4, t2, i4)
```

ia will be printed from position 10 and ja will be printed from position 2. It may be noted that though in the list ia appears before ja, ja will be printed before ia (nearer to the left margin) because of the t edit descriptor.

Case II: The Tab left edit descriptor has the form tln, where n is the position relative to the current position to the left.

```
      read 10, i, j
10    format (t6, i1, tl3, i1)
```

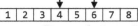

FIGURE 8.3
Tab left.

If the data is 12345678, i will be assigned to 6 and j will be assigned to 4 (Figure 8.3). After reading 6 (t6, 6th position), the current position is 7. With respect to this position, 3 positions to the left (tl3) is the 4th position from the left. For output also, position is calculated with respect to the current position to the left. The tl edit descriptor cannot go beyond position 1.

Case III: The Tab right edit descriptor has the form trn, where n is the position relative to the current position to the right. If the format statement is changed to

```
10    format (i1,tr3,i2)  ! Figure 8.4
```

FIGURE 8.4
Tab right.

i and j will be assigned to 1 and 56, respectively. After reading 1, the current position is 2, and 3 places to the right is the position 5. In all three cases mentioned here, n should be positive and greater than 0.

8.19 X Descriptor

This edit descriptor is used to skip few positions of the input field depending on the repeat factor of 'x' or to insert blanks (horizontally) in an output record. The form of the x edit descriptor is nx, where n is a positive integer greater than 0.

```
      read 95, ia, ib
95    format (2x, i3, 3x, i4)
```

First two positions of the input record are skipped, and contents of the next three locations become the value of ia. Next three positions are skipped, and ib is read from the next four positions of the input field. If the input is 123456789997, ia will be 345 and ib is set to 9997. When used with an output format, the x edit descriptor inserts blank(s) in the output record. As old Fortran compilers (prior to Fortran 2003) used to interpret the first character of an output record as carriage control character, '1x' is used with an output format to insert a blank at the beginning of a record so that this may be interpreted as carriage control character (single space) whenever required. The present version of Fortran does not support this feature.

```
      print 400, a, b, c
400   format(1x,3f10.3)
```

The peculiarity of x edit descriptor is that it must have a repeat factor, even if it is one (gfortran allows format (i2,x, i2) for both the input and the output, where x is taken as 1x; however, both NAG and ifort do not).

8.20 Slash Descriptor

If a new record is desired at any point, slash (/) is used as edit descriptor. On output, a slash causes the current line (record) to be terminated, and subsequent output starts from a new line (record). On input, the current record is terminated, and input begins from the next record.

```
      read 93, n, (a(i), i=1,n)
93    format(i5/(5e16.6))
```

n is read from the first record, and the a's are read from the subsequent records—5 from each record. Note the importance of the inner brackets. Because of this pair of brackets, when the format is exhausted, a's are read according to the edit descriptor 5e16.6 until the list is exhausted. Without this pair of brackets, the edit descriptor i5/5e16.6 will give runtime format error.

Slash can be repeated, or repeat count may be placed before the slash.

```
       print 73, a, b
73     format(f10.3 /// f8.3)
```

Two blank lines will be inserted between the values of a and b. The preceding format may be written as:

```
73     format(f10.3, 3/, f8.3)
```

In this case, the comma after f10.3 is essential. However, it is better to use comma both before and after slash (/).

In general, for output operation, if there are n number of slashes between two edit descriptors, (n-1) blank lines are inserted. The slash can also be placed at the beginning or at the end of the edit specification.

```
73     format (//f10.3///f8.3//)
```

The output will be two blank lines followed by the value of a according to f10.3 edit descriptor, and two more blanks lines followed by the value of b according to f8.3 edit descriptor followed by two blank lines.

8.21 Embedded Blanks

An embedded blank with an input field may be treated in two different ways. By default, it is treated as null. However, with an appropriate edit descriptor, it can be interpreted as zero.

```
       read 100, ia
100    format (i3)
```

If the data is 1b2 (b stands for blank), by default, ia will be set to 12. The default can be changed by bz edit descriptor. If the format statement is modified as

```
100    format (bz, i3)
```

blanks will be treated as zero, and so i will be read as 102. Once set, this will remain effective until the end of the format statement. This can be forced back to the default value by bn edit descriptor.

```
       read 400, ia, ja
400    format (bz, i3, bn, i3)
```

If the data is 1b23b4, ia will be set to 102 and j to 34. If the embedded or the trailing blanks are treated as zero, one should be extremely careful while reading data. Trailing blanks after the exponent in scientific notation (e edit descriptor) would make e+01b to e+10.

If the field is totally blank, the value is taken as zero irrespective of the interpretation of the blank (bn or bz). Another important point may be noted. If the blanks are interpreted as null, still they take part in determining the width of the respective fields.

```
        integer:: ia,ib
        read(5,10)ia,ib; print *, ia,ib
   10   format(bn,2i5)
        end
```

If the input is 1b23456bb9 (where b stands for blank), ia is available between positions 1 and 5. As blanks are treated as null, the value assigned to ia is 1234. Similarly, ib is available between positions 6 and 10 and is assigned to 569. If bn is replaced by bz, ia and ib are assigned to 10234 and 56009, respectively.

The edit descriptors bn and bz do not have any effect on output. These descriptors can be used only with numeric editing (i, b, o, z, f, e, en, es, ex, d and g edit descriptors).

8.22 Apostrophe and Quote Descriptors

Any character string may be displayed by enclosing the string within apostrophes or quotes.

```
        print 100, a
   100  format (' The result is = ', f8.3)
```

The print statement will display the string followed by the value of 'a' according to the edit descriptor f8.3. To print an apostrophe, two successive apostrophes must be used.

```
        print 110
   110  format (' don''t')
```

The same thing can be done by enclosing the string within quotes.

```
   110  format (" don't")
```

There is another way of printing a string, which is now considered a deleted feature.

```
        print 67
   67   format(8hiacs cal)
```

This print statement will display iacs cal on the screen. The disadvantage of the method (h edit descriptor) is that one has to count the exact number of characters and that becomes the coefficient of 'h'. Since this is an obsolete feature, it will not be discussed in detail.

8.23 Colon Descriptor

Colon editing is useful when it is necessary to terminate the format control because there is no further item in the list. This is illustrated with an example.

```
        a=32.25;  b=43.25
        print 215, a
 215    format (' a= ',  f5.2,' b= ',  f8.3)
```

The `print` statement of the program will generate an output as follows:

```
        a= 32.25 b=
```

Since the list element does not contain b, it is desirable that the string `'b= '` should not be in the output record. This can be achieved with the help of the colon (:) edit descriptor.

```
 215    format (' a= ',  f5.2 : ' b= ',  f8.3)
```

Though the example shows colon editing with the output statement, it can also be used to terminate an input statement. Note that a comma is not required between other edit descriptors and the colon. If the list is not exhausted, the colon edit descriptor is ignored.

8.24 Decimal Editing

This edit descriptor temporarily changes decimal editing mode. The descriptors act on the d, e, f, en, es and g edit descriptors.

The edit descriptor dc changes decimal to comma, and dp reinstates comma to decimal.

```
        f=2.3; g=5.3; write(6, 10)f, g
 10     format(dc, f5.2, dp, f5.2)
```

The output will be b2,30b5.30 (b stands for blank).

8.25 Rounding Modes

The rounding mode may be changed during the input/output statement using the edit descriptors ru, rd, rz, rn, rc and rp. They are, respectively, rounding up, down, zero, nearest, compatible and processor dependent. The edit descriptors used along with the read or the write statement override the rounding options, if used, with the open statement (Chapter 9).

If the mode is rounding up (ru), the resultant value is the smallest representable value greater than or equal to the original value. When the mode is rounding down (rd), the resultant value is the largest representable value less than or equal to the original value. If the mode is rounding zero (rz), the resultant value is as closest to the original value but certainly not greater than the original value. When the mode is rounding nearest (rn), the resultant value is taken as the value of the closest neighbor. If the neighbors are equidistant from the original value, the resultant value is processor dependent. If the mode is compatible (rc), the resultant value is taken as the value of the closest neighbor. If the neighbors are equidistant from the original value, the value away from zero is taken. When the mode is processor dependent (rp), the resultant value is processor dependent, which is actually one of the five modes mentioned earlier.

The following program is tested with Intel ifort compiler.

```
real:: x=3.1415926
print 10, x,x,x,x,x,x ! ifort compiler
10   format(ru,f9.6,rd,f9.6,rz,f9.6,rn,f9.6,rc,f9.6,rp,f9.6)
end
```

The results are as follows:

```
3.141593 3.141592 3.141592 3.141593 3.141593 3.141593
```

8.26 Variable Format

In place of a format, a character variable containing the edit descriptors may be used.

```
character (len=20) :: form
form='(3i5)'
read form, ia, ib, ic; print form, ia, ib, ic
end
```

Three integer variables ia, ib and ic will be read according to the editing specification contained in the character variable form, which in this case is 3i5.

The editing specification can be read from the external device and subsequently can be used as a format for the read or the print statement.

```
      character (len=20) :: form
      read 110, form ! read edit descriptor
110   format(a)
      read form, ia, ib, ic
      print form, ia, ib, ic
      end
```

The first read statement will read the editing specification to be used for the next read and print statement. If the variable form is read as (3i3) (note that the parentheses must be present along with the edit descriptor), the next read statement, which uses this character variable in place of format, will use this edit descriptor while reading ia, ib and ic. In the preceding program, the character variable is used to display the numbers. Note that the length of the character variable must be sufficient to accommodate all the edit descriptors including the left and right brackets.

8.27 Memory to Memory Input/Output

Usually, in any input/output operation, there is a transfer of data between the memory and some external device. However, it is possible to have a situation where one part of the memory may behave as an external device. Data from one portion of the memory may be transferred to another portion of the memory under the control of a `format`. Both `read` and `write` operations are possible.

```
        character (len=8) :: cbuf='12345678'
        integer :: ia, ib, ic
        read (cbuf, 10) ia, ib, ic
  10    format (i1, 1x, i2, 1x, i3)
        print 10, ia, ib, ic
        end
```

The results are as follows:

```
  1 34 678
```

The character variable, in this case, behaves as input device. Data is transferred to `ia`, `ib` and `ic` under the control of the `format` statement from the character variable `cbuf`. ASCII characters from the location `cbuf` are converted into binary through the `format` statement similar to the conversion of ASCII characters supplied from the keyboard. That is character '1' becomes integer 1 (binary number).

In a similar way, it is possible to 'write' in memory under the control of a `format`.

```
        character (len=12) :: ckuf
        integer :: ia=1, ib=23, ic=567
        write(ckuf, 20) ia, ib, ic
  20    format (3i4)
        print *, ckuf
        end
```

The output from the preceding program will be `bbb1bb23b567`, where 'b' denotes blank. It is evident that under the control of the `format`, binary numbers are converted into ASCII and stored in the character variable `ckuf`. As already mentioned, the length of the character variable should be sufficient to accommodate the output. Edit descriptor is `3i4`, that is, it will require at least `12` (3 × 4) characters. The length of the corresponding character variable, in this case `ckuf`, should be at least `12`.

8.28 NAMELIST

`Namelist` does not require any input or output list; it does not require any `format` statement either. Yet, it is grouped with the `format` statements, as it is very similar to formatted input/output statements; namelist uses a namelist block, which is a declarative statement. However, there is a difference between the list elements of a namelist block and standard list elements associated with the read/write statement. When the namelist block is used with the read statement, it is not necessary to specify all the list elements present in the namelist block. The elements that have not been supplied from the input device retain their current (existing) value. A namelist block is defined in the following way:

```
namelist/block-name/ list_elements
```

As an example, a namelist block blk may be defined as follows:

```
namelist/blk/a, b, c, d
```

where blk is the name of the block and a, b, c and d are list elements.

The read statement associated with the namelist block blk is as follows:

```
        read (5, blk)
or,     read (5, nml=blk)
```

The data supplied from the keyboard must have one '&' character in the first position followed by the block name and the variable name=value of the variable in any order. The data items are separated by blank, comma, tab or linefeed. It is terminated by a slash ('/'). For example, if one wants to modify b and d, the data to be supplied is as follows:

```
        &blk b=10.0, d=20.0 /
or,     &blk d=20.0, b=10.0 /
or,     &blk b=10.0
        d=20.0 /
```

Data supplied from the keyboard must start with an ampersand (first non-blank character) and ends with a slash as shown in the preceding program. Note further that in this case only the values of b and d have been changed, and therefore, a and c will retain their old (existing) values.

The write statement associated with the namelist block is as follows:

```
        write (5, blk)
or,     write (5, nml=blk)
```

The same namelist block may be used for both the input and output statement. The following program will illustrate this point.

```
        integer ::ia, ib, ic, id
        namelist/blk/ia, ib, ic, id
        ia=100; ib=200; ic=300; id=400
        read(5, blk); write(6, blk)
        end
```

If the data supplied through the keyboard is:

```
        &blk ia=7, id=27 /
```

the output will be as follows:

```
&BLK IA=7, IB=200, IC=300, ID=27/
```

The output will have the `namelist` block name attached to it.

The `read` statement with `namelist` is useful when it is necessary to change some and not all the list elements of an input list.

Real numbers may be supplied in normal decimal form or in scientific notation. However, `es` and `en` descriptors are not allowed. Character data is to be enclosed within quotes or apostrophes. The data corresponding to complex variable must be enclosed within brackets containing two numbers—one for the real and another for the imaginary part. For logical variable, the data must be `t` or `f` or `.t` or `.f` (uppercase is also allowed). Double precision variable must have `'d'` as exponent in the data.

```
integer :: i
real :: r
double precision :: d
complex :: c
logical :: l
character(len=4) :: s
namelist /nlt/ i, r, d, c, l, s
read(5, nlt); write(6,nlt)
end
```

The inputs may be as follows:

```
&nlt i=2 r=10.7 d=3.14159265d0 c=(2.0, 3.0) l=.t s="iacs" /
```

Blanks within the character string are treated as part of the character string. Blank is not allowed within a numeric or logical value. Integer must be supplied in decimal form: binary, octal and hex numbers are not allowed.

The program unit may have any number of `namelist` blocks, and no two blocks can have same block name. If the list element is an array, values may be supplied for the whole array or for the individual elements.

```
integer, dimension(5) :: p=999
namelist /n2/ p
read (5, n2); write (6, n2)
end
```

The data may be of the following type:

```
&n2 p=1, 2, 3, 4, 5/ ! all the elements are assigned, p(1)=1, p(2)=2
                     !... p(5)=5
&n2 p(1:3)=10, 20, 30/      ! only elements 1, 2 and 3 are assigned
&n2 p(1)=7, p(4)=20/        ! only p(1) and p(4) are assigned.
&n2 p(1:5:2)= 10, 20, 30/   ! p(1), p(3) and p(5) are assigned
```

In case the inputs are:

```
&n2 p=1 2 3 /
```

only p(1), p(2) and p(3) are assigned to 1, 2 and 3, respectively, and p(4) and p(5) are not affected. If the inputs are &n2 p=1 2 3 4 5 6 /, that is, more than the required number of data, runtime error will terminate the program.

If some of the consecutive array elements are to be assigned with the same value, a repeat factor r may be used in the form of r*value, where r is a positive integer.

```
&n2 p=5*100 ! all the array elements are assigned to 100
&n2 p=2*100, 2*200, 300 ! p(1)&p(2)=100,p(3)&p(4)=200,p(5)=300
```

When the namelist block contains a user-defined type variable like:

```
type point
 real :: a, b
end type point
type(point) :: twod
namelist/n2/twod
read(5, n2); write(6, n2)
end
```

the data may be supplied as shown next:

```
       &n2 twod%a=7.0 twod%b=8.0 /
or,    &n2 twod= 7.0 8.0/
```

Now, consider the following program:

```
type point
 real :: a, b
end type point
type vector
 type(point):: xy
 real :: z
end type vector
type(vector) :: threed
namelist/n1/threed
read(5, n1); write(6, n1)
end
```

In this case, the data may be supplied as follows:

```
       &n1 threed= 2.0 3.0 4.0 /
or,    &n1 threed%xy%a= 2.0
       threed%xy%b= 3.0
       theeed%z= 4.0 /
or,    &n1 threed%xy= 2.0 3.0
       threed%z= 4.0/
```

The same procedure is followed when a user-defined type inherits properties from its parent (discussed in Chapter 19).

```
type point
 real :: x, y
```

```
end type point
type, extends (point) :: vector
 real :: z
end type vector
type(vector) :: threed
namelist/nl/threed
read(5, nl)
write(6, nl)
end
```

The inputs may be as follows:

```
        &nl threed= 2.0 3.0 4.0 /
or,     &nl threed%point= 2.0 3.0
        threed%z= 4.0 /
or,     &nl threed%point%x= 2.0
        threed%point%y= 3.0
        threed%z= 4.0 /
```

8.29 NAMELIST Comment

`Namelist` comment starts with an exclamation sign (`'!'`) after the value separator. The comment continues until the end of the input line. As such a slash (`'/'`) after the exclamation sign in the same line does not terminate the input record; rather it becomes a part of the comment.

```
        &nl a=2 b=3 ! this is a comment
        /
or,     &nl ! this is a comment
        a=2 b=3 /
```

Comments are ignored and are used for documentation.

8.30 Rules for NAMELIST

There are some restrictions on the type of the list elements used with the `namelist`. Some of them have already been discussed.

- It cannot be an allocatable array (Chapter 15).
- It cannot be a pointer (Chapter 16).
- It cannot be dummy array with non-constant bound.

- It cannot be a character array with non-constant character length.
- If one of the `namelist` group-variable is public, all other variables must be public (Chapter 12).

8.31 Processor Dependency

The results of list-directed and namelist input/output are processor dependent. In addition, the width of the field, number of decimal places and the width of the exponent field for g0 edit descriptor are processor dependent.

9

Auxiliary Storage

The output of a program may be the input of another program. In such cases, it is convenient to store the output of the first program to some external device (or disk) so that it is available even after the end of the job. Subsequently, the same can be read by the second program from the same external device knowing the arrangement of the data on the output device. Sometimes, it is not convenient to supply the input to a program from the keyboard during the execution. In such cases, the data may be supplied from some external device. The creation of such data on the external device may be accomplished by some other software, not necessarily by Fortran. We first describe a few terms that are required for handling external files.

9.1 Record

A record is a collection of related objects. For example, an employee record may consist of a name of a person, his address, age, sex, salary, date of joining and date of retirement. There are three types of records: formatted, unformatted and endfile.

9.2 File

A file is a collection of records. For example, a collection of employees' records containing information about all the employees of an institution makes an employee file. Three types of files are available in Fortran: sequential, direct and stream.

9.3 Formatted Record

A formatted record consists of a string of characters as permitted by the system (usually ASCII characters). The record length of such a record is the total number of characters present in the record. Within a Fortran program, formatted records are generated through a formatted output (write/print) statement. They are read by a formatted input (read) statement. A formatted record may be created by a program other than a Fortran program. A record having a length equal to zero is permitted.

9.4 Unformatted Record

Unformatted records are processor dependent; they are read or written by an unformatted `read` or `write` Fortran statement. The length of the unformatted record is usually calculated as the total number of bytes required to store the record. A record having a length equal to `zero` is permitted.

9.5 Endfile Record

This is a special type of record. It signals the end of a file of a sequential file. The end of file record is the last record of a file. There is no length property associated with the `endfile` record.

9.6 Sequential File

A sequential file can be read sequentially. It is not possible to read the 10th record without reading the first 9 records. It is also necessary to create a file sequentially. All the records are either formatted or unformatted. These two forms of records cannot be mixed. The last record may be an end of file record.

9.7 Direct File

A direct file can be read in any order, that is, one can read directly the 10th record without having to read the first 9 records. Similarly, it is possible to create the 10th record without creating the first 9 records. The record size of all records of a direct file is the same. List-directed input/output, namelist input/output and non-advancing input/output are not allowed in direct files. All the records are either formatted or unformatted. These two forms of records cannot be mixed.

9.8 Stream File

In a steam-oriented input/output operation, a record is read or written in term of bytes or some integer multiples of bytes. This is discussed in Section 9.25.

9.9 Unit Number

A file is accessed within a Fortran program through a unit number. The unit number can be any positive integer between 1 and 99.

9.10 Scratch and Saved Files

A file created within a program may be temporary or permanent. A temporary file, normally called a scratch file, is deleted from the system when the job ends. A scratch file does not require any name. A permanent file, on the other hand, remains available on the external device after the job ends. This file must have a name—either supplied by the programmer or automatically supplied by the system.

9.11 OPEN, CLOSE and INQUIRE Statements

In this section, we discuss three statements: `open`, `close` and `inquire`.

An `open` statement makes a connection between the unit number and the external file. Some files such as keyboard (input), screen (output) and error file are usually pre-connected to some fixed unit numbers. These files are not to be opened explicitly. Usually, unit 5 is connected to the keyboard and unit 6 is connected to the screen.

A `close` statement is used to terminate in a tidy manner the connection between the unit number and the external file.

An `inquire` statement is used to make an inquiry about a file.

The syntax of all the three statements is similar:

```
open(unit=unit no, olist=value)
close(unit=unit no, olist=value)
inquire(unit=unit no, olist=value)
```

where `olist` is an optional specifier. The optional specifiers are discussed next. Not all the optional specifiers are available to all the three statements. Again, some specifiers are dependent on other specifiers.

9.12 Optional Specifiers

`UNIT=unit-number (open, inquire, close)`

As already mentioned, the unit number establishes a connection between the Fortran program and the file. All transactions take place through the unit number; `unit=unit-number` may be replaced by just the unit number. Normally, the unit number is the first item of `olist`.

```
open(unit=10, .....)
close(unit=20)
inquire (30, ....)
```

NEWUNIT=integer (open)

The processor can choose a suitable unit number while opening an external file.

```
open(newunit=integer variable, ....)
```

The integer variable is assigned to an integer by the processor, which becomes the unit number of the external file if the open statement is successfully executed.

```
integer :: iu
open(newunit=iu, ....)
```

FILE=char-variable (open, inquire)

The character string is the name of the file to be connected with the unit. If the status= 'scratch', this specifier is not required—the system uses a processor-dependent file name.

```
open(unit=10,file='a.dat')
```

For a successful execution of the open statement, unit 10 will be connected with the file 'a.dat'. For an inquire statement, file= is used to make an inquiry about a file. It is not necessary to connect the file and open the file either.

```
inquire (file='t37.f95', ...)
```

Both file= and unit= cannot be present in the inquire statement.

status=char-variable (open, close)

The character string can be 'old', 'new', 'scratch', 'replace' or 'unknown' for an open statement. If 'new' is specified, the file must not be present. If 'old' is specified, the file must be present. If 'scratch' is specified, the file is created and subsequently deleted at the end of the job. In this case, file= must not be present. If 'replace' is specified and the file is present, the existing file is deleted and a new file is created; if the file is not present, a new file is created like 'new'. If 'unknown' is specified, the status becomes processor dependent.

```
open(unit=10, file='a.dat', status='replace')
```

For the close statement, the character string may be either 'keep' or 'delete'. If the status='keep', the file is saved after the unit is closed. If the status='delete', the file is deleted when the unit is closed. As the scratch file is always deleted when the unit is closed, it should not have a status attribute during the close operation. The default status for this statement is 'keep'. It is obvious that for a scratch file, the default status is 'delete'.

ACCESS= char-variable (open, inquire)

The character string can be 'sequential', 'direct' or 'stream' depending on the access during the opening of the file. The default is 'sequential'. For an existing file, the access must be an allowed value. For an `inquire` statement, if no file is connected, it returns `'UNDEFINED'`.

```
character (len=10) :: acc
open (20, access='sequential')
inquire (20, access=acc)
```

If the unit `20` is opened in `sequential` mode, the `inquire` statement will set the character variable `acc` to `'SEQUENTIAL'`.

ACTION= char-variable (open, inquire)

The character string can be `'read'`, `'write'` or `'readwrite'` during the opening of the file. If the action is `'read'`, the file is opened in read-only mode, and as such `write`/`print` and `endfile` statements cannot be used on that file. If the action is `'write'`, a read operation is not permitted for such connection. If `'readwrite'` is the value of the action, there is no restriction. If the action is omitted, the default value is processor dependent. The default is usually `'readwrite'`.

```
character (len=10) :: acc
open(10,action='write')
inquire (10, action=acc)
print *, acc
```

If the file is not connected, the `inquire` statement returns a value `'UNDEFINED'`.

FORM= char-variable (open, inquire)

The character string can be `'formatted'` or `'unformatted'` during the opening of the file. If this is omitted, it is `'formatted'` for a sequential file and `'unformatted'` for the direct or stream access.

```
character (len=10) :: acc
open(10,access='sequential',form='formatted',file='f.dat',status &
     ='new')
inquire(10, form=acc)
print *, acc
end
```

The `inquire` statement returns `'FORMATTED'` and stores this value in `acc`.

ASYNCHRONOUS= char-variable (open, inquire)

The character string can be `'yes'` or `'no'` during the opening of the file. If it is `'yes'`, asynchronous input/output is allowed in the corresponding unit. The default is `'no'`. The `inquire` statement returns the status `YES` or `NO` depending on the status of the unit.

BLANK= char-variable (open, inquire)

The character string is either 'null' or 'zero'. The default is 'null'. The specifier blank can be used only for the formatted input. It determines how blanks during the numeric input are to be interpreted. When the value is 'null', blanks are ignored. However, if the field is completely blank, it is taken as zero. When the value is 'zero', blanks will be treated as zero. This has no effect on the output. The inquire statement returns the status NULL or ZERO depending on the setting.

```
character (len=10) :: acc
open(10,blank='zero')
inquire (10, blank=acc)
print *, acc
end
```

DECIMAL=char-variable (open, inquire)

The character string is either 'comma' or 'point'. This specifier can be used only with the formatted input/output. The default value is 'point'. If the file is opened with decimal= 'comma', the decimal points are replaced by a comma. If the unit 10 is opened with decimal= 'comma', the file created by the following program

```
     real :: r=10.2
     open (10, file='a.dat',status='new', decimal='comma')
     write(10, 20)r
20   format(f8.3)
     end
```

will be bb10,200, where b represents the blank.

However, if decimal= 'point' is absent or omitted, the file will contain bb10.200.

In the preceding program, if an inquiry is made about the specifier DECIMAL, COMMA will be returned.

```
     real :: r=10.2
     character (len=15) :: acc=" "
     open (10, file='a.dat',status='new', decimal='comma')
     write(10, 20)r
20   format(f8.3)
     inquire(10, decimal=acc)
     print *, acc
     end
```

Consider the following two programs:

```
complex :: c
namelist/nl/c
open(unit=5, decimal='comma')
read(5,nl)
write(6,nl)
end
```

If the file is opened with decimal='comma', the input to the read statement for a complex variable under the control of namelist should have a semicolon as the separator:

```
&nl c=(2,0 ; 3,0)/
```

Similarly, for a list-directed input statement,

```
real :: a,b,c
open(unit=5, decimal='comma')
read *, a,b,c
print *, a,b,c
end
```

the values corresponding to the read statement are separated by semicolons instead of commas (blanks, newlines, etc. may also be used). The input to the preceding program may be as follows:

```
      2,0 ; 3,0; 4,0
or    2,0 3,0 4,0
or    2,0
      3,0
      4,0
```

If there is no connection between the unit number and the file or the file is not opened for formatted input/output, inquire will return the value of the specifier as 'UNDEFINED'.

DELIM= char-variable (open, inquire)

The character string may be 'apostrophe', 'quote' or 'none'. The specifier is permitted for formatted output statements. This specifier is the current value of the delimiter that will be used to delimit character constants for the list-directed or namelist output operation. If 'none' is specified, the characters will not be delimited by any character (apostrophe or quote). The default value is 'none'.

```
character (len=10) :: c='BENGAL'
character (len=10) :: acc
open(10, access='sequential', form='formatted', &
    delim='quote', file='delim.dat', status='new')
write(10,*)c
inquire(10, delim=acc)
print *, acc
end
```

The output file delim.dat will have a record 'BENGALbbbb', where b represents the blank.

The inquire statement with delim=char-variable returns the value of the delim specifier. If there is no connection between the unit and the file or form is not 'formatted', 'UNDEFINED' is returned. In the preceding program, the value of acc will be 'QUOTE' as the file has been opened with delim='quote'.

DIRECT= char-variable, STREAM=char-variable (inquire)

The character string can have the value 'yes' or 'no' depending on whether a direct access mode or a stream access mode, respectively, is allowed for the file or not. It returns 'UNKNOWN' if the processor cannot determine the exact status.

ENCODING= char-variable (open, inquire)

This specifier is allowed only for formatted input/output. It can assume two values: 'utf-8' (Unicode) and 'default'. If the file is opened without this specifier, the default value is 'default'.

ERR= label (open, inquire, close)

If any error condition occurs during the execution of an open, inquire or close statement, the control is passed to the statement label mentioned along with err.

```
open (10, err=100, ....)
```

Control will be passed to the statement label 100, if an error is detected while opening the file. If err=label (or iostat=) is not present, the job is terminated.

EXIST=logical variable (inquire)

This checks whether the file specified by the specifier file is available on the disk. It returns true if the file exists and false otherwise.

FORMATTED= char-variable, UNFORMATTED= char-variable (inquire)

On inquiry, the character string would contain 'YES' or 'NO' depending on the form specifier and 'UNKNOWN' if the processor is unable to determine the status.

ID= integer (inquire)

It is the identifier of a pending data transfer. It interacts with the specifier pending.

IOMESG= char-variable (open, inquire, close)

In the case of any error or end of file or end of record occurring during the execution of an I/O statement or with an open, inquire or close statement, the processor sets some explanatory message to this variable; it remains unaltered if there is no error (eof or eor). Consider the following program:

```
character (len=40) :: im='xxxxx'
integer :: is
open(10, file='form.dat', status='new', iostat=is, iomsg=im)
print *, im
end
```

Let us assume that a file form.dat already exists in the disk. The open statement would then be unsuccessful because status= 'new' is expected to create a new file. If the file exists, an error condition results. Now, im is initialized to 'xxxxx'. When there is an error condition, the processor will change this string of characters by some explanatory message, which is, in this case, "new file already exists." Note that in this case, the iostat= specifier must be present.

IOSTAT= integer variable (open, inquire, close)

During the opening of a file, 0 is returned if no error condition occurs. The specifier becomes a processor-dependent positive number if an error occurs. In case of an error, if neither `iostat` nor `err` is present, the job is terminated. For the successful execution of `inquire` and `close` statements, `iostat` returns 0.

NAME= char-variable (inquire)

On inquiry, the character string would contain the name of the file or `'UNKNOWN'`.

```
character(len=15) :: ch
open(25,file='seqf.dat',status='replace', form='formatted')
inquire (25,name=ch)
```

NAMED=logical variable (inquire)

On inquiry, the logical variable is T (`true`) if the file has a name. It is F (`false`) otherwise.

NEXTREC= integer variable (inquire)

This is used for direct I/O. Its value is `last record` +1. If the file does not contain any record, it is 1. If it is not a direct file, it is undefined.

NUMBER= integer variable (inquire)

The unit number assigned to a file (`file=`) is returned to the integer variable. If no unit is connected to the file, -1 is returned (the file name may be case sensitive) for the NAG compiler. The file name written in the `open` statement must be exactly the same as the file name given in the `inquire` statement.

```
integer :: n
open(25,file='sEQF.DAT', status='replace', form='formatted')
.
inquire(file='sEQF.DAT', number=n)
```

Here, n will be 25. However, if `file= 'SEQF.DAT'`, n will be -1 (note that in this case, `'sEQF.DAT'` would not be considered the same as `'SEQF.DAT'`).

OPENED= logical variable (inquire)

On inquiry, the logical variable is true if the corresponding file is open and false otherwise.

```
logical :: l
open(25,file='seqf.dat', status='replace', form='formatted')
.
inquire(25, opened=l)
print *, l
```

Here, l is true. The `inquire` statement may of the form

```
inquire(file='seqf.dat', opened=l)
```

PAD=char-variable (open)

The character string can be `'yes'` or `'no'`. This is available only for formatted I/O. The default value is `'yes'`. If the value is `'yes'`, blank characters are used for padding, and the list and format specification combination requires more characters than the number of characters available in the record. If the value is `'no'`, the number of characters in the record must not be less than the number of characters required for the list and format specification. (The exception is when `advance= 'no'` is specified or `eor=` or `iostat=` specification is present.)

PENDING (inquire)

This specifier is used to find out whether previously initiated asynchronous data transfer is over (complete). The logical variable associated with `pending` is true if the transfer is not over and false if the transfer is complete.

POS=integer (inquire)

This is used with stream I/O. It returns the number of file storage units immediately following the current position. If the file is not connected, it is undefined.

POSITION=char-variable (open, inquire)

The character string can be `'asis'`, `'rewind'` or `'append'`. This is available for sequential or stream access. When a file is created, it is positioned at the beginning; `'rewind'` positions the file at the beginning of the file, and `'append'` positions the existing file just before the end-of-file mark. If there is no end-of-file mark, the file is positioned at the end of the last record. If `position= 'asis'` and the file exists and is connected, the position remains unaltered. However, if the file exists and is not connected, `'asis'` leaves the file position unspecified. `Inquire` returns `position= 'UNDEFINED'` if the file is not connected.

READ=char-variable, WRITE=char-variable, READWRITE=char-variable (inquire)

All these three specifiers are similar in nature. If the `access` matches with the specifier, the value of the character string is `'YES'`; if it does not match, the value is `'NO'`. The value is `'UNKNOWN'` if the processor is unable to determine.

```
character(len=15) ::ch
open(25,file='s.dat',status='replace',form='formatted', &
   action='write')
inquire(25, write=ch)
print *, ch
```

Since `action= 'write'`, the value of `ch` will be `'YES'`.

RECL=integer (open, inquire)

The positive integer indicates the size of each record of a direct I/O. For sequential access, if this is not specified, the default value is dependent on the processor. If there is no connection, the `inquire` statement returns `-1`, and for stream access, it is `-2` in Fortran 2018.

ROUND=char-variable (open, inquire)

The character string can be `'up'`, `'down'`, `'zero'`, `'nearest'`, `'compatible'` or `'processor_defined'`. This specifier is applicable for formatted I/O (just for I/O). This specifies the I/O rounding mode.

SEQUENTIAL=char-variable (inquire)

The char-variable is set to 'YES' if the file is opened with access= 'sequential', 'NO' otherwise and 'UNKNOWN' if the processor is unable to determine.

SIGN=char-variable (open, inquire)

The character string can be 'plus', 'suppress' or 'processor_defined'. This is available for formatted I/O. If it is 'plus', a '+' sign is included for positive numbers. The default is 'processor_defined', which is usually 'suppressed', that is, '+' sign is suppressed when the number is positive. For inquire, if the unit is not connected, the char-string is set to 'UNDEFINED'.

END=label

If an end of file is detected during the input, the control is passed to the statement label mentioned along with end. If end of file occurs and the end=label is omitted, the job is terminated.

SIZE=integer (inquiry)

On inquiry, the integer contains the size of an existing disk file. If the processor cannot determine the size of the file, -1 is returned.

```
integer :: n
inquire (file='xxa.dat', size=n)
print *, n
```

Before we end this section, here is a complete list of specifiers for open, inquire and close statements.

The I/O control list of the open statement is summarized as follows:

Unit = unit number of the file, integer	Form = type of format, character
Access = access mode, character	Iomsg = additional error message related to file opening, character
Action = action on the file, character	Iostat = status of the open operation, integer
Asynchronous = asynchronous I/O, char	Newunit = unit number assigned by the system, integer variable
Blank = blank handling during input, character	Pad = status of pad mode in formatted I/O, character
Decimal = decimal edit mode, character	Position = file position, character
Delim = delimiter for character, character	Recl = length of the record, integer
Encoding = character encoding—UTF-8 or ASCII	Round = rounding mode, character
Err = label	Sign = handling of leading sign, character
File = name of the file, character	Status = file status, character.

The I/O control list of the `inquire` statement is summarized as follows:

`Unit` = unit number of the file, integer	`Named` = returns if the file has a name, logical
`File` = file name, character	`Nextrec` = next record number in direct I/O, integer
`Access` = access mode, character	`Number` = unit number assigned to a file, integer
`Action` = action on the file, character	`Opened` = checks if the file is open. logical
`Asynchronous` = asynchronous I/O, character	`Pad` = pad mode in formatted I/O, character
`Blank` = blank handling during input, character	`Pending` = asynchronous data transfer pending, logical
`Decimal` = decimal edit mode during I/O, character	`Pos` = current position in stream I/O, integer
`Delim` = delimiter for character, character	`Position` = file position, character
`Direct` = direct file indicator, character	`Read` = read allowed on the file, character
`Encoding` = character encoding—UC or ASCII	`Readwrite` = read and write allowed on the file, character
`Err` = label	`Recl` = record length, integer
`Exist` = existence of file, logical	`Round` = rounding mode, character
`Form` = type of format, character	`Sequential` = sequential mode permitted, character
`Formatted` = check for formatted file, character	`Sign` = mode of handling of leading sign, character
`Id` = identifier for pending data transfer, integer	`Size` = size of file in terms of storage unit, integer
`Iomsg` = extra error message related to I/O, character	`Stream` = check for stream mode, character
`Iostat` = status of I/O operation, integer	`Unformatted` = check unformatted I/O, character
`Name` = returns the name of the file, character	`Write` = check for the `write` operation on the file, character

The I/O control list of the `close` statement is summarized as follows:

`Unit` = unit number of the file, integer
`Iostat` = status of I/O operation, integer
`Iomsg` = additional error message related to I/O, character
`Err` = label
`Status` = disposition of the file, character

9.13 Kind Type Parameters of Integer Specifiers

Fortran allows integer specifiers such as nextrec and number to be of any type (kind) of integer.

9.14 ENDFILE Statement

The syntax of an endfile statement is as follows:

```
        endfile unit-no
like,   endfile 10
```

This statement writes a special record called 'end-of-file' on a file opened for sequential access. The file is positioned after the end of file record. The end-of-file record is generated if the statement after the write operation is rewind or backspace or the file is closed.

9.15 REWIND Statement

The rewind statement brings the record pointer to the beginning of a file. If the record pointer is at the beginning, the statement has no effect. The syntax of rewind is as follows:

```
        rewind unit-no
like,   rewind 25
```

9.16 BACKSPACE Statement

The backspace statement backspaces one record of the specified unit. If the file pointer is at the beginning of the file, no action is taken. If the file pointer is just after the end of file mark, it is positioned before the end-of-file mark. Sometimes, the backspace statement writes an end-of-file record before backspacing. In that case, the record pointer is positioned before the record that precedes the end-of-file record. An example of backspace statement is as follows:

```
        backspace unit-no
like,   backspace 10
```

where 10 is the unit number. It is prohibited to backspace a file that is connected but does not exist; backspace is not allowed for a unit that is being written through namelist of list-directed formatting.

9.17 Data Transfer Statement

The read statement is used to read data from a file, and the print or write statement is used to write data on a file. The syntax of the read statement is as follows:

```
        read(unit-no, io-control-list) list
or,     read format, list
```

The syntax of print and write statements is as follows:

```
        print format, list
        write(unit-no, io-control-list) list
```

9.18 READ/WRITE Statement

The read format, list and print format, list have already been discussed in Chapter 8. The following are the io-control-list of read and write statements:

UNIT= unit-no, integer (read, write)

This is used to indicate the unit number on which the input/output operation will be performed. The file is opened by either an open statement or a write statement, and the unit number is pre-connected with some file. If the unit number is '*', it is usually 5 for read statement and 6 for write statement. However, just a '*' without any format specifier indicates a list-directed input/output operation.

FMT=char-variable/expression or statement label or '*' (read, write)

This specifies the format of the input/output statement. It can be a statement number corresponding to the format statement or a character variable containing the format specification or a character constant or a '*' indicating free-formatted input/output.

```
        read (unit=5, fmt=10) a, b, c
        read (unit=5, fmt='(i6)') ia
        write(unit=6, fmt=*) a, b, c
```

NML=char-variable, NAMELIST group name (read, write)

This specifies the namelist group name. This must be associated with the input/output statement involving namelist.

ASYNCHRONOUS= char-variable

This has been discussed in section 9.19.

ADVANCE=char-string (read, write)

This specifier can have a value 'yes' or 'no'. During input, if advance='no', the leftover data in the input buffer is read by the next statement. The default is 'yes'. In that case, the leftover data in the input buffer is ignored.

```
      read(5,10,advance='no')i,j
      read(5,10)k,l
10    format(4i1)
      print *, i,j,k,l
      end
```

If four numbers 1234 are given as input, the first read statement sets i=1 and j=2. As advance='no', the next read statement reads 3 and 4 from the same input buffer and assigns k to 3 and l to 4.

If advance='no' is used with the write statement, the next write statement does not start a fresh line—instead, it uses the same line. Suppose it is desired to have an output like the following:

```
9 9 9 9 9 9 9 9 9
8 8 8 8 8 8 8 8
7 7 7 7 7 7 7
6 6 6 6 6 6
5 5 5 5 5
4 4 4 4
3 3 3
2 2
1
1
2 2
3 3 3
4 4 4 4
5 5 5 5 5
6 6 6 6 6 6
7 7 7 7 7 7 7
8 8 8 8 8 8 8 8
9 9 9 9 9 9 9 9 9
```

Using advance='no' with the write statement, it is easy generate such output as follows:

```
      implicit none
      integer :: i,j
      do i=9,1,-1
        do j=1,i
          write(6,10,advance='no')i ! carriage motion is suppressed
10        format(i2)
        end do
      write(6,10) ! dummy write. go to the next line
      end do
      do i=1,9,1
        do j=1,i
          write(6,10,advance='no') i
        end do
      write(6,10)
      end do
      end
```

BLANK=char-string (read)

The interpretation of blank in the input (numeric) is controlled by this specifier. It can assume a value 'null' or 'zero'.

DECIMAL=char-string (read, write)

This has already been discussed.

DELIM=char-string (write)

This has already been discussed.

END=label (read, write)

If an end-of-file condition is sensed, the program branches to the label specified with the end=. If this specifier or iostat is not present, the job is aborted.

EOR=label (read, write)

If an end-of-record condition is sensed, the program branches to the label specified with the eor=. If this specifier or iostat is not present, the job is aborted.

ERR=label (read, write)

If an error condition is sensed, the program branches to the label specified with the err=. If this specifier or iostat is not present, the job is aborted.

ID=integer variable (read, write)

The processor allocates an identifier id to the integer variable, which may subsequently be used to test the status of the asynchronous data transfer using the pending specifier.

IOMSG=char-variable (read, write), IOSTAT=integer-variable (read, write), PAD=char-string (read), ROUND=char-string (read, write), SIGN=char-string (write), SIZE=integer (read):

These have already been discussed.

POS=integer (read, write)

This specifies the file position in the file storage unit (specified on iso_fortran_env). The file must be opened with access='stream'.

REC=integer (read, write)

This specifier is the record number to be read or written for a direct access file.

The specifiers associated with the read and write statements are summarized as follows:

- UNIT = unit no, integer
- FMT = format
- NML = namelist group name

- ADVANCE = char-string
- ASYNCHRONOUS = char-string
- BLANK = char-string (read only)
- DECIMAL = char-string
- DELIM = char-string (write only)
- END = label
- EOR = label
- ERR = label
- ID = integer variable
- IOMSG = char-variable
- IOSTAT = integer variable
- PAD = char-string (read only)
- POS = integer
- REC = integer
- ROUND = char-string
- SIGN = char-string (write only)
- SIZE = integer variable (read only)

9.19 Asynchronous Input/Output

In the synchronous input/output operation, the system waits until the operation is completed. The asynchronous input/output, on the other hand, continues to execute following instructions, and input/output operations proceed asynchronously. It is the responsibility of a programmer not to use the data before the transfer is over, and the programmer is supposed to check the status of the I/O operation before using the data.

To transfer data in asynchronous mode, the file must be opened with asynchronous= 'yes'.

```
open(unit=20, asynchronous='yes',.........)
```

The read and write statements should also have the asynchronous='yes' specifier to initiate the asynchronous mode of data transfer.

```
read(unit=10,asynchronous='yes',...)
write(unit=10,asynchronous='yes',...)
```

Once this type of data transfer is selected, the program does not wait, but it continues its execution. It is thus necessary to ensure that, say, the input operation is completed before the data is used. The statement wait does not allow the program to proceed until the asynchronous data transfer is over.

```
read(unit=10,asynchronous='yes',...)
    .
wait(10)
```

The statement after wait will be executed only after the data transfer is over. The wait statement may have a number of specifiers: unit, end, eor, err, id, iomsg and iostat. These have been discussed in earlier sections. The file-positioning statements, close and inquire, may have an implied wait statement.

There is another way of making an inquiry about the status of the asynchronous data transfer. The read or the write statement may contain an additional specifier id that receives a unique identification (integer) corresponding to the read or write statement.

```
read(unit=10, asynchronous='yes',id=myid,...)
```

Subsequently, through the inquire statement, the status of the data transfer may be determined as follows:

```
inquire(unit=10, id=myid, pending=pnd)
```

where pnd is a logical variable. The inquire statement sets pnd to true if the input/output operation is pending; otherwise, it sets pending to false.

The following program segments summarize all the aforementioned concepts:

```
      integer, dimension(10000,10000) :: buf=100
      integer :: ist
      logical :: pnd
      open (unit=10, asynchronous='yes', action='write', &
      status='new', file='async.dat', access='sequential', iostat=ist)
      write(unit=10, asynchronous='yes', fmt=100)buf
100   format(12i10)
         .
      wait(10)
      close(10)
      end
```

Although it is not mandatory, one may inform the compiler that a variable will take part in the asynchronous input/output data transfer by using the asynchronous attribute with the variable declaration.

```
      integer, asynchronous, dimension(10000,10000) :: buf=100
```

The preceding program creates a file async.dat. The statement after the wait statement cannot be executed until the data transfer is completed. The next program uses the inquire statement to monitor the status.

```
      integer, dimension(10000,10000) :: buf
      integer :: ist, myid
      logical :: pnd
      open (unit=10, asynchronous='yes', action='read', &
        status='old', file='async.dat', access='sequential',&
        iostat=ist)
      read (unit=10, asynchronous='yes', fmt=100, id=myid)buf
100   format(12i10)
         .
```

```
inquire(unit=10, id=myid, pending=pnd)
if(pnd) then
  wait(unit=10)
endif
.
end
```

If pnd is true, the program waits, as the data transfer is not yet over; otherwise, the program continues.

9.20 FLUSH Statement

The flush statement acts on an external file and causes all data to be written on the external file to make it available to other processes. Essentially, flush flushes input/output buffer.

```
flush unit-number
flush 10 or flush(10)
```

The flush statement may have usual specifications, such as iostat, iomesg and err. Once the flush statement is executed, the most recent data is available from the external file.

9.21 Rules for Input/Output Control List

A particular I/O control specifier may appear only once.

- unit = unit number may be written without unit=. If unit= is omitted, the corresponding unit number must be the first item in the I/O control specification list.
- delim and sign can be used only with the read statement.
- blank, pad, err, eor and size can be used only with the read statement.
- The label (statement number) associated with err, eor and end must be in the same scoping unit as the read statement.
- If nml= namelist-group-name is replaced by just the namelist-group-name, it should be the second item in the I/O control list.
- If the unit number is other than a file number, rec and pos, the I/O control must not be present.
- If rec is present, end and namelist-group-name must not be present, and format must not be list directed (asterisk).
- If advance is used for sequential or stream with the format statement, the unit must be an integer (not internal I/O).
- If eor or size is used, advance must appear in the control list.

- If asynchronous='yes', the unit number must be an integer.
- If id appears, asynchronous must be 'yes'.
- If pos is present, io-control-list should not contain rec.
- decimal, blank, pad, sign and round specifiers can only be used with format or namelist-group-name.
- size and eor can be used only with advance='no'.
- For advance, asynchronous, decimal, blank, delim, pad, sign and round, trailing blanks in the list of I/O control are ignored. In addition, the list elements are case independent.

9.22 IS_IOSTAT_END

This function [is_iostat_end(i)] returns true if the end-of-file condition is reached; otherwise, it returns false. Here, i is an integer variable associated with iostat=i.

9.23 IS_IOSTAT_EOR

This function [is_iostat_eor(i)] returns true if the end-of-file record is reached; otherwise, it returns false. Here, i is an integer variable associated with iostat=i.

9.24 Examples of File Operations

1. **Creation of a formatted sequential file**

```
        integer :: a=2
        real :: b=20.5
        character(len=5) :: c='india'
        double precision :: d= 3.1415926589d0
        complex :: e=(2.0,3.0)
        open(25,file='seqf.dat',status='replace',form='formatted')
        write(25,100)a,b,c,d,e
100     format(i5,f8.3,a5,d20.10,2f8.3)
        close(25)
        end
```

2. **Reading a formatted sequential file**

```
        integer :: a
        real :: b
        character(len=5) : :c
        double precision :: d
        complex :: e
```

```
      open(25,file='seqf.dat',status='old',form='formatted')
      read(25,100)a,b,c,d,e
100   format(i5,f8.3,a5,d20.10,2f8.3)
      print *,a,b,c,d,e
      close(25)
      end
```

3. Creation of unformatted sequential file

```
      integer :: a=2
      real :: b=20.5
      character(len=5) :: c='india'
      double precision :: d= 3.1415926589d0
      complex :: e=(2.0,3.0)
      open(25,file='seqf.dat',status='replace',form='unformatted')
      write(25)a,b,c,d,e
      close(25)
      end
```

4. Reading unformatted sequential file

```
      integer :: a
      real :: b
      character(len=5) :: c
      double precision :: d
      complex :: e
      open(25,file='seqf.dat',status='old',form='unformatted')
      read(25)a,b,c,d,e
      print *,a,b,c,d,e
      close(25)
      end
```

5. Creation and reading of formatted direct file

```
      integer :: a=2,p
      integer :: i,irec=1
      real :: b=20.5,q
      character(len=5) :: c='india',r
      double precision :: d=3.1415926589d0,s
      complex::e=(2.0,3.0),t
      open(25,access='direct',file='seqf.dat',status='replace',&
        form='formatted',recl=54)
      do i=1,10
       write(25,100,rec=irec)a+i,b,c,d,e
       irec=irec+1
      enddo
100   format(i5,f8.3,a5,d20.10,2f8.3)
      irec=7
      read(25,100,rec=irec)p,q,r,s,t
      print *,p,q,r,s,t
      close(25)
      end
```

6. **Creation and reading of unformatted direct file**

```
integer :: a=2,p,n
integer :: i,irec=1
logical :: ll
real :: b=20.5,q
character(len=5) :: c='india',r
character(len=30) :: ch1
double precision :: d=3.1415926589d0,s
complex :: e=(2.0,3.0),t
open(25,access='direct',file='seqf.dat',status='replace',&
  form='unformatted',recl=29)
inquire(25,name=ch1)
inquire(25,named=ll)
inquire(file='seqf.dat',number=n)
print *,n; print *,ll; print *,ch1
do i=1,10
   write(25,rec=irec)a+i,b,c,d,e
   irec=irec+1
enddo
inquire(25,nextrec=n)
print *,n
inquire(25,formatted=ch1)
print *,"==========",ch1
inquire(25,size=n)
print *,n
irec=3
read(25,rec=irec)p,q,r,s,t; print *,p,q,r,s,t; close(25)
end
```

No explanations are given for these programs. Readers are requested to read these programs; they can type and run these programs and try to explain the results.

9.25 Stream Input/Output

Input/output operations of standard Fortran are basically record oriented. Although for the direct I/O, the records can be accessed randomly; yet, the record size for this type of I/O is fixed.

Often the situation is quite different. If the file is generated by a non-Fortran program or by some package or by some measuring instrument, the structure of records may not be similar to the standard Fortran. In such a case, the input/output statements discussed in the previous sections are not suitable for data transfer. Fortunately, Fortran provides an alternative way, called stream I/O, for accessing these types of records.

9.26 Storage Unit of Stream Input/Output

In case of stream I/O, the basic storage unit is a byte. The system keeps track of the number of bytes required for a particular type of variable—four for standard integer and real, eight for complex variable and so on. An internal counter is maintained containing the byte position of the next item to be read/written on/from the file. This is incremented automatically when a variable is written to the file, and the increment takes place according to the type of the variable. The next item is written at the current position of the record pointer. Similarly, by setting a value to the counter, the corresponding item may be fetched randomly from the file. When the file is opened, the counter is set to 1. The concept of record ceases to exist for files created by stream I/O.

9.27 Stream Input/Output Type

Stream I/O can be of two types: formatted and unformatted. Unformatted stream I/O is more convenient than formatted stream I/O. We will first discuss the unformatted stream I/O.

9.28 Stream File Opening

The file to be created for stream I/O is opened in the following manner:

```
open(10,file="stream.dat",access="stream",status="new",form="unformatted")
```

If an existing file created earlier is to be opened, the corresponding open statement is

```
open(10,file="stream.dat",access="stream",status="old", form="unformatted").
```

Access is specified as "stream", meaning that the file will be accessed in stream mode. The default form of stream I/O is unformatted, so in the preceding case, form="unformatted" is redundant.

9.29 Unformatted Stream File

This is illustrated with the following example:

```
program streamio
real :: r=2.0,r1
integer :: a=10,ppos
character(len=7) :: ch="SUBRATA"
character(len=4) :: ch1
double precision :: d=3.141592658d0
complex :: c=(2.0,3.0)
open(10,file="stream.dat",access="stream",status="replace")
write(10)r,a,ch; write(10)d,c
inquire(10,pos=ppos); inquire(10,access=ch)
print *,ch; print 20, ppos-1
```

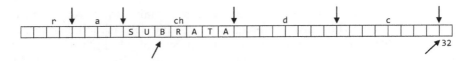

FIGURE 9.1
Output of stream `write`.

```
20      format('Total no of bytes written  ',i3)
        read(10,pos=11)ch1; print *,ch1; read(10,pos=28)r1; print *,r1;
        close(10)
        end
```

The outputs from the program are as follows:

```
STREAM
Total no of bytes written   31
BRAT
3.00000000
```

Let us try to understand the output.

The `write` statement will create a file containing 4 bytes each for `r` and `a`, 7 bytes for `ch`, 8 bytes of for `d` and 8 bytes for `c`. This is depicted in Figure 9.1.

The `inquire` statement will return 32 as it is the next position where writing can take place. Having set `pos` to 11, the `read` statement reads `"BRAT"` and stores it in the location `ch1` (the length of the `ch1` being 4); `r1` reads the last four bytes whose value is 3.0, the imaginary part of `c`.

9.30 Formatted Stream I/O

In the formatted stream I/O, the file must be opened with `form="formatted"`, and I/O operations are performed with the `format` statement. The difficulty in the formatted stream I/O is that each `write` statement adds a record terminator, which is machine dependent. It may be a carriage return or line feed or both or some other character. Therefore, the so-called random reading/writing is difficult to perform as counting of bytes may not be easy because of the presence of additional record terminating character(s). However, `pos=1` when the file is opened, and the value of `pos` may be determined by an `inquire` statement. Formatted stream file I/O is rarely used.

```
        real :: r=2.0
        integer :: a=10,ppos
        integer*1 y
        character(len=7) :: ch="SUBRATA"
        character (len=1) :: x
        double precision :: d=3.141592658d0
        complex :: c=(2.0,3.0)
        open(10,file="stream.dat",access="stream",form="formatted",&
         status="replace")
```

```
       write(10,40)r,a,ch
40     format(f8.3,i3,a7)
       write(10,21)d,c
21     format(d20.10,2e16.6)
       inquire(10,pos=ppos)
       print 20, ppos-1
20     format('Total no of bytes written   ',i3)
       ppos=1         ! start from the beginning
       do i=1,74
         read(10,23,pos=ppos)x ! read one character at a time in a format
         ppos=ppos+1 ! increment the record pointer
23       format(a1)
         write (6,24,advance='no')x ! print the character in a format
24       format(a1)
       enddo
       close(10)
       end
```

The outputs are as follows:

```
Total no of bytes written    74
2.000 10SUBRATA        0.3141592658D+01      0.200000E+01      0.300000E+01
```

The output shows 74, that is, 4 additional characters have been added $(8+3+7+20+2*16=70)$. We have read the same file byte by byte and have printed the record character by character using 'a' edit descriptor.

9.31 Rule of Thumb

The choice of file structure obviously lies with the programmer. However, a rule of thumb may be formulated. Formatted I/O requires conversion (binary to ASCII and ASCII to binary), whereas unformatted I/O does not require any conversion because the data as it is laid out in the computer memory is written directly to disk.

If the output of a program becomes the input to another program and that too in a different machine, formatted I/O should be used. It is difficult to read the binary file if it is created in a machine of different type, for example, the representation of a number may be laid out in a different arrangement of bits. Unformatted I/O is recommended so long as I/O operations are performed within the same machine. It is faster because no conversion takes place. All scratch files normally use unformatted I/O.

The choice of sequential or direct (random) I/O is governed by the nature of the problem. If the data is to be read sequentially, a sequential file is used. When the records are accessed in random, direct file is preferred.

The choice between sequential, direct and stream I/O is dictated by the nature of the file. If the file is created by a Fortran program, sequential or direct I/O is preferred. If the file is created by non-Fortran program/package/some instrument having a record structure different from Fortran, most of the time, such file cannot be properly read by the standard Fortran read statement. In such a case, stream I/O may be an alternative and should be tried although it is necessary to understand the structure of the data to make use of it later.

9.32 Recursive Input/Output

If an input/output operation initiates another input/output operation, it is called a recursive input/output operation. Under certain conditions, Fortran permits recursive I/O.

9.33 Processor Dependencies

The default values of action=specifier, encoding='default', file connected for status='scratch', record=specifier, status='unknown', recl=specifier in the open statement are processor dependent.

The name of the file may be case sensitive under Unix (or Linux) even under Windows. The default rounding mode, sign mode, the effect of rounding mode when it is equal to 'processor_defined' are not specified in the Fortran Report. In addition, it is not specified when the rounding mode is nearest, and the left-value and right-value are equidistant which value is taken by the processor. The value assigned to the iostat=specifier or to the iomsg=specifier when an error is detected is processor dependent.

10

Numerical Model

This chapter deals with the numeric model of Fortran. A thorough understanding of the numeric model and the intrinsics associated with it is required to write truly processor-independent computer programs. First, we define a few terms that will be used to define the model.

Bit: Bit stands for binary digit. It can have a value either zero or one.

Byte: Byte is a collection of bits. Eight bits make one byte.

Storage: A variable or constant may occupy one byte or more than one bytes depending upon its type. Each byte has a unique address within the system.

Bit Numbering: Bits are numbered from right to left within a byte or combination of bytes. The rightmost bit of a byte or a combination of bytes is numbered as zero. If four contiguous bytes are joined to store a real number, bits are numbered 0 to 31 from right to left.

10.1 Numerical Model for Integers

An integer is stored as a binary number. The value of such a binary number is given by

$$n = s \sum (k=1,n) \; w_k \; . \; b^{(k-1)}$$

where k is the total number of bits required to represent the integer, b is the radix or base that is 2 (binary) and w's are integers such that 0 <= w <b. Naturally, where b is 2, w can be either 0 or 1, and s is either +1 or -1, depending on the sign of the number.

10.2 BASE

The base or radix can be determined by intrinsic `radix`. The argument is either an integer or a real number. It may be a scalar or an array valued. The base corresponding to an integer is returned by the intrinsic.

```
integer :: i
print *, radix(i)    ! prints 2 as the base of the number system
```

10.3 Largest Integer

Intrinsic huge returns the largest integer available corresponding to the type of the argument. The largest integer for a particular kind is $2^{(k-1)}-1$, where k is the total number of bits allocated to the integer (the leftmost bit is a sign bit; it is 0 if the number is positive and 1 if the number is negative). The print statements print *, huge(0_1); print *, huge(0_2); print *, huge(0_3); print *, huge(0_4) will print 127, 32767, 2147483647 and 9223372036854775807 corresponding to kind=1, 2, 3 and 4 for the NAG compiler. These numbers correspond to 2^7-1, $2^{15}-1$, $2^{31}-1$ and $2^{63}-1$. A better way is to use the integer parameters corresponding to different kinds indirectly.

```
integer, parameter :: vsmall=selected_int_kind(2)
integer, parameter :: small=selected_int_kind(3)
integer, parameter :: medium=selected_int_kind(5)
integer, parameter :: big=selected_int_kind(10)
integer(vsmall) :: a
integer(small) :: b
integer(medium) :: c
integer(big)::d
print *,huge (a); print *,huge (b); print *,huge (c); print *,huge (d)
print *,huge(0_vsmall); print *, huge(0_small)
print *,huge(0_medium); print *, huge(0_big)
end
```

10.4 DIGITS for Integers

Intrinsic digits returns the number of bits (without the sign bit) the system allocates to represent an integer. Again, using the variables defined earlier, we find that

```
print *,digits(a); print *,digits(b); print *,digits(c)
print *,digits(d)
```

displays 7, 15, 31 and 63, respectively.

10.5 RANGE for Integers

Range returns the exponent range that may be fully represented in this model. Using the variables defined in Section 10.3, print *, range(a); print *, range(b); print *, range(c); print *, range(d) display 2, 4, 9 and 18, respectively. We have already noted that huge returns the largest integer available in the system corresponding to its argument. Now consider huge(c). This number is 2147483647. Thus,

```
10ˣ = 2147483647
```
or
```
x = log₁₀ 2147483647
```

which is 9 when truncated to the nearest integer. Similarly, $\log_{10}127$, $\log_{10}32767$ and $\log_{10}9223372036854775807$, when integerized by truncation, result in 2, 4 and 18, respectively. In other words, range returns int($\log_{10}$(real(huge(x)))).

10.6 Real Numbers

A real number is represented as

```
X = 0
```

or

```
= s.bᵉ. ∑ (k=1,p) fₖ.b⁽⁻ᵏ⁾ = s.bᵉ. (1/2+∑(k=2,p)) fₖ.b⁽⁻ᵏ⁾
[f₁ is always 1 which is 1.2⁽⁻¹⁾ = 1/2]
```

where b and p are integers and are greater than 1, f_k is a non-negative integer and is less than b, f_1 is non-zero, s is +1 for a positive number and -1 for a negative number and e (integer) satisfies $e_{min} \le e \le e_{max}$, where e_{min} and e_{max} are system-dependent minimum and maximum values of the exponent. This can be achieved by adjusting the exponent such that f_1 is always 1 (except for x=0). For x=0, both e and f_ks are 0. For a real number, bit 31 is the sign bit, which is 0 when the number is positive and 1 when the number is negative. Bits 30 to 23 are used to store the exponent, and bits 22 to 0 contain the fraction.

10.7 FRACTION and EXPONENT

The numerical model indicates that the real number consists of two parts: a fraction and an exponent. Intrinsics fraction and exponent return the fractional and exponent parts of a real number, respectively.

```
print *, exponent (7.0); print *, fraction (7.0)
```

It will display 3 and 0.875, respectively. Let us try to understand these two numbers. First, we have to express 7.0 according to the prescription given earlier: $7.0 = 2^3 \times (0.875)_{10} = 2^3 \times (.111)_2$.

We consider a few more examples:

```
0.25 = (.01)₂ = 2⁻¹ (.1)₂ = 2⁻¹ . (0.5)₁₀    !fraction=0.5,   exponent=-1
0.75 = (.11)₂ = 2⁰ (.11)₂ = 2⁰ . (0.75)₁₀    !fraction=0.75,  exponent=0
0.0625 = (.0001)₂ = 2⁻³ (.1)₂ = 2⁻³.(0.5)₁₀  !fraction=0.5,   exponent=-3
1.0 = 2¹ (.1)₂ = 2¹.(0.5)₁₀                  !fraction=0.5,   exponent=1
```

10.8 MAXEXPONENT and MINEXPONENT

These two intrinsics return the maximum and minimum of the exponent values permitted by the numerical model for a particular kind of real number.

```
real :: r
double precision :: r1
print *, maxexponent(r); print *, minexponent(r)
print *, maxexponent(r1); print *, minexponent(r1)
end
```

These will print 128, -125, 1024 and -1021 (2 to the power), respectively.

10.9 Largest and Smallest Real Numbers

The computer is a finite bit machine, and it obviously has a limit for storing real numbers. The largest and smallest positive numbers that the machine can store are returned by intrinsics huge and tiny.

```
real :: r
print *, huge (r); print *, tiny (r)
```

This will display the largest and smallest real numbers (single precision) that the processor can handle. The values that are returned are 3.4028235e+38 and 1.1754944e-38. If, however, r is declared as a double precision variable, the corresponding numbers are 1.7976931348623157E+308 and 2.2250738585072014E-308, respectively.

10.10 DIGITS for Real Numbers

This intrinsic returns the number of significant digits corresponding to a particular kind of a real number.

```
print *, digits (0.0); print *, digits (0.0d0)
```

It displays 24 and 53, respectively. The result is explained in Chapter 14.

10.11 RANGE for Real Numbers

The range of a real number is defined as the decimal exponent range in a numerical model. For a real number, it is given by int (min (log$_{10}$(huge (x)), -log$_{10}$(tiny (x)))). Therefore, print *, range (0.0) displays 37 as the range of the standard real number and print *, range (0.0d0) displays 307 as the range of the double precision number.

10.12 PRECISION

This intrinsic returns the decimal precision of a particular `kind`. It is less than or equal to

```
int ((p-1) * log₁₀(b)) + k
```

where p and b have the usual meaning, and k is 1 if b is an integral power of 10; otherwise it is 0.

```
print *, precision (0.0)
```

It displays 6. The calculation is shown next:

```
int ((23-1)*log₁₀(2.0))+k = int(22 * 0.3010) = int(6.6) = 6
```

A similar calculation would show that `print *, precision (0.0d0)` will display 15.

10.13 SCALE

This intrinsic takes two arguments: real x and integer i. It returns $x.b^i$ (in our system b=2). For example, `scale (5.0, 2)` returns 20.

10.14 SET_EXPONENT

This exponent takes two arguments: x (real) and i (integer). The intrinsic takes the fractional part of x according to the numerical model and multiplies it by 2^i.

```
print *, set_exponent(7.0, 2)
```

It displays 3.5. Let us try to understand the number. As the factional and exponent parts are, respectively, 0.875 and 3, therefore $0.875 * 2^2 = 3.5$.

10.15 EPSILON

The intrinsic takes one real number, x, as its argument and returns a real number of the same type and kind as x that is almost negligible compared to 1. For real numbers of standard precision, `print *, epsilon (1.0)` will return a real number with magnitude 2^{-23} (1.1920929E-07). It is the smallest number that will change the result of the computation.

10.16 NEAREST

Intrinsic `nearest` returns the nearest machine-representable number of its first argument in a given direction determined by the second argument. Only the sign of the second argument is considered in determining the direction.

If the sign of the second argument is positive, the nearest number is just greater than the first argument, and if it negative, it is just less than the first argument. Consider the following program:

```
real :: r
integer :: i
equivalence (r,i) ! r and i share same location
r=1.0; print *, i
r=nearest(r,1.0); print *, i
r=1.0; r=nearest(r,-1.0); print *, i
end
```

The output looks as follows (shown in the same line):

```
1065353216 1065353217 1065353215.
```

If the bit pattern corresponding to real number 1.0 is treated as an integer, the value of the corresponding integer is `1065353216`. If `1` is added or subtracted, the number becomes `106535217` and `1065353215`, respectively.

10.17 SPACING

The function takes one real argument x. The function returns a real number of the same type and kind of its argument, which is the absolute spacing in the numerical model. It is $2^{(e-p)}$ when x is not equal to zero. If x is zero, it is `tiny(x)`. Actually, spacing corresponding to a real number x is defined as the difference between x and the next floating point number having the same exponent as that of x. It is the smallest gap between two numbers of a given type.

$$x=b^e \sum (k=1, p) \ f_k \ b^{-k}$$

Therefore, the spacing is $b^e b^{-p} = b^{e-p}$.

10.18 RRSPACING

This intrinsic takes one real argument x and returns the reciprocal of the relative spacing of its argument. It returns a real number of the same type and kind as x. The relative spacing near a real number x is as follows:

$$|x.b^{-e}| \ b^p$$

10.19 Programming Example

Using one of the intrinsics mentioned earlier, we try to find the root of equation

```
f(x) = x² - 25 = 0
```

using Newton–Raphson method. The technique is to start with a guess value xg. Obviously, f(xg) is not equal to zero as xg is not a root of the equation. However, one can choose 'h' such that f(xg+h) = 0 with h = -f(xg)/fp(xg) where fp(xg) is the derivative of f(xg) with respect to xg. The new value of x (actually xg) which is xg+h is then used and the process is repeated. This process continues until the absolute value of h becomes very small and in that case xg is the root of the equation. Note that the convergence is tested with the intrinsic spacing (shown in bold letters within the program).

```
real :: x, xg, fx, fpx, h
integer, parameter::limit=100
integer :: index=1
xg=10.0 ! guess value
do while (index <= limit)
   fx=xg**2 -25.0; fpx=2.0**xg; h=-fx/fpx;x=xg+h
   if(abs(x-xg) .le. spacing(x)) then
      exit
   else
      index=index+1
      xg=x
   endif
enddo
if(index <= limit ) then
   print *, 'Convergence achieved ', xg
else
   print *, 'Convergence not achieved ', xg
endif
end
```

The check for convergence may be replaced by the following:

```
if(xg>=nearest(x,-1.0) .and. xg <=nearest(x,1.0)) then
```

It is perhaps apparent that a processor-independent program can be developed easily with the help of the aforementioned intrinsics. Often the convergence criterion of a numerical problem is dependent on the processor, the smallest and largest values it can handle, the spacing between two adjacent real numbers, etc. Instead of using processor-dependent parameters, one can use these intrinsics and make the program more general and portable.

11

Library Functions

Library functions or intrinsics were introduced in earlier chapters. This chapter is a general introduction to library functions.

11.1 Generic Names

A group of library functions may be referred to by their family name and the compiler can substitute the family name by an appropriate particular name depending upon the type of argument. The family name is called the generic name of the group.

11.2 Intrinsic Procedures

The Fortran library contains both functions and subroutines. User-written subprograms (functions and subroutines) are introduced in Chapter 12.

11.3 Pure Procedures

The main property of a `pure` procedure is that it has no side effects. This is discussed in Chapter 12.

11.4 Elemental Procedures

The dummy arguments of these procedures contain only scalar arguments. However, they can be called with arrays also. Except `mvbits` and `move_alloc`, which are subroutines, all elemental procedures are functions. When the elemental procedures are called with scalars, the returned value is also a scalar. When the elemental procedure is called with an array-valued actual argument, the procedure acts on every element of the array and returns an array having the same shape and size as the actual argument. In case there is more than one array-valued argument, all these arguments must be of the same shape and size. For example, library function `sin` takes one argument (in radian) and returns sine of the argument. The argument can be an array and library function `sin` returns an array of the same type and size as the input array containing the sine of the corresponding elements of the input array.

```
real, dimension(4) :: a,b
real, parameter :: pi=3.1415926
a=[pi/6.0,pi/4.0,pi/3.0,pi/2.0]; b=sin(a); print *, b
end
```

This is equivalent to `b(1)=sin(a(1)); b(2)=sin(a(2)); b(3)=sin(a(3)); b(4)=sin(a(4))`. The output corresponds to `sin` of $\pi/6$, $\pi/4$, $\pi/3$ and $\pi/2$, respectively. By default, elemental procedures are pure. Fortran allows defining impure elementary procedure by appropriate declaration (Chapter 12).

11.5 Enquiry Functions

The enquiry function may or may not have any argument. It makes enquiry about certain things and returns an answer. The result does not depend on the value of the argument, if present. For example, `bit_size(i)` returns the number of bits needed to store variable `i`.

11.6 Transformational Functions

These types of functions have one or more arrays as the argument. They do not act on the array elements like the elemental procedure. They may return a scalar or an array created from the input array; `maxval` is one such function.

11.7 Non-elemental Procedures

These types of procedures are called with scalar arguments, and scalars are returned. Except subroutine `mvbits`, all library subroutines are non-elemental.

11.8 Argument Keywords

Intrinsic procedures can be called with argument keywords. If the actual arguments are in proper order, keywords may be omitted. However, when the argument keyword is used, the actual keyword may be supplied in any order. For example, function `ibset (i, pos)` may be invoked as `j = ibset (j, 3)`, `j = ibset (i=j, pos=3)` or `j = ibset (pos=3, i=j)`.

If one of the arguments is used with the keyword, the rest must also be used with the keyword; `j=ibset (i=j, 3)` is not acceptable, as the first argument is used with a keyword but the second argument does not contain keyword `pos`.

11.9 Variable Number of Arguments

Sometimes the number of essential arguments is not fixed. Library function `max` returns the maximum of its arguments. The number of arguments is at least two, but it can be any number greater than two; `max (a1, a2)` returns the maximum of `a1` and `a2`. However, `max (a1, a2, a3)` returns the maximum of `a1`, `a2` and `a3`.

11.10 Optional Arguments

Some of the arguments may be optional. They may be omitted, if they are not required. For example, intrinsic `minval` contains an optional argument `dim`.

11.11 Types of Available Intrinsics

The intrinsics can be divided into several groups. The important intrinsics and their usage are discussed in different chapters of this book. Last three items of the following list are not part of Fortran 2018 standard. Here procedures include both functions and subroutines.

- Numeric procedures
- Mathematical procedures
- Character procedures
- Kind procedures
- Miscellaneous type conversion procedures
- Numeric inquiry procedures
- Array inquiry procedures
- Other inquiry procedures
- Bit manipulation procedures
- Floating-point manipulation procedures
- Vector and matrix multiply procedures
- Array reduction procedures
- Array construction procedures
- Array location procedures
- Null procedures

- Allocation transfer procedure
- Random number subroutines
- System environment procedures
- IEEE procedures
- Atomic subroutines
- Coarray routines
- Procedures to handle strings of arbitrary length
- OpenMP routines
- MPI routines

A list of all library functions is available as Annexure C.

11.12 Intrinsic Statement

This statement is used to specify the intrinsic procedure.

```
intrinsic :: intrinsic-name-list
```

where intrinsic-name-list is the name of the intrinsics. The name cannot be repeated in the list.

```
intrinsic :: sqrt
real :: x= 2.0; print *, sqrt(x)
end
```

Note that without the intrinsic statement, the program would work. However, it is a good programming practice to include this statement to increase the readability of the program.

11.13 Processor Dependencies

The values returned by some of the intrinsics may be processor dependent.

11.14 Final Word

If a library function or subroutine is available for performing a particular task, readers are advised to use the same instead of depending on his or her own code. The use of library functions/subroutines helps the compiler to generate good optimized codes suitable for target machines.

12

Subprograms

Perhaps the most attractive feature of Fortran is the concept of subprogram. A subprogram is just like a main program, which can be compiled independent of the main program. It can be debugged and tested by some driver programs (the actual main program is not required). Different users can share a subprogram, which can be ported from one program to another. A complicated task is often divided into several subtasks having a specific objective for each such subtask. Program development becomes easy. For example, a program may require calculation of factorial and hypergeometric functions. Instead of writing a single program unit, it is always better to design separate subprograms to calculate factorial and hypergeometric functions for a given input. These subprograms can be thoroughly tested. Later on, all these subprograms may be combined with the actual main program to form a load module (the module that the computer can execute). Until now all our examples contained only one program unit—called the main program. Every complete load module must have only one main program and may have one or more subprograms. It is not possible to execute a subprogram directly. A subprogram must be invoked by a main program or another subprogram.

There are two types of subprograms: function and subroutine. Statement function, which is not a subprogram but may be considered somewhat similar to the internal subprogram, has not been discussed as it is not considered as an obsolete feature now.

A subprogram may be an external subprogram or an internal subprogram and can be a part of a module.

An external subprogram is a subprogram that is not contained within a main program or another subprogram or a module. However, an internal subprogram is contained within a main program or a subprogram or a module. The situations are summarized as follows:

Situation 1: There is only one main program. No subprogram is present.

Situation 2: There is a main program with one or more external subprograms.

Situation 3: There is a main program with or without an internal subprogram plus an external subprogram with or without an internal subprogram.

Situation 4: There is a module with or without a subprogram and a main program with or without an internal subprogram plus an external subprogram with or without an internal subprogram that uses this module.

Situation 5: Any other combination involving a main program, internal subprograms, external subprograms and modules.

These are discussed in detail in subsequent sections. In this book, we use the word procedure to mean a function, a subroutine or both. The external subprogram may be written using other programming languages also.

12.1 FUNCTION Subprogram

A function subprogram starts with keyword `function` followed by the name of the function and the arguments of the function within parentheses. If there is more than one argument, the arguments are separated by a comma. If a function does not require any argument, a left parenthesis and a right parenthesis with nothing in between are placed after the function name. The function returns a value to the calling program in two different ways—through its name or through a `result` clause. If the `result` clause is present, it is placed after the right parenthesis. A variable is attached to the `result` clause, and this follows `result` and is enclosed within parentheses.

```
function function-name(arg1,...,argn) result(res)
```

where `function-name` is the name of the `function`; `arg1`, `arg2`, etc. are the arguments of the function and `res` is the variable through which the function returns a value to the calling program. If the `result` clause is absent, the function definition is given as follows:

```
type function function-name(arg1, arg2, ..., argn)
```

where `type` is a type declaration that determines the type of value (`real`, `integer`, `character`, etc.) the function is supposed to return (default `i-n` rule may also be used).

Although in Fortran a function may return one or more values to the calling program, we use only functions where one value is transferred from the function to the calling program. Subroutine subprograms will be used to transfer more than one value or no value at all to the calling program.

Usually, the data is exchanged between the calling program and the called program through the arguments of the subprogram (also called dummy parameters). When the subprogram is called, the actual arguments are supplied. For the time being, we assume that the number and type of the actual and dummy arguments are the same. This will be relaxed as we proceed.

After the function declaration, usually all the arguments of the functions are declared. Next, all the local variables that are required as temporary storage within the function are declared. The body of the function contains Fortran statements. The last statement of the function is the `end` statement, which may contain `function` and the function name.

```
function func-name(arg1,...,argn) result(res)
<declaration>
<Fortran statement>
end function func-name
```

For example, function `sqr` takes one argument x—a real number—and returns the square of x.

```
function sqr(x) result(sq_x)
real :: x, sq_x
sq_x = x*x
end function sqr
```

The name of the function is sqr, and its argument is x. The result is returned through the real variable sq_x. The argument x is called the dummy parameter or argument. The function does not know what the value of x is. It only knows that it is a scalar and of type real. When the function is invoked, the dummy parameter gets the value to be squared (in this case from the calling program). Note that the function is quite general— for any x it returns the square of x. It may be said that when the function is invoked, the dummy parameter points to the actual parameter. The name x is quite arbitrary. When the function is invoked, the corresponding actual parameter must be a real scalar quantity. As the function returns value through sq_x, the variable must appear at least once on the left-hand side of the assignment sign within the function body.

The function is invoked by writing its name and supplying the actual argument. The following main program may call the preceding function sqr:

```
program main
real :: y=40.0,z,sqr
z= sqr(y)
...
end
```

The function along with its actual argument must appear on the right-hand side of an assignment sign in the calling program. In this case, when the function is invoked, the dummy argument x of function sqr points to the actual variable y of the calling program. The actual argument y could be a number or expression. If it is an expression, it is first evaluated.

Function sqr may also be written without the result clause. In this case, the function returns the value through its name, and naturally, the function name (without the argument) must appear at least once on the left-hand side of the assignment sign within the function body. The function returns a value either through the result clause or through the function name.

```
function sqr(x)
real :: x, sqr
sqr = x*x
end function sqr
```

or

```
real function sqr(x)
real :: x
sqr = x*x
end function sqr
```

A function may call another function. Let us define function cube, which, using function sqr, calculates the cube of its argument.

```
function cube(x) result(cb)
real :: x, cb, sqr
cb = sqr(x) * x
end
```

Note that function cube invokes function sqr. Some more examples of functions are given as follows:

```
function fact(n) result(facto)  ! calculates factorial n
integer ::n
```

```
   real :: facto
   real :: temp        ! local variable
   integer ::i
   if(n .lt. 0) then
     facto = -1.0        !error
   else if(n.eq.0) then
     facto = 1.0          ! factorial zero
   else
     temp = 1.0
     do i =2, n
      temp = temp *i
     enddo
     facto = temp
   endif
  end function fact
```

The function can be invoked as x = fact (n), where n is defined (say, 5) before the function is called.

Although we can write a function without the result clause, it is better to use the result clause always. If the type of the function name is not explicitly specified, the default i-n rule applies. That is, if the function name starts with i, j, k, l, m or n, it returns an integer value; otherwise, it returns a real value. The following is an example of a logical function; the function returns a logical quantity (true or false).

Without the result clause:

```
logical function odd_even(n)        ! returns logical value
integer :: n ! true if n is even otherwise false
if(mod(n,2).eq.0) then
 odd_even = .true.
else
 odd_even = .false.
endif
end function odd_even
```

With the result clause:

```
function odd_even1 (n) result(res)! same as above with result clause
logical:: res
integer ::n
if(mod(n,2).eq.0) then
 res = .true.
else
 res = .false.
endif
end function odd_even1
```

The preceding functions may be invoked as follows:

```
logical odd_even, odd_even1
integer ::a=2
logical:: result
result = odd_even(a)
```

```
result = odd_even1(a)
     .
end
```

Some more examples of functions are given as follows:

```
integer function fact(n)
complex function cfun(a,b,c)
character(len=20) function convert(ch)
```

The last function declaration tells us that function `convert` returns a character of size 20. It may be noted that the type of the arguments and the value returned by the function need not be the same. In the examples shown previously, function `cfun` returns a complex number, but its arguments, `a`, `b` and `c`, need not be complex numbers; in this case, without any declarations, they are real variables.

We close this section with one complete program containing a function that takes one character variable as an argument and returns a character variable with the characters in the reverse order. Note that in the function, the size of the returned character is specified as the assumed size by means of an asterisk.

```
character (len=10) :: ch, reverse
read *, ch
print *, ch
ch=reverse(ch)
print *, ch
end
function reverse(ch) result(res)
character(len=*) :: ch, res
integer :: length, i, it
length = len(ch)
do i=1, length
 it = length-i+1 !character from the right
 res(i:i) = ch(it:it)
enddo
res = adjustl(res)
end
```

12.2 SUBROUTINE Subprogram

A subroutine subprogram may return any number of values including zero (no value at all) to the calling program. The subroutine subprogram starts with keyword `subroutine` followed by the name of the subroutine and its arguments separated by a comma within parentheses.

```
subroutine sub (arg1, arg2,..., argn)
```

where `sub` is the name of the subroutine and `arg1`, `arg2`, . . . , `argn` are dummy arguments of the subroutine. A subroutine may not have any argument at all; in that case, the parentheses may be omitted.

```
subroutine sub
```

However, in this case the subroutine may be defined with the name of the subroutine followed by a left and a right parenthesis: `subroutine sub()`.

The subroutine returns the values through its arguments, and naturally, the name of the subroutine does not have any relation with the type of the values the subroutine would return.

After the subroutine declaration, usually the arguments are defined. This is followed by the declarations of the local variables that the subroutine may need. These are followed by Fortran statements. The subroutine is terminated by the end statement.

```
subroutine mysub(a,b)
integer :: a
real :: b
real :: t   ! local variable
<fortran statements>
    .
end subroutine mysub
```

The portion `'subroutine mysub'` after the end statement is optional but can greatly help future maintainers read the program. We illustrate the use of subroutine by means of the following example. This subroutine takes two integer arguments and interchanges these two arguments.

```
subroutine exchange(a,b)
integer :: a,b
integer :: t        ! local variable
t = a; a=b; b=t
end subroutine exchange
```

12.3 CALL Statement

The subroutine is invoked from the calling program by a call statement. The syntax of the call statement is as follows:

```
call subroutine-name (arg1,...,argn)
```

where `subroutine-name` is the name of the subroutine and the arguments are actual arguments.

```
call exchange(ia,ib)
```

When the subroutine is invoked, the dummy arguments of subroutines a and b point to the actual arguments ia and ib, respectively. Any modifications made to the dummy variables a and b are normally reflected in the corresponding actual variables ia and ib in the calling program. This is discussed in detail in subsequent sections. Also note that in this case when the subroutine is entered, the dummy variables a and b contain the current values of the actual arguments ia and ib, respectively.

```
program main
integer :: ia, ib
ia = 10
ib = 20
```

```
call exchange(ia,ib)
...
end program main
```

The next program shows how different types of intrinsic variables are passed as dummy arguments to a subroutine.

```
program main_prog
integer:: p=2
real:: q=3.2
double precision:: r=2.0d0/3.0d0
complex:: s=(2.0,3.0)
logical:: t=.true.
character(len=100):: u= &
 "Association of Voluntary Blood Donors, West Bengal"
call mysub(p,q,r,s,t,u)
end
subroutine mysub(a,b,c,d,e,f)
integer:: a
real:: b
double precision:: c
complex:: d
logical:: e
character(len=*):: f
print *, a,b,c,d,e; print *, len_trim(f)
end
```

Dummy parameters are placeholders—only the type and the number of parameters must match; the actual name does not matter. For example, no harm is done if the actual parameters of the preceding program (p, q, r, s, t and u) are replaced by a, b, c, d, e and f, respectively.

```
subroutine fact(f,n)
real :: f
integer ::n
integer :: i
if(n .lt. 0) then
   f = -1.0              ! error
 else if (n.eq.0) then
      f = 1.0
     else
     f = 1.0
     do i =1, n
     f = f*i
     enddo
endif
end subroutine fact
```

Subroutine fact may be called by the following program segment:

```
real :: factorial
integer :: n=5
call fact (factorial, n)
...
end
```

The first argument (actual parameter) factorial contains the value returned from the subroutine; the second argument, n, is an integer and the factorial of this number is returned. The corresponding dummy arguments of fact are scalars of type real and integer, respectively.

A subroutine may call another subroutine or function. The cube function discussed in Section 12.1 may be recast as a subroutine. This subroutine now calls function sqr.

```fortran
function sqr(n)
integer :: n,sqr
sqr = n*n
end function sqr
subroutine cube(n,c)
integer :: n,t,sqr,c
t = sqr(n); c = t*n
end subroutine cube
```

A subroutine need not have any argument.

```fortran
subroutine messg
print *, "i am here"
end
```

When this subroutine is called as

```fortran
call messg
```
or
```fortran
call messg()
```

the message is displayed on the screen. In this case, the subroutine neither takes any input from the calling program nor returns anything to the calling program. Subroutine messg may be defined as subroutine messg() as there is no argument associated with it. Similarly, it may be called as call messg().

A subroutine may take an input from the calling program, but it might not return anything to the calling program.

```fortran
subroutine message(n)
integer ::n
integer :: i
do i=1,n
 print *, "i am here"
enddo
end
```

When this subroutine is called with some value greater than 0 corresponding to the dummy parameter n, the message is printed n times. The example may seem pointless, but one could envisage cases where some external devices like an audible alarm of a power switch are activated.

12.4 INTENT

The arguments of a subprogram may be of three types: input only, output only and both input and output. For an input-only argument, the subprogram is not supposed to modify the arguments. For an output-only argument, the subprogram returns the value through it. Similarly, for an argument, which is of the input–output type, the subprogram takes input through the argument from the calling program, modifies it and returns the same to the calling program. Fortran provides an additional attribute `intent` for the arguments of the subprogram. As expected, `intent` may be `in`, `out` or `inout` (or `in out`). The purpose of this attribute is to declare the intention of a programmer clearly, thereby helping the compiler to perform some additional checks on the arguments. For example, an argument declared with `intent(in)` cannot be modified within the subprogram; that is, it cannot appear on the left-hand side of the assignment sign within the subprogram. We rewrite the subroutines `exchange` and `cube` with the `intent` attribute.

```
subroutine exchange (a,b)
integer, intent(inout) :: a, b
...
end subroutine exchange
subroutine cube(n,c)
integer, intent(in) ::n
integer, intent(out)::c
...
end
```

Using subroutine `mysub` of section 12.3, the usefulness of `intent` is demonstrated. If the comment from the statement shown in bold is removed, there will be a compilation error as the `intent` of variable e is `in`, and as such, it cannot be modified.

```
subroutine mysub(a,b,c,d,e,f)
integer, intent(in):: a
real, intent(in):: b
double precision, intent(in):: c
complex, intent(in):: d
logical,intent(in):: e
character(len=*), intent(inout):: f
print *, a,b,c,d,e
print *,trim(f), len_trim(f)
f="west bengal" ! no error as the intent is inout
!   e=.false. ! if the comment is removed the compiler will object
end
```

12.5 Internal Procedure

The subprograms discussed so far are external procedures. An internal procedure is a subprogram that is contained within a main program or another subprogram (modules are discussed later). The internal procedures are accessible from the unit that contains the internal procedure. We consider again our old examples `sqr` and `cube` to describe the internal subprogram. The internal subprograms are placed immediately before the end statement of their mother (host). Before the internal subprogram, keyword `contains` must be present. The presence of a `contains` statement indicates that the unit contains internal procedures. It separates the internal subprograms from their host. The program does not fall through into the internal subprogram. It must be called in the usual manner. After executing the statement just before the `contains` statement, and after skipping all the statements belonging to the internal subprograms, the control is passed to the end statement. The end statement may have a statement number, and one can jump to the end statement from an executable statement to terminate the program. An internal subprogram cannot contain another (nested) internal subprogram. Internal subprogram may be defined within a main program, within an external procedure and within a module. The internal subprogram may call other internal subprograms, but it cannot call itself.

```
program cone
integer :: n=2
integer :: s,t
s=sqr(n); t=cube(n)
print *, s,t
contains
   function sqr(a) result(res)
   integer ::res
   integer, intent(in)::a
   res=a*a
   end function sqr
   function cube(b) result(rc)
   integer, intent(in)::b
   integer ::rc
        .
   end function cube
end program cone
```

There may not be any statement between `contains` and end. However, this is meaningless. For internal subprograms, the type of the function should not be declared within the calling program. For example, the compiler will flag an error if `integer::sqr`, cube statement is introduced before the first executable statement of program cone. Internal subprograms can share variables with their mother. Note that if an external subprogram sub is declared, this subprogram cannot call `sqr` or `cube`. Consider the following program:

```
program cone
integer :: n=4
integer :: t, cube
t=cube(n)
print *, t
end program cone
function cube(b) result(rc)
integer, intent(in)::b
integer ::rc
rc=sqr(b)*b
contains
   function sqr(a) result(res)
   integer ::res
   integer, intent(in)::a
```

```
  res=a*a
  end function sqr
end function cube
```

The main program cannot call sqr. Only cube can call sqr because sqr is an internal function of cube. Note that cube is an external function to cone, and as such, it has been declared as an integer within cone; sqr is an internal function to cube, and so this has not been declared as an integer within cube. Another example of the internal subprogram is given as follows:

```
program main
integer :: sum, num !  add 1+2+3+...   num
read *, num
sum=myadd(num)
print *, sum
contains
  function myadd(n) result (res)
  integer, intent (in) :: n
  integer :: res
  integer :: temp, i
  temp=0
  do i=1, n
   temp=temp+i
  enddo
  res=temp
 end function myadd
end program main
```

The next program again uses the internal procedure. It calculates the days of the week using the fact that January 1, 1900, was a Monday. The technique is to calculate the number of days passed since January 1, 1900. Once this is obtained, the next thing is to take a modulo with 7. Care is to be taken to handle the leap years.

```
implicit none
character(len=3),dimension(0:6)::day=[ &
    "mon","tue","wed","thu","fri","sat","sun"]
integer,dimension(12)::month=[ &
 31,28,31,30,31,30,31,31,30,31,30,31]
integer:: iy,no_of_days=0 ! jan 1, 1900 was a Monday.
integer:: dd, mm, yy
print *, "type the date  dd mm yyyy"
read *, dd,mm,yy
if(yy<1900) then   ! check for valid data
 print *, yy, " ... invalid year ... exit"
 stop
 endif
 if(mm<1 .or. mm>12) then
  print *, mm, " ... invalid month ... exit"
  stop
 endif
 if(dd<1 .or. dd>31) then
  print *, dd, " ... invalid date ... exit"
  stop
 endif
```

```
   if(is_leap_year(yy)) then
    month(2)=29 ! Leap year
   endif
 if( dd > month(mm)) then
   print *, dd, " ... invalid date ... exit"
   stop
 endif
 do iy=1900, yy-1
   if(is_leap_year(iy)) then
     no_of_days=no_of_days+366 ! Leap year
   else
    no_of_days=no_of_days+365 ! Normal year
   endif
 enddo
 do iy=1,mm-1
   no_of_days=no_of_days+month(iy)
 enddo
 no_of_days=no_of_days+dd
 print *, day(mod(no_of_days-1,7))
 contains
  logical function is_leap_year(yr)
  integer,intent(in)::yr
  if(mod(yr,4).eq.0 .and.   &
    (mod(yr,100).ne.0 .or. mod(yr,400).eq.0)) then
   is_leap_year=.true.
  else
   is_leap_year=.false.
 endif
 end function is_leap_year
 end
```

12.6 Character Type Argument

The size of a dummy character argument may be declared with an asterisk (*). The length of such a dummy argument is the length of the actual argument. Therefore, the length of the dummy character argument becomes a 'variable.' The same subprogram may be called with different actual character arguments having different sizes. The following subroutine calculates the number of vowels within the character string:

```
implicit none
character(len=10):: c1
character(len=20):: c2
integer :: nv
read *, c1              ! read the first string
call nvowel(nv,c1)
print *, nv            ! no of vowels in c1
read *, c2             ! read the second string
call nvowel(nv,c2)
print *, nv            ! no of vowels in c2
end
```

```
subroutine nvowel(num,string)
character(len=*),intent(in):: string
integer, intent(out)::num
integer::l, i
l=len(string)
num=0
do i=1, l
  select case(string(i:i))
    case('a','e','i','o','u','A','E','I','O','U')
      num=num+1
  end select
enddo
end
```

The next program converts Rupees in figure to Rupees in words using an external subprogram containing an internal subprogram. The logic is explained through in-line comments.

```
subroutine fig_to_word(r,buf)
implicit none
integer, parameter::sz=selected_int_kind(9)
integer(kind=sz)::rs, paise, r1, r2
double precision:: r
character (len=*)::buf ! upper limit is rs 999999999.99
character (len=10), dimension(19):: &
 onetwo=[character(len=10):: &
 "one", "two", "three", "four", "five", "six", "seven",&
 "eight", "nine", "ten", "eleven", "twelve", "thirteen", &
 "fourteen", "fifteen", "sixteen", "seventeen", "eighteen", &
 "nineteen"]
character(len=10), dimension(2:9):: &
tentwenty=[character(len=10)::&
 "twenty", "thirty", "forty", "fifty", "sixty", "seventy", &
 "eighty", "ninety"]
character(len=15)::tbuf
buf="Rupees "
rs=r ! rupees = whole number
paise=(r-rs)*100.0+0.5 ! paise = fraction
print *, "Rupees ",rs, " and Paise ", paise
if (rs == 0) then      ! successive div and mod (integer) operations
 buf= trim(buf)// " " //"zero"  ! are used to extract the digits
else
r1=rs ! extraction of digits
r2=r1/10000000_sz
 if (r2 .gt. 0) then
  call convert(r2, tbuf)
  buf= trim(buf)//" " //trim(tbuf)//" crore "
 endif
r1=mod(r1,10000000_sz)
r2=r1/100000_sz
 if (r2 .gt. 0) then
  call convert(r2, tbuf)
  buf= trim(buf)//" " //trim(tbuf)//" lakh "
 endif
r1=mod(r1, 100000_sz)
```

```fortran
r2=r1/1000_sz
 if (r2 .gt. 0) then
 call convert(r2, tbuf)
 buf= trim(buf)//" " //trim(tbuf)//" thousand "
 endif
r1=mod(r1,1000_sz)
r2=r1/100_sz
 if (r2 .gt. 0) then
 call convert(r2, tbuf)
 buf= trim(buf)//" " //trim(tbuf)//" hundred "
 endif
r1=mod(r1,100_sz)
 if (r1 .gt. 0) then
  call convert(r1, tbuf)
 buf= trim(buf)//" " //trim(tbuf)
 endif
endif
buf=trim(buf) //" and Paise "
 if (paise .eq. 0) then
 buf=trim(buf) //" zero only"
 else
  call convert(paise, tbuf)
 buf= trim(buf)//" " //trim(tbuf)//" only."
 endif
 contains
  subroutine convert(n,tbuf)
  integer(kind=sz), intent(in)::n
  character(len=*), intent(out)::tbuf
  character(len=10):: t1,t2 ! local variables
  integer (kind=sz):: temp, nt
  t1=""; t2=""; nt=n; tbuf=""
  if(nt>19) then
   temp=nt/10_sz
   t1=trim(tentwenty(temp))
   nt=mod(nt,10_sz)
  endif
  if(nt>=1 .and. nt<=19) then
   t2= trim(onetwo(nt))
  endif
  if(len_trim(t1) >0) then ! if t1 is not empty "add" t1 and t2
   tbuf=trim(t1)//" "//t2
  else
   tbuf=t2      ! t1 is empty, return only t2
  endif
  end subroutine convert
end subroutine fig_to_word
 program rs_to_word ! Main program
 double precision :: rupees ! single precision may not produce
 character(len=120)::buf=" " ! correct rounding
 print *, "enter data ... read rupees.paise "
 read *, rupees
 call fig_to_word(rupees,buf)
 print *, buf
 end program rs_to_word
```

12.7 Argument Types

We have already mentioned that while calling a subprogram, the number and the type of the arguments of the actual parameters must match with the corresponding dummy arguments of the subprogram. There is, usually, one-to-one correspondence between the actual arguments and the dummy arguments. This essentially means that the first dummy argument points to the first actual argument, the second dummy argument points to the second actual argument and so on. There are two ways by which parameters are transferred between the calling program and the called program. They are known as call by reference and call by value.

12.8 Call by Reference

If the actual argument is a variable name, the corresponding argument is said to have been passed by reference. When an argument is passed by reference, any allowed modifications done within the subprogram are reflected in the corresponding actual argument in the calling program. Consider subroutine exchange once again.

```
subroutine exchange(a,b)
integer :: a,b
integer :: t
...
end subroutine exchange
program main
integer ::ia, ib
ia = 3; ib = 4
call exchange(ia,ib)
print *,ia,ib
end program main
```

When subprogram exchange is called, the dummy arguments a and b point to the corresponding actual arguments ia and ib, respectively. Any modifications done on a and b within the subprogram are carried over to the calling program and are reflected in the actual arguments ia and ib, respectively. Note that within the subprogram, a and b are interchanged, and since a and b, respectively, point to ia and ib, the changes are also reflected in the calling program. Arguments passed in this manner are said to be "call by reference."

12.9 Call by Value

The actual argument may be a constant or an expression. If it is an expression, it is evaluated and the dummy arguments are initialized with these values. The same thing happens if the actual expression is a constant. For example, if subroutine exchange is called as call exchange (10,20), the dummy arguments a and b are initialized with 10 and 20, respectively. Now the variables a and b are interchanged within the subprogram, and after the interchange is done, the value of a is 20 and that of b is 10 as expected. However, when the subprogram returns control to the calling program, the modifications done on a and b within the subprogram are not carried over to the calling program. Arguments passed in this manner are said to be "call by value."

If the actual argument (variable) is enclosed within parentheses, it is (assumed to have been) passed by value; call exchange ((ia), (ib)) will not interchange ia and ib, as the arguments, though not an expression, are passed by value. The situation will be interesting when only one of the actual arguments is passed by value, call exchange ((ia), ib). If ia=3 and ib=4 before the call, ia will remain 3 and ib will be 3 after the call. This is due to the fact that the dummy arguments a and b, corresponding to the actual arguments ia and ib, are interchanged within the subroutine exchange. As the first is called by value, the modification done on the corresponding dummy argument within the subroutine is not carried over to the calling program. As the second argument is called by reference, the modification done on the corresponding dummy argument is reflected in the calling program.

12.10 RETURN Statement

Control from the called program is transferred to the calling program after processing the last executable statement within the subprogram. This is called normal return. The subprogram may contain a statement called return that when executed transfers the control back to the calling program. A subprogram may have more than one return statement. In fact, return is always assumed before the end statement of the subprogram. The following function subprogram calculates factorial n and uses the return statement:

```
function fact(n) result(res)
integer, intent(in) :: n
integer :: res
integer:: i
if(n < 0) then
 res = -1        ! error
 return
endif
if (n == 0) then
 res=1           !factorial 0 is 1
 return
endif
res = 1
do i =2, n
 res = res *i
enddo
end function fact
```

For n < 0, factorial is not defined, so res is set to -1 (error). As no further processing is required, control is passed to the calling program by executing the return statement. The return statement is not allowed within the main program.

12.11 INTERFACE Block

An `interface` block is a place where information about an external subprogram, that is, its name, type and number of its arguments, is kept for the compiler. The `interface` block starts with keyword `interface` and ends with keyword end `interface`. Between these two lines, the name of the subprogram along with the definitions of its arguments is placed.

```
interface
   subroutine exchange(a,b)
   integer, intent(inout) :: a, b
   end subroutine exchange
end interface
```

The following example does not really require an `interface` block. Later we will see that there are situations where an `interface` block is mandatory.

```
program main
interface
   subroutine exchange(a,b)
   integer, intent(inout) :: a,b
   end subroutine exchange
end interface
integer :: ia=10, ib=20
call exchange (ia,ib)
...
end program main
subroutine exchange(p,q)
integer, intent(inout) :: p,q
...
end subroutine exchange
```

We rewrite one of our old programs with the `interface` block, which is actually not necessary for this program. It is a good programming practice to use the `interface` block with every external subprogram.

```
program main_prog
interface
 subroutine mysub(a,b,c,d,e,f)
 integer:: a
 real:: b
 double precision:: c
 complex:: d
 logical:: e
 character(len=*):: f
 end subroutine mysub
end interface
integer:: a=2
real:: b=3.2
.
call mysub(a,b,c,d,e,f)
end
```

```
subroutine mysub(a,b,c,d,e,f)
integer:: a
real:: b
double precision:: c
complex:: d
logical:: e
character(len=*):: f
.
end
```

12.12 Array as Arguments

An array can be an argument to a subprogram. The dimension of the array (rank) must be
specified within the subprogram. It is not necessary to specify the actual size within the
subprogram.

```
subroutine zero(a)
integer, dimension(:)::a
a=0
end subroutine zero
```

However, an appropriate `interface` block must be declared. In the previous case, the
following `interface` block in the calling program is required.

```
interface
 subroutine zero(a)
 integer, dimension(:)::a
 end subroutine zero
end interface
```

For an array of rank 2, two colons separated by a comma are necessary to specify the
dimension within the subprogram. Subprogram `set_value` stores a particular value in
all the locations of a two-dimensional array.

```
subroutine set_value(x,value)
integer, dimension(:,:)::x
integer, intent(in):: value
x=value
end subroutine set_value
```

The corresponding `interface` block in the calling program looks like

```
interface
  subroutine set_value (x,value)
  integer, dimension(:,:) :: x
  integer, intent(in)::value
  end subroutine set_value
end interface
```

The calling program will have an appropriate declaration for the actual parameter corresponding to x.

```
integer, dimension (3,3) :: y
```

An internal subprogram, in an identical fashion, may use an array as its argument.

The lower bound of the dummy argument of a subprogram may be different from the corresponding actual argument.

```
program diff_lb
integer::i
integer, dimension(10)::a=[(i, i=1,20,2)]
integer, dimension(0:4)::b=[(i, i=20,36,4)]
integer, dimension(1:4)::c=[(i, i=30, 33)]
integer, dimension(4:6)::d=[(i, i=40,60,10)]
interface
  subroutine sub(x)
   integer, dimension(0:)::x
   end subroutine sub
end interface
call sub(a)
call sub(b)
call sub(c)
call sub(d)
end program diff_lb
subroutine sub(x)
integer, dimension(0:):: x
print *, x(2)
end subroutine sub
```

As the lower bound of x is 0 and the upper bound of the corresponding actual argument, say, a, is 9, array a can have subscripts between 0 and 9 within the subroutine. Similarly, the bounds of other arrays b, c and d inside the subroutine are 0:4, 0:3 and 0:2, respectively. The outputs are 5, 28, 32 and 60 (shown in the same line). However, if the bounds of the array within the interface block and the subroutine is replaced by '*' or ':', the lower bound of array x within the subroutine takes the default value 1, and as such, the output will be 3, 24, 31 and 50, respectively.

An array within a subprogram may be dimensioned through one of its arguments. This is also called variable dimension.

```
program main
integer, dimension(10)::a
.
call sub(a,10)
.
end program main
subroutine sub(a,n)
integer, dimension(n)::a
.
end subroutine sub
```

Using Fortran-77 style, a rank 1 array may be given a dimension of just one within the subprogram.

```
program main
integer, dimension(10)::a
call sub(a)
...
end program main
subroutine sub(a)
integer, dimension(1)::a
...
end subroutine sub
```

12.13 User Defined Type as Argument

User-defined variables (structures) may be an argument to a subprogram. For an internal procedure, the type declaration in the main unit is sufficient for the subprogram.

```
program main
type emp
 sequence
 integer ::id
 character(len=20) ::name
end type emp
type(emp) :: p
call set_emp(p)
...
contains
 subroutine set_emp(q)
 type(emp) ::q
 q%id=10; q%name="P K Mukherjee"
 end subroutine set_emp
end program main
```

For external subprograms, the type declaration must also be present in the subprogram.

```
program main
type emp
 sequence
 integer ::id
 character(len=20) ::name
end type emp
type(emp) :: p
call set_emp(p)
...
end program main
subroutine set_emp(q)
```

```
type emp
  sequence
  integer ::id
 character(len=20) ::name
end type emp
type(emp) ::q
 .

end subroutine set_emp
```

12.14 MODULE

We have just seen that data is exchanged between a main program and the subprograms or between subprograms through arguments. There are two other ways of sharing data among different units of programs: through the common statement and through module. The common statement is declared as obsolete, but still we discuss it in this book.

A module is a unit that contains constants, variables and procedures to be shared by different units of a program. A module, like functions or subroutines, may be compiled independent of the main program. Later it may be used with other units by a suitable linker.

A module starts with keyword module followed by the module name. It ends with the end module statement.

```
module mycons
integer :: ia,ib
real, parameter :: pi = 3.1415926
end module mycons
```

The mycons module may be used by a main program or a subprogram through the use statement.

```
program main
use mycons
ia = 10
call sub
print *, ia
end program main
subroutine sub
use mycons
print *, ia
ia=20
end subroutine sub
```

The constants, variables or subprograms defined within a module are available within the units where the module has been used. For example, in program main, ia is set to 10. In subroutine sub, the print statement displays 10, because of the use statement; variable ia of main and variable ia of subroutine sub are the same. Subsequently, ia is set to 20 in subroutine sub. Therefore, the print statement of main displays 20.

A module may contain internal procedures.

```
module mod_name
<declaration>
contains
  subroutine sub
    .
  end subroutine sub
  function fun
    .
  end function fun
end module mod_name
```

If subroutine exchange is included in a module, an appropriate use statement can access it.

```
module swap
contains
subroutine exchange(a,b)
  integer, intent(inout)::a,b
  integer :: t
  t = a; a=b; b=t
  end subroutine exchange
end module swap
program main
use swap
integer :: ia=10, ib=20
call exchange(ia,ib)
print *, ia, ib
end program main
```

It is seen that in this way one can create a program library containing several subprograms and use them as and when needed. A module may use another module but cannot use itself directly or indirectly.

The parameters and variables defined in the module can be selectively used, and even can be renamed, through the use statement.

```
module select
real, parameter :: pi=31415926
real(kind=kind(0.0d0)), parameter :: dpi=3.1415926589793d0
end module select
program main
use select, only : dpi
...
end program main
```

The only clause, placed after the module name, allows the access of a particular item (in this case dpi) or a number of items mentioned with the clause. In the aforementioned case, only variable dpi is available in program main—the other variable pi, declared in module select, is not available in main. Note the comma after the module name and single colon after only. While using a module the variables declared in it may be renamed, if necessary.

```
program main
use select, mypi =>dpi
...
end program main
```

Variable dpi, declared in the module select, is used in main; however, in this routine, variable dpi is called as mypi (renamed). The symbol to rename an item is the equal sign followed by the greater than sign.

The next example shows a module containing user-defined type declaration.

```
module emptype
  type emp
   sequence
   integer ::id
   character(len=20) :: name
  end type emp
end module emptype
program main
use emptype
type(emp) :: p
.
call set_type(p)
.
end program main
subroutine set_type(q)
use emptype
type(emp) :: q
.
end subroutine set_type
```

12.15 MODULE PROCEDURE

A module may contain procedures, which may be used by other units. All the procedures are placed within the module after statement contains.

```
module mylib
...
contains
  subroutine sub
  ...
  end subroutine sub
  function func (arg1,..., argn)
  ...
  end function func
end module mylib
program main
use mylib
...
end program main
```

Unit `main` may use subprograms `sub` and `func` defined in module `mylib`. It is apparent that a module may contain both variable declarations and subprograms.

The concerned unit may use all the variables and subprograms defined in the module, or it may choose variables and subprograms from the module selectively. It is also possible to rename a subprogram within the unit like a variable. We use the following module to demonstrate all the features mentioned previously:

```
module mylib
real, parameter :: pi=3.1415926
contains
  subroutine iexchange(a,b)
  integer, intent(inout):: a,b
  .
  end subroutine iexchange
  subroutine rexchange(p,q)
  real, intent(inout) :: p,q
  .
  end subroutine rexchange
end module mylib
```

Case I: Use of whole module.

```
program main
use mylib
...
end
```

In this case, both subprograms `iexchange` and `rexchange` are available within `main`. In addition, `main` can access `pi`.

Case II: Use statement containing the `only` attribute.

```
program main
use mylib, only : iexchange
...
end
```

Unit `main` can use only subprogram `iexchange` defined in module `mylib`.

Case III: Use statement containing the `only` attribute with renaming of a subprogram.

```
program main
use mylib, only : myexchange => iexchange
.
end
```

Unit `main` can use only subprogram `iexchange` defined in the module. However, this subroutine `iexchange` is called `myexchange` within `main`.

Case IV: Use of whole module with renaming of a subprogram.

```
program main
use mylib, myexchange => iexchange
...
end
```

Unit main may use both subprograms iexchange and rexchange as well as the named constant pi. However, subprogram iexchange is called as myexchange within program main. Since subprogram rexchange is not renamed, it is used as rexchange within main.

12.16 PUBLIC and PRIVATE Attributes

The variables and procedures defined within a module are, by default, public. This indicates all the variables and procedures are available to the unit that uses this module. By declaring some variables and subprograms as private, the programmer may restrict the use of a variable or subprogram from outside the module. Statement public sets the default to public accessibility. Similarly, statement private sets the default to private accessibility.

The access specification, public or private, may be used along with the variable, subprogram, user-defined operator (will be discussed soon), etc.

```
module mymod
public             ! default
private :: x,y,z
...
end module mymod
```

The private declaration in the preceding module restricts the use of x, y and z outside the module. The creator of the module now has a means of hiding some of the items from the units using the module.

Consider the following program:

```
module sq_cube
private:: sqr        ! can be called only from within the module
public:: cube        ! can be called from outside the module if
contains             ! the module is used
  real function sqr(x)
  real, intent(in)::x
    sqr=x*x
  end function sqr
  real function cube(x)
  real, intent(in)::x
    cube=sqr(x)*x
  end function cube
end module sq_cube
```

```
program main
use sq_cube
real:: y=3.0
print *, cube(y)
end
```

Function cube calls sqr. This is allowed. However, the main program cannot call sqr directly as sqr is declared private. In the next example, the module contains a procedure cube, which in turn contains an internal procedure sqr.

```
module sq_cube
public :: cube
contains
  real function cube(x)
  real, intent(in)::x
  cube=sqr(x)*x
  contains
    real function sqr(x)   ! internal to cube
    real, intent(in)::x
    sqr=x*x
    end function sqr
  end function cube
end module sq_cube
program main
use sq_cube
real:: y=3.0
print *, cube(y)
end
```

Obviously, the program main can call cube. It cannot call sqr directly. Now consider the following:

```
module mypriv
private
real:: val
public:: set_val, get_val
contains
  subroutine set_val(x)
  real, intent(in)::x
  val=x
  end subroutine set_val
  subroutine get_val(x)
  real, intent(out)::x
  x=val
  end subroutine get_val
end module mypriv
```

In this module, unless otherwise mentioned, all the variables and procedures are by default private because of the private declaration (shown in bold). Following this logic, val is a private variable, and procedures set_val and get_val are public functions.

However, if declaration private is changed to public and the line public:: set_val, get_val is removed, val becomes private if the declaration real, private:: val is introduced, procedures set_val and get_val become public by default.

Suppose we have a module that is used to diagonalize a matrix. The module may contain more than one subprogram—one of them takes the input from outside, and the results of the computation it returns should be made available. Other subprograms that are called by the aforementioned subprograms need not be made available to an user because these subprograms are not to be called from outside, and as such they may be made `private` routines. This ensures that even accidentally these `private` subprograms cannot be called from outside the module. Thus, we see that it is possible to restrict the use of a variable or subprogram in two different ways—at the user level by using `only` or at the developer's level by using `private`.

We give some examples of the access specifications: `public` and `private`.

```
module mypriv
real, private:: val
public:: set_val, get_val
contains
  subroutine set_val(x)
  real, intent(in)::x
  val=x
  end subroutine set_val
  subroutine get_val(x)
  real, intent(out)::x
  x=val
  end subroutine get_val
end module mypriv
program main
use mypriv
real::x,y
read *, x
call set_val(x)
print *, "x= : ",x
val=37.2
call get_val(y)
print *, "y= : ",y
print *, "val= : ", val
end
```

We first examine the output of the program when the data corresponding to the `read` statement is 25.5.

```
x= :     25.5000000
y= :     25.5000000
val= :     37.2000008
```

The value of variable (third line) `val` is 37.2. This indicates that this variable `val` is not the same as variable `val` defined in module `mypriv`. Variable `val` defined within the module is a `private` variable, and as such it cannot be accessed within the `main` program. This type of problem can be avoided by using statement `implicit none` in the `main` program, and this will force the compiler to treat the undeclared variable `val` (the private variable `val` declared within the module is not visible within the main program) in the `main` program as an undefined variable resulting in a compilation error.

The individual component of a derived type may be `public` or `private`. Unless otherwise stated, the default is `public`. If any individual component is `private`, this

elementary item of the corresponding variable can be accessed only within the module.
This is illustrated with the following example:

```
        module mymod
        public
        type :: twod
         real:: x
         real, private:: y
        end type twod
        type(twod):: point
        contains
         subroutine set_val(a,b)
          real, intent(in)::a
          point%x=a
          point%y=b
         end subroutine set_val
         subroutine get_val(a,b)
          real,intent(out)::a,b
          a=point%x
          b=point%y
         end subroutine get_val
        end module mymod
        use mymod
        real::p=2.0,q=3.0,r,s
        call set_val(p,q)
        call get_val(r,s)
        print *,r,s
        print *, point%x
!       print *, point%y  !error
        end
```

If the comment from the line print *, point%y is removed, there will be a compilation error because of the presence of real, private:: y in the definition of the user-
defined type. However, statement print *, point%x does not give any compilation error
as real::x is equivalent to real, public::x. Similarly, if td of type twod is defined in
the main program, td%y=20.0 (when the comment is removed) would generate a compilation error due to the reason stated previously.

```
        use mymod
        type(twod)::td
        td%x=10.0
!       td%y=20.0 !error
        end
```

The variable itself can be a private variable.

```
        module mymod
        public
        type :: twod
         real:: x
         real:: y
        end type twod
        type(twod), private:: point  ! point is private
        contains
         subroutine set_val(a,b)
          real, intent(in)::a
```

```
        point%x=a
        point%y=b
        end subroutine set_val
        subroutine get_val(a,b)
        real, intent(out)::a,b
        a=point%x
        b=point%y
        end subroutine get_val
      end module mymod
      use mymod
      type(twod)::td ! ok
      real::p=2.0,q=3.0,r,s
      call set_val(p,q)
      call get_val(r,s)
      print *,r,s
      print *, point%x        ! error
      print *, point%y        ! error
      td%x=100.0              ! ok
      td%y=200.0              ! ok
      end
```

In-line comments indicate the acceptable and illegal statements. Finally, the type itself could be private. Thus, no variable of this particular type can be defined outside the module.

```
      module mymod
      public
      type, private :: twod ! private type
       real:: x
       real:: y
      end type twod
      type(twod):: point
      contains
       subroutine set_val(a,b)
        real, intent(in)::a
        point%x=a
        point%y=b
       end subroutine set_val
       subroutine get_val(a,b)
        real,intent(out)::a,b
        a=point%x
        b=point%y
       end subroutine get_val
      end module mymod
      use mymod
      type(twod)::td    ! error
      real::p=2.0,q=3.0,r,s
      call set_val(p,q)
      call get_val(r,s)
      print *,r,s
      td%x=100.0         ! error
      td%y=200.0         ! error
      end
```

Again in-line comments indicate the illegal statements.

12.17 PROTECTED Attribute

The public variables of a module can be protected from modification using this protected attribute. These variables may be modified only within the module. This attribute is similar to attribute intent(in) of a dummy argument of a subprogram; the protected attribute can be specified only within a module. The protected variable cannot appear on the left-hand side of the assignment sign outside the module.

```
      module mymod
      integer, protected::a=30,b=40,c
      contains
        subroutine myadd(a,b,c)
          integer:: a,b,c
          c=a+b
        end subroutine myadd
      end module mymod
      program main
      use mymod
      call myadd(a,b,c)
   !  c=100                  ! error if the comment is removed
      print *, a,b,c         ! allowed a,b and c are protected variable
      end
```

The protected variable 'c' is modified inside the subroutine declared within the module. However, the protected variable 'c' cannot be modified within program main. For example, statement c=100 within program main (shown as comment and in bold letters) would generate a Fortran error when the comment is removed. We now give another example of the protected variable.

```
      module mypriv
      implicit none
      real, protected:: val
      public:: set_val, get_val
      contains
        subroutine set_val(x)
          real, intent(in)::x
          val=x
        end subroutine set_val
        subroutine get_val(x)
          real, intent(out)::x
          x=val
        end subroutine get_val
      end module mypriv
      use mypriv
      implicit none
      real::x,y
      read *, x
      call set_val(x)
      print *, "x= : ",x
      call get_val(y)
      print *, "y= : ",y
```

```
!  val=37.5  not allowed as val is a protected variable
   print *, "val= : ", val
   end
```

The outputs are as follows:

```
x= :    25.5000000
y= :    25.5000000
val= :    25.5000000
```

The output shows that the protected variable val is visible in the main program, where module mypriv has been used. As the protected variable is a read-only variable, any attempt to set this variable to some value within the main program will be considered as an illegal operation and therefore will generate an error message during compilation.

Attribute protected cannot be used with the definition of a derived type.

```
module mymod
public
type :: twod
 real, protected:: x    ! illegal
 real, protected:: y    ! illegal
end type twod
...
end module mymod
```

However, a user-defined type variable can be declared as protected.

```
module mymod
 public
 type :: twod
  real:: x
  real:: y
 end type twod
 type(twod), protected:: point
 contains
   subroutine set_val(a,b)
    real, intent(in)::a
    point%x=a
    point%y=b
   end subroutine set_val
   subroutine get_val(a,b)
    real,intent(out)::a,b
    a=point%x
    b=point%y
   end subroutine get_val
 end module mymod
 use mymod
 real::p=2.0,q=3.0,r,s
 call set_val(p,q)              ! valid
 call get_val(r,s)              ! valid
 print *,r,s
 print *, point%x              ! valid
 print *, point%y              ! valid
```

```
point%x=20.0                  ! invalid
point%y=30.0                  ! invalid
end
```

As point is a protected variable (i.e., it is a read-only variable), statements print *, point%x and print *, point%y are valid Fortran statements. However, statements point%x=20.0 and point%y=30.0 would generate a compilation error as point cannot be modified outside the module.

12.18 Scope Rules

The variable defined within a unit is called a local variable for the unit. A variable available in a unit but not defined within the same unit is called a global variable for the unit. At a particular point of time, all local and global variables are available for manipulation. This fact is illustrated with the help of examples.

Case I: When only the main program, but no subprogram, is present, the variables declared within the main program are available within the main program unless they are deallocated (Chapter 15).

```
program main
integer :: ia
real :: r
...
end program main
```

Variables ia and r are available within program main. These locations are given back to the system once the job is over. They are available so long as the job runs.

Case II: When a main program and one or more external subprograms are present, the local variables declared within the main program and subprogram are quite independent of each other. For example, in the following program segment:

```
program main
integer ::ia
...
end program main
subroutine sub
integer :: ia
.
end subroutine sub
```

variable ia of the main program and variable ia of subroutine sub are different. They are stored in different locations. Moreover, when the subroutine is exited, variable ia of the subroutine is deallocated, and when the subprogram is reentered, locations are assigned to the variable—the last value does not reappear unless it is a saved variable (will be discussed shortly). Note that the subprogram can only access the variable declared in the main or other subprograms through arguments or a common block (will be discussed shortly) or through the module.

Case III: Internal procedures do have access to all the variables declared within the unit that contains the internal procedure.

```
program abc
integer :: ia=10
real :: r=20.0
...
call sub
contains
   subroutine sub
   ...
   print *, ia, r
   end subroutine sub
end program abc
```

Variables ia and r are global variables for internal subroutine sub. Therefore, print statement within the subroutine prints 10 and 20.0, respectively. Any modifications done on ia and r within the subroutine are carried over to program abc. However, if the internal subprogram contains a declaration having the same name as that of its 'mother', the local variable is only accessible within the subprogram and the corresponding global variable is masked from the subprogram.

```
program pqr
integer :: ia=10
...
call sub
print *, ia ! ia=10
contains
   subroutine sub
   integer :: ia=20              ! local variable
   print *, ia
   ia=30                         ! ia=30
   end subroutine sub
end program pqr
```

Within the subroutine, the global variable ia, that is, variable ia declared in program pqr, is unavailable, as there is a name conflict. The subroutine can access only the 'local' ia declared within it. Note that the modification of ia (i.e., 30) will not disturb the 'global' ia. When the subroutine is exited, the print statement within program pqr will display the value of ia as 10. The important point is that only the name matters—the type of the variable does not. For example, if integer :: ia=20 within the subroutine is replaced by real :: ia=20.0, the result is the same—namely, the local real variable ia will mask the global integer variable ia.

Case IV: If a module is 'used,' all the variables and subprograms declared within the module are normally available to the unit where the module has been used.

```
module mylib
integer :: ia=10
real :: r=20
end module mylib
```

```
program test
use mylib
print *, r
...
end program test
```

This `print` statement within program `test` will display `20` because the use statement will make variable `r` available to unit `test`.

Finally, it may be noted that the same statement label may be used in the `main` program and subprograms, or among the subprograms, there will not be any conflict.

```
     program main
     ...
10   ...
     end program main
     subroutine sub1(..., ...)
     ...
10   ...
     end subroutine sub1
     subroutine sub2(..., ...)
     ...
10   ...
     end subroutine sub2
```

The same statement number may be used within a program unit and its internal subprograms.

It has already been mentioned that using attribute `only` or `private`, variables or subprograms declared may be made unavailable to a unit containing the `use` statement.

12.19 Generic Subprograms

We have already encountered generic subprograms in connection with the library functions. For example, `sqrt` is the generic name of the square root family. If this is used, the compiler from the nature of its argument can substitute the appropriate routine when the argument is, say, a complex variable by the appropriate function `csqrt`. This facility is directly available in Fortran for the user. If a generic subprogram is defined, the compiler by examining the arguments of the `'call'` statement may substitute the generic name by an appropriate name commensurate with the argument. To use user-defined generic subprograms, the interface block must be present.

Suppose we want to exchange two real numbers and two complex numbers. One solution could be to define two subroutines—`rexchange` for real numbers and `cexchange` for complex numbers—and call explicitly the appropriate routines depending on the type of arguments. The other solution is to define a generic subprogram and let the compiler substitute the appropriate subprogram after analyzing the arguments.

```
program main
interface swap
   subroutine cexchange(a,b)
   complex, intent(inout) ::a, b
   end subroutine cexchange
   subroutine rexchange(p,q)
   real, intent(inout) :: p,q
   end subroutine rexchange
end interface
complex :: ca=(10.0, 20.0), cb=(40.0, 50.0)
real :: a=100.0, b=200.0
call swap(ca,cb)
...
call swap(a,b)
...
end program main
subroutine cexchange(a,b)
complex, intent(inout) :: a, b
complex :: t
t=a; a=b; b=t
end subroutine cexchange
subroutine rexchange(p,q)
real,intent(inout) :: p,q
real :: t
t=p; p=q; q=t
end subroutine rexchange
```

The `interface` block has a name `swap`, and this is the generic name for the subroutines `iexchange` and `rexchange`. When `swap` is called with complex arguments, the compiler generates a call to subroutine `cexchange`, and similarly, when `swap` is called with real arguments, the compiler substitutes the generic name by the appropriate name `rexchange`. The arguments to the generic name `swap` determine the routine to be used by the program. All the subprograms mentioned within this type of `interface` block must be of the same type; that is, all of them must be either a subroutine subprogram or a function subprogram.

It is possible to include the `private` subprograms within the module. The subprogram becomes local to the module. In this case, the subprograms will be referred to within the `interface` block as `module procedure`. So, the previous program would look like the following:

```
module gen_proc
 public:: swap
 private:: cexchange, rexchange
 interface  swap
  module procedure:: cexchange, rexchange
 end interface
contains
 subroutine cexchange(a,b)
 complex, intent(inout) :: a, b
 complex :: t
 t=a; a=b; b=t
 end subroutine cexchange
 subroutine rexchange(p,q)
```

```
   real,intent(inout) :: p,q
   real :: t
   t=p; p=q; q=t
   end subroutine rexchange
 end module gen_proc
 program test_gen
 use gen_proc
 complex :: ca= (10.1, 15.2), cb=(20.3, 25.4)
 real :: a=100.0, b=200.0
 call swap(a,b)          ! interchange two real no; call to rexchange
 print *, a,b            ! interchange two complex no; call to cexchange
 call swap(ca,cb)
 print *, ca,cb
 end program test_gen
```

The `interface` block does not contain the complete executable code (in fact if the complete executable code is present, it will generate a Fortran error). The executable code for the subroutines are present either as external subprograms or as module procedures. Note that private procedures `cexchange` and `rexchange` are not directly accessible from program `test_gen`. They can be accessed indirectly through the generic name `swap`. The word `module` may be dropped from `module procedure`.

12.20 ABSTRACT Interface

Often an `interface` block requires a number of declarations of subroutines or functions with identical types of arguments. All such subroutines/functions are to be declared individually within the `interface` block.

```
   interface
     subroutine sub1(a,b)
       integer, intent(in):: a,b
     end subroutine sub1
     subroutine sub2(a,b)
       integer, intent(in):: a,b
     end subroutine sub2
     ...
     subroutine subn(a,b)
       integer, intent(in):: a,b
     end subroutine subn
   end interface
```

This can be avoided by declaring an `abstract interface` with a template of the subprograms and associating the subprograms with the template through the `procedure` statement.

```
   abstract interface
     subroutine subtemp(a, b)
       integer, intent(in) :: a, b
     end subroutine subtemp
```

```
  end interface
  procedure(subtemp):: sub1, sub2, sub3, ..., subn
```

These subroutines `sub1`, `sub2`, ..., `subn` do not have to be declared explicitly within the `interface` block, and this saves a lot of typing. One example is given as follows:

```
program abstract_demo
abstract interface
 subroutine subint(a, b, c)
  real, intent(in) :: a, b
  real, intent(out):: c
 end subroutine subint
end interface
procedure(subint):: sub1, sub2, sub3, sub4
real :: a=100.0, b=200.0, c
call sub1(a,b,c)
print *, a,b,c
call sub2(a,b,c)
print *, a,b,c
call sub3(a,b,c)
print *, a,b,c
call sub4(a,b,c)
print *, a,b,c
end program abstract_demo
subroutine sub1(a,b,c)
real, intent(in) :: a, b
real, intent(out):: c
c=a+b
end subroutine sub1
subroutine sub2(a,b,c)
real, intent(in) :: a, b
real, intent(out):: c
c=a-b
end subroutine sub2
subroutine sub3(a,b,c)
real, intent(in) :: a, b
real, intent(out):: c
c=a*b
end subroutine sub3
subroutine sub4(a,b,c)
real, intent(in) :: a, b
real, intent(out):: c
c=a/b
end subroutine sub4
```

The procedure statement can be used to associate the `interface` block of another procedure having explicitly a declared `interface` block.

```
program proc_test
 interface
  integer function myfun1(a,b)
   integer,intent(in):: a,b
  end function myfun1
 end interface
```

```
integer:: a=20,b=10,c
procedure (myfun1):: myfun2 ! interface of myfun2 is same as myfun1
c=myfun1(a,b); print *, a,b,c
c=myfun2(a,b); print *, a,b,c
end program proc_test
integer function myfun1(a,b)
integer,intent(in):: a,b
myfun1=a+b
end function myfun1
function myfun2(a,b)
integer,intent(in):: a,b
myfun2=a*b
end function myfun2
```

These are just for demonstration. It may be noted that in this case the `interface` block is really not required.

12.21 Keyword Arguments

It has already been pointed out that there is one-to-one correspondence between the arguments of the calling program and the called program. Calling by keyword is a very useful facility of Fortran, where the arguments are passed through their name defined in the `interface` block; the ordering of the arguments is not relevant. For example, if function `trap` is defined as

```
function trap(a,b)
```

it can be called as

```
z = trap(a=10,b=20.0)
```
or
```
z = trap(b=20.0,a=10)
```

provided an `interface` block exists. In the first case, the dummy parameter 'a' gets a value 10, and the second dummy parameter 'b' gets a value 20.0. The example demonstrates the real power of keyword method of argument passing. Note that in the actual call, the second argument, b, appears before a. However, because of the keyword assignment, 'a' gets a value 10 and 'b' gets a value 20.0. The following also gives an identical result:

```
z = trap(10.0,b=20.0)
```

The dummy parameter 'a' here gets 10.0 and the second parameter 'b' gets 20.0. The last example illustrates both the use of the natural method of using actual parameters and the use of keywords. Once the keyword is used, all the subsequent parameters must be used with keywords. This is illustrated next. Suppose we have defined a function as follows:

```
function func(x1,x2,x3,x4)
```

This can be invoked as follows:

```
z = func(10.0, 20.0, 30.0, 40.0)
z = func(p1, p2, p3, p4) [p1, p2, p3 and p4 are real]
z = func(x4=40.0, x1=10.0, x2=20.0, x3=30.0)
z = func(10.0, 20.0, x4=40.0, x3=30)
z = func(x1=p1, x2=p2, x3=p3, x4=p4)
z = func(p1, p2, x4=p4, x3=p3)
```

However, z = func(x1=10, x2, x3, x4) is an invalid function call. The reason is that once the keyword argument is used, all the subsequent arguments must be specified in terms of keywords.

If a mixed type of arguments are used, the arguments before the keyword argument must have one-to-one correspondence with the corresponding dummy arguments. The following example will display the value of trap_a as 7.5 in all the three cases:

```
program key
interface
  function trap(p,q) result(area)
  real, intent(in)::p,q
  real ::area
  end function trap
end interface
real :: a=10.0, b=5.0, trap_a
trap_a=trap(a,b); print *, trap_a  ! normal call
trap_a=trap(p=10.0,q=5.0); print *, trap_a ! through keyword
trap_a=trap(q=5.0,p=10.0); print *, trap_a
end program key
function trap(a,b) result(area)
real, intent(in) :: a,b
real :: area
area = 0.5*(a+b)
end function trap
```

In this example, in the interface block, function trap is defined in terms of p and q. If the keyword argument is used, variable 'a' within the function will be substituted by the value of p irrespective of its position as actual parameter in the calling sequence. This is true for the other parameters too.

12.22 Operator Overloading

In earlier chapters, various kinds of operators were introduced—arithmetic, relational, logical, etc. The operators have a predefined meaning to the compiler. We have already seen that some of the operators can have different types of arguments. The symbol for unary minus and binary minus is the same; from the context, the compiler can determine the meaning of the operator. In addition, the same binary operator, say, addition operator, can be used to add two integers or two real numbers or two complex quantities. This process is known as operator overloading. In this case, the operators are intrinsically overloaded. A user can overload an operator of his or her choice. For example, one can overload the multiplication operator (*) in such a way that it performs the multiplication operations between two matrices. However, when the same multiplication operator gets two integer or real or complex quantities as operands, the normal multiplication operations are performed. While overloading any operator, it is illegal to modify the basic characteristics of the concerned operator. The intrinsic meaning of the multiplication operator in the case of integer, real or complex quantities cannot be altered. The multiplication operator is intrinsically binary in nature. It is not possible to overload this binary operator such that it becomes a unary operator. No two overloaded operators can have identical types of arguments. In addition to overloading the intrinsic operator, Fortran allows user-defined operators too.

If a relational operator, say, .eq., is overloaded, its equivalent symbol '==' gets automatically overloaded in an identical fashion. This is true for other relational operators too.

The operators are overloaded through the `interface operator` statement. Suppose we want to overload the unary plus and minus operators such that when the operator operates on a character, it changes the lowercase letter to the corresponding uppercase letter and the uppercase letter to the corresponding lowercase letter, respectively.

The `interface` block also contains the definition of the subprogram, which the compiler will 'call' automatically when the overloaded operator is called into play. The `function` subprogram corresponding to the unary operator has one non-optional argument having `intent(in)`; for binary operators two non-optional arguments are required with `intent(in)` for both the arguments.

```
interface operator(+)
 function low_to_up(ch) result(res)
 character, intent(in)::ch
 character :: res
 end function low_to_up
end interface
interface operator(-)
 function up_to_low(ch) result(res)
 character,intent(in) :: ch
 character :: res
 end function up_to_low
end interface
```

When these overloaded operators are used, appropriate subprograms are called, which in turn perform the operations that the operators are supposed to perform.

```
character :: ch1
read *, ch1          ! read a lowercase letter
print *, ch1
ch1 = +ch1           ! uppercase through the overloaded unary + operator
print *,ch1
ch1 = -ch1           ! lowercase through overloaded unary - operator
print *, ch1
end
```

Now we have to design functions low_to_up and up_to_low.

```
function low_to_up(ch) result(res)
character, intent(in) :: ch
character :: res
```

```
res=achar(iachar(ch)-iachar('a') + iachar('A'))
end function low_to_up
function up_to_low(ch) result(res)
character, intent(in) :: ch
character :: res
res=achar(iachar(ch)-iachar('A') + iachar('a'))
end function up_to_low
```

The next example shows how to overload the addition operator to 'add' two strings.

```
program stringadd
interface operator(+)
 function str_add(s1,s2) result(res)
 character (len=*), intent (in) :: s1,s2
 character (len=len(s1)+len(s2)) :: res
 end function str_add
end interface
character (len=10) :: s1
character (len=15) :: s2
character (len=30) :: s3= ' '
s1 = 'debasis'
s2 = 'sengupta'
s3 = s1+s2                  ! this will call function str_add
print *, s3
end program stringadd
function str_add(s1,s2) result(res)
character(len=*), intent(in) ::s1,s2
character(len=len(s1)+len(s2)) ::res
res = trim(s1) // " " // trim(s2)
end function str_add
```

The length of the character variable res is the 'sum' of s1 and s2. The next program shows overloading of the relational operator .gt. , which will be used to find whether one user-defined variable is greater than or not greater than another user-defined type variable according to the prescription of the programmer.

```
module mymod
implicit none
type :: point
 integer::x,y,z
end type point
end module mymod
program vecgr
use mymod
interface operator(.gt.)
 function vgt(a,b) result(res)
 use mymod
 type(point),intent(in)::a,b
 logical::res
 end function vgt
end interface
type(point)::z1,z2
z1=point(3,7,4)
z2=point(2,4,11)
```

```
   if (z1>z2) then ! according to the prescription of the programmer
    print *, "z1 is greater than z2"
   else
    print *, "z1 is not greater than z2"
   endif
   end program vecgr
   function vgt(a,b) result(res)
   use mymod
   type (point),intent(in)::a,b
   logical::res
   res=sqrt(real(a%x**2+a%y**2+a%z**2)) > &
        sqrt(real(b%x**2+b%y**2+ &
        b%z**2)) !arbitrary condition to test greater
   end function vgt
```

Note that we have overloaded the relational operator .gt.. This has resulted in overloading the corresponding symbol '>'.

Next, we give an example to add English distance using addition operator (+).

```
   program distance ! add english distance
   type dist
   sequence
    integer :: mile
    integer :: yds
   end type dist
   interface operator (+)
    function adddist(d1,d2) result(res)
    import dist ! discussed in chapter 19
    type(dist), intent(in)::d1,d2
    type(dist)::res
    end function adddist
   end interface
   type(dist):: d1,d2,d3
   read *, d1%mile, d1%yds, d2%mile, d2%yds! elementary items of d1 and d2
   d3=d1+d2 ! calls function addlist
   print *," d1.miles = ", d1%mile, " d1.yds = ",d1%yds
   print *," d2.miles = ", d2%mile, " d2.yds = ",d2%yds
   print *," d3.miles = ", d3%mile, " d3.yds = ",d3%yds
   end !data validation not done for simplicity yds must be less than 1760
   function adddist(d1,d2) result(res)
   type dist
   sequence
    integer :: mile
    integer :: yds
   end type dist
   type(dist), intent(in)::d1,d2
   type(dist)::res,d3
   d3%mile=0
   d3%yds=d1%yds+d2%yds
   if (d3%yds >=1760) then   ! 1760 yds = 1 mile
    d3%mile=d3%mile+1
    d3%yds=d3%yds-1760
   endif
```

```
d3%mile=d3%mile+d1%mile+d2%mile
res=d3   ! d1+d2 is returned
end function adddist
```

A new statement `import` is used in this example. This statement imports the definition of type `dist` from its host. This is discussed in Chapter 19.

Finally, the preceding program is shown using `module procedure`.

```
module mymod
type dist
 integer :: mile
 integer :: yds
end type dist
interface operator (+)
 module procedure adddist
end interface
contains
 function adddist(d1,d2) result(d3)
 type(dist), intent(in)::d1,d2
 type(dist)::d3
 d3%mile=0
 d3%yds=d1%yds+d2%yds
 if (d3%yds >=1760) then   ! 1760 yds = 1 mile
  d3%mile=d3%mile+1
  d3%yds=d3%yds-1760
 endif
 d3%mile=d3%mile+d1%mile+d2%mile   ! d1+d2 is returned
 end function adddist
end module mymod
program distance_add
use mymod
type(dist):: d1,d2,d3
read *, d1%mile, d1%yds, d2%mile, d2%yds
d3=d1+d2 !  add "english" distance
print *," d1.miles = ", d1%mile, " d1.yds = ",d1%yds
print *," d2.miles = ", d2%mile, " d2.yds = ",d2%yds
print *," d3.miles = ", d3%mile, " d3.yds = ",d3%yds
end
```

12.23 Overloading of Assignment Operator

The assignment operator can also be overloaded. This is illustrated with an example. In the programming language `'C'`, anything that is nonzero is considered as `true`, and if it is zero, then it is `false`. We shall now overload the assignment operator such that we will be able to write the following:

```
logical variable = integer variable
```

If the integer variable is zero, the logical variable will be set to `false`; otherwise, it will be set to `true`. Overloading the assignment operator is again done through the `interface` statement.

```
interface assignment(=)
...
end interface
```

Between these two statements, the name and the type of the parameters of the subprogram (in this case it is a subroutine and not a function) that the system will call when the overloaded assignment operator is used are placed. The subroutine has two non-optional arguments. The `intent` of the first one is `out`, and the `intent` of the second one is `in`.

In the following program, when an integer variable `index` is equated to a logical variable, the system will automatically generate a call to subroutine `ftn_to_c`.

```
program forttoc
interface assignment(=)
  subroutine ftn_to_c(l,i)
  integer,intent(in) ::i
  logical, intent(out)::l
  end subroutine ftn_to_c
end interface
integer :: num
logical :: index
read *, num
index = num
print *,index
end program forttoc
subroutine ftn_to_c(l,i)
integer,intent(in) :: i
logical, intent(out) :: l
if(i .eq. 0) then
 l=.false.
else
 l=.true.
endif
end subroutine ftn_to_c
```

12.24 Overloading of Standard Library Functions

The standard library functions, also called intrinsics, expect a certain type of arguments. For example, the library functions `sqrt` and `log10` do not allow integer arguments. Compiler generates an error message if an integer argument is used. It is true that integer arguments can be converted to real numbers using the library function `real`. However, another solution could be to overload the library function so that depending on the argument, the standard library function or the overloaded library function is called. We demonstrate this using the library function `sin`, which expects a real argument in radian.

Using the following module, it is possible to calculate the `sin` function when the argument is in degrees (integer, say, `30` degrees):

```
module sin_in_deg
implicit none
real, parameter:: pi=3.14159265
real, parameter::piby180=pi/180.0
intrinsic:: sin
public:: sin
private:: sin_integer
interface sin
 module procedure sin_integer
end interface
contains
 elemental function sin_integer (deg) result(sin_in_deg)! see section 12.45
 integer, intent(in):: deg
 real:: sin_in_deg
 sin_in_deg = sin(deg*piby180)
end function sin_integer
end module sin_in_deg
program calc_sin_deg
use sin_in_deg
implicit none
integer:: degree
degree=30
print *, sin(degree)     !30 deg, calls sin_integer(30) = 0.5
print *, sin(pi/2.0)     ! pi/2 radian as real no, calls intrinsic sin
end
```

12.25 User Defined Operators

A user may define his or her own operator. Such operators are enclosed within periods. The name of the user-defined operator cannot contain digits or underscore—it can contain only letters. The length is less than or equal to `63`. There should not be any space between the period and the operator-name. The name cannot be `.true.` or `.false.`. Suppose we wish to define an operator `myand`. This will be a binary operator whose operands are integers. If both the operands are nonzero, it will return `true`; otherwise, it returns `false`. Again, the user-defined operator requires `interface`.

```
program userop
  interface operator(.myand.)
    function testand(i1,i2) result(res)
    integer,intent(in) :: i1,i2
    logical :: res
    end function testand
end interface
integer :: num1, num2
logical :: logi
read *, num1, num2
logi = num1 .myand. num2  ! calls testand
```

```
print *, logi
end program userop
function testand(i1, i2) result (res)
integer,intent(in) :: i1,i2
logical :: res
if(i1.ne.0.and. i2.ne. 0) then
 res=.true.
else
 res=.false.
endif
end function testand
```

Another example of user-defined unary operator is shown. The operator is used to reverse a character string.

```
program userop
interface operator(.reverse.)
 function revstr(ch) result(res)
 character(len=*),intent(in)::ch
 character(len=len(ch))::res
 end function revstr
end interface
character (len=11)::c1="noitaicossa"
c1=.reverse.c1
print *,c1
end
function revstr(ch) result(res)
character(len=*),intent(in)::ch
character(len=len(ch))::res
integer::s,i,j
s=len(ch)
do i=1,s
 j=s-i+1
 res(i:i)=ch(j:j)
enddo
return
end function revstr
```

The output is a character string 'association', which is the reverse of 'noitaicossa'.
   A module may contain user-defined operators. The function associated with the user-defined operator may be either an internal or an external procedure. These are shown with examples. The priority of the user-defined operator is less than that of the arithmetic operator.

```
module mymod
interface operator(.mymul.)
 function mul(a,b) result (res)
 integer,intent(in)::a,b
 integer::res
 end function mul
end interface
end module mymod
program modop
```

```
use mymod
integer::x=10,y=20,z=30,a
a=x.mymul.y + z        ! y+z is performed first, priority of addition is
print *, x,y,z,a       ! more than the user defined operator
end
function mul(a,b) result (res)
integer,intent(in)::a,b
integer::res
res=a*b
end function mul
```

Function mul is, in this case, outside the module. The next example shows function mul as an internal procedure to the module.

```
module mymod
  interface operator(.mymul.)
  module procedure mul
  end interface
contains
    function mul(a,b) result (res)
    integer,intent(in)::a,b
    integer::res
    res=a*b
  end function mul
end module mymod
program modop
use mymod
integer::x=10,y=20,z=30,a
a=x.mymul.y + z        ! this is x*(y+z)
print *, x,y,z,a       ! output 10 20 30 500
end
```

The interface must contain statement module procedure mul or just procedure mul.

12.26 Use Statement and Renaming Operators

A user-defined operator defined within a module may be renamed through the use statement.

```
use mymod, operator (.yourand.) => operator (.myand.)
```

Operator myand defined in module mymod can now be accessed as yourand in the following program. However, this is not allowed for intrinsic operators.

```
module mymod
interface operator(.myand.)
  function testand(i1,i2) result(res)
  integer,intent(in) :: i1,i2
  logical :: res
  end function testand
```

```
      end interface
      end module mymod
      program userop
      use mymod, operator (.yourand.)=> operator(.myand.)
      integer :: num1, num2
      logical :: logi
      read *, num1, num2
      logi = num1 .yourand. num2
      print *, logi
      end program userop
      function testand(i1, i2) result (res)
      integer,intent(in) :: i1,i2
      logical :: res
      if(i1.ne.0.and. i2.ne. 0) then
       res=.true.
      else
       res=.false.
      endif
      end function testand
```

12.27 Precedence of Overloaded Operators

The priority of an overloaded operator is the same as that of the corresponding operator defined in Fortran; that is, if a multiplication operator (*) is overloaded, the priority of this overloaded operator will be the same as that of the multiplication operator.

12.28 Precedence of User Defined Operators

The priority of all user-defined binary operators is the same and is less than that of all intrinsic operators. This may be demonstrated very easily. Consider the following expression: a=x.mymul.y + z, where x=10, y=20 and z=30, and x.mymul.y calculates x*y. However, the multiplication operator is replaced by a user-defined operator .mymul., which performs the same multiplication. As the priority of the user-defined operator is less than that of the intrinsic operator (in this case addition operator), the addition (y+z) will be performed before the user-defined operator .mymul.. Thus, the expression is equivalent to x .mymul. (y+z). It may be verified that the result is indeed 10*(20+30)=500.

The priority of all user-defined unary operators is the same and is greater than that of all intrinsic operators. This may be verified as follows:

```
      program testprio
      interface operator(.mychsign.)
       function chsign(a) result(res)
       integer,intent(in)::a
       integer::res
       end function chsign
```

```
end interface
integer::b=2
print *, .mychsign. b**2
print *, -b**2
end
function chsign(a) result(res)
integer,intent(in)::a
integer::res
res=-a
end function chsign
```

The outputs are 4 and -4, respectively. The first number confirms that the priority of the user-defined unary operator .mychsign. is higher than that of the exponentiation operator (**). Therefore, (-b)**2 is 4. On the other hand, the priority of the exponentiation operator (**) is higher than that of the unary minus (-) Thus, -(b)**2 is -4.

12.29 OPTIONAL Arguments

Arguments to a subprogram may be optional. For example, a call to the subprogram may or may not contain all the actual parameters if the corresponding dummy parameters have been declared as optional. Obviously, the subprogram must have some way of assigning some values to these optional parameters, if they are not supplied as arguments or the other parameters are such that the optional parameters are not required for this particular call. To use a particular parameter as optional, keyword optional is used.

```
interface
 function op_demo(a1,a2,a3,a4) result (res)
 integer, intent(in):: a1,a2
 integer, intent(in), optional:: a3,a4
 integer :: res
 end function op_demo
end interface
```

The interface block in the preceding program indicates that the first two parameters a1 and a2 must be supplied while calling—the third and fourth parameters may or may not be present. Therefore, the function may be invoked in the following manner. Each parameter can appear only once.

```
k = op_demo(10,20,30,40)   ! all the arguments are supplied
                           ! call by value
k = op_demo(m1,m2,m3,m4)   ! call by reference
k = op_demo(m1,m2)
k = op_demo(m1,20,m3)
k = op_demo(m1,20)
```

It is also possible to invoke the function through keyword arguments.

```
k = op_demo(a1=10,a2=20,a3=30,a4=40)
k = op_demo(a2=20,a1=20,a4=40)
k = op_demo(a3=10,a1=20,a2=30)
```

12.30 PRESENT Intrinsic

This intrinsic is used to check the presence of an actual argument within the subprogram body. It was already mentioned that the subprogram must somehow handle the situation when an optional argument is not present.

```
if (present(a3)) then
   ...                  ! optional argument is present
else
   ...                  ! optional argument is not present
endif
```

The present intrinsic returns true if the corresponding argument is present in the call to the subprogram—otherwise, it returns false. The following program illustrates this feature. The program calls a function with two to four arguments. The first and second arguments must be present, and the other two are optional arguments. The function calculates the average of all the arguments present in the call. If any optional argument is absent, it is not considered for calculating the average.

```
program calc_average
interface
 real function average(a1,a2,a3,a4)
 integer, intent(in) :: a1,a2
 integer, intent(in), optional :: a3,a4
 end function average
end interface
integer :: a1,a2,a3,a4
real :: ave
read *, a1,a2,a3,a4
ave = average(a1,a2,a3,a4); print *, ave
ave = average(a1,a2,a3); print *, ave
ave = average(a1,a2); print *, ave
ave = average(a1=10,a2=20,a4=40); print *, ave
ave = average(a1=20,a4=40,a2=30); print *, ave
end program calc_average
real function average(a1,a2,a3,a4)
integer, intent(in) :: a1,a2
integer, intent(in), optional :: a3,a4
integer :: number, total
number = 4 ! this should be 4 each time when the function is entered
total = a1 + a2
if(present(a3)) then ! optional arg a3 present?
 total = total+a3
else
 number = number -1 ! optional arg a3 not present, decrement number
endif
if(present(a4)) then ! optional arg a4 present
 total = total + a4
else
 number = number -1 ! optional arg a4 not present, decrement number
endif
average = real(total)/real(number)
end function average
```

The logic of the program is easy to follow. If an argument is not present, the total number of items is decremented by 1.

12.31 Assumed Rank of Dummy Arguments

The rank of a dummy argument within a subprogram may assume different values depending on the rank of the actual calling argument. The dimension of the dummy argument within the subprogram is specified with two dots (. .). The dimension of the argument within both the `interface` block and the subprogram must be declared with two dots (. .) shown in bold letters.

```
program assumed_rank
integer, dimension(10)::a=10
integer, dimension(3,3)::b=20
integer, dimension(2,2,2)::c=30
integer, dimension(2,2,2,2,2,2,2)::d=40
interface
  subroutine sub(x)
  integer, dimension(..)::x
  end subroutine sub
end interface
call sub(a)
call sub(b)
call sub(c)
call sub(d)
end program assumed_rank
subroutine sub(x)
integer, dimension(..):: x ! at present available with ifort and
                           ! gfortran
print *, "Rank is ", rank(x)
end subroutine sub
```

The output are as follows:

```
Rank is          1
Rank is          2
Rank is          3
Rank is          7
```

The **select rank** construct allows to execute a block of code depending on the rank of the selector (array).

```
subroutine sub(a)
integer, dimension(..):: a
select rank(a)
 rank(0)        ! here if rank is zero
   .
 rank(1)        ! here if rank is 1
   .
```

```
rank(*)          ! here for assumed size
   .
rank default   ! none of the above, default
   .
end select
end subroutine sub
```

12.32 Array-Valued Functions

A user-defined function may return an array.

```
function arr(a1) result(res)
integer :: a1
integer, dimension(4) ::res
res(1)=a1+10; res(2)=a1+20; res(3)=a1+30; res(4)=a1+40
end function arr
program main
interface
 function arr(a1) result(res)
 integer :: a1
 integer, dimension(4) ::res
 end function arr
end interface
integer :: x=100
integer, dimension(4):: y
y=arr(x)
print *, y ! output: 110, 120, 130, 140
end
```

When this function is invoked, it returns an integer array of rank 1 and size 4.

12.33 SAVE Variables

The local variables of any subprogram are lost on exit from the subprogram, and when the subprogram is entered again, it does not remember the last values. If the save declaration is used, the corresponding variable reappears with the last value when the subprogram is entered again. In addition, if a local variable of a subprogram is initialized either along with its declaration or by a data statement, it automatically becomes a save variable.

For example if the following two statements from Section 12.30

```
integer :: number
number = 4
```

are merged into a single statement: `integer :: number=4`, variable `number` becomes a `'save'` variable, and we will see that this will invite trouble in this particular case. Although it is not available outside the subprogram, yet when the function is entered again, the `number` reappears with its last value. The initialization is done only once. The program logic expects that the value of variable `number` should be 4 each time the function is entered (total number of arguments). When the function is called for the second time with three arguments, `number` becomes 3, and when the function is called again with 4 arguments, variable `number` with value 3 reappears. Naturally, the calculation of average in this case fails as the wrong value of `number` is used to calculate the average.

A variable may be forced to become a `save` variable through the `save` declaration.

```
function func(a1) result(res)
integer , save :: num ! num is now a save variable
integer :: kunm
```

Variable `num` is a `'saved'` variable but `kunm` is not. `Save` without any argument makes all the local variables as `'saved'` variable.

```
function func(a1) rsult(res)
...
save
integer :: num ! both num and r are saved variable
real :: r
...
end function func
```

A simple program to know how many times a subprogram is entered is as follows:

```
program main
integer::i
do i=1,3
 call sub
end do
end
subroutine sub
integer::count=0 ! count is now a save variable
count=count+1
print *, "subroutine called ", count, " time(s)"
return
end
```

Being a save variable, `count` reappears with the last value each time the subroutine is entered. The outputs of the program are thus as follows:

```
subroutine called 1 time(s)
subroutine called 2 time(s)
subroutine called 3 time(s)
```

12.34 COMMON Statement

Two or more units may share locations through the common statement. The common statement is hardly used nowadays as a better alternative—the module—is now available. The common statement had been a part of almost every program in earlier years.

There are two types of common statements: blank and block. In a blank common no name is given to the common block.

```
program main
common a,b,c
...
end program main
subroutine sub
common p,q,r
...
end subroutine sub
```

In the common statement of main, there are three real scalars: a, b and c. In the subroutine, the corresponding items (real) are p, q and r, respectively. The variables p and a, q and b and r and c share the same location. Therefore, if a is set to 20.0 in main, this value is reflected through p in subroutine sub. Similarly, if p is set to 100 in the subroutine, it is reflected through a in the main unit. Note that the names do not matter. The relative position of the variables within the common block actually matters. So the names of the variable within the subprogram may be anything, say, x, y and z. The first variable of the common within the subroutine will share the same location with the first variable a of the common block of the main unit. It is, of course, assumed that the number and type of the variable in both the common blocks are the same so that a one-to-one correspondence is maintained.

If common is of the following form,

```
real, dimension(3)::a
common a
...
end
subroutine sub
common p,q,r
...
end
```

then a(1) and p, a(2) and q and a(3) and r will share the same locations. A blank common may be shared between the main program and the subprograms or between the subprograms. Earlier, we have seen that through arguments, data can be transferred between different units of a program; common is another way of sharing data among different units of a program. A particular variable appearing in common cannot be passed as an argument to a subprogram. In addition, the variable cannot appear twice in the common block.

A blank common may also be represented as follows:

```
common // a, b, c
```

where two successive slash symbols indicate the blank common.

There is another type of common, called a block common. In this case, a name is attached to a common block, and there may be several such common blocks in a unit. In blank common, whether one needs them or not, all the variables must be present in every unit. In block common, only the relevant block is required to be present in the unit.

The block name common is placed after keyword common and is enclosed within two slashes (/).

```
program main
common /blk1/ a,b,c
common /blk2/ p,q,r
 . . .
end
subroutine sub1
common /blk1/a,b,c
 . . .
end
subroutine sub2
common/blk2/p,q,r
 . . .
end
```

Since subroutine sub1 contains only block blk1, it can use and modify variables a, b and c of the main unit, but it cannot access variables p, q and r as blk2 is not present in sub1. Similarly, through the blk2 subroutine sub2 may use and modify p, q and r of the main unit. It cannot touch the variables a, b and c of the main unit.

12.35 BLOCK DATA

A block common may be initialized through a special subprogram called block data. The block data subprogram contains only the declarations and the initialization statements. It cannot contain any executable statements. It is not necessary to call the block data subprogram.

```
program main
common /blk1/ a,b,c
common /blk2/ p,q,r
 . . .
end
subroutine sub1
common /blk1/a,b,c
 . . .
end
subroutine sub2
common/blk2/p,q,r
 .
end
block data
common /blk1/ a,b,c
common /blk2/ p,q,r
```

```
data a,b,c/2.0,3.0,4.0/
data p,q,r/100.0,200.0,300.0/
end
```

A block data subprogram can have a name. For a named block data, end block data must have the same name. However, this is optional.

```
block data fb
common /blk2/ p,q,r
data p,q,r/100.0,200.0,300.0/
end block data fb
```

The end block data fb in the preceding program may be replaced just by end block data. There can be any number of named block data subprogram to initialize common blocks; it is not necessary to initialize all the common blocks in a single block data subprogram. No two block data subprograms can have the same name in a single load module. In addition, no common block can appear in two block data subprograms.

```
program main
common /blk1/ a,b,c
common /blk2/ p,q,r
print *, a,b,c,p,q,r
call sub1
call sub2
end program main
subroutine sub1
common /blk1/a,b,c
print *, a,b,c
end subroutine sub1
subroutine sub2
common/blk2/p,q,r
print *, p,q,r
end subroutine sub2
block data fb
common /blk2/ p,q,r
data p,q,r/100.0,200.0,300.0/
end block data fb
block data sb
common /blk1/ a,b,c
data a,b,c/2.0,3.0,4.0/
end block data sb
```

Readers are encouraged to guess the output of this program and verify their observation by executing the program.

The usual rules to initialize variables through the data statement are applicable within block data. A block data can contain only the definitions of the user-defined type declaration statements. It can also contain asynchronous, bind, common, data, dimension, equivalence, implicit, parameter, pointer, save, target, use and volatile statements. However, it cannot contain allocatable, external and bind attribute specifiers.

The final point to be noted is that there can be one blank block data subprogram (block data without a name) in a load module.

12.36 COMMON and DIMENSION

Dimension of a variable may be given along with common:

```
dimension a(10)
common/blk1/a
```
and
```
common/blk1/a(10)
are equivalent.
```

12.37 COMMON and User Defined Type

The common statement works on the assumption that variables are stored sequentially within a block. This may not be true for user-defined variables, and as such, the user-defined variables should be forced to occupy consecutive locations.

If a user-defined type variable is initialized through block data, sequence must be present in the type definition.

```
type:: t
 sequence
 integer:: a
 real:: b
end type t
type(t)::x
common/blk1/x
print *, x%a, x%b
end
block data
type:: t
 sequence
 integer:: a
 real:: b
end type t
type(t)::x
common/blk1/x
data x/t(10,20.0)/
end
```

12.38 COMMON and EQUIVALENCE

When two variables share a location within a program unit, the equivalence statement is used. When two variables share the same locations between two different units, the common statement is used.

A variable, declared in a common block, may be made equivalent to a local variable declared in the same unit. However, there are certain restrictions.

Rule I: Variables declared in a particular common block cannot be made equivalent to one another. In addition, variables declared in two different common blocks cannot be made equivalent.

```
common /blk1/a,b,c
equivalence (a,b)            ! not allowed
common/blk2/p,q,r
common/blk3/x,y,z
equivalence (p,x)            ! not allowed
```

Rule II: A common block cannot be extended beyond its lower bound through the equivalence statement.

```
dimension a(10)
common/blk1/ b(10)
equivalence(b(1),a(2))       ! not allowed
```

Rule III: A common block can be extended in the direction of its upper bound.

```
equivalence (b(2),a(1))      ! allowed
```

However, this will increase the size of the common block to 11. This procedure is discouraged.

12.39 EXTERNAL Statement

We have seen that the name of a subprogram can be anything including the name of an intrinsic function. If the name of an intrinsic is used as the name of a user-defined subprogram, it is necessary to pass this information through an `external` statement. In that case, compiler will correctly use the user-defined subprogram.

```
program main
external sqrt
real:: sqrt
real:: x,y=2.0
x=sqrt(y)        ! library function sqrt will not be called.
print *, x       ! user defined sqrt will be called
end
function sqrt(x)
real, intent(in):: x
sqrt=x*x ! this sqrt calculates square of x and not square root of x
end              ! but this is certainly not advisable
```

Without the `external` statement, the library function `sqrt` will be used in `main`. The `external` statement prevents the use of the library function; instead, the user-defined function sqrt is used. However, if the user-defined function sqrt is inside a module, an external statement is not required.

```
module ext_sq
contains
 function sqrt(x)
 real, intent(in):: x
 sqrt=x*x ! this sqrt calculates square of x and not
          ! square root of x
 end function sqrt ! however this is not a good
                   ! programming practice
end module ext_sq
program main
use ext_sq
real:: x,y=2.0
x=sqrt(y)  ! library function sqrt will NOT be called
print *, x ! user defined sqrt will be called
end
```

The next example shows how the intrinsics are passed as arguments to a subprogram. In this case, an intrinsic declaration is required.

```
intrinsic:: sin, cos
real,parameter::pi=3.14159265
real:: myfunc, result
result=myfunc(sin,pi/6)      ! result is 0.5
print *, result
result=myfunc(cos,pi/3)      ! result is 0.5
print *, result
end
real function myfunc(f,x)
external f
myfunc=f(x)
end function myfunc
```

If the subprogram is an internal one, an external statement is not required.

```
intrinsic:: sin, cos
real,parameter::pi=3.14159265
real :: result
result=myfunc(sin,pi/6); print *, result
result=myfunc(cos,pi/3); print *, result
contains
 real function myfunc(f,x) ! external not required
  myfunc=f(x)
 end function myfunc
end
```

Again, if the subprogram is within a module, the external statement is optional.

```
module mymod
real,parameter::pi=3.14159265
contains
 real function myfunc(f,x)
```

```
 external f   ! optional
 real, intent(in):: x
 myfunc=f(x)
 end function myfunc
end module mymod
use mymod
intrinsic:: sin, cos
real :: result
result=myfunc(sin,pi/6); print *, result
result=myfunc(cos,pi/3); print *, result
end
```

There is, of course, no harm even if all the external subprograms are declared as `external`.

```
program main
external sub1, sub2
...
end
subroutine sub1
...
end
subroutine sub2
...
end
```

If a subprogram name is to be used as an argument of a function, it must be declared as `external` in the calling unit. The following program integrates a function using the trapezoidal rule. One of the arguments to function `trap` is a function name, and so it has been declared as `external`.

```
program main
real :: a=0.0,b=90,h=0.1,f,result
external f
result = trap(f,a,b,h)
print *, result
end program main
function trap(f,a,b,h)
real, intent(in)::a,b,h
interface
 real function f(x)
 real, intent (in)::x
 end function f
end interface
real :: res
integer :: i,np
np=nint((b-a)/h); res=0
do i=1, np-1
 res=res+f(a+i*h)
enddo
res=h/2.0*(f(a)+f(b)+2.0*res); trap=res*3.14149265/180.0
end function trap
function f(x)
real f
real,intent(in)::x
```

```
real,parameter::pi=3.1416926/180.0
f=sin(x*pi)
end function f
```

If an intrinsic is passed as an argument, generic name should not be used—the particular name must be passed as follows:

```
intrinsic:: csqrt        ! not sqrt
complex:: myfunc, result
complex::z=cmplx(2.0,2.0)
result=myfunc(csqrt,z)   ! not sqrt
print *, result; print *, csqrt(z)
end
complex function myfunc(f,x)
complex :: f, x
external f
myfunc=f(x); end function myfunc
```

Both the print statements should display the identical result.

12.40 Recursion

A function or a subroutine can call itself. This is called recursion. A recursive procedure (function or subroutine) must have one exit point; otherwise, recursion becomes infinite—a never-ending process.

A recursive procedure can always be converted into an iterative procedure. For example, calculation of the factorial can be done with a recursive procedure—it can also be performed using the iterative method. A recursive procedure consumes more system resources than the corresponding iterative counterpart does; the recursive procedure takes more time to perform an identical computation than an iterative procedure. However, as some of the algorithms are intrinsically recursive, it is sometimes convenient to express such an algorithm as a recursive procedure.

12.41 RECURSIVE FUNCTION

A recursive function, say, factorial, is declared as follows:

```
recursive function factorial(n) result(res)
```

As the recursive function calls itself, the name of the function cannot be used to return the value calculated from the function—the result clause is required to transfer the value calculated by the function. The complete declaration is given as follows:

```
recursive function factorial(n) result(value)
integer, intent(in) :: n
integer :: value
```

In each recursive call, new local variables n and `value` are created. As the recursive function returns value through the `result` clause, variable `value` must appear at least once on the left-hand side of the assignment sign. A complete recursive function to calculate factorial is as follows:

```
recursive function factorial(n) result(value)
integer,intent(in) :: n ! n >= 1
integer :: value
if(n==1) then
 value=1 ! exit point
else
 value=n*factorial(n-1)
 ! calls itself
endif
end function factorial
```

| 4*f(3) | 3*f(2) | 2*f(1) | f(1)= 1 |

FIGURE 12.1
Recursive call (n=4).

We consider a case when the recursive function is called with n=4, that is, to calculate factorial 4. As n is not equal to 1 when the function is called, the 'else' part of the if statement is executed, and `value` can be obtained only when `factorial(n-1)`, that is, `factorial(3)`, is known. In fact, `factorial(3)` is a call to the recursive factorial function with n=3. This process continues until the `factorial` is called with n=1. Now the 'then' part, that is, the exit part of the recursive function, is reached, and 1 will be returned, which will, in turn, enable to calculate `factorial(2)`. Having obtained `factorial(2)`, now `factorial(3)` can be calculated in an identical manner. In a similar manner, ultimately `factorial(4)` will be obtained (Figure 12.1).

The main program to call the recursive `factorial` function is as follows:

```
program recur
interface
 recursive function factorial(n) result(val)
 integer, intent(in) :: n
 integer:: val
 end function factorial
end interface
integer :: f,n
read *, n ! n=4 (say)
f=factorial(n)
print *, f
end program recur
```

A recursive function within a module does not require an `interface` block.

```
module rec_fun
contains
  recursive function factorial(n) result(value)
  integer,intent(in) :: n
  integer :: value
  if(n==1) then
   value=1 ! exit point
  else
   value=n*factorial(n-1) ! calls itself
  endif
```

```
  end function factorial
end module rec_fun
use rec_fun
integer :: x=5
print *, factorial(5)
end
```

Example of another recursive function is given as follows (calculation of GCD):

```
recursive function gcd(n,m) result(gasagu) ! calculation of gcd.
integer,intent(in) :: n,m
integer :: gasagu
integer :: r ! local variable
r=mod(n,m)
if(r==0) then
 gasagu=m ! exit point
else
 gasagu=gcd(m,r)
endif
end function gcd
```

The main program to call the gcd function is as follows:

```
integer :: n=80, m=50
interface
 recursive function gcd(n,m) result(gasagu)
 integer,intent(in) :: n,m
 integer :: gasagu
 end function gcd
end interface
print *, gcd(n,m)
end
```

12.42 RECURSIVE SUBROUTINE

A subroutine also can call itself. As the subroutine returns the value through its argument, it must not contain any result clause. If the recursive subroutine is an external subprogram, an interface block is required. However, if it is inside a module, an interface block is not required (will give error if used). The following program uses a module and calculates the factorial through a recursive call.

```
module rec_sub
contains
 recursive subroutine fact(n,value)
 integer, intent(in) :: n ! n >=1
 integer, intent(out) :: value
 if(n==1) then
  value=1 ! exit point
 else
```

```
    call fact((n-1),value)
     value=n*value
    endif
    end subroutine fact
   end module rec_sub
   program recsub
   use rec_sub
   integer :: num, f
   read *, num
   call fact(num,f)
   print *, f
   end program recsub
```

The program logic is similar to that of the recursive function. However, the difference is that the subroutine returns the value through its argument, and in this case, it returns the value through variable value.

The next program sorts an array using the quicksort algorithm. The algorithm, based on a technique called divide and conquer, may be found in any standard textbook on data structures.

```
   program quick_sort
   interface
    recursive subroutine quick(a,first,last)
    integer, dimension(:), intent(inout)::a
    integer, intent(in)::first, last
    end subroutine quick
   end interface
   integer, dimension(10)::a=[32,17,31,6,51,-26,19,12,-2,77]
   integer::first=1
   integer::last=10
   print *, a
   call quick(a,first,last)
   print *,a
   end program quick_sort
   recursive subroutine quick (a,first,last)
   integer, dimension(:),intent(inout)::a
   integer, intent(in)::first, last
   integer::mid,high,low,t
   mid=a((first+last)/2)
   high=last
    low=first
    do
      do while (a(low)<mid)
       low=low+1
      enddo
      do while(a(high)>mid)
        high=high-1
      enddo
      if (low<=high) then
       t=a(low)
       a(low)=a(high)
       low=low+1
       a(high)=t
       high=high-1
```

```
      endif
      if(low>high) then
        exit
      endif
    enddo
    if(low<last) then
      call quick(a,low,last)
    endif
    if(first<high) then
      call quick(a,first,high)
    endif
  end subroutine quick
```

12.43 PURE Procedure

A pure procedure is a type of procedure that does not have any side effect. For a pure function, the procedure just returns a value; for a pure subroutine, it can modify only the intent(out) or intent(inout) parameter only. Unless otherwise specified all elemental procedures are pure. All the intrinsic functions, subroutines mvbits and move_alloc, are pure. A statement function (obsolete feature, not discussed in the book) is considered as pure if all the functions it uses are pure.

A forall (also obsolete feature) statement can call a pure procedure directly. A pure procedure can call another pure procedure; also, it can be passed as an argument to a pure procedure.

A pure procedure is declared with suffix pure:

```
pure integer function func (a, b)
integer, intent (in) :: a, b
...
end function func
pure subroutine sub(a, b, c)
real, intent (in) :: a
integer, intent(out) :: b
integer, intent (inout) :: c
...
end subroutine sub
```

12.44 Rules for PURE Procedure

- All non-pointer dummy arguments of the pure function must be of intent (in).
- All the non-pointer dummy arguments of the subroutine must have their intent specified.
- A local variable of a pure procedure cannot have a save attribute attached to it.

- A local variable cannot be initialized along with its declaration or through a data statement within a pure procedure.
- All the internal subprograms of the pure procedure must be pure.
- The statements open, close, inquire, print, endfile, backspace, rewind, flush, wait, read, print, write, pause, error stop and stop are not allowed within a pure procedure.
- A pure procedure cannot be a recursive procedure.
- A pure procedure cannot contain an image control statement (Chapter 20).
- If the dummy argument of a pure procedure is the name of another procedure, the corresponding actual procedure (parameter) must be a pure procedure.
- A forall statement can 'call' a pure procedure directly.
- Arguments transferred through the common block cannot be modified inside a pure procedure. The following program would generate a compilation error:

```
integer :: a, b
common/blk1/ c
call sub(a, b)
...
end
pure subroutine sub (x, y)
integer, intent (in) :: x, y
common/blk1/ z
...
z=20.0   ! not allowed
...
end
```

The pure procedure has been introduced with an eye to facilitate parallel programming.

12.45 ELEMENTAL Procedure

An elemental procedure is a function or a subroutine whose arguments are all scalars, but it can be called with conformable arrays. If the arguments are scalar, the result is also scalar. For functions, the result must be scalar. However, if the procedure is called with arrays, in such a case the procedure acts on every element of the array in an identical manner. This is best explained with the help of an example. An elemental procedure starts with keyword elemental. All the elemental procedures are, by default, pure so the rules stated previously are applicable to the elemental procedure. It is permitted to have elemental procedures that are not pure (see Section 12.46).

```
module ele_proc
contains
  elemental subroutine exchange(a, b)
    real, intent(inout) :: a, b
    real :: t
    t=a; a=b; b=t
  end subroutine exchange
```

```
end module ele_proc
program main
use ele_proc
real, dimension (3) :: p, q
p=[10.0, 20.0, 30.0]; q=[100.0, 200.0, 300.0]
call exchange (p, q); print *, p, q
end
```

The elemental subroutine is declared with scalar arguments a and b. However, it is called with arrays p and q. The subroutine will interchange p(1) with q(1), p(2) with q(2) and p(3) with q(3).

An elemental procedure cannot be recursive or vice versa.

12.46 IMPURE ELEMENTAL Procedure

By default, all elemental procedures are pure. An elemental procedure may be declared impure by using the prefix impure. Like pure elemental procedure, the arguments are scalar.

```
impure elemental subroutine sub(...)
```

If the procedure is called with arrays, the procedure acts on every element of the array according to the order of the array elements. Moreover, it may have a side effect. Consider the following program:

```
module mymod
integer,save::num=0
contains
  impure elemental subroutine sub(x,y,z)
  integer,intent(inout)::x,y,z
  z=x+y
  num=num+1
  end subroutine sub
end module mymod
use mymod
integer, dimension(10)::a,b,c
a=4; b=5; call sub(a,b,c); print *, num
end
```

Subroutine sub is an elemental subroutine. Therefore, when it is called with an array, the subroutine acts on every element of the array, that is, c(1)=a(1)+b(1), c(2)=a(2)+b(2), ... As the subroutine is an impure subroutine, it is permitted to modify variable num, and each time the subroutine is called, variable num is incremented by 1. Note that num is a saved variable. The size of a, b and c arrays is 10, and therefore, the subroutine will be called 10 times. Thus, the value of num becomes 10 after the subroutine call. It is needless to point out that the compiler will generate a compilation error if the word impure is removed from the subroutine declaration (shown in bold).

12.47 SUBMODULE

A change in the module procedure body requires compilation of the whole module, which is not very convenient especially when the module is very large. Normally, a well-designed module interface block does not change with time. It would be convenient if the interface block and the procedure body could be separated so that a change in the mode of computation need not require the compilation of the whole module. This is precisely done with the help of a submodule. The interface block stays with the module, and the body of the procedure goes with the submodule. Thus, a change in the submodule requires recompilation of the submodule only. We illustrate this with the help of simple programs.

```
module m1
  interface
   module subroutine sub1 (a,b,c)
   integer, intent(in):: a,b
   integer, intent(out):: c
   end subroutine
  end interface
end module
submodule(m1) m1_s   ! compiled with ifort compiler
contains
 module subroutine sub1(a,b,c)
 integer, intent(in):: a,b
 integer, intent(out):: c
 c=a+b ! if this instruction is changed only submodule m1_s
 ! is to be recompiled as the interface block within
 ! module is not changed.
 end subroutine
end submodule
program test_submodule
use m1
integer:: x=20, y=10, z
call sub1(x,y,z)
print *, "x, y, z = ",x,y,z
end
```

In the preceding program, the interface block is within the module, and the actual calculation is performed within submodule. Keeping the interface block same if the method of calculation is changed, say, c=a+b is changed to c=a-b, the submodule m1_s is to be recompiled only. This example is very trivial. In actual practice, the module may contain many statements, and recompilation may take a considerable amount of time.

```
module m1
  interface
   module subroutine sub1 (a,b,c)
   integer, intent(in):: a,b
   integer, intent(out):: c
   end subroutine
  end interface
end module
submodule(m1) m1_s ! submodule of module m1
contains
```

```
    module subroutine sub1(a,b,c)
    integer, intent(in):: a,b
    integer, intent(out):: c
    c=a-b ! this instruction is changed so submodule m1_s
! is to be recompiled as the interface block within
! module is not changed
    end subroutine
  end submodule
  program test_submodule
  use m1
  integer:: x=20, y=10, z
  call sub1(x,y,z)
  print *, "x, y, z = ",x,y,z
  end
```

Now consider the following program:

```
  module smod
  type :: dis
   real:: a,b,c
  end type dis
  interface
   module real function discr(z)
   type(dis), intent(in):: z
   end function discr
  end interface
  end module smod
  submodule(smod) submodf1
  contains
   module real function discr(z)
   type(dis),intent(in):: z
   discr=(z%b**2-4.0*z%a*z%c)
   end function discr
  end submodule submodf1
  program test
  use smod
  type(dis):: z
  real:: d
  z%a=2.0; z%b=6.0; z%c=3.0; d=discr(z); print *, d
  end
```

The prefix module (shown in bold) to function discr within the interface block indicates that this interface is related to the module procedure and is not related to any external procedure. Moreover, the operating part of the module procedure is within the submodule.

A submodule can also have a submodule. The modules and submodules maintain a parent–child–grandchild relation. If a submodule is changed, only the submodule and its descendants are to be recompiled.

A submodule may contain its own variables. These are available to the submodule and its descendants. These are not accessible outside the submodule. If there is a name conflict with the parent, the corresponding variable of the parent is masked—it is not available within the submodule where it is defined and to its descendants.

The next program shows two submodules m1_s1 and m1_s2 of module m1.

```
module m1
 interface
  module subroutine sub1 (a,b,c)
  integer, intent(in):: a,b
  integer, intent(out):: c
  end subroutine
  module subroutine sub2 (a,b,c)
  integer, intent(in):: a,b
  integer, intent(out):: c
  end subroutine
 end interface
end module
submodule(m1) m1_s1
contains
 module subroutine sub1(a,b,c)
 integer, intent(in):: a,b
 integer, intent(out):: c
 c=a+b
 end subroutine
end submodule
submodule(m1) m1_s2
contains
 module subroutine sub2(a,b,c)
 integer, intent(in):: a,b
 integer, intent(out):: c
 c=a-b
 end subroutine
 end submodule
program test_submodule
use m1
integer:: x=20, y=10, z
call sub1(x,y,z)
print *, "x, y, z = ",x,y,z
call sub2(x,y,z)
print *, "x, y, z = ",x,y,z
end
```

If m1_s2 is a descendant of m1_s1, it is defined as submodule (m1:m1_s1) m1_s2.

```
module m1
 interface
  module subroutine sub1 (a,b,c)
  integer, intent(in):: a,b
  integer, intent(out):: c
  end subroutine
  module subroutine sub2 (a,b,c)
  integer, intent(in):: a,b
  integer, intent(out):: c
  end subroutine
 end interface
end module
submodule(m1) m1_s1
contains
 module subroutine sub1(a,b,c)
```

```
 integer, intent(in):: a,b
 integer, intent(out):: c
 c=a+b
 end subroutine
end submodule
submodule(m1:m1_s1) m1_s2 ! submodule of m1_s1
contains
 module subroutine sub2(a,b,c)
 integer, intent(in):: a,b
 integer, intent(out):: c
 c=a-b
 end subroutine
end submodule
program test_submodule
use m1
integer:: x=20, y=10, z
call sub1(x,y,z)
print *, "x, y, z = ",x,y,z
call sub2(x,y,z)
print *, "x, y, z = ",x,y,z
end
```

12.48 EQUIVALENCE and MODULE

If a variable is declared within a module, the same cannot be made equivalent with a local variable.

```
module mymod
  integer :: i
end module mymod
program main
use mymod
integer :: k
equivalence (i,k) ! not allowed
...
end
```

The rule is that objects belonging to different scoping units cannot be made equivalent.

12.49 Function Calls and Side Effects

When two logical conditions are connected by logical operators such as and/or, evaluation of both the conditions is not always required to arrive at the final result. For example, if two relational expressions are connected by an 'and' operator and if the first relation is false, there is no need to evaluate the second expression. Similarly, if they are connected by 'or' and if the first condition is true, there is no need to evaluate the second condition.

```
if (a>b .and. f(q)) then
...
endif
```

Suppose f is a logical function (returns true or false) and the program logic demands that it is to be evaluated irrespective of the result of the first condition. If a>b is false, it is likely that function f will not be invoked and this might create a logical problem. This is called a side effect. In such a situation, the function should be called before the if statement as follows:

```
logical :: l
...
l=f(q)
if(a>b .and. l) then
...
endif
```

As function f is called before the if statement, the logical problem is avoided, and at the same time the returned value is used in the if statement, if required. A similar problem may arise with other logical operators too.

12.50 Mechanism of a Subprogram Call

When a subprogram is called, the control is transferred to the calling program from the called program. In addition, the call statement carries with it the return address. The return address is used when the control is transferred back to the calling program from the called program. The control is returned to the statement following the call statement. In case of a function call like c=func(x), the next instruction after returning from the function call is the assignment of the returned value from the function to variable c.

12.51 Recursive Input/Output

Input/output statement can be recursive, that is, when one input/output statement is in operation, it can initiate another input/output statement.

```
implicit none
integer:: x
print *, 'type 1 or any other integer'
read *, x
print *, fun(x)
contains
 real function fun(n)
 integer, intent(in):: n
```

```
    if (n.eq.1) then
     print *, 'n is equal to 1'
    fun=10.0
    else
     print *, 'n is not equal to 1'
     fun=20.0
     endif
     end function fun
    end
```

The outputs for two sets of input, 1 and 3, are given as follows:

Input
1

Output

n is equal to 1   ⟸   Generated due to recursive call
10.0   Generated due to normal function call in
                         the main program

Input
3

Output

n is not equal to 1   ⟸   Generated due to recursive call
20.0   Generated due to normal function call in
                           the main program

12.52 Programming Examples

The following program solves a differential equation using the Runge–Kutta method. It is actually a predator–prey problem, say, tiger and deer. When the tiger population goes up, the deer population goes down. Again when the deer population goes down, the tigers do not have enough food, and as such, the tiger population diminishes. Then when the tiger population goes down, the deer population goes up, as there are not many tigers to kill the deer. There will be a periodic oscillation of the population of the tiger and the deer. This is graphically shown side by side. The following differential equations are solved numerically using the standard Runge–Kutta method:

```
dx/dt=ax-bxy
dy/dt=cxy-dx
```

where a, b, c and d are constants.

```
program plot
implicit none
integer:: i
real, parameter::a=4.0,b=2.0,c=1.0,d=3.0
```

```
character, parameter::dot='.', cros='x', star='*',blk=' '
character, dimension(51)::line=blk
real, dimension(421)::x1,y1
real:: xx,yy,ak1,ak2,ak3,ak4,bk1,bk2,bk3,bk4,xxt,t,xm,ym,gm
real:: h,hf,h6
integer::ir,if
x1(1)=8.0
y1(1)=2.5
h=0.01
hf=0.5*h
h6=h/6.0 ! runge-kutta
do i=2,421
 xx=x1(i-1)
 yy=y1(i-1)
 ak1=f1(xx,yy)
 bk1=f2(xx,yy)
 xxt=xx+hf
 ak2=f1(xxt,yy+hf*ak1)
 bk2=f2(xxt,yy+hf*bk1)
 ak3=f1(xxt,yy+hf*ak2)
 bk3=f2(xxt,yy+hf*bk2)
 xxt=xx+h
 ak4=f1(xxt,yy+ak3*h)
 bk4=f2(xxt,yy+bk3*h)
 x1(i)=xx+h6*(ak1+2.0*(ak2+ak3)+ak4);
 y1(i)=yy+h6*(bk1+2.0*(bk2+bk3)+bk4)
end do
 xm=maxval(x1); ym=maxval(y1)! find maximum
 line=dot ! plot y-axis  -------->
 print 20, line
20       format(1x,19x,51a)
 gm=max(xm,ym); gm=1.0/gm; t=0.0
 do i=1,421,12
  line=blk ! fill the line with blank
  if=50.0*(y1(i)*gm)+1.5 ! the values of if and ir lie between
                          ! 1 and 51
  ir=50.0*(x1(i)*gm)+1.5
  line(1)=dot ! dot for the x-axis
  line(if)=star
  line(ir)=cros
  print 10,t,x1(i),y1(i),line
10       format(1x,3f6.2,1x,51a)
  t=t+0.12
 end do
 contains
  real function f1(x,y)
  real::x,y
  f1=a*x-b*x*y
  end function f1
  real function f2(x,y)
  real:: x,y
  f2=c*x*y-d*y
```

```
      end function f2
   end program plot
```

The output of the program is shown next. The width of the paper is used as y-axis. The motion of the "carriage" is the x-axis.

```
                  . . . . . . . . . . . . . . . . . . . . . . . . . . . . . . . . .
0.00   8.00   2.50 .                    *                            X
0.12   6.01   4.18 .                           *            X
0.24   3.16   5.09 .                   X            *
0.36   1.53   4.69 .          X                   *
0.48   0.88   3.77 .    X                    *
0.60   0.63   2.88 . X              *
0.72   0.56   2.16 . X         *
0.84   0.57   1.61 . X      *
0.96   0.66   1.21 . X    *
1.08   0.82   0.92 .    X*
1.20   1.08   0.72 .    *X
1.32   1.48   0.58 .   *      X
1.44   2.09   0.50 .   *        X
1.56   2.96   0.47 . *           X
1.68   4.20   0.50 .  *               X
1.80   5.82   0.63 .  *                    X
1.92   7.67   0.98 .     *                        X
2.04   8.91   1.86 .        *                          X
2.16   7.77   3.67 .               *             X
2.28   4.28   5.40 .                     X    *
2.40   1.82   5.40 .          X              *
2.52   0.88   4.41 .    X                *
2.64   0.55   3.35 . X                 *
2.76   0.44   2.48 . X          *
2.88   0.42   1.82 . X        *
3.00   0.46   1.34 . X     *
3.12   0.57   0.99 .   X *
3.24   0.74   0.75 .    X
3.36   1.02   0.58 .   * X
3.48   1.44   0.47 . *      X
3.60   2.07   0.40 . *        X
3.72   3.01   0.38 . *           X
3.84   4.35   0.40 . *               X
3.96   6.17   0.52 .  *                   X
4.08   8.32   0.86 .     *                          X
4.20   9.88   1.80 .        *                              X
```

The next example uses a two-dimensional array to plot a graph of sin(x), where x is between 0 and 360 degrees.

```
   subroutine xyplot(x,y,lines,graph,width,last,symb)
   implicit none
   real, dimension(:)::x,y
   integer, intent(in)::lines,width,last
```

```
      character, dimension(:,:)::graph
      character, intent(in)::symb
      character, parameter::border="i", blank=" "
      real::yl,ys,xs,xl,xscale,yscale,yv
      real, dimension(6)::lx
      integer::i,j,ix,iy
      yl=maxval(y); ys=minval(y); xl=maxval(x); xs=minval(x)
      graph=blank
      xscale=(xl-xs)*0.02; yscale=(yl-ys)/float(lines-1)
      graph(1,:)=border; graph(width,:)=border
      do i=1,last
       ix=(x(i)-xs)/xscale+1.5; iy=(y(i)-ys)/yscale+0.5 ; iy=lines-iy
       graph(ix,iy)=symb
      enddo
      do i=1,6
       lx(i)=10.0*(i-1)*xscale+xs
      enddo
      print 200
200   format('      y values ',11('i       '))
      yv=yl+yscale
      do i=1,lines
       yv=yv-yscale
       print 400, yv,(graph(j,i),j=1,width)
400    format(e12.2,1x,51a1)
      enddo
      print 600
600   format(13x,10('i....'),'i')
      print 700, (lx(i),i=1,6)
700   format(9x,f7.2, 5(3x,f7.2))
      print 800, xscale,yscale
800   format(/8x,' xscale= ',e10.2,4x,'yscale= ',e10.2)
      end subroutine xyplot
      program xy_plot
      implicit none
      interface
       subroutine xyplot(x,y,lines,graph,width,last,symb)
       real,dimension(:)::x,y
       integer,intent(in)::lines,width,last
       character,dimension(:,:)::graph
       character,intent(in)::symb
       end subroutine xyplot
      end interface
      real, dimension(73)::x,y
      character, dimension(51,51)::graph
      real, parameter::factor=3.1415926/180.0
      integer::i,k
      k=1
      do i=1,361,5
        x(k)=i-1; k=k+1
      enddo
      y=sin(factor*x)
      call xyplot(x,y,51,graph,51,73,'*')
      end program xy_plot
```

The output of the preceding program is given as follows:

```
Y VALUES I    I    I    I    I    I    I    I    I    I
0.10E+01 I         ****                                  I
0.96E+00 I          *      *                             I
0.92E+00 I         **     **                             I
0.88E+00 I        *          *                           I
0.84E+00 I                                               I
0.80E+00 I       *           *                           I
0.76E+00 I      *              *                         I
0.72E+00 I     *                *                        I
0.68E+00 I                                               I
0.64E+00 I     *                 *                       I
0.60E+00 I                                               I
0.56E+00 I    *                   *                      I
0.52E+00 I   *                      *                    I
0.48E+00 I                                               I
0.44E+00 I  *                        *                   I
0.40E+00 I                                               I
0.36E+00 I  *                         *                  I
0.32E+00 I                                               I
0.28E+00 I                                               I
0.24E+00 I *                           *                 I
0.20E+00 I                                               I
0.16E+00 I*                             *                I
0.12E+00 I                                               I
0.80E-01 I*                              *               I
0.40E-01 I                                               I
-0.24E-06 *                              *              *
-0.40E-01 I                                              I
-0.80E-01 I                              *             *I
-0.12E+00 I                                             I
-0.16E+00 I                             *              *I
-0.20E+00 I                                             I
-0.24E+00 I                          *             *  I
-0.28E+00 I                                            I
-0.32E+00 I                                            I
-0.36E+00 I                        *              *   I
-0.40E+00 I                                           I
-0.44E+00 I                       *               *   I
-0.48E+00 I                      *                    I
-0.52E+00 I                                      *    I
-0.56E+00 I                    *                  *   I
-0.60E+00 I                                           I
-0.64E+00 I                  *                 *       I
-0.68E+00 I                                           I
-0.72E+00 I                 *                 *        I
-0.76E+00 I                *                 *         I
-0.80E+00 I               *                 *          I
-0.84E+00 I                                           I
-0.88E+00 I              *                *            I
-0.92E+00 I             **              **             I
-0.96E+00 I             *               *              I
-0.10E+01 I              ****                          I
         I....I....I....I....I....I....I....I....I....I
         0.00    72.00   144.00   216.00   288.00   360.00
         xscale=  0.72E+01    yscale=   0.40E-01
```

13

String with Variable Length

The character variable (string) was introduced in Chapter 4. It was pointed out that the length of the character variables needs to be specified at the compilation time. The present version of Fortran supports character strings of variable length by using a module called iso_varying_string. The character variable that will have this property (variable length) must be of type varying_string defined in the module iso_varying_type. Strictly speaking, it is not a feature of the language as it is introduced through a module written in Fortran 95, and this is not a part of Fortran 2018.

```
program varingstr
use iso_varying_string
type (varying_string) :: name, surname
type (varying_string), dimension (10) :: fname
...
```

Variables name and surname are of the type varying_string, and the variable fname is an array of rank 1 of size 10 of the type varying_string. These variables may change their length dynamically during the execution of the program. The length must be non-negative, and there is no restriction on the size; there may be some processor-dependent upper bound. The characters are numbered 1, 2, 3, ... n, where n is the length of the string. The declarations of the varying_string type variables do not contain any reference to their size. In all subsequent discussions, v refers to a variable string, and c refers to a standard character string. This chapter is based on a report ISO/IEC 1539-2: 2000. An implementation of the module iso_varying_string is available from ftp.nag.co.uk/sc22wg5/ISO_VARYING _STRING.

13.1 Assignment

The assignment sign can be used to assign a constant or variable of the usual character type or a varying_string type to a varying_string type variable. The following operations are allowed:

```
v = constant
c = constant
v = c
v = v
c = v
```

As indicated, v and c stand for varying_string type and standard character type variables.

```
use iso_varying_string
type(varying_string) :: v, v1
character(len=10) :: c
v="kolkata"
c=v
v1=v
print *, c
end
```

Case I: When two sides of the assignment sign contain variables of type varying_ string, the length of the variable on the left-hand side of the assignment sign becomes equal to the length of the variable on the right-hand side of the assignment sign.

In the preceding example, the length of v is 7, and when the variable v1 is equated to v, the length of v1 becomes also 7. Again, if the statement v1 = "IACS" is executed, the length of v1 becomes 4.

Case II: When a varying_string type is equated to a character variable, the length of the varying_string type variable becomes the length of the character variable. Thus, if one writes v1 = c, the length of v1 becomes 10 as the length of c is 10.

Case III: When an ordinary character variable is equated to a varying_string type variable, three cases may arise:

a. If the length of the variables on both sides of the assignment sign is same, it is a simple assignment.

```
type(varying_string) :: v
character(len=4) :: c
v='iacs'
c=v
```

b. If the length of the ordinary character variable is more than the varying_ string type variable, blanks are added at the end.

```
type(varying_string) :: v
character(len=10) :: c
v="iacs"
c=v
```

c becomes "iacsbbbbbb", where 'b' indicates blank.

c. If the length of the ordinary character variable is less than the varying_ string type variable, truncation from the right takes place.

```
type(varying_string) :: v
character(len=10) :: c
v="iacs, kolkata"
c=v
```

As the length of c is 10 and the length v is 13, due to truncation from the right, c will be set to "iacs, kolk".

13.2 Concatenation

Concatenation may be performed between constants and variables of the standard character type and variables of `varying_string` type. The resultant string has a length equal to the sum of the length of each string.

```
use iso_varying_string
type (varying_string) :: v, v1, v2
character (len=4) :: c="iacs"
v= "kolkata "
v2= " india"
v1= v // "700032"              ! "kolkata 700032"
v1= c //','// v                ! "iacs,kolkata"
v1= v // v2                    ! "kolkata india"
end
```

All the preceding concatenation operations are valid.

13.3 Comparison

Two strings may be compared with standard relational operators. The result of comparison is either `true` or `false`. The first non-matched character decides the issue. This was discussed in detail in Chapter 4. If the strings are of unequal length, blanks are added to the smaller string to make the length same as the bigger one. The comparison may be performed between v and v, c and v, v and c and c and c.

```
use iso_varying_string
type (varying_string) :: v, v1
character (len=4) :: c="iacs"
logical :: l
v="abcd"
v1="efgh"
l= v <= v1
l= v == v1
l= v > c
if (v .eq. v1) then
.
else
.
endif
```

13.4 Extended Meaning of Intrinsics

The intrinsics `adjustl`, `adjustr`, `char`, `iachar`, `ichar`, `index`, `len`, `len_trim`, `lge`, `lgt`, `llt`, `lle`, `repeat`, `scan`, `trim` and `verify` were discussed in Chapter 4. With the introduction of the module `iso_varying_string`, these intrinsics can accept `varying_string` variables as argument also. In other words, the capabilities of the intrinsics have been extended to include variables of type `varying_string`. There is no point in repeating the old story again. We give example of intrinsics with a `varying_string` variable as its argument. We may use the following:

```
i=len (v)  ! v is a varying string: "abcd"
v1=repeat(v,4)
```

It is thus seen that all the discussions of Chapter 4 are equally applicable for strings with variable length except that appropriate `varying_string` variable argument is to be used.

13.5 PUT

The subroutine `put` is used to write a record to an external file `call put (unit, string)`, where string is a `varying_string` type variable. If no unit is mentioned, the default unit is 6—the screen. The unit is an integer (scalar): `call put (string)`; put can have one optional argument `iostat`. It is a scalar and an integer. It returns zero when the `put` operation is successful. It returns some positive number in case of any error. If `iostat` is absent, in case of any error, the job is terminated. The `put` subroutine writes the string at the current position of the record. For example, `call put (v)` ; `call put(v1)` will write the value of v followed by the value of v1 as one record.

13.6 PUT_LINE

This takes identical arguments to `put`. The only difference is that after execution it terminates the current record, and a new record (new line) begins.

```
call put_line(6,v)
call put_line(v1)
call put_line(v1, iostat)
call put_line(6, v1, iostat)
```

13.7 GET

The subroutine `get` can be used to input a `varying_string` variable. In its simplest form, it is `call get (v)`, where `v` is variable of the type `varying_string`.

```
use iso_varying_string
type (varying_string) :: v
call get(v)
```

This reads `v` from the keyboard. There are other forms of `get`; `call get(unit, string)`. If unit (scalar) is omitted, it is assumed to be the keyboard (unit 5); `call get (string, maxlen)`, `maxlen` is an integer (scalar), which, if used, specifies maximum number of characters to be read from the input device. It is optional; `call get (v, 4)` will read only 4 characters from the keyboard and store in the location `v`. If `maxlen` is zero or negative, nothing will be read, and the string will be a null string. If `maxlen` is omitted, it is assumed to be `huge(1)`. Another form of `get` is `call get (string, set)`. Here `set` is a scalar of either of the type standard character or of the type `varying_string`. This variable contains a set of characters that may be used as the terminator. The reading is terminated when the input matches any member from the set of characters. The terminal character is not stored in the string. Next, `call get (string, set, separator)`, where separator is a scalar and of `varying_string` type. It is also optional. When separator is used, the actual character that separated (terminated) the record is stored in the location separator. Finally, `call get (string, set, separator, maxlen, iostat)`, where `iostat`, an optional parameter, is a scalar and an integer. After the read operation is performed, it is zero if read operation is valid and end of record is not encountered. A positive value indicates an error condition, and a negative value indicates that either an end of file or an end of record is reached. If `iostat` is absent, when end of file or end of record is reached, the job is terminated.

```
use iso_varying_string
type (varying_string) :: v,separator,set
integer :: maxlen=20, iostat
set="#"
call get(v,set,separator,maxlen,iostat)
print *, len(v)
call put_line(separator)
call put_line(v)
print *, iostat
end
```

The outputs from the program when the input is "ASSOCIATION#OF" are as follows:

```
11
#
ASSOCIATION
0
```

The output tells us that 11 characters have been stored in the location v. The character "#" acted as a separator, which was taken from the variable set, and it was stored in the location separator; as the get operation is successful, the value of iostat is zero.

13.8 EXTRACT

This function extracts a string from a varying_string or a character string. It can have optional start and finish parameters; both are integers. It returns a varying_string from the position start to finish. If start is omitted or is less than 1, it is taken to be 1. If finish is not present or is more than the length of the string, it is assumed to be len(string). If finish is less than start, a null string is returned. If finish is less than start, no character is extracted.

```
use iso_varying_string
type (varying_string) :: v, v1
v="indian association"
v1=extract(v,3,6)
call put_line(v1)
end
```

The output is "dian" (character positions 3-6).

13.9 REMOVE

The arguments are same as extract. It removes the string from start to finish.

```
use iso_varying_string
type (varying_string) :: v, v1
v="indian association for the cultivation of science"
v1=remove(v,19,38)
call put_line(v1)
end
```

The output is "indian association of science" after removing characters 19 to 38 from the string.

13.10 REPLACE

This function replaces a string by a substring between positions start and finish. If start is not supplied or it is less than 1, it is assumed to be 1. If finish is not supplied or is greater than the length of the string, finish is assumed to be len(string).

```
use iso_varying_string
type (varying_string) :: v, v1
v="indian association for the cultivation of science"
v1=replace(v,8,38,"institute")
call put_line(v1)
end
```

The output is "indian institute of science". The function replaces the substring "association for the cultivation" by "institute". There is another form of replace as follows:

```
replace(string, target, substring, every, back)
```

The string is searched for the occurrence of the string present in the target. When a match is found, it is replaced by substring. If every is true, all such occurrences are replaced; otherwise only the first occurrence is replaced. The default value of every is false. If back is true, searching takes place in the backward direction. If it is false, searching takes place in the forward direction. The default is false.

```
use iso_varying_string
type (varying_string) :: v, v1, v2
v="indian association for the cultivation of science"
v1="ti"
v2=replace (v,v1, "xx",.true.,.false.)
call put_line(v2)
end
```

The output is "indian associaxxon for the culxxvaxxon of science". As every is true, all occurrence of "ti" is replaced by "xx".

13.11 SPLIT

This subroutine splits a string into two substrings. The splitting is done through a set of "splitting characters."

In expression call split (string, word, set), string and word are of type varying_string, and set is of type character or varying_string. The subroutine divides the string into two when a character from the set of characters matches with any character of the string. The characters that are passed over are transferred to the variable word and removed from the string. If a fourth optional argument separator is present, the matched character is stored in that location; otherwise it is dropped. There is a fifth optional argument back; the default value of back is false. If it is true, searching starts from the end of the string. Obviously, back is a logical variable; set and separator are of either type character or varying_string.

```
use iso_varying_string
type (varying_string) :: v, word, set, separator
set="fc"
```

```
v="indian association for the cultivation of science"
call split(v,word,set,separator)
call put_line(word)
call put_line(v)
call put_line(separator)
end
```

The outputs are as follows:

```
indian asso
iation for the cultivation of science
```

It is not difficult to interpret the output. The string v is scanned. When it hits the
character 'c', one of the characters from the variable set, all characters from the first
character of v up to the character just before the character 'c' are transferred to the
variable word, and all these characters including the matched character 'c' are deleted
from the variable v. However, because of the presence of the variable separator, the
matched character is transferred to the variable separator.

14

IEEE Floating Point Arithmetic and Exceptions

Most of the modern computers follow IEEE (Institute of Electrical and Electronics Engineers) standards for storing and using floating point numbers. During the execution of a program, a situation may arise when computation cannot proceed due to some unusual behavior of the program. In such a case, an exception signal is raised, and usually the job is aborted. In certain cases, it is possible to catch the exception signal, and the control may be passed to the programmer to enable the programmer to take some corrective measure so that the program can continue. For example, during the execution of a program, the programmer can trap the division by zero exception, and then it becomes the responsibility of the programmer to devise an algorithm to handle such a situation. Based on this standard, the current version of Fortran provides a number of functions and subroutines to handle exception conditions. These routines help the programmers to develop efficient numerical software. In this chapter, we shall first examine how floating point numbers are stored according to the IEEE standard.

14.1 Representation of Floating Point Numbers (IEEE Standard)

The floating point numbers are represented as

```
        x = 0
and     x = s.bᵉ *∑(k=1,p) f_k * b⁻ᵏ
```

where s represents the sign plus or minus, b=2 for binary representation and f_ks satisfies $0 <= f_k < b$.

For binary b=2 and the f_ks is either 0 or 1.

14.2 Single Precision 32-Bit Floating Point Numbers (IEEE Standard)

A single precision 32-bit real number is stored in the following way according to the IEEE standard:

Bit 31: sign bit, 0 for positive number and 1 for negative number

Bits 30-23: exponent bits (biased, to be discussed shortly)

Bits 22-0: used to store the fraction

The smallest and largest numbers that can be stored in eight bits (30-23) are 0 [(00000000)$_2$] and 255 [(11111111)$_2$], respectively. These two numbers are reserved (to be discussed later). Consequently, the smallest and largest numbers that are stored in these bits are 1 through 254 [(11111110)$_2$]. These are not the actual exponents of a binary number. In order to handle negative exponents efficiently (i.e., without wasting a bit for the sign of the exponent), the number 127 is added to the exponent; that is, the exponent is biased with 127. Thus, exponent 3 is stored as 127 + 3=130, and –3 is stored as 127-3 = 124. The maximum and minimum values of the exponent bits (after adding 127) are 254 and 1, respectively, corresponding to actual exponents 127 and –126, respectively.

The fraction is stored in normalized form. The exponent is so adjusted that the number becomes 2^e × 1.yyy..., where yyy... are binary digits—either 0 or 1. For example, for the decimal system, 234.56 may be written as 2.3456 × 10^2. Similarly, a binary number 110101.111 may be written as 2^5 × 1.10101111. A single precision real number having biased exponent between 1 and 254 and the fraction containing 1 before the binary decimal point is termed a normalized real number. Therefore, a floating point normalized number looks like the following:

```
x = s * 2ᵉ * 1. (normalized binary fraction)
```

The advantage of storing a fraction in its normalized form is that the most significant bit, which is $2^0 = 1$ for any fraction, need not be stored—it is implied. Therefore, the fraction actually uses 24 bits to store its value, though only 23 bits are allocated for the fraction.

We shall now show the bit pattern of a few real numbers.

1. 1.25 = 2^0 × (1.01)$_2$

   Exponent = 0 + 127 = 127 = (01111111)$_2$

   Fraction = (1) 010 0000 0000 0000 0000 0000

   1.25 = 0 01111111 010 0000 0000 0000 0000 0000 [hidden bit is not shown]

   = (07750000000)$_8$ = (3FA00000)$_{16}$

2. 0.0625 = (0001)$_2$ = 2^{-4} × (1.0)$_2$

   Exponent = −4 + 127 = 123 = (01111011)$_2$

   Fraction = (1) 000 0000 0000 0000 0000 0000

   0.0625 = 0 01111011 000 0000 0000 0000 0000 0000 [hidden bit not shown]

   = (07540000000)$_8$ = (3D800000)$_{16}$

3. $-5.0 = 2^2 \times (1.01)_2$ (value only)

   Sign bit = 1

   Exponent = 2 + 127 = 129 = 10000001

   Fraction = (1) 010 0000 0000 0000 0000 0000

   -5.0 = 1 10000001 010 0000 0000 0000 0000 0000 [hidden bit not shown]

         = $(30050000000)_8 = (C0A00000)_{16}$

4. $-17.625 = 2^4 \times (1.1015625)_{10} = 2^4 \times (1.0001101)_2$ (value only)

   Sign bit = 1

   Exponent = 4 + 127 = 131 = 10000011

   Fraction = (1) 0001101 0000 0000 0000 0000

   -17.625 = 1 10000011 000 1101 0000 0000 0000 0000 [hidden bit not shown]

         = $(30143200000)_8 = (C18D0000)_{16}$

A double precision number requires 8 bytes (64 bits). For such numbers, arrangements are shown next:

Bit 63: Sign bit

Bits 62:52 exponent

Bits 51:0 fraction

The exponent is biased by 1023. The biased exponent can have a value between 1 and 2046, corresponding to the actual exponent -1022 and 1023. We shall now find out the internal representation of -5.0D0.

Bit 63: Sign bit 1

Bit 62:52 Exponent 10000000001

Bit 51:0 Fraction (1)0100 0000 0000 0000 0000 0000 0000 0000 0000 0000 0000 0000 0000

$-5.0D0$ = 1 10000000001 0100 0000 0000 0000 0000 0000 0000 0000 0000 0000 0000 0000 0000

      = $(1400240000000000000000)_8$

      = $(C014000000000000)_{16}$

These results may be verified using boz edit descriptors.

```
        real:: r
        read *, r
        print 10, r ! Ok with ifort and gfortran
        print 20, r ! NAG compiler gives runtime error
        print 30, r
10      format(b32.32)
20      format(o11.11)
30      format(z8.8)
        end
```

14.3 Denormal (Subnormal) Numbers

Denormal numbers (also called subnormal numbers) or denormalized numbers are those that are less than the smallest normalized floating point number available in the system. They actually fill the gap between zero and the smallest floating point number. As these numbers are smaller than the smallest floating point number, these numbers are also called subnormal numbers. In IEEE floating point representation, denormal numbers are stored with bias 126 and with all the exponent bits switched off. In addition, there is no hidden bit associated with a denormal number. The smallest non-zero positive denormal number is as follows:

0000 0000 0000 . . . 0001 (total 32 bit, single precision)

Exponent is biased by 126, so the actual exponent is 0-126=-126. Rightmost bit is on, and its value is 2**(-23). Therefore, the smallest non-zero positive denormal number is as follows (Figure 14.1):

2**(-126) ×2**(-23)=2**(-149) = 1.4012985E-45 (approximately)

Similarly, the largest denormal positive number is one with all bits 0 to 23 switched on. It is as follows (Figure 14.2):

2**(-126) × (1 - 2**(-23)) = 1.1754942E-38

Readers are encouraged to find out the smallest and largest double precision denormal numbers.

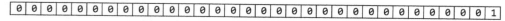

FIGURE 14.1
Smallest positive denormal number.

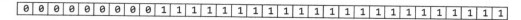

FIGURE 14.2
Largest positive denormal number.

14.4 Representation of Zero

There are two zeros in this representation: positive zero (+0) and negative zero (-0). For positive zero, all bits are switched off. For negative zero, sign bit is on and all other bits are zero (Figures 14.3 and 14.4).

Though the bit patterns of positive and negative zeros are different, the numerical values of -0.0 and 0.0 are the same. For the purpose of comparison, they are treated as equal. However, the print statement displays the negative sign of the negative zero. The library function `sign` (which takes two arguments, and the sign of the second argument is transferred to the first argument) respects the sign of the negative zero. The following program illustrates these points.

```
        real:: r1,r2,r3 = 5.0,  r4=-5.0
        r1=-0.0;  r2 = 0.0
        r3=sign(r3,r1) ! r1 = -0.0,  r3 = -5.0
        r4=sign(r4,r2) ! r2 = 0.0,  r4 = 5.0
        print 10, r1,r2,r3,r4
 10     format(4f8.1)
        end
```

The output of the program is $-0.0\ 0.0\ -5.0\ 5.0$.

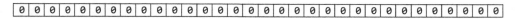

FIGURE 14.3
Positive zero.

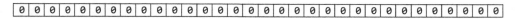

FIGURE 14.4
Negative zero.

14.5 Representation of Infinity

There are two infinities: positive $(+\infty)$ and negative $(-\infty)$. Positive `infinity` is represented by the following bit combinations: sign bit off, all exponent bits on and all other bits zero. Negative `infinity` is same as positive `infinity` except the sign bit is on. The infinite values are created by division by zero or by arithmetic overflow (Figures 14.5 and 14.6).

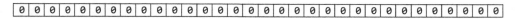

FIGURE 14.5
Positive `infinity`.

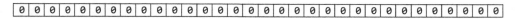

FIGURE 14.6
Negative `infinity`.

14.6 Representation of NaN (Not a Number)

Not a number (NaN) is created by an invalid operation like 0/0 or 0x∞. NaN is represented with all bits of the exponent as 1 and the fractional parts as non-zero (different compiler represents in different ways). There are two types of NaN: signaling and quiet. If an operand contains signaling NaN, the invalid exception signal is generated, and the result becomes a quiet NaN. Quiet NaN as an operand does not raise any exception. For a quiet NaN, the most significant bit of the fractional part is 1, and for a signaling NaN, the most significant bit of the fractional part is zero (Figures 14.7 and 14.8).

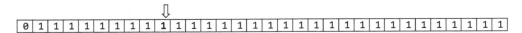

FIGURE 14.7
Quiet NaN.

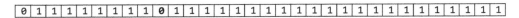

FIGURE 14.8
Signaling NaN.

14.7 Summary of IEEE "Numbers"

Denormal values, infinite values (both positive and negative) and NaN are known as exceptional floating point values. A value that does not belong to these three categories is known as a normal value (Table 14.1).

IEEE has defined operations of these so-called special numbers. If one of the operand is NaN, then the result is NaN. Table 14.2 (using NAG compiler) summarizes the result of operations of other special numbers.

Table 14.3 summarizes arithmetic operations (using NAG compiler) when one of the operand is a special number and the other operand is a normal real number.

TABLE 14.1

Single Precision IEEE "Numbers"

Type	Sign	Exponent	Mantissa
Positive zero	0	0	0
Negative zero	1	0	0
Denormalized number	0 or 1	0	Non-zero
Positive infinity	0	255	0
Negative infinity	1	255	0
NaN (quiet)	0	255	Non-zero, bit 22 is on
NaN (signaling)	0	255	Non-zero, bit 22 is off
Normalized number	0 or 1	1 to 254	Any

TABLE 14.2

Operations Involving "Special" Numbers

$0 + 0 = 0$	$\infty + \infty = \infty$
$0 - 0 = 0$	$\infty - \infty = NaN$
$0 * 0 = 0$	$\infty * \infty = \infty$
$0/0 = NaN$	$\infty/\infty = NaN$
$0 ** 0 = NaN$	$\infty ** \infty = \infty$
$0 + (-0) = 0$	$\infty + (-\infty) = NaN$
$0 - (-0) = 0$	$\infty - (-\infty) = \infty$
$0 * (-0) = -0$	$\infty * (-\infty) = -\infty$
$0/(-0) = NaN$	$\infty/(-\infty) = NaN$
$0 ** (-0) = NaN$	$\infty ** (-\infty) = 0$
$-0 + 0 = 0$	$-\infty + \infty = NaN$
$-0 - 0 = -0$	$-\infty - \infty = -\infty$
$-0 * 0 = -0$	$-\infty * \infty = -\infty$
$-0/0 = NaN$	$-\infty/\infty = NaN$
$-0 **0 = NaN$	$-\infty ** \infty = NaN$
$-0 + (-0) = -0$	$-\infty + (-\infty) =-\infty$
$-0 - (-0) = 0$	$-\infty - (-\infty) = NaN$
$-0 * (-0) = 0$	$-\infty * (-\infty) = \infty$
$-0/(-0) = NaN$	$-\infty/\infty = NaN$
$-0 **(-0) = NaN$	$-\infty ** (-\infty) = NaN$
$0 + \infty =\infty$	$-0 + \infty = \infty$
$0 - \infty = -\infty$	$-0 - \infty = -\infty$
$0 * \infty = NaN$	$-0 * \infty = NaN$
$0/\infty = 0$	$-0/\infty = -0$
$0 ** \infty = 0$	$-0 ** \infty = 0$
$0 + (-\infty) = -\infty$	$-0 + (-\infty) = -\infty$
$0 - (-\infty) = \infty$	$-0 - (-\infty) = \infty$
$0 * (-\infty) = NaN$	$-0 * (-\infty) = NaN$
$0/(-\infty) = -0$	$-0/(-\infty) = 0$
$0 ** (-\infty) = \infty$	$-0 ** (-\infty) = \infty$

TABLE 14.3

Operation Involving "Special Number"
and Normal Real Number

$\infty + x = \infty$	$-\infty ** x = -\infty$
$\infty - x = \infty$	$x/-\infty = -0.0$
$\infty * x = \infty$	$NaN + x = NaN$
$\infty/x = \infty$	$NaN - x = NaN$
$\infty ** x = \infty$	$NaN * x = NaN$
$x/\infty = 0.0$	$NaN/x = NaN$
$-\infty + x = -\infty$	$NaN ** x = NaN$
$-\infty - x = -\infty$	$x/NaN = NaN$
$-\infty * x = -\infty$	$NaN ** x = NaN$
$-\infty/x = -\infty$	

14.8 Divide by Zero

This exception is raised when the processor tries to divide a non-zero numerator (real or complex) by zero denominator. The result of such an operation is signed `infinity`: `1.0/0.0` is $+\infty$ and $-1.0/0.0$ is $-\infty$.

14.9 Overflow

This exception is raised when the absolute value of the result of a computation (real or complex) is larger than a processor-dependent limit.

14.10 Underflow

This exception is raised when the absolute of the result of a computation (real or complex) is smaller than the processor-dependent limit or the number cannot be represented with full precision.

14.11 Inexact Computation

This exception is raised when the result of a computation (real or complex) is not exact. This may happen when during rounding the number overflows or underflows. The resultant number is not equal to the original number. This exception is also raised when a real number is converted into an integer and the resultant integer cannot be stored as it is too large.

14.12 Invalid Arithmetic Operation

This exception is raised when arithmetic operation (real or complex) is not valid. For example, square root of a real negative number is not a valid operation. Also, an expression like (−a)**b where a and b are both real numbers (a is positive) involves evaluating logarithm of a real negative quantity.

14.13 IEEE Modules

IEEE exception handling in Fortran is performed through three modules: `ieee_exceptions`, `ieee_arithmetic` and `ieee_features`. The module `ieee_exceptions` is used for exceptions, the module `ieee_arithmetic` is used for IEEE arithmetic and the module `ieee_features` is used to control the features supported by the compiler. The `ieee_arithmetic` contains a use statement for `ieee_exceptions`; that is, `ieee_arithmetic` modules appear to contain a use `ieee_exceptions` statement. All functions of IEEE modules are pure, and procedure names are generic. An exception may or may not cause program termination. Normally, the job is aborted when, say, division by zero occurs. However, by suitable instructions, the programmer may allow the program to continue.

14.14 IEEE Features

This module defines IEEE features. The module defines a data type `ieee_feature_type`. The following named constants are defined in this module:

`ieee_datatype`, `ieee_denormal` (also known as `ieee_subnormal`), `ieee_divide`, `ieee_halting`, `ieee_inexact_flag`, `ieee_inf`, `ieee_invalid_flag`, `ieee_nan`, `ieee_rounding`, `ieee_sqrt`, `ieee_underflow_flag`.

For a particular processor, some of the IEEE features are available by default; some others, being inefficient and perhaps slow, are not available by default, and support for some may not be available at all. When the IEEE features are not included by the use statement, the processor will support only the default features. When IEEE features are used, support for all the available IEEE features will be provided. It is possible to use an only clause with the use statement to provide support for a particular feature. For example, suppose the default feature does not support denormalized numbers. In such a case, all the underflowed values are treated as zero. There is no smooth transition from the tiny (smallest normalized floating point number) number to zero. However, if the compiler is forced to compile a program with denormalized support (assuming such support exists, not as default), the transition from tiny number to zero becomes smooth through the denormalized numbers. This is achieved with the appropriate use statement as follows:

```
use ieee_features ! all the features are available
use ieee_features, only: ieee_denormal
```

14.15 IEEE FLAGS

There are five IEEE flags: `ieee_overflow`, `ieee_divide_by_zero`, `ieee_invalid`, `ieee_underflow` and `ieee_inexact`.

IEEE_overflow: Overflow is a condition when the number is too large to be represented in a given machine. The flag is set when overflow occurs.

IEEE_divide_by_zero: When division by zero is attempted, this flag is set.

IEEE_invalid: This flag is set when the arithmetic operation is invalid.

IEEE_underflow: This happens when the value of a real number is too small and it cannot be represented with its full precision (there is loss of precision).

IEEE_inexact: This occurs when the result of computation is not exact.

14.16 Derived Types and Constants Defined in the Modules

The module `ieee_exceptions` defines following derived data types:

1. **IEEE_FLAG_TYPE:** The named constants defined in this module are `ieee_invalid`, `ieee_overflow`, `ieee_divide_by_zero`, `ieee_underflow` and `ieee_inexact`. Also, this module defines array constants `ieee_usual` and `ieee_all` with the following:

   ```
   ieee_usual=[ieee_overflow, ieee_divide_by_zero, ieee_invalid] ! dimension 3
   ieee_all = [ieee_usual, ieee_underflow, ieee_inexact] ! dimension 5
   ```

2. **IEEE_STATUS_TYPE:** This is used for representing the current floating point status.

3. **IEEE_MODES_TYPE:** This is used for representing the floating point modes.

The module `ieee_arithmetic` defines the following derived data types:

1. **IEEE_CLASSTYPE:** This is used to identify the class of a floating point number. The named constants defined in this module are `ieee_signaling_nan`, `ieee_quiet_nan`, `ieee_negative_inf`, `ieee_positive_inf`, `ieee_negative_normal`, `ieee_negative_denormal`, `ieee_negative_zero`, `ieee_positive_zero`, `ieee_positive_denormal`, `ieee_positive_normal` and `ieee_other_value`. The named constants `ieee_negative_denormal` and `ieee_positive_denormal` have the same value as `ieee_negative_subnormal` and `ieee_positive_subnormal`, respectively.

2. **IEEE_ROUND_TYPE:** This is used to define a rounding mode. The named constants defined in the modules are `ieee_nearest`, `ieee_to_zero`, `ieee_up`, `ieee_down`, `ieee_away` and `ieee_other`.

   `ieee_nearest`: The result is rounded to the exact result to the nearest representable value.

   `ieee_to_zero`: The result is rounded to the exact result toward zero to the next representable value.

   `ieee_up`: The result is rounded to the exact result toward `+infinity` to the next representable value.

   `ieee_down`: The result is rounded to the exact result toward `-infinity` to the next representable value.

   `ieee_away`: The result is rounded away from zero to the next representable value.

   `ieee_other`: It is processor dependent.

14.17 IEEE Operators

Only the relational operators, equal (`.eq.`, `==`) and not equal (`.ne.`, `/=`), can be used to compare class and round type. The result is either `true` or `false`.

14.18 Inquiry Functions (Arithmetic Module)

In all subsequent discussions of this section, if the argument x of the function is absent, the argument is assumed to be any kind of real number. Also, all these program segments contain two statements:

```
use ieee_arithmetic
real:: x
```

IEEE_SUPPORT_DATATYPE([X]): This function returns true if the processor supports the variable of type (kind) x; otherwise it returns false:

```
if (ieee_support_datatype(x)) then
  print *, 'x is supported datatype'
else
  print *, 'x is not supported datatype'
endif
end
```

IEEE_SUPPORT_DENORMAL([X]): This function returns true if the processor supports IEEE denormalized numbers. It returns false otherwise. The optional argument x is real of any kind. It may be scalar or array valued (this routine is also called ieee_support_subnormal).

IEEE_SUPPORT_DIVIDE([X]): This function returns true if the processor supports divide with the accuracy specified in the IEEE standard for real variable of same kind type parameter as x. Otherwise it returns false.

IEEE_SUPPORT_INF([X]): This function returns true if the processor supports IEEE infinity for same kind type as x. It returns false otherwise.

IEEE_SUPPORT_NAN([X]): This function returns true if NAN is supported by the processor for the same kind variable x. It returns false otherwise.

IEEE_SUPPORT_HALTING(FLAG): This function returns true if the processor has the ability to control program termination for these exception conditions: ieee_invalid, ieee_overflow, ieee_divide_by_zero, ieee_underflow, ieee_inexact. It returns false otherwise.

```
if (ieee_support_halting(ieee_divide_by_zero)) then
 print *, &
  'program termination can be controlled for divide by zero exception'
else
 print *, &
 'program termination cannot be controlled for divide by zero exception'
endif
```

IEEE_SUPPORT_ROUNDING (ROUNDVALUE, [X]): This function returns true if the round value of type type(ieee_round_type) is supported for the same kind type parameter x. The round value must be one of the type: ieee_nearest, ieee_to_zero, ieee_up, ieee_down, ieee_away.

```
if (ieee_support_rounding (ieee_nearest, x)) then
 print *, 'rounding according to ieee nearest mode is supported'
else
 print *, 'rounding according to ieee nearest mode is not supported'
endif
```

IEEE_SUPPORT_FLAG(FLAG, [X]): It returns `true` if a flag mentioned subsequently is supported for the same kind type as x. It returns `false` otherwise. The flags may be one of the following: `ieee_overflow, ieee_divide_by_zero, ieee_invalid, ieee_underflow, ieee_inexact.`

IEEE_SUPPORT_SQRT([X]): This function returns `true` if the `sqrt` function has been implemented according to the IEEE standard for the same kind type variable x. It returns `false` otherwise.

The preceding statement implies `sqrt(+0.0)=0.0`, `sqrt(-0.0)=-0.0`, `sqrt(+inf)=+inf`, `sqrt(other negative number)=NaN`.

IEEE_SUPPORT_STANDARD([X]): This function returns `true` if all inquiry functions mentioned earlier, namely, `ieee_support_datatype, ieee_support_denormal, ieee_support_divide, ieee_support_flag (for every valid flag), ieee_support_halting (for every valid flag), ieee_support_inf, ieee_support_nan, ieee_support_rounding (for every round value)` and `ieee_support_sqrt` return `true`. It returns `false` otherwise.

IEEE_SUPPORT_UNDERFLOW_CONTROL([X]): It returns `true` if the processor supports underflow control mode for variables of same kind type as x. It returns `false` otherwise.

IEEE_SUPPORT_IO([X]): This function returns `true` if the processor supports base conversion (described as `ieee_up, ieee_down, ieee_zero` and `ieee_nearest`) during formatted input/output.

14.19 IEEE_CLASS

This function returns the class of the variable used as its argument. The result is of derived type `ieee_class_type` defined in the `ieee_arithmetic` module. The result is of one of the following:
`ieee_signaling_nan, ieee_quiet_nan, ieee_negative_inf, ieee_positive_inf, ieee_negative_normal, ieee_negative_denormal, ieee_negative _zero, ieee_positive_zero, ieee_positive_denormal, ieee_positive_normal, ieee_other_value.`

```
use ieee_arithmetic
type (ieee_class_type):: res
real:: r=20.0
res=ieee_class(r)
```

```
if (res == ieee_positive_normal) then
 print *, 'positive normal'
else
 print *, 'not a positive normal'
endif
end
```

Since r is a positive number, the 'then' path will be followed, and the print statement with 'positive normal' will be executed. If r is changed to -20.0, the res will not be equal to ieee_positive_normal, and as such the 'else' path will be followed.

14.20 IEEE_COPY_SIGN

This function takes two arguments, x and y (both real). It returns the absolute value of x with the sign of y.

```
use ieee_arithmetic
real:: a=5.0, b=-10.0
if (ieee_support_datatype(a).and. ieee_support_datatype(b)) then
 a=ieee_copy_sign(a, b)
endif
print *, 'a = ', a, 'b = ', b
end
```

The value of a will be absolute value of a with the sign of b; that is, -5.0.

14.21 IEEE_VALUE

This intrinsic takes two arguments: the first one, x, is a real variable, and the second one is of type ieee_class_type. It returns a real value of the same kind as x with the value set by the second argument.

```
use ieee_arithmetic
real:: x
x=ieee_value(x, ieee_positive_inf)
print *, x
end
```

The variable x is set to IEEE positive infinity. The print statement will display infinity.

14.22 IEEE_IS_FINITE

This function takes a real argument. It returns `true` if the value of the argument is finite; that is, it is neither IEEE `infinity` nor IEEE NaN. If the argument is not `finite`, it returns `false`.

14.23 IEEE_IS_NAN

This function takes one real argument and tests whether the value of the argument is NaN. It returns `true` if the argument is a NaN and returns `false` otherwise.

14.24 IEEE_IS_NEGATIVE

This function takes one real argument. It returns `true` if the argument is a negative number; otherwise it returns `false`. If the value of `r` is set to negative `infinity`, then the number is still considered to be a negative number. Also, if `r` is set to `ieee_negative_zero`, `ieee_is_negative` treats the number as a negative number.

14.25 IEEE_IS_NORMAL

A real variable is considered to be normal if it belongs to one of the following categories: `ieee_positive_normal`, `ieee_negative_normal`, `ieee_positive_zero`, `ieee_negative_zero`. This function takes one real argument and returns `true` or `false` depending on the argument.

14.26 IEEE_INT

This function converts the first argument `'a'` (real) to an integer according to the second argument round (`ieee_round_type`). The returned value of type kind is specified by the optional third argument. If the third argument `kind` is absent, the value returned is of default type integer. The flag `ieee_inexact` is signaled if the returned value is not exact; `ieee_int(9.6, ieee_up)` returns `10`.

14.27 IEEE_REAL

This routine converts the first argument `'a'` (real or integer) to a real number according to the existing rounding mode. If the optional second parameter `kind` is present, the returned value is of type kind; otherwise the subroutine returns a real number of default kind; `call ieee_real(456)` returns `456.0`.

14.28 IEEE_SIGNBIT

This function takes a real number as its argument. The returned value is `true` if the number is negative; otherwise it returns `false`.

14.29 IEEE_MAX_NUM and IEEE_MIN_NUM

Both these functions take two real arguments, x and y; `ieee_max_num` returns `max(x, y)`. If one or both are signaling NaN, the result is NaN and `ieee_invalid` is signaled; `call ieee_max_num(20.0, 30.0)` returns `30.0`; `ieee_min_num` is same as `ieee_max_num` except it returns the minimum value.

14.30 IEEE_MAX_NUM_MAG and IEEE_MIN_NUM_MAG

The two functions `ieee_max_num_mag` and `ieee_min_num_mag` are identical to the corresponding `ieee_max_num` and `ieee_min_num` routines except that they return the maximum of minimum value after calculating the absolute values of the arguments. This means that the maximum and minimum are determined between `abs(x)` and `abs(y)` (x and y are arguments). Thus, `ieee_max_num_mag(2.5,-3.6)` returns `-3.6` as `abs(-3.6) > abs(2.5)`. Similarly, `ieee_min_num_mag(2.5,-3.6)` returns `2.5` as `abs(2.5) < abs(-3.6)`. No exception is raised unless one or both the arguments are signaling NaN.

14.31 IEEE_FMA

This function takes three arguments a, b and c. All parameters are real. The arguments b and c are of same kind as the argument a. The routine returns `(a*b)+c` using the current rounding mode. The flags overflow, underflow, and inexact flags could be signaling depending on the final result of computation.

14.32 IEEE_LOGB

This function takes one real number as argument and returns the unbiased exponent.

Case I: If the value is neither `infinity` nor `NaN`, the returned value is unbiased exponent.

Case II: If `x=0`, the result is negative `infinity` and the flag `ieee_divide_by_zero` is set.

Case III: If `x=NaN`, the result is also `NaN`.

Case IV: If x is either `+infinity` or `−infinity`, the result is `+Infinity` (for `−Infinity` gfortran returns `−Infinity`).

Assuming one instruction (which will be discussed shortly) that allows the program to continue even if divide by zero occurs, the following program allows us to test all the cases mentioned earlier.

```
use ieee_arithmetic
real:: r, e
call ieee_set_halting_mode(ieee_all,.false.)   ! assume now
read *,r
e=ieee_logb(r)
print *, e
end
```

14.33 IEEE_NEXT_AFTER, IEEE_NEXT_DOWN and IEEE_NEXT_UP

This function `ieee_next_after` takes two arguments x and y (both real) and returns the next representable neighbor of x according to the direction of y. A positive y indicates the neighbor to the right (greater than x), and a negative y indicates the neighbor to the left (less than x).

Case I: If x is equal to y, the result is same as x, and no exception flags are signaled.

Case II: When x is not equal to y, the intrinsic returns the next representable neighbor of x, the direction being determined by the sign of y. Flags `ieee_overflow` and `ieee_inexact` are signaled when x is finite but `ieee_next_after(x, y)` is infinite. Flags `ieee_underflow` and `ieee_inexact` are signaled when `ieee_next_after(x,y)` is either zero or denormalized.

Case III: If x or y is quiet NaN, the result is one of the input NaN values.

The routine `ieee_next_down (x)` returns the adjacent next lower machine number; `ieee_next_up(x)` returns the higher adjacent machine number.

14.34 IEEE_REM

This is a remainder function independent of rounding mode. It takes two real arguments, x and y. The function returns x-y*n, where the integer n is nearest to the exact value of x/y satisfying the condition:

```
abs(n - x/y)=1/2, n is even.

use ieee_arithmetic
print *, ieee_rem(5.0, 3.0) ! prints -1
print *, ieee_rem(4.0, 3.0) ! prints 1
print *, ieee_rem(3.0, 2.0) ! prints -1
print *, ieee_rem(4.0, 2.0) ! prints 0
end
```

14.35 IEEE_SCALB

This function takes two arguments: real x and integer i. It returns a real of the same kind as x whose value is (2**i)*x. If x is finite but x*(2**i) exceeds the capacity of the machine, the overflow flag is signaled, and the result is infinity with the sign of x. If x*(2**i) is too small, underflow flag is set, and the result is nearest representable number with the sign of x. If x is infinite, the result is x, but no exception flags are set.

```
use ieee_arithmetic
print *, ieee_scalb(2.0,2)  ! 8.0
print *, ieee_scalb(5.0,3)  ! 40.0
print *, ieee_scalb(-2.0,2) ! -8.0
print *, ieee_scalb(-5.0,2) ! -20.0
end
```

14.36 IEEE_GET_ROUNDING_MODE

This subroutine takes a variable of type ieee_round_type as its argument. It returns the current rounding mode. The rounding mode can be one of ieee_nearest, ieee_to_zero, ieee_up, ieee_down, ieee_away and ieee_others types.

```
use ieee_arithmetic
type (ieee_round_type) :: round
call ieee_get_rounding_mode(round)
if (round==ieee_nearest) then
 print *, 'nearest'
```

```
      else if (round==ieee_to_zero) then
       print *, 'to zero'
      else if (round==ieee_up) then
       print*, 'up'
      else if (round==ieee_down) then
       print *, 'down'
      else if (round==ieee_away) then ! ieee_away is not yet supported
       print *, 'away'
      endif
      end
```

In this case, the `print` statement displays 'nearest' because this is the default rounding mode. The rounding mode `ieee_others` is processor dependent.

14.37 IEEE_SET_ROUNDING_MODE

This subroutine also takes one argument of type `ieee_round_type` and sets the rounding mode to one of the permitted values, that is, `ieee_nearest`, `ieee_to_zero`, `ieee_up`, `ieee_down`, `ieee_away` or `ieee_others`.

```
      use ieee_arithmetic
      type (ieee_round_type):: round
      call ieee_set_rounding_mode(ieee_up)
      call ieee_get_rounding_mode(round)
      if (round == ieee_up) then
       print *, 'ieee_up'
      else
       print *, 'ieee_up not set'
      endif
      end
```

The program segment displays 'ieee_up' because the rounding mode is set to `ieee_up`.

14.38 IEEE_RINT

This function takes one real variable x as its argument and rounds to integer according to the current mode of rounding. The returned value is of same kind type as x. If x is either infinity or NaN, it returns infinity or NaN.

```
      use ieee_arithmetic
      call ieee_set_rounding_mode(ieee_nearest); print *, ieee_rint(1.8)
      call ieee_set_rounding_mode(ieee_up); print *, ieee_rint(1.8)
      call ieee_set_rounding_mode(ieee_to_zero); print *, ieee_rint(1.8)
      call ieee_set_rounding_mode(ieee_down); print *, ieee_rint(1.8)
      end
```

The outputs are, respectively, 2.0, 2.0, 1.0 and 1.0

14.39 IEEE_UNORDERED

This function takes two real arguments, x and y. It returns true if x or y or both are NaN; otherwise it is false.

```
use ieee_arithmetic
real:: x, y=0.0
x=ieee_value(x, ieee_quiet_nan)
print *, ieee_unordered(x, y)
end
```

The result returned is true as x is NaN.

14.40 IEEE_GET_HALTING_MODE

This subroutine takes two arguments: the first one is of type ieee_flag_type and the second one is logical. The logical returns true if the exception specified by the first argument would cause halting; otherwise it returns false.

```
use ieee_arithmetic
logical::l ! NAG compiler
call ieee_get_halting_mode(ieee_overflow, l); print *, l
call ieee_get_halting_mode(ieee_divide_by_zero, l); print *, l
call ieee_get_halting_mode(ieee_invalid, l); print *, l
call ieee_get_halting_mode(ieee_underflow, l); print *, l
call ieee_get_halting_mode(ieee_inexact, l); print *, l
end
```

If l returns true, the corresponding halting mode would halt the program. If it returns false, the program would not terminate in case the corresponding flag is raised. The default values of these flags for the NAG compiler are T T T F F. These values are compiler dependent. For ifort and gfortran compilers, all the default values are false.

14.41 IEEE_SET_HALTING_MODE

This subroutine can set or reset the halting mode. It takes two parameters: the first one is of type ieee_flag_type and the second one is logical. If the logical halting parameter is true, the program will halt when the exception corresponding to the flag occurs. If the logical parameter is false, the program does not stop when the exception corresponding to the flag occurs.

The program will not terminate when the divide by zero takes place if the following instruction is executed before the flag is set.

```
call ieee_set_halting_mode(ieee_divide_by_zero,.false.)
```

All halting modes can also be set or reset by a single instruction as follows:

```
call ieee_set_halting_mode(ieee_all,.false.)      ! reset
call ieee_set_halting_mode(ieee_all,.true.)       ! set
```

14.42 IEEE_GET_MODES and IEEE_SET_MODES

These two modes subroutines are used to get and set floating point modes, respectively. Both take a single argument modes of type `ieee_modes_type`. The following examples demonstrate these two routines:

```
use ieee_arithmetic
type (ieee_modes_type):: mymodes
.
call ieee_get_modes(mymodes)                  ! save all modes
call ieee_set_rounding_mode(ieee_nearest) ! change rounding_mode
                                          ! do some calculation
.
call ieee_set_modes(mymodes)                  ! restore all modes
```

14.43 IEEE_GET_STATUS and IEEE_SET_STATUS

These two subroutines take a single argument of type `ieee_status_type`.

The `ieee_get_status` gets the current value of the floating point status, and the `ieee_set_status` restores the floating point status. An example will make this point clear:

```
use ieee_arithmetic
type (ieee_status_type):: stat
call ieee_get_status(stat)             ! floating status is stored in stat
call ieee_set_flag(ieee_all,.false.) ! flags are changed
.
call ieee_set_status(stat)             ! flags are restored
end
```

The call to the subroutine `ieee_get_status` stores the floating point status in the variable `stat`. The subroutine `ieee_set_flag` changes the floating point status. After calculations are over, the original status is restored through the call to the subroutine `ieee_set_status`.

14.44 IEEE_GET_FLAG and IEEE_SET_FLAG

Both these subroutines take two arguments: the first one is `flag` of type `ieee_flag_type` (`intent in`). The second argument `flagvalue` is a logical variable having `intent out`.

The subroutine `ieee_get_flag` returns the status of `flag`: `true` if it is set and `false` if it is not set.

```
call ieee_get_flag(ieee_overflow, flagvalue)
```

If overflow flag is signaling, `flagvalue` is `true`; if it is not signaling, `flagvalue` is `false`.

The subroutine `ieee_set_flag` can be called if `ieee_support_halting(flag)` is true. The `flagvalue` true sets the corresponding flag to signaling mode, and `false` sets the corresponding flag to non-signaling mode.

```
call ieee_set_flag(ieee_divide_by_zero,.false.)
```

This call will make the `divide_by_zero` flag non-signaling. The following example shows the use of `ieee_get_flag` and `ieee_set_flag`.

```
program flag
use ieee_arithmetic
logical, dimension(5):: fv    ! gfortran compiler
call ieee_get_flag(ieee_all, fv)
print *, 'Initial status: ', fv
call ieee_set_flag(ieee_divide_by_zero,.true.)
call ieee_get_flag(ieee_all, fv) ! NAG compiler will abort the job unless
print *, 'Status after modification: ',fv ! it is compiled with -ieee=full
end
```

The outputs of the program are as follows:

```
Initial status: F F F F F
Status after modification: F T F F F
```

The expressions `fv(1)`, `fv(2)`,..., `fv(5)`, respectively, are the statuses of `ieee_overflow`, `ieee_divide_by_zero`, `ieee_invalid`, `ieee_underflow` and `ieee_inexact`.

14.45 IEEE_GET_UNDERFLOW_MODE

This subroutine takes a logical variable as its argument. If the returned value is `true`, the present underflow mode is gradual. If it is `false`, the present underflow mode is abrupt.

```
use ieee_arithmetic
logical:: gradual
call ieee_get_underflow_mode(gradual)
 if(gradual) then
  print *, 'underflow mode is gradual'
else
 print *,'underflow mode is abrupt'
endif
end
```

14.46 IEEE_SET_UNDERFLOW_MODE

For ieee_set_underflow_mode if the argument is true, gradual underflow is set; if it is false, gradual underflow is removed. These subroutines can be called if ieee_support_underflow(x) is true for the particular x.

```
call ieee_set_underflow_mode(.true.)     ! set
call ieee_set_underflow_mode(.false.)    ! reset
```

14.47 IEEE_SELECTED_REAL_KIND

This function takes three integers p, r and radix as its argument. All the arguments are scalar. Both p and r are assumed to be zero if they are absent. At least one argument must be specified. If there is no requirement for radix for a particular selected kind, it may be absent. The result is a kind parameter (integer) for the IEEE real data type having decimal precision at least p digits and an exponent r.

If no kind parameter corresponding to 'p' is available with a particular radix, it returns –1. If kind corresponding to 'r' with a particular radix is not available, it returns –2. If kind corresponding to both 'p' and 'r' for a particular radix is not available, it returns –3. If kind corresponding to both 'p' and 'r' is supported but not both together for a particular radix, it returns –4, and –5 is returned if the processor does not support IEEE real type with a particular radix. If more than one kind parameter is supported, the kind parameter corresponding to the smallest decimal precision is returned.

```
use ieee_arithmetic
integer:: s
s=ieee_selected_real_kind(6,35); print *, s
s=ieee_selected_real_kind(12); print *, s
s=ieee_selected_real_kind(12,400); print *, s
s=ieee_selected_real_kind(50,300); print *, s
s=ieee_selected_real_kind(12,300); print *, s
s=ieee_selected_real_kind(50,400); print *, s
s=ieee_selected_real_kind(6); print *, s
end
```

The outputs are 1, 2, –2, –1, 2, –3, 1, respectively.

14.48 Arithmetic IF and IEEE_VALUE

Though `arithmetic if` has been declared as an obsolete feature of the language, it is necessary to mention a few words when the argument of `arithmetic if` is, say, `ieee_quiet_nan`.

It may be noted that `ieee_quiet_nan` is not a number—it is not a negative, or a positive or a zero. Therefore, the syntax of `arithmetic if` is not valid in this case. The path chosen by the code is compiler dependent. The `arithmetic if` treats both `ieee_positive_zero` and `ieee_negative_zero` as zero, `ieee_positive_inf` as a positive number and `ieee_negative_inf` as a negative number.

14.49 IEEE_QUIET Compare Routines

There are six routines in this category. They are `ieee_quiet_eq`, `ieee_quiet_ge`, `ieee_quiet_gt`, `ieee_quiet_le`, `ieee_quiet_lt` and `ieee_quiet_ne`. These routines take two reals (say, a and b) as arguments. The arguments are of same kind. These routines compare a and b, respectively, for being equal, greater than or equal, greater than, less than or equal, less than and not equal. Accordingly, the routines return `true` or `false`. If one of the arguments is NaN, the returned values are `false`, `false`, `false`, `false`, `false` and `true`, respectively. If a or b is a signaling NaN, `ieee_invalid` signal is raised.

14.50 IEEE_SIGNALING Compare Routines

There are six routines in this category. They are `ieee_signaling_eq`, `ieee_signaling_ge`, `ieee_signaling_gt`, `ieee_signaling_le`, `ieee_signaling_lt` and `ieee_signaling_ne`. These routines are almost same as the quiet routines discussed in the previous section. If either a or b is NaN, `ieee_invalid` signal is raised.

14.51 NaN, Infinity and Format

The edit descriptors `f`, `e`, `d` and `g` may be used to assign NaN and `infinity` (both positive and negative) to a real variable through `read` statement:

```
      use ieee_arithmetic   ! this may be omitted
      real::x
      read 10,x
      print *,x
10    format(f9.3)
      end
```

The input may be NaN or INF or +INF, INFINITY or +INFINITY or -INF or -INFINITY (lowercase letters are also allowed). It may be noted that if it is desired to input +INFINITY or -INFINITY, the width of the field must be at least 9. The data should be within the field—leading and trailing blanks are ignored. However, if +INF or -INF is used, the width should be greater than or equal to 4. The output is always NaN or Infinity or -Infinity depending on the input. These discussions are based on NAG Fortran compiler. It may be different for other compilers.

14.52 Relational Operators, Infinity and NaN

NaN is not a number. If it is compared with any normal number, the result is false. However, if NaN is tested for non-equality with another NaN, the result is true.

```
use ieee_arithmetic
real::x, y=10.0
x=ieee_value(x, ieee_quiet_nan)
print *, x>y, x>=y, x<y, x<=y, x==y, x==x, x/=x
end
```

The output is as follows:

```
F F F F F F T
```

Positive infinity is greater than any number including negative infinity.

```
use ieee_arithmetic
real:: x, y=10.0,z
x=ieee_value(x, ieee_positive_inf)
z=ieee_value(x, ieee_negative_inf)
print *, x>y, x>=y, x<y, x<=y, x==y, x==x, x /= x, x>z
end
```

The output is as follows:

```
T T F F F T F T
```

Similarly, negative infinity is less than any positive or negative number.

```
use ieee_arithmetic
real:: x, y=10.0,z
x=ieee_value(x, ieee_negative_inf)
z=ieee_value(z, ieee_positive_inf)
print *, x>y, x>=y, x<y, x<=y, x==y, x==x, x/=x, x<z
end
```

The result is as follows:

```
F F T T F T F T
```

When NaN is compared with a positive or negative `infinity`, the result is always `false` as NaN is not a number.

```
use ieee_arithmetic
real:: x, y
x=ieee_value(x, ieee_positive_inf)
y=ieee_value(y, ieee_quiet_nan)
print*, x>y, x>=y, x<y, x<=y, x==y
end
```

The result is as follows:

```
F F F F F
```

Also for the following:

```
use ieee_arithmetic
real:: x, y
x=ieee_value(x, ieee_negative_inf)
y=ieee_value(y, ieee_quiet_nan)
print*, x>y, x>=y, x<y, x<=y, x==y
end
```

The result is as follows:

```
F F F F F
```

14.53 Exception within a Procedure

If a flag is set within a procedure, it is not reset on exit from the procedure.

```
use ieee_arithmetic
real:: r=1.0, s=0.0,t
logical:: flagvalue
call ieee_set_halting_mode(ieee_divide_by_zero,.false.)
call ieee_get_flag(ieee_divide_by_zero, flagvalue)
if(flagvalue) then
  print *, 'Divide by zero flag is on (1)'
else
  print *, 'Divide by zero flag is off (1)'
endif
call sub(r,s,t)
call ieee_get_flag(ieee_divide_by_zero, flagvalue)
if(flagvalue) then
  print *, 'Divide by zero flag is on (3)'
else
  print *, 'Divide by zero flag is off (3)'
endif
```

```
      end
      subroutine sub(r, s,t)
      use ieee_arithmetic
      real:: r, s,t
      logical:: flagvalue
      t=r/s
      call ieee_get_flag(ieee_divide_by_zero, flagvalue)
      if(flagvalue) then
        print *, 'Divide by zero flag is on (2)'
      else
        print *, 'Divide by zero flag is off (2)'
      endif
      t=r+s
      end
```

The outputs are as follows:

```
Divide by zero flag is off (1)
Divide by zero flag is on (2)
Divide by zero flag is on (3)
Warning: Floating-point divide by zero occurred
```

The output shows that the division by zero flag set inside the subroutine is not reset when the subroutine returns the control to the calling program.

14.54 Exception Outside a Procedure

If a flag is set outside the procedure, this value of the flag is not carried over when the procedure is called.

```
      use ieee_arithmetic
      real:: r=1.0, s=0.0,t
      logical:: flagvalue
      call ieee_set_halting_mode(ieee_divide_by_zero,.false.)
      t=r/s
      call ieee_get_flag(ieee_divide_by_zero, flagvalue)
      if(flagvalue) then
       print *, 'Divide by zero flag is on (1)'
      else
       print *, 'Divide by zero flag is off (1)'
      endif
      call sub(r, s,t)
      call ieee_get_flag(ieee_divide_by_zero, flagvalue)
      if(flagvalue) then
       print *, 'Divide by zero flag is on (3)'
      else
       print *, 'Divide by zero flag is off (3)'
      endif
```

```
      end
      subroutine sub(r, s,t)
      use ieee_arithmetic
      real:: r, s,t
      logical:: flagvalue
      call ieee_get_flag(ieee_divide_by_zero, flagvalue)
      if(flagvalue) then
       print *, 'Divide by zero flag is on (2)'
      else
       print *, 'Divide by zero flag is off (2)'
      endif
      t=r+s
      end
```

The outputs are as follows:
```
Divide by zero flag is on (1)
Divide by zero flag is off (2)
Divide by zero flag is on (3)
Warning: Floating-point divide by zero occurred
```

14.55 Programming Examples

The following program shows transition of a number from normal to denormal when the number is repeatedly divided by 10.0; it ultimately becomes zero.

```
      use ieee_arithmetic
      real:: x=10.0 ! NAG compiler
      do
       if (x.eq.0.0) then
       stop 'zero reached'
       endif
       if(ieee_is_normal(x)) then
        print *, x, 'Normal'
       else
        print *, x, 'Denormal'
       endif
      x=x/10.0
      enddo
      end
```

The last line of the output is 1.4012985E-45 Denormal. Readers are advised to run this program to have a feeling for the denormal number.

The second program demonstrates the use of floating overflow. It calculates factorial, and after the calculation, check is made for overflow. The program is terminated when overflow occurs. However, at the beginning of the program, halting due to overflow is disabled, and the control is given to the programmer.

```
use ieee_arithmetic ! calculation of factorial 1,2,3,4....
real:: f1, f2
integer:: n=2
logical::flag
if(.not. (ieee_support_halting(ieee_overflow))) then
 print *, 'program termination cannot be controlled &
  & by overflow condition'
 stop
endif
call ieee_set_halting_mode(ieee_overflow,.false.)
f1 = 1.0 ! job will not be aborted when overflow occurs
f2=f1
do
 f1=n*f1 ! 1 × 2x3 × 4....
 call ieee_get_flag(ieee_overflow, flag)
 if(flag)then
  exit ! exit when there is overflow, last valid factorial is printed
 endif
f2=f1 ! f2 contains the last valid result
n=n+1
enddo
 print *, 'Factorial up to ', n-1, 'is possible, Value... ', f2
end
```

The output of the program is as follows:

```
Factorial up to 34 is possible, Value ... 2.9523282E+38
```

The output tells that this particular processor can calculate up to factorial 34 (normal precision). The processor flags overflow (if it is signaling) if an attempt is made to calculate factorial 35.

The next program shows a method to prevent program termination due to an invalid arithmetic operation. The example calls the library function square root (sqrt) with a negative real number as its argument.

```
use ieee_arithmetic
real:: x, y
x=-4.0
call ieee_set_halting_mode(ieee_invalid,.false.)
y=sqrt(x)
print *, y
end
```

Square root of a real negative number is an invalid arithmetic operation. If the call to the subroutine ieee_set_halting_mode is initiated with the first parameter as ieee_ invalid and the second parameter as .false., the job will not be aborted when the statement y=sqrt(x) is executed with negative real number as argument. It is, of course, the responsibility of the programmer to handle such a situation. The print statement will display NaN. Note that the program will be aborted with arithmetic exception if the call to the subroutine is removed because the default action is abort. Similarly, if x=-2.1 and y=4.5, execution of (x)**y will set the ieee_invalid flag as this involves calculation of logarithm of a negative number.

14.56 Out of Range

The intrinsic out_of_range is used to test whether a real or an integer can safely be converted to real or integer of different kind or type. This intrinsic takes three arguments. The first one x is either real or integer, and the second argument mold is of type real or integer. It is a scalar, and it is not necessary to assign a value to mold if it is a variable. The third optional argument (scalar) round is of type logical if the first argument is real and the second argument is an integer. The argument round can be true or false, and it controls rounding during conversion. The function returns true if the first argument can safely be converted to mold type; otherwise it returns false.

14.57 Processor Dependencies

A particular processor may not support all IEEE features. The initial rounding mode, underflow mode and halting mode are processor dependent. In addition, the floating point exception flags are processor dependent when non-Fortran procedures are entered.

The interpretation of a sign in the input field is processor dependent. The output corresponding to IEEE NaN is usually NaN. It may be different.

15

Dynamic Memory Management

Dynamic memory management, that is, managing the memory during the execution of a program, is one of the new features of Fortran. It was mentioned in Chapter 6 that the size of the array is to be declared at the compilation time. Therefore, if the array size changes from one run to another, it is necessary to recompile the program with this changed value of the array size. The alternative is to define an array of sufficiently large size so that the program can handle all kinds of problems. The first one is clearly not convenient and the second one sometimes may block a large amount of memory even when the requirement is small. Fortran allows an array to be allocatable, and thus, its size may be specified during the execution of the program. This enables the user to grab memory dynamically and release the same to the system when it is no longer required. The initial status of an allocatable array is unallocated. This means that no memory locations have been assigned to the allocatable array. It is to be allocated during the execution of the program. It is also possible to allocate location to a scalar during execution.

15.1 ALLOCATABLE Arrays

An array can be declared "allocatable" by specifying only the rank. The size of the array is specified during the execution.

```
integer, allocatable, dimension(:) :: a
```

This declares a as an allocatable array of rank 1. The size of the array is specified later through the allocate statement.

```
allocate(a(1000))
```

Memory locations sufficient to store 1000 such 'a' are obtained from the system through this allocate statement. The array is then accessed in the usual manner. The argument of allocate need not be constant. It may be a variable or even an expression.

```
read *, n
allocate (a(n))
allocate (a(2*n+1000))
allocate (a(0)) ! allowed, size=0
```

While allocating an array, both the lower and upper bounds can be specified. Of course, the lower bound is 1, if it is not specified.

```
allocate (a(0:1999))
```

The subscript of a array may vary between 0 and 1999.

The following are examples of allocatable arrays of rank more than 1. The ranks are specified by colons separated by a comma.

```
real, allocatable, dimension(:,:) :: b !  2-d allocatable array
real, allocatable, dimension(:,:,:) :: c ! 3-d allocatable array
integer, allocatable, dimension(:,:,:,:,:,:,:,:,:,:,:,:,:,:,:) :: x ! 15-d
```

These allocatable arrays are given memory locations with the allocate statement.

```
allocate (b(100,100))
allocate (b(0:99,0:99))
allocate (b(-10:10,10:20))
allocate (c(5,5,5))
allocate (c(0:5,-1:10,10:20))
allocate (c(0:5,10,20))
allocate(x(2,2,2,2,2,2,2,2,2,2,2,2,2,2,2))
```

There can be no more than 15 colons in the allocatable statement. Two or more allocatable variables may be allocated locations by a single allocate statement.

```
allocate (a(1000),b(10,10))
```

Some compilers do not allow the following statements:

```
allocate (a(1000), b(size(a),size(a))
```

This is possible, if the allocate statement is broken into two allocate statements.

```
allocate (a(1000))
allocate (b(size(a),size(a))
```

If, for some reason, the allocate statement fails to allocate locations, the job is aborted. This can be prevented if a status variable stat=integer-variable is used with the allocate statement, where integer-variable is set to zero if the system could assign the requisite number of locations mentioned in the allocate statement. It returns non-zero if the requisite number of locations cannot be made available to the program. If stat is used, the program continues even if the system fails to allocate the requisite number of locations. It is the responsibility of the programmer to take necessary action should such a situation arise.

```
allocate (a(10000), stat=istat)
.
if(istat .eq. 0) then    !    locations allocated
.
 else                    !    locations not allocated
.
endif
```

If more than one variable is assigned locations through a single allocate statement, the status indicates the results of all allocations together.

```
allocate (a(10000), b(10), stat=istat)
```

Successful execution of an `allocate` statement sets `istat` to zero. Unsuccessful execution of an `allocate` statement sets `istat` to a processor-dependent positive integer. The status variable cannot be allocated in the same `allocate` statement.

The `allocate` statement can have an optional parameter `errmsg` of type character. In case the `allocate` statement fails, this variable will contain some additional information. Consider the following program:

```
integer, dimension(:), allocatable :: a
integer :: istat
character(len=60) :: err=" "
allocate(a(100))
allocate(a(200),stat=istat,errmsg=err) !   intentional error
print *, "Status : ",istat; print *, "Errmsg : ",err !NAG compiler
end
```

Array `'a'` has been allocated and needs to be deallocated before it could be allocated again. Thus, the second allocate statement is bound to generate an error message (compiler dependent), and because of the presence of `stat`, the job will not be aborted. The outputs of the program are as follows:

```
Status :  2
Errmsg : Allocatable variable is already allocated
```

The Variable `err` is not modified if the processor could allocate required number of locations specified in the allocate statement. Allocated array may be of size zero.

`Allocate` with the `source` attribute is used to produce a clone. In the example given next, the allocated array `'a'` gets the upper and lower bounds and values of the elements from array b.

```
integer, allocatable, dimension(:) :: a
integer, dimension(-1:3) :: b=[1,2,3,4,5]
allocate(a, source=b)
print *,a; print *, lbound(a), ubound(a) ! ifort compiler
end
```

The output is as follows:

```
1       2           3           4           5
 -1      3
```

The `print` statement will display 1 2 3 4 5. The extents of arrays a and b must be same. Array `'a'` gets the lower and upper bounds from array b. For example, the following statements

```
allocate(a(3), source=b)
allocate(a(10), source=b)
```

are not acceptable.

Allocation with `mold` produces an array having the same size and same upper and lower bounds as `mold`. However, no value is assigned to the allocated array.

```
integer, allocatable, dimension(:) :: a
integer, dimension(-1:3) :: b=[1,2,3,4,5]
allocate(a, mold=b)
print *,a  ! this will print garbage
print *, lbound(a), ubound(a)
end
```

The output is as follows:

```
1633772636   1634348652   1761608554        23488     -23920304
        -1                3
```

The `allocate` statement only takes the information related to the size of `mold` (in this case b array). As array a only gets the upper and lower bounds of b, `print *,a` displays garbage (shown in bold; this may vary from one run to another) as the array has not been assigned to any value. In other words, the result is same if array a is just defined without initializing.

Automatic allocation is possible through assignment statement if the two arrays have the same rank. If the array appearing on the left-hand side of the assignment sign is not yet allocated, it is allocated with the same size as the array that appears on the right-hand side of the assignment sign. However, if the array appearing on the left-hand side of the assignment sign is already allocated, it is first deallocated and then the assignment operation takes place as mentioned earlier.

```
integer, allocatable, dimension(:) :: a
integer, dimension(5) :: b=[1,2,3,4,5]
integer, dimension(8) :: c=[1,2,3,4,5,6,7,8]
a=b    ! allocates 5 locations and sets b to a
a=c    ! first deallocates a; then allocates 8 locations and
       ! sets c to a
end
```

15.2 DEALLOCATE Statement

The allocated arrays may return locations back to the system by using the `deallocate` statement.

```
deallocate(a)
deallocate(a,b)
```

Once an array is deallocated, the array is no longer available, and if the variable is allocated again, old values do not reappear.

Like the `allocate` statement, unsuccessful deallocation causes the job to be terminated. Again, `stat =integer-variable` may be used to monitor the status of the deallocation process. Successful deallocation causes the integer variable to be returned with zero; unsuccessful deallocation sets the integer variable to be returned with a processor-dependent positive value. Note that if `stat` is present, the job is not aborted even if the system fails to deallocate an array. The integer variable associated with `stat` cannot be deallocated in the same `deallocate` statement. Like the `allocate` statement, the `deallocate` statement also takes an optional parameter `errmsg` of type character. This works exactly like the `allocate` statement.

```
integer, dimension(:), allocatable :: a
integer :: istat
character(len=50) :: err= " "   ! NAG compiler
allocate(a(100))
a=200
deallocate(a)
deallocate(a,stat=istat,errmsg=err) ! already deallocated.
print *, "Stat : ",istat; print *, "Errmsg : ",err
end
```

As expected the outputs will indicate the error condition.

```
Stat :  3
Errmsg : Allocatable variable is not allocated
```

15.3 ALLOCATED Intrinsic

This intrinsic is used to test whether an allocatable array has been allocated by `allocate` statement. This intrinsic returns `true` if the array has been allocated; otherwise it returns `false`.

```
if (allocated(a)) then     !  a has been allocated
.
else                       !  a has not been allocated
.
endif
if(allocated(a)) then
   deallocate(a)
endif
```

15.4 Derived Type and ALLOCATE

A user-defined type may be made allocatable.

```
type t
  integer :: a
  real :: b
end type t
type(t), allocatable :: x
allocate(x)
x%a=2
x%b=3.5
print *, x%a
```

```
print *, x%b
deallocate(x)
end
```

Like standard variables, a user-defined type variable may be an allocated array.

```
type mytype
  integer :: a
  real :: b
end type mytype
type(mytype),allocatable,dimension(:) :: c
allocate(c(2))                    ! note this statement
c(1)%a=2; c(1)%b=2.5; c(2)%a=10; c(2)%b=10.5
print *,c(1)%a,c(1)%b; print *,c(2)%a,c(2)%b
deallocate(c)
end
```

The elementary items of a user-defined variable may also be an allocated array:

```
type mytype
  integer :: a
  real :: b
  integer, allocatable, dimension(:) :: e
end type mytype
type(mytype), allocatable, dimension(:) :: c
allocate(c(2)); allocate(c(1)%e(2)); allocate(c(2)%e(3))
c(1)%a=2; c(1)%b=2.5; c(2)%a=10; c(2)%b=10.5
c(1)%e(1)=27; c(1)%e(2)=37; c(2)%e(3)=100
print *,c(1)%a, c(1)%b; print *,c(2)%a, c(2)%b
print *,c(1)%e(1), c(1)%e(2); print *,c(2)%e(3)
end
```

Variable c is an allocatable array of rank 1. It has been allocated, and its size has become 2. It contains an integer allocatable elementary item. This has been allocated individually for c(1) and c(2). The elementary items corresponding to c(1) are c(1)%e(1) and c(1)%e(2) and corresponding to c(2) are c(2)%e(1), c(2)%e(2) and c2%e(3).

15.5 Allocated Array and Subprogram

The local variable of a subprogram may be an allocated array. On exit from the subprogram, the locations are returned to the system. Moreover, when the subprogram is entered again, the locations allocated earlier are not available to the program. This is illustrated with an example.

```
program alloc_demo
integer :: i
do i=1,2
  call sub(i)
```

```
    .
    enddo
    end program alloc_demo
    subroutine sub(i)
    integer :: i
    integer, allocatable, dimension(:) :: a
    if(i.eq.1) then
       allocate(a(5))
       a=[1,2,3,4,5]
    endif
    print *, a
    end subroutine sub
```

When the subprogram is entered for the first time, locations are allocated—the `print` statement works correctly. When the subroutine is entered for the second time with `i=2`, the `print` statement tries to access an array that is not available, and therefore, it is not legal. In the earlier version of Fortran (Fortran 90), if the local array is not deallocated explicitly, it used to create problems. Let us see what happens when the `if` statement of the subprogram is removed and the number of times the do loop to be executed is increased in the main program (assume that the earlier version of Fortran compiler is being used).

```
    do i=1,10000
     call sub(i)
    enddo
    end
    subroutine sub(i)
    integer :: i !    i is not used
    integer, allocatable, dimension(:) :: b
    allocate(b(10000))
    b(1)=100
    end subroutine sub
```

Each time the subroutine is entered, b array is allocated. Since the b array is not deallocated in the subprogram, the locations, though unavailable on exit from the subroutine, are not released to the system. The locations remain attached somehow to the program without any facility of being accessed by the program. Each time the subroutine is called, fresh locations are assigned to b and unavailable locations allocated during the earlier call remain attached to the program. This goes on, and the subroutine grabs memory locations in each call and does not release the same on exit from the subroutine, and so more and more memory locations continue to be attached to the program. Ultimately, after several such calls, when no more locations can be allocated, the allocate statement within the subroutine fails and the job is aborted.

The current version of Fortran automatically deallocates the allocated array on exit from the subprogram.

If the allocatable array is declared with `save` attribute and allocated only once, the locations reappear with the last value on subsequent entry to the subroutine. The second method is to declare the allocatable array within a module and use the module within the subroutine. Allocatable arrays declared within a module are not released when the subroutine is exited. Both these methods are illustrated next.

Method I: Allocatable array with `save` attribute

```
integer :: i
do i=1,100
  call sub(i)
enddo
end
subroutine sub(i)
integer :: i
integer, allocatable, save, dimension(:) :: c
if (i .eq. 1) then
  allocate(c(5))
  c=[10,20,30,40,50]
endif
print *, c
end
```

The `print` statement will not fail when the value of i is greater than 1 (it is assumed that the first call to the subroutine is executed with i = 1). This is because of the `save` attribute; the c array will reappear with its last value when the subroutine is entered again (i greater than 1). However, it is a fact that the c array is unavailable when the subroutine is exited.

Method II: Allocatable array within a module

The allocatable array is declared within a module so that the same array is available when the subroutine is entered again.

```
module myalloc
   integer, allocatable, dimension(:) :: d
end module myalloc
program main
integer :: i
do i=1, 100
  call sub(i)
enddo
end
subroutine sub(i)
use myalloc
integer :: i
if(i.eq.1) then
  allocate(d(5))
  d=[100,200,300,400,500]
endif
.
end
```

15.6 ALLOCATE and Dummy Parameter

An allocatable array may be a dummy argument of a subprogram. In such a case, both the actual and dummy parameters must be declared as an allocatable array with identical rank. The array is allocated either in the calling program or in the called program depending on the `intent`. Interface block must be present.

Case I: The `intent` is `out`. In this case, array is allocated within the subprogram.

```
program casei
integer, allocatable, dimension(:) :: a
interface
  subroutine sub(a)
  integer, intent (out), allocatable, dimension(:) :: a
  end
end interface
call sub(a); print *,a; deallocate (a)
end
subroutine sub(b)
integer, intent(out), allocatable, dimension(:) :: b
allocate (b(5))
b=[1,2,3,4,5]
end
```

The outputs are as follows:

```
1 2 3 4 5.
```

Case II: The `intent` is `in`. In this case, the array is allocated within the calling program.

```
program caseii
integer, allocatable, dimension(:) :: a
interface
   subroutine sub(a)
   integer, intent(in), allocatable, dimension(:) :: a
end
end interface
allocate (a(5))
a=100
call sub(a); print *, a; deallocate (a)
end
subroutine sub(b)
integer, intent(in), allocatable, dimension(:) :: b
print *, b
end
```

The outputs are as follows:

```
100    100  100  100   100
100    100  100  100   100
```

Case III: The `intent` is `inout`. The array is allocated either in the calling or within the called program. Since the `intent` is `inout`, the array is supposed to carry some value from the calling program. This implies the array is allocated in the calling program. However, despite the fact that the `intent` is `inout`, if the array does not carry any input from the calling program, the array may be allocated within the called program. However, this not proper. Improper use of `intent` seems to be a bad programming practice.

```
program caseiii
integer, allocatable, dimension(:) :: a
interface
  subroutine sub(a)
  integer, intent(inout), allocatable, dimension(:) :: a
end
end interface
allocate (a(5))
a=100
call sub(a); print *, a; deallocate (a)
end
subroutine sub(b)
integer, intent(inout), allocatable, dimension(:) :: b
print *, b
b=[1,2,3,4,5]
end
```

The outputs are as follows:

```
100   100   100   100   100
1      2     3     4     5
```

15.7 Allocatable Character Length

While declaring a character variable, the length may be specified as " : ". When this variable is assigned to a value or allocated, processor assigns locations for the variable.

```
character (len=: ), allocatable :: ch
ch = "saha institute of nuclear physics"; print *, len(ch)
ch = "university of calcutta"; print *, len(ch)
end
```

The output is as follows (shown in the same line):

```
33 22
```

The locations can also be allocated using the `allocate` statement.

```
allocate (character(len=20) :: ch)
```

Now, we add a few more statements just before end statement to the preceding program.

```
deallocate(ch)
allocate (character(len=20) :: ch)
print *, len(ch); ch = "iacs, kolkata"; print *, len(ch)
```

The output (shown in the same line):33 22 20 13.

It is clear that the allocate statement allocates locations for storing 20 characters, so that statement print *, len(ch) displays 20. However, when ch is set to another value, the length of ch changes dynamically that is displayed as 13—the length of the string "iacs, kolkata".

15.8 Character and Allocatable Arrays

Since the size of a character variable can be specified in the character declaration, allocatable character arrays need to be mentioned separately.

```
character(len=30),allocatable,dimension(:)::ch
    .
allocate(ch(100))
```

The preceding allocate statement will allocate sufficient memory locations to accommodate a character array of size 100 with each element of size 30 characters.

Fortran allows specification of length of a character variable to be deferred, and it can be allocated by an allocate statement.

```
character(len=:),dimension(:),allocatable :: ch
    ...
allocate(character(len=10) :: ch(20))
```

The declaration may also be written as character(:),allocatable:: ch(:).

Array ch will have 20 elements, each having a size of 10 characters.

15.9 Allocatable Scalar

Memory may be allocated to a scalar during the execution of a program. Subsequently, the same may be deallocated in the usual way.

```
integer, allocatable::x
allocate(x)
x=10
    .
deallocate(x)
    .
end
```

Similarly, other variables such as real, logical, complex, double precision and character can be allocated and deallocated in the same manner.

15.10 Allocatable Function

A function returns value through its name or through the `result` clause. The result of the function may have an allocatable attribute. The following program demonstrates this feature.

```
program main
interface
  function ac(x,n)
  real, allocatable, dimension(:) :: ac ! note this declaration
  real, dimension(:),intent(in) :: x
  integer, intent(in) :: n
  end function ac
end interface
real, allocatable, dimension(:) :: y,z
real, parameter::fac=3.1415926/180.0 ! degree to radian
integer::i,n
print *,'enter the value of n'
read *,n
allocate(y(n),z(n))
do i=1,n
    y(i)=30.0*fac*i ! 30, 60, 90 degree etc.
enddo
print *,sin(y)+cos(y)
z=ac(y,n)
print *,z
end
function ac(x,n)
real, allocatable, dimension(:) :: ac
real, dimension(:), intent(in) :: x
integer, intent(in) :: n
allocate(ac(n)) ! note this statement
ac=sin(x)+cos(x)
end
```

The two `print` statements should give identical result.

If the function returns the value through the `result` clause, the preceding program is to be modified to take into account this fact.

```
program main
interface
  function ac(x,n) result(res)
  real, allocatable, dimension(:) :: res
  real, dimension(:), intent(in) :: x
  integer, intent(in) :: n
  end function ac
end interface
real, allocatable, dimension(:) :: y,z
real, parameter :: fac=3.1415926/180.0
integer :: i,n
    .
z=ac(y,n)
```

```
print *,z
end
function ac(x,n) result (res)
real, allocatable, dimension(:) :: res
real, dimension(:), intent(in) :: x
integer, intent(in) :: n
allocate(res(n))
res=sin(x)+cos(x)
end
```

This subprogram also calculates $sin(x)+cos(x)$ for n values and returns the values through the `result` clause. The current version of Fortran is capable of automatically allocating required number of locations to variable z depending upon the size of the returned array and would deliver correct result if the statement `allocate(y(n),z(n))` is replaced by `allocate (y(n))`.

15.11 Allocation Transfer

Subroutine `move_alloc` is used to move allocation including the values of different locations from one array (`from`) to another array (`to`). The syntax of `move_alloc` is as follows:

```
call move_alloc(from, to)
```

After the subroutine is called, the `from` array is deallocated; the `to` array is allocated obtaining all values from the corresponding locations of `from` array.

```
integer, allocatable :: a(:), b(:)
allocate(a(5))
a=[1,2,3,4,5]
call move_alloc(a,b); print *, b
end
```

The `print` statement will display 1 2 3 4 5 because while moving the allocation, the subroutine has set b=a and deallocated array a.

This prescription may be used conveniently to increase the size of an allocatable array without destroying the current content of the array.

```
integer, allocatable, dimension(:) :: x,y
integer :: sz
x=[1,2,3,4,5]
sz=size(x)                        ! size of array x; 5 in this case
allocate(y(2*sz))                 ! size of the array y is 10
y(1:sz)=x                         ! y(1)... y(5)= x(1)...x(5)
call move_alloc(from=y, to=x)
x(sz+1:2*sz)=100                  ! size of x is now 10 with
print *, x                        ! x(1), ... x(5) same as old x
end
```

The size of array y is two times the size of array x; y(1: sz)=x ensures that x(1), x(2), ..., x(5) are copied on to y. Subroutine move_alloc reallocates x and copies the content of y to x. It also deallocates y. The next statement stores 100 in locations x(6), x(7), ..., x(10).

15.12 Restriction on Allocatable Arrays

The common statement works on the assumption that the storage is assigned in a sequential fashion; moreover, the size of the common block must be available during compilation. An allocatable array whose size will be available only during execution, therefore, cannot appear in a common block. However, this is not a serious limitation as different units may share locations through a module.

15.13 Programming Example

In the programming example, a 3×3 linear simultaneous equation will be solved by the Gauss method. This method solves linear simultaneous equation by setting elements below the diagonal to zero by arithmetic manipulation. If the equations are as follows:

$$a_{11}x_1 + a_{12}x_2 + a_{13}x_3 = b_1 \tag{1}$$
$$a_{21}x_1 + a_{22}x_2 + a_{23}x_3 = b_2 \tag{2}$$
$$a_{31}x_1 + a_{32}x_2 + a_{33}x_3 = b_3 \tag{3}$$

By arithmetic manipulation this can be changed to the following:

$$a_{11}x_1 + a_{12}x_2 + a_{13}x_3 = c_1 \tag{4}$$
$$d_{22}x_2 + d_{23}x_3 = c_2 \tag{5}$$
$$d_{33}x_3 = c_3 \tag{6}$$

From equation (6) one can get x_3, and substituting this value in (5), the value of x_2 is obtained from (5). Having obtained the value of x_2 and x_3, it is easy to find the value of x_1 from (4).

The following program uses allocate, deallocate, maxloc and maxval. Any standard textbook on numerical method may be consulted for the Gauss algorithm.

```
program gausstest
real, dimension(3,4) :: a
real, dimension(3) :: x
interface
   subroutine gauss(a,x,n,np1)    !  the equation is ax=b
   real, intent(inout), dimension(:,:) :: a
   real, intent(out), dimension(:) :: x
   integer, intent(in) :: n, np1
   end subroutine gauss
end interface ! a array, rightmost column contains b.
a=reshape([1,3,-1,1,2,1,1,7,1,5,32,1],[3,4])
call gauss(a,x,3,4);  print *, x ! the roots are returned through x
end
subroutine gauss(a,x,n,np1)
implicit none
integer, intent(in) :: n, np1
real, intent(inout), dimension(:,:) :: a
real, intent(out), dimension(:) :: x
real, dimension(:), allocatable :: t
real :: f,sum
integer :: l, kp1,nm1,i,j,k, ip1
allocate(t(np1))
nm1=n-1
do k=1, nm1 !  by arithmetic manipulation set the
   kp1=k+1      ! triangle below the diagonal to zero
   l=maxval(maxloc(abs(a(k:n,k))))+k-1 ! largest element in each column
   if(l.ne.k) then !   interchange the row if necessary
     t(k:np1)=a(k,k:np1); a(k,k:np1)=a(l,k:np1); a(l,k:np1)=t(k:np1)
   endif
   do i=kp1,n
      f=a(i,k)/a(k,k)
      do j=kp1, np1
       a(i,j)=a(i,j)-f*a(k,j)
      enddo
   enddo
enddo
x(n)=a(n,np1)/a(n,n)     !  back substitution
do i = nm1,1,-1
  ip1=i+1; sum=0.0
  do j=ip1,n
   sum=sum+a(i,j)*x(j)
  enddo
 x(i)=(a(i,np1)-sum)/a(i,i)
enddo
deallocate(t)
return
end
```

The program is used to solve the following equations:

$$x+y+z=5, \quad 3x+2y+7z=32, \quad -x+y+z=1.$$

The roots of the equations are `2.0`, `-1.0`, `4.0`.

16

Pointers

A pointer is a variable that during execution can point to another variable (variable, array, derived type, etc.); it can dynamically associate itself with another object. The pointer becomes the alias of such a variable.

When a variable is declared, the declaration informs the compiler about the variable and, if it is initialized, the initial value. For an array, its type, rank and size are specified. The compiler allocates enough memory for the object accordingly. The information about the object is called the descriptor, which describes the object and memory locations that contain the value of the object. For a pointer variable only, the descriptor without the memory location is specified, and during the execution of the program, it is made to point to a memory location. In addition, a pointer can dynamically change its association during the execution of the program.

A pointer must point to the right kind of object. For example, an integer pointer can point to an integer quantity and a real pointer can point to a real quantity. An integer pointer cannot point to a real quantity.

16.1 POINTER Declaration

A pointer is declared as follows:

```
integer, pointer :: ip          !  ip can point to an integer
real, pointer :: rp             !  rp can point to an real quantity
character(len=30), pointer :: cp !  cp is a character pointer
```

Pointer `ip` can point to an integer. Similarly, pointer `rp` can point to a real quantity. Pointer `cp` can point to a character variable of length 30. It can, however, be made to point 30 characters from a character variable having a length of greater than 30 characters.

16.2 TARGET

A pointer can point to a variable declared with an attribute `target`.

```
integer, target :: a, b
real, target :: r, s
character(len=30), target :: c
```

If the variable is not declared with the `target` attribute, a pointer cannot point to such a variable. A pointer can, however, point to another pointer.

A variable declared with the `target` attribute cannot have the `pointer` attribute. The compiler treats the `target` variables in a special way. It ensures that the `target` variable is always available within its scope. Thus, the compiler prevents the `target` variable from disappearing under some temporary unnamed variable created by the compiler during optimization of the program to increase the computing speed. A pointer could be associated with a target, which is still in an undefined state.

16.3 POINTER Status

A pointer may exist in three different states: (a) undefined, (b) null and (c) associated. At the beginning of the program, the pointer is in the undefined state unless it is initialized along with its declaration. In the null state, the pointer does not point to anything but it is not in an undefined state. In the associated state, the pointer points to some object.

16.4 POINTER Initialization

A pointer can be initialized to `null` along with its declaration.

```
real, pointer :: rp=>null()
```

Pointers can be initialized through `data` statements too.

```
data rp/null()/
```

16.5 POINTER Assignment

Pointer `ip`, for instance, may be made to point to an integer variable through pointer assignment. There is a special symbol for this purpose. It is equal sign followed by a greater than sign (`=>`).

```
ip => a
```

Now, `ip` points to a. Thus, `ip` is an alias to a; a = 2 and `ip` = 2 are the same (Figure 16.1). Subsequently, `ip` may point to, say, b.

```
ip => b
```

At this moment, `ip` becomes alias to b—and it no longer points to a. However, this dissociation of the pointer with 'a' does not disturb the value of a. Consider the following program segment:

```
integer, pointer :: ip, iq
integer, target :: a=10, b=20
ip => a      ! ip points to a
iq => b      ! iq points to b
```

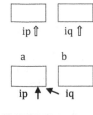

FIGURE 16.1
Pointer assignment.

```
ip = iq        ! same as a=b
iq => ip       ! iq points to a
```

After the first two statements are executed, ip points to a and iq points to b. The statement

```
ip = iq
```

is equivalent to

```
a = b
```

Both ip and iq still point to a and b, respectively. The final statement

```
iq => ip
```

forces pointer iq to point to a. Now both the pointers point to a. Pointer assignments are illustrated with the help of the following program:

```
integer, pointer :: p1,p2,p3
integer, target :: x=10, y=20, z
p1=>x
p2=>y
p3=>z
z=x*y
print *, "x,y,z     : ",x,y,z
p3=p1*p2
print *, "p1,p2,p3 : ",p1,p2,p3 ! z=x*y
p1=>y
p3=p1*p2
print *, "p1,p2,p3 : ",p1,p2,p3 ! z=y*y
p1=p2
print *, "p1,p2     : ",p1,p2     ! x=y
end
```

The outputs of the program are as follows:

```
x,y,z      :   10 20 200
p1,p2,p3 :   10 20 200
p1,p2,p3 :   20 20 400
p1,p2    :   20 20
```

It is not difficult to interpret the output. Now consider the following program segment:

```
character(len=5), pointer :: cp
character(len=18), target :: ch = "Indian Association"
cp => ch            ! not allowed
cp => ch(1:5)       ! allowed
cp => ch(14:18)     ! allowed
```

The first statement will give a compilation error, as the size of the pointer and the target are different. The second statement will make the pointer to point to the string 'India'. Finally, the last statement will make the pointer to point to the string 'ation'. Following the same logic, the next program segment will give a compilation error.

```
integer, parameter :: s1=selected_real_kind(4,307)
integer, parameter :: s2=selected_real_kind(6,37)
integer (kind=s1), target :: a=10.0
integer (kind=s2), pointer :: ia
ia=>a              ! not of same type
```

If a user-defined type variable has the `target` attribute, all its non-pointer components acquire the same attribute by default.

```
type :: t
 integer :: a=10
 real :: r=20.0
end type t
type(t), target :: myt
type(t), pointer :: pt
integer, pointer :: pi
real, pointer :: pr
pt=>myt
pi=>myt%a            ! a acquires target attribute by default
pr=>myt%r            ! r acquires target attribute by default
print *, pi, pr
print*, myt%a
print *, myt%r
end
```

16.6 NULLIFY

This intrinsic is used to dissociate a pointer from a target.

```
nullify(pa)
```

where pa is the pointer pointing to a target. After `nullify` is executed, the pointer does not point to anything. It will be in a null state.

16.7 POINTER and Array

A pointer can point to an array of the right type. The pointer declaration must contain the rank of the target array.

```
integer, pointer, dimension(:) :: pt
integer, target, dimension(10) :: a
.
```

```
a = [1,2,3,4,5,6,7,8,9,10]
pt => a
```

The array elements may be accessed through pointers pt (1), pt (2), ...
More examples of pointers and arrays are given as follows:

```
integer, pointer, dimension(:) :: p1
real, pointer, dimension(:,:) :: p2
complex, pointer, dimension(:,:,:) :: p3
integer, dimension(5), target :: a=10
real, dimension(2,2), target :: b
complex, dimension(2,2,2), target :: c=cmplx(3.0,4.0)
b=reshape([11.0,12.0,13.0,14.0],[2,2])
p1=>a
p2=>b
p3=>c
print *, p1(1),p2(2,1),p3(2,2,2)
end
```

It is obvious that the output of the preceding program is 10, 12.0 and (3.0, 4.0).

16.8 POINTER as Alias

A section of an array may be referred to with the help of a pointer. The pointer acts as an alias. Two examples are given as follows:

```
integer, dimension(10),target :: a=[(i,i=10,100,10)]
integer, pointer, dimension(:) :: p1,p2
p1=>a(2:ubound(a,dim=1):2) ! dim=1 is required to get a scalar
print *, p1
p2=>p1(1:ubound(p1,dim=1):2)
print *, p2
end
```

The first pointer assignment statement makes the following:

```
p1(1) points to a(2)   ! 20
p1(2) points to a(4)   ! 40
p1(3) points to a(6)   ! 60
p1(4) points to a(8)   ! 80
p1(5) points to a(10)  ! 100
```

The next one makes the following:

```
p2(1) points to the target pointed by p1(1) ! 20
p2(2) points to the target pointed by p1(3) ! 60
p3(3) points to the target pointed by p1(5) ! 100
```

The two statements shown in bold in the previous program may be replaced by the following statements:

```
p1=>a(2 :: 2)
p2=>p1(1 :: 2)
```

In this case, the upper bound of the respective items is assumed. Next, consider a 4 × 4 (rank 2) array. Suppose it is necessary to access a section of the array marked with bold digits with a pointer.

TABLE 16.1a

Locations

1,1	1,2	1,3	1,4
2,1	**2,2**	**2,3**	2,4
3,1	**3,2**	**3,3**	3,4
4,1	4,2	4,3	4,4

```
integer, target, dimension(4,4) :: x
integer, pointer, dimension(:,:) :: pa
integer :: i,j
do i=1,4
 do j=1,4
  x(i,j)=10*i+j
  enddo
enddo
print *,x
pa=>x(2:3,2:3)
print *,pa
end
```

TABLE 16.1b

Pointer as Alias

11	12	13	14
21	22	23	24
31	32	33	34
41	42	43	43

x Array

The outputs are as follows:

```
11 21 31 41 12 22 32 42 13 23 33 43 14 24 34 44
22 32 23 33
```

The statement pa=>a(2:3, 2:3) makes the pointers to point to
the array section such that pa(1,1) points to a(2,2), pa(2,1) to a(3,2), pa(1,2) to a(2,3) and pa(2,2) to a(3,3). As a result the print statement displays the content of location a(2,2), a(3,2), a(2,3) and a(3,3) (Table 16.1).

16.9 ALLOCATE and POINTER

A pointer can be made to point to an object without the pointer assignment. This is done through the allocate statement.

```
integer, pointer :: ip
.
allocate (ip)
```

One can use optional arguments stat and errmsg with the allocate statement. Now the system will allocate some memory location that can store only integer and that can be accessed only through pointer ip. There is no name attached to the location.

16.10 POINTER and ALLOCATABLE Array

A pointer may point to an allocatable array.

```
integer, allocatable, target, dimension(:) :: a
integer, pointer, dimension(:) :: ap
    .
allocate(a(100))
ap => a
```

One hundred locations are allocated to 'a', and pointer ap is made to point to a.

A pointer may be made to point to an unnamed array created by the allocate statement.

```
integer, pointer, dimension (:) :: p1
    .
allocate(p1(100))
p1=200
```

The allocate statement creates an unnamed array of size 100 and the pointer points to that array. These locations are accessed only through pointer p1 in the usual manner; that is, p1(1) will point to the first integer and so on.

The behavior of stat and errmsg is slightly different when a pointer is allocated and a nameless array is created. However, the basic philosophy remains the same. When a pointer is allocated

```
allocate(p1(10), stat=istat, errmsg=err)
```

the value of istat is 0 and the character array err is not modified if the allocation is successful. Subsequently, if pointer p1 is allocated again, fresh unnamed locations are allocated without deallocating the locations already allocated by the previous allocate statement. These locations, being nameless, remain inaccessible unless they are assigned to another pointer. This is illustrated with the help of the following program:

```
integer, dimension(:), pointer :: p1, p2
integer :: istat
character(len=50) :: err="no error "
allocate(p1(5),stat=istat, errmsg=err)
print *, 'stat : ',istat
print *, 'err  : ',err
p1=10
p2=>p1
print *, 'p1   : ',p1
print *, 'p2   : ',p2
allocate(p1(6),stat=istat, errmsg=err )
print *, 'Stat : ',istat
print *, 'Err  : ',err
p1=20
print *, 'p1   : ',p1
print *, 'p2   : ',p2
end
```

The output is as follows:

```
Stat :   0
Err  :   No Error
p1   :   10 10 10 10 10
p2   :   10 10 10 10 10
Stat :   0
Err  :   No Error
p1   :   20 20 20 20 20 20
p2   :   10 10 10 10 10
```

Note that when the instruction p2=>p1 is executed, both p2 and p1 point to the same name-less locations (say, n1) created earlier. However, when p1 is reallocated, p1 is now pointing to other nameless locations (say, n2). When locations pointed by p1 are set to different values, p2 is still pointing to the old nameless locations (n1). Therefore, print *, p1 displays the contents of new location (n2), and print *, p2 displays the contents of the old location. Note that no error is committed in the allocation process, and as such, istat is still 0 and err is not modified.

In the array pointer assignment, the lower bound may be set to any value. Consider the following program:

```
real, dimension(-5:5), target :: a=100.0
real, pointer :: p1(:),p2(:), p3(:)
p1=>a
p2=>a(-5:5)
p3(10:)=>a(-5:5)
print *,lbound(p1), ubound(p1)
print *,lbound(p2), ubound(p2)
print *,lbound(p3), ubound(p3)
end
```

The outputs are as follows:

```
-5 5
1 11
10 20
```

This indicates that the statement p1=>a sets the lower and upper bounds of p1 to -5 and 5, respectively. On the other hand, p2=>a(-5:5) sets the lower and upper bounds of p2 to 1 and 11, respectively. The statement

```
p3(10:)=>a(-5:5)
```

sets the lower and upper bounds of p3 to 10 and 20, respectively. This statement may also be written as follows:

```
p3(10:)=>a
```

The target of a multidimensional array pointer could be a single-dimensional array pointer.

```
integer, pointer :: a(:), b(:,:), c(:)
integer :: n=5, i
```

```
allocate(a(n*n))          ! n*n=25
a=[(i, i=1,25)]   ! array elements are 1 to 25 column-wise
b(1:n, 1:n)=> a
c=>a(1:n*n:n+1)
print *, c
end
```

Readers may verify that the c array will contain the diagonal elements of b. Another way of writing the same statement is

```
c=>a( :: n+1)
```

where both the lower and upper bounds of a are replaced by colons (:).

16.11 DEALLOCATE

Like an allocatable array, pointer p1 allocated earlier, may be deallocated by a deallocate statement.

```
deallocate (p1)
```

The deallocate statement may also have stat and errmsg. In the next program p1 is deallocated and cannot be deallocated without allocating again. Thus, the second deallocate statement in the following program sets istat to some nonzero value, and now the character array err contains the error message.

```
integer, dimension(:),pointer :: p1
integer :: istat
character(len=50) :: err="no error" ! NAG compiler
allocate(p1(5),stat=istat,errmsg=err)
print *, 'Stat : ',istat
print *, 'Err  :  ',err
p1=10
print *, 'p1    : ',p1
deallocate(p1) ! no error
deallocate(p1,stat=istat,errmsg=err)! error. err contains error
                                    ! message
print *, 'stat : ',istat
print *, 'err  :   ' ,err
end
```

The output is as follows:

```
Stat :   0
Err  :   No Error
p1   :   10 10 10 10 10
Stat :   4
Err  :   Pointer is not associated
```

16.12 Unreferenced Storage

Again consider the following program segment:

```
integer, pointer, dimension(:) :: ptr
 .
allocate(ptr(100))
```

If pointer `ptr` is made to point to some other location or nullified, the memory locations created by the `allocate` statement in the above program are not available afterward; moreover, these locations are not given back to the system—they remain hanging within the program. The correct strategy would be to deallocate pointer `ptr` first and subsequently modify its association with the target or set to null.

16.13 ASSOCIATED Intrinsic

This intrinsic tests for the association of a pointer with some target or some particular target. It returns `true` if it is associated with a target or with a particular target; otherwise, it returns `false`.

```
integer, pointer :: pa
integer, target :: a,b
pa=>a
```

Now, `print *, associated (pa)` or `print *, associated(pointer=pa, target=a)` displays true. However, `print *, associated (pointer=pa, target=b)` displays `false` as pointer pa is not associated with target b.

16.14 Dangling Pointer

Consider the following program segment:

```
integer, pointer :: ptr1, ptr2
...
allocate(ptr1)
...
ptr2 => ptr1
...
```

The `allocate` statement allocates locations in memory that can be accessed through `ptr1`. The pointer assignment statement makes both `ptr2` and `ptr1` as alias of the same location. If `ptr1` is deallocated and the location is returned to the system, pointer `ptr2` is still pointing to some non-existent location. So, if `ptr2` is used, the result may be unpredictable. This is called a dangling pointer. One should set `pt2` to `null()` before deallocating pointer `pt1`.

16.15 POINTER within Subprogram

Pointers declared within a subprogram become undefined on exit from the subprogram. This can be prevented if the pointer is declared with the `save` attribute.

```
real, pointer, save :: ptr
```

The `save` attribute ensures that when the subprogram is entered for the second time, the pointer does not become undefined.

```
subroutine sub(i)
integer :: i
integer, pointer, save :: ptr
.
if(i.eq.1) then
  allocate(ptr)
  ptr=10
endif
print *, ptr
end
```

The pointer is allocated when the subroutine is entered for the first time (i=1). Subsequently, the subroutine is entered with the value of i not equal to 1. This ensures that the condition for the `if` statement is `false`. However, because of the `save` attribute, the pointer allocated earlier reappears with its last value. The attribute `save` is not required if pointer `ptr` is a part of a module and subroutine `sub` uses the module.

16.16 POINTER and Derived Type

A pointer may point to a derived type variable too.

```
type emp
 integer :: id
 character(len=30) :: name
end type emp
type(emp), pointer :: dtype
type(emp), target :: employee
.
dtype => employee
dtype%id=100
dtype%name="Sankar Chakravorti"
```

Memory space may be created for a derived object through a pointer.

```
allocate(dtype)
dtype%id=200
```

The `allocate` statement with pointer `dtype` of type `emp` will create space for one such object.

16.17 Self-Referencing Pointer

A derived type may have an elementary item that can point to a variable of similar type (derived type).

```
type emp
  integer :: id
  character(len=30) :: name
  type(emp), pointer :: next
end type emp
```

The elementary item next is a pointer that can point to a record of type emp.

```
type (emp), target :: e1, e2
e1%id=10
e1%name="some name"
e2%id=20
    .
e1%next=>e2
```

The elementary item id of e2 can be accessed as e2%id, or with reference to e1, it is e1%next%id.

  With this kind of facility, one can easily build up a linked list.

16.18 FUNCTION and POINTER

The following program sorts an array using a function and returns the array through the result clause.

```
program sortx
interface
  function sort(a) result(res)
  integer, dimension(:), target :: a
  integer, pointer, dimension(:) :: res
  end function sort
end interface
integer,target,dimension(10) :: a=[10,-5,2,23,-10,6,100,0,-1,1]
print *, sort(a)
end program sortx
function sort(a) result(res)
integer,target, dimension(:) :: a
integer, pointer, dimension(:) :: res
integer :: i,j,length, temp
allocate (res(size(a)))
res=>a
length=size(a)              ! not an efficient sort
do i=1, length-1
```

```
  do j=i+1,length
   if(res(i) .gt. res(j))then
    temp=res(i) !  interchange
    res(i)=res(j)
    res(j)=temp
   endif
  enddo
 enddo
end function sort
```

16.19 POINTER and Subprogram

The argument of a subprogram may be a pointer. Two cases may arise.

Case I: The actual argument is a pointer, but the dummy argument is not a pointer. In this case, the actual argument must point to a target.

```
program main
integer, target :: a=10
integer, pointer :: pa
pa=>a
call sub(pa)
print *, pa        ! a is now 100
end
subroutine sub(b)
integer :: b
b=100
end
```

Case II: The dummy argument is a pointer. In this case, the actual argument must be a pointer and an explicit interface must be used. The dummy parameters must not have any intent.

```
program main
interface
  subroutine sub(a)
   integer, pointer :: a
   end subroutine sub
end interface
integer, target :: b=100
integer, pointer :: pb
pb=>b
call sub(pb)
print *, pb        ! pb is now 200
end
subroutine sub(b)
integer, pointer :: b
b=200
end subroutine sub
```

16.20 POINTER INTENT

The dummy pointer argument of a subprogram may have an `intent` attribute attached to it. The intent used in this context refers to its association and not to the target variable. If the `intent` is `in`, the pointer cannot be modified; that is, it cannot be reassociated with another target. It cannot be set to `null` either. If the `intent` is `out`, the association is not defined when the subprogram is entered. It has an association on exit from the subprogram.

```
integer, target :: x=10
integer, pointer :: ptr
 interface
    subroutine testptr(p)
       integer, pointer, intent(in) :: p
    end subroutine testptr
 end interface
ptr=>x
print *, ptr
call testptr(ptr)
print *, x
end
subroutine testptr(p)
integer, pointer, intent(in) :: p
p=27
end subroutine testptr
```

We reiterate that `intent(in)` refers to the association of the pointer with the target. This pointer p cannot point to any other target inside the subroutine `testptr`. Nor it can be set to null by `p=>null()`. However, `intent(in)` has no connection with the target that p is pointing to. The target can be modified; x was 10 when the subroutine is called, but it is 27 on exit from the subroutine. If the `intent` is set to `inout`, pointer p may be made to associate itself with another target within the subprogram, if necessary. It can also be set to `null()`.

```
subroutine testptr(p)
integer, pointer, intent(inout) :: p
p=27
p=>null()
end subroutine testptr
```

Needless to say, the corresponding `interface` block in the `main` program must be modified—`intent(in)` is to be changed to `intent(inout)`.

16.21 PROCEDURE and POINTER

A pointer can point to a procedure. The interface may be either explicit or implicit.

```
procedure(sub), pointer :: p=>null()
```

The pointer is initialized to null. Pointer p can point to any procedure having similar calling parameters (same number, type and intent) as sub, and subsequently, the pointer may be used in place of the procedure name. The pointer can dynamically change its association with the procedure having the same calling parameters. Individual interface may be used; however, in this case abstract interface is preferred (Chapter 19). The procedure pointer p is made to point to procedure proc through the statement p=>sub1, and the subsequent call uses this p in place of the procedure name. Here the interface is explicit.

```
program proctest2
interface
   subroutine sub(a,b,c)
   integer,intent(in) :: a,b
   integer,intent(out) :: c
   end subroutine sub
   subroutine sub1(a,b,c)
   integer,intent(in) :: a,b
   integer,intent(out) :: c
   end subroutine sub1
   subroutine sub2(a,b,c)
   integer,intent(in) :: a,b
   integer,intent(out) :: c
   end subroutine sub2
end interface
integer :: a=20,b=10,c
procedure(sub),pointer :: p=>null()
p=>sub1
call p(a,b,c)
print *, a,b,c
p=>sub2
call p(a,b,c)
print *,a,b,c
end
subroutine sub1(a,b,c)
integer,intent(in) :: a,b
integer,intent(out) :: c
c=a+b
end subroutine sub1
subroutine sub2(a,b,c)
integer,intent(in) :: a,b
integer,intent(out) :: c
c=a-b
end subroutine sub2
```

The outputs are as follows:

```
20 10 30
20 10 10
```

The next example shows an implicit interface. Pointers with implicit interface are defined as follows:

```
procedure(), pointer :: p=>null()
```

Note that for implicit interface, the interface name is absent; however, the left and right parentheses are present as follows:

```fortran
program proctest3
integer :: a=20,b=10,c
procedure(),pointer :: p=>null()        ! implicit interface
p=>sub1
call p(a,b,c)
print *, a,b,c
p=>sub2
call p(a,b,c)
print *,a,b,c
contains
 subroutine sub1(a,b,c)
  integer,intent(in) :: a,b
  integer,intent(out) :: c
   c=a+b
  end subroutine sub1
  subroutine sub2(a,b,c)
  integer,intent(in) :: a,b
  integer,intent(out) :: c
  c=a-b
  end subroutine sub2
  end
```

The component of a derived type may be a procedure pointer.

```fortran
abstract interface
 subroutine sub(a,b,c)
 integer,intent(in) :: a,b
 integer,intent(out) :: c
 end subroutine sub
end interface
type proctest
 integer :: i,j,k
 procedure(sub), nopass, pointer :: p1 ! nopass is discussed in ch 19
end type proctest
type(proctest) :: x
procedure(sub) :: sub1,sub2
x%i=20
x%j=10
x%p1=>sub1
call x%p1(x%i, x%j, x%k)
print *, x%i, x%j, x%k
x%p1=>sub2
call x%p1(x%i, x%j, x%k)
print *, x%i, x%j, x%k
end
subroutine sub1(a,b,c)
integer,intent(in) :: a,b
integer,intent(out) :: c
c=a+b
end subroutine sub1
subroutine sub2(a,b,c)
```

```
integer,intent(in) :: a,b
integer,intent(out) :: c
c=a-b
end subroutine sub2
```

Another way of writing the preceding program is shown next. Here a local procedure pointer (p1) is defined, and the call to the procedure is made through p1.

```
abstract interface
  subroutine sub(a,b,c)
  integer,intent(in) :: a,b
  integer,intent(out) :: c
  end subroutine sub
 end interface
 type proctest
  integer :: i,j,k
  procedure(sub), nopass, pointer :: p1 ! nopass discussed in chapter 19
 end type proctest
 type(proctest) :: x
 procedure(sub) :: sub1,sub2
 procedure(sub), pointer :: p1
 x%i=20
 x%j=10
 p1=>sub1
 call p1(x%i, x%j, x%k)
 print *, x%i, x%j, x%k
 p1=>sub2
 call p1(x%i, x%j, x%k)
 print *, x%i, x%j, x%k
 end
 subroutine sub1(a,b,c)
 integer,intent(in) :: a,b
 integer,intent(out) :: c
 c=a+b
 end subroutine sub1
 subroutine sub2(a,b,c)
 integer,intent(in) :: a,b
 integer,intent(out) :: c
 c=a-b
 end subroutine sub2
```

16.22 ALLOCATE with SOURCE

The allocate statement with source attribute may use a pointer in place of an array.

```
integer, pointer, dimension(:) :: a
integer, dimension(5) :: b=[1,2,3,4,5]
allocate(a(5), source=b)
print *,a
end
```

The print statement will display 1 2 3 4 5.

16.23 ALLOCATE with MOLD

If `mold` is used in place of `source`, the dimension of the allocated array is controlled by `mold`. It is not necessary to define the array associated with `mold`.

```
integer, allocatable, dimension(:) :: a
integer, dimension(20) :: b
allocate(a, mold=b)
```

This will allocate 20 locations for 'a'. Note that even if b is defined, no value is transferred from b. At this point, no value is assigned to a.

16.24 CONTIGUOUS

This attribute is used with array pointers and assumed-shape dummy arrays. This indicates that the pointer points to a continuous target. A continuous array is defined as an array, where the elements are contiguous; that is, they are not separated by other elements. Two examples of `contiguous` attributes are given as follows:

```
      program contig_1
      integer, contiguous, dimension(:),pointer :: p
      integer, dimension(10), target :: a=100
  !   p=>a(1:10:2)            ! ERROR, can point to a contiguous target
      p=>a(1:10)              ! correct, no error
      .
      end program contig_1
```

In the preceding example, if the comment is removed, the compiler will flag an error. This is due to the fact that p cannot point to a non-contiguous array a(1:10:2). However, the next instruction p=>a(1:10) is correct as now p is pointing to a contiguous target. The next example uses an assumed-shape dummy array. Here, an `interface` block is required.

```
program contig_2
integer, dimension(10) :: a=[1, 2, 3, 4, 5, 6, 7, 8, 9, 10]
interface
 subroutine sub(a)
 integer, intent(inout), contiguous, target, dimension(:) :: a
 end subroutine sub
end interface
call sub(a)
print *, a
end program contig_2
subroutine sub(x)
integer, intent(inout), target, contiguous :: x(:)
integer, contiguous, pointer, dimension(:) :: p
p=>x
print *, p
p=[10, 20, 30, 40, 50, 60, 70, 80, 90, 100]
end subroutine sub
```

16.25 IS_CONTIGUOUS

This function tests the contiguity of an array. The array may be of any data type. If it is a pointer, it must be associated with a target.

```
integer, dimension(:),pointer :: p
integer, dimension(30), target :: a=100
p=>a
if (is_contiguous(p)) then
  print *, "Contiguous"
else
  print *, "Not contiguous"
endif
end
```

In the preceding case, 'true' path is chosen. If p=>a is replaced by p=>a(2:30:2), the 'false' path will be chosen, and "Not contiguous" will be displayed. Also, is_contiguous(a) or is_contiguous(a(1:30:1)) or is_contiguous(a(1:30)) returns the 'true' value. However, is_contiguous(a(1:30:2)) returns the 'false' value.

16.26 Programming Example

We now give an example of doubly linked list. In a doubly linked list, each record has two referencing pointers. In our case, they are prior and next. The prior pointer points to the previous record, and the next pointer points to the next record in the list. The prior pointer corresponding to the first record is nullified as there is no record before the first record. The next pointer corresponding to the last record is also nullified as there is no record after the last record. Two more pointers start and last always point to the first and the last record, respectively. The program will create a telephone directory in the memory.

Records can be added to the list, and they can also be deleted from the list. When a record is added, a space is created in the memory for it; similarly, when a record is deleted, the space is returned to the system. The adjustments of various pointers are taken care of by the program logic.

The module link_list contains the description of the record and the definition of the pointers start and last. The module is used by various subprograms.

```
module link_list
  type addr
    character(len=30) :: name
    character(len=40) :: address
    integer :: pin
    type(addr), pointer :: next
    type(addr), pointer :: prior
  end type addr
type(addr), pointer :: start
```

```
    type(addr), pointer :: last
    end module
! the main program is l_list. depending upon the input
! it performs various operation, like - insertion, deletion,
! listing and searching
    program l_list
    use link_list
    logical :: bool=.true.
    integer :: choice
    nullify(start)
    do while(bool) !  make a selection
    call menu(choice)
    select case (choice)
     case(1) !  insert record
     call enter
     case(2) !  delete record
     call delrec
     case(3) !  list records
     call list
     case(4) !  search a name
     call search
     case(5) ! exit from the program
     bool=.false.
     case default
     print *, "the input should be between 1 & 5"
    end select
    enddo
    end program l_list
    subroutine menu(choice) !this subprogram menu displays a menu
    integer, intent(out) :: choice ! and reads an integer from the keyboard
    print *
    print *,   "********* menu *********"
    print *,   "1. enter a name"
    print *,   "2. delete a name"
    print *,   "3. list the names"
    print *,   "4. search a name"
    print *,   "5. quit"
    print *,   "********* menu *********"
    read *, choice
    end subroutine menu
    subroutine enter ! this subroutine enter adds a record to the list and
    use link_list ! '@' is typed as name the routine is exited.
    type(addr), pointer :: info
    do !  create a space for a record
    allocate(info)
     print *
    print *, "enter name"
    read *, info%name
    if(info%name(1:1) .eq. "@") then
     exit
    endif
    print *, "enter address"
    read *, info%address
    print *, "enter pin"
```

```
    read *, info%pin
    if(associated(start)) then !  list not empty
     last%next => info
     info%prior => last
     last => info
     nullify(last%next)
    else !  list empty
     start => info
     nullify(start%next)
     last => start
     nullify(start%prior)
    endif
    enddo
    end subroutine enter
    subroutine list! to display the list on the screen
    use link_list
    interface
     subroutine show(info)
     use link_list
     type(addr), pointer :: info
     end subroutine show
    end interface
    type(addr), pointer :: info
    info => start !  start from the beginning
    do while (associated(info))
     call show(info)
     info => info%next ! fetch the next item from the list
    enddo
    end subroutine list
    subroutine show(info) ! displays the list on the screen
    use link_list !after getting the record from the subroutine list
    type(addr), pointer :: info
    print *
    print *, info%name
    print *, info%address
    print *, info%pin
    print *, "------------------------"
    print *
    end subroutine show
! this subroutine is used to delete a record; the routine
! can delete any record from the list; note that the first
! and other records are handled in different ways;
! after the deletion the record pointers are modified accordingly
    subroutine delrec
    use link_list
    type(addr), pointer :: info
    character(len=30) :: nam
    logical :: index
    interface
     subroutine find(info,nam,index)
     use link_list
     type(addr), pointer :: info
      character(len=30), intent(in) :: nam
     logical :: index
```

```
  end subroutine find
 end interface
 index=.false.
 print *
 print *, "type the name to be deleted"
 print *
 read *, nam
 call find(info,nam,index)
 if(index) then
  if(associated(info,start)) then ! first record
     if (.not.(associated(info%next))) then ! first record and only
                                              ! record
        nullify(start)
    else ! first record but not only record
    start=>info%next
      nullify(start%prior)
   endif
  else ! other than first record
   info%prior%next => info%next
   if(associated(info,last)) then ! last record
    last =>info%prior
   else
    info%next%prior=info%prior
   endif
  endif
  deallocate(info) !  release the location
  print *
    print *, trim(nam), " deleted from the list"
  print *
 else
  print *, "name not present in the list"
 endif
 end
 subroutine search ! to display a particular name
 use link_list
 type(addr), pointer :: info
 character(len=30) :: nam
 logical :: index
 interface
  subroutine show(info)
  use link_list
  type(addr), pointer :: info
  end subroutine show
  subroutine find(info,nam,index)
  use link_list
  type(addr), pointer :: info
  character(len=30), intent(in) :: nam
  logical :: index
  end subroutine find
 end interface
 index=.false.
 print *, "type the name to be searched"
 read *, nam
 call find(info,nam,index)
```

```
if(index) then !  record found
 call show(info)
else
 print *, "name not found"
endif
end subroutine search
subroutine find(info,nam,index)
use link_list
type(addr), pointer :: info
character(len=30), intent(in) :: nam
logical :: index
info => start
do while(associated(info))
 if(nam .eq. info%name) then
  index = .true.
  exit
 else
  info => info%next
 endif
 enddo
end subroutine find
```

The logic of this program is slightly difficult. This section may be skipped during the first reading.

17

Bit Handling

Fortran contains a number of intrinsics to manipulate a single bit or a group of bits. These include bit testing, bit setting, shifting of bits, logical operations such as 'and' and 'or' and transferring bits from one location to another. In our subsequent discussion, whenever the type of any array is not mentioned, it is assumed as an integer array. To keep the diagrams simple, only the right-most 8 bits are shown—left-most 24 bits (which are zero because of choice of value for the variable or constant) are not shown in the diagram.

17.1 BIT_SIZE

This library function takes an integer as argument and returns the size of the variable; that is, the number of bits used to store the integer. The returned integer is the same type as the argument.

```
integer, parameter :: s1=selected_int_kind(2)
integer, parameter :: s2=selected_int_kind(4)
integer, parameter :: s3=selected_int_kind(9)
integer, parameter :: s4=selected_int_kind(12)
integer (kind=s1) :: k1
integer (kind=s2) :: k2
integer (kind=s3) :: k3
integer (kind=s4) :: k4
integer :: l1, l2, l3, l4
l1=bit_size(k1)          ! l1=8
l2=bit_size(k2)          ! l2=16
l3=bit_size(k3)          ! l3=32
l4=bit_size(k4)          ! l4=64
print *, l1, l2, l3, l4
end
```

This program shows the number of bits required to represent integers of various kinds in the system. These are processor dependent.

17.2 BTEST

Case I: This intrinsic takes two integers as arguments: the first one is an integer variable and the second one is an integer or integer expression. The function tests the bit of the first argument whose position is specified by the second argument pos (if the second argument is an expression, it is first evaluated). It returns `true` if the bit in question is on and returns `false` if the bit is off (Figure 17.1).

```
integer :: a=10
logical :: l1, l2
l1=btest(a,3)    ! true
l2=btest(a,4)    ! false
print *, l1, l2
end
```

(only rightmost eight bits are shown)

0	0	0	0	1	0	1	0

FIGURE 17.1
Bit pattern of 10.

The value of a is 10. It is internally stored as $(1010)_2$ with 28 leading zeros. Bits 1 and 3 are on, and the rest are zero. Therefore, `btest(a, 3)` is `true`, and `btest(a, 4)` is `false`. The second argument `pos` satisfies this inequality: $0 <= pos < bit\_size(a)$. If the condition is not satisfied, the result is `false`. Note that `bit_size` returns the number of bits in a, which is 32 in this case. Bits are numbered from 0 to 31, and hence, `pos` has to be less than `bit_size (a)`.

Case II: The first argument to the intrinsic is an array. The second argument is a scalar. The intrinsic tests the bit of the elements of the first argument whose position is specified by the second argument (`pos`). The intrinsic returns a logical array of the same shape and size as the first argument (Figure 17.2).

```
integer, dimension (3) :: a =[11, 26, 31]
logical, dimension (3) :: x
x=btest(a,0)
print *, x
end
```

a(1)

0	0	0	0	1	0	1	1

a(2)

0	0	0	1	1	0	1	0

a(3)

0	0	0	1	1	1	1	1

FIGURE 17.2
a = [11, 26, 31].

`x(1)` will be `true` as bit 0 of `a(1)` is on. Similarly, `x(2)` is `false`, and `x(3)` is `true`. For a two-dimensional integer array `b(3x3)` having elements shown in Figure 17.3, `btest(b,1)` returns a two-dimensional logical array having its elements `f, t, t, f, f, t, t , f` and `f` (column wise).

1	4	7
2	5	8
3	6	9

f	f	t
t	f	f
t	t	f

FIGURE 17.3
Array b and `btest(b,1)`.

Case III: Both the first and second arguments may be the arrays of the same shape and size. The intrinsic tests the bit of all the elements of the first array whose position is stored in the second array (Figure 17.4).

```
integer, dimension (3) :: a=[21, 36, 53]
integer, dimension (3) :: pos=[0, 5, 7]
logical, dimension (3) :: x
x=btest(a, pos); print *, x
end
```

a(1)

0	0	0	1	0	1	0	1

a(2)

0	0	1	0	0	1	0	0

a(3)

0	0	1	1	0	1	0	1

FIGURE 17.4
a = [21, 36, 53].

`x(1)` will be `true` as the bit 0 of `a(1)` is on. Similarly, `x(2)` is `true`, and `x(3)` is `false`.

17.3 IBSET

Case I: This intrinsic also takes two arguments: the first one is an integer variable and the second one (pos) is an integer or integer expression. The function returns an integer of the same kind as the first argument with the bit whose position is specified by the second argument set to 1 (Figure 17.5).

	Before k1=13						
0	0	0	0	1	1	0	1

	After k2=15						
0	0	0	0	1	1	1	1

FIGURE 17.5
Example of ibset.

```
integer :: k1=13
integer :: k2
k2=ibset(k1,1); print *, k2
end
```

When bit 1 of k1 is set, the number becomes 15. The second argument pos satisfies the following inequality:

```
0 <= pos <bit_size(k1)
```

Case II: The first argument to the intrinsic is an integer array. The second argument is a scalar (non-negative integer). The intrinsic sets the bit of the elements of the first argument whose position is specified by the second argument. The intrinsic returns an array of the same type as the first argument (Figure 17.6).

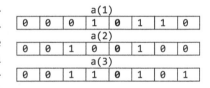

	a(1)						
0	0	0	1	0	1	1	0

	a(2)						
0	0	1	0	0	1	0	0

	a(3)						
0	0	1	1	0	1	0	1

FIGURE 17.6
a = [22, 36, 53].

```
integer, dimension (3) :: a=[22, 36, 53]
integer, dimension (3) :: b
b=ibset(a,3); print *, b; end
```

Thus, b(1), b(2) and b(3) become 30, 44 and 61, respectively, as bit 3 of each element is switched on, and it adds $2^3=8$ to all the elements of a.

Case III: Both the first and second arguments may be the arrays of the same shape and size. In this case the intrinsic sets a bit of each element of the array a by the corresponding elements of the second array whose position is stored in the second array (Figure 17.7).

	a(1)						
0	0	0	1	0	1	0	0

	a(2)						
0	0	1	0	0	1	0	0

	a(3)						
0	0	1	1	0	1	0	1

```
integer, dimension (3) :: a=[20,36,53]
integer, dimension (3) :: pos=[1,0,3]
integer, dimension (3) :: b
b=ibset(a, pos); print *, b
end
```

FIGURE 17.7
a = [20, 36, 53].

Thus, b(1), b(2) and b(3) become 22, 37 and 61, respectively. The bits that will be switched on are indicated by bold letters.

```
b(1)=a(1)+2 [bit 1 is switched on]
b(2)=a(2)+1 [bit 0 is switched on]
b(3)=a(3)+8 [bit 3 is switched on]
```

17.4 IBCLR

Case I: This intrinsic also takes two arguments: the first one is an integer variable and the second one (pos) is an integer or integer expression. The function returns an integer of same kind and type as first argument with bit position as specified by the second argument set to 0, keeping all other bits unchanged (Figure 17.8).

```
integer :: a=14
integer :: b
b=ibclr(a,1)
print *, b
end
```

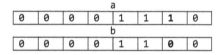

FIGURE 17.8
Arrays a and b.

As bit 1 of a is set to 0, the number becomes 12, and thus, b becomes 12. The second argument pos satisfies the inequality: 0 <= pos < bit_size(a).

Case II: The first argument is an integer array, and the second argument is a scalar (non-negative integer). The intrinsic clears the bit of all the elements of the first argument whose position is specified by the second argument (pos). The intrinsic returns an array of the same type as the first argument (Figure 17.9).

```
integer, dimension (3) :: a=[22,36,53]
integer, dimension (3) :: b
b=ibclr(a,2)
print *, b
end
```

Thus, b(1), b(2) and b(3) will become 18, 32 and 49, respectively.

a array

0	0	0	1	0	1	1	0
0	0	1	0	0	1	0	0
0	0	1	1	0	1	0	1

b array

0	0	0	1	0	0	1	0
0	0	1	0	0	0	0	0
0	0	1	1	0	0	0	1

FIGURE 17.9
a = [22, 36, 53], b = [18, 32, 49].

Case III: Both the first and second arguments may be integer arrays of the same shape and size. It returns an array of the same type as the input array. The intrinsic clears a bit of each element of the input array by the value stored in the corresponding elements of the second array.

```
integer, dimension (3) :: a=[22,36,53]
integer, dimension (3) :: pos=[1,2,0]
integer, dimension (3) :: b
b=ibclr(a, pos)
print *, b
end
```

The elements of the b array become 20, 32 and 52, respectively (Figure 17.10).

a array									b array							
0	0	0	1	0	1	1	0		0	0	0	1	0	1	0	0
0	0	1	0	0	1	0	0		0	0	1	0	0	0	0	0
0	0	1	1	0	1	0	1		0	0	1	1	0	1	0	0

FIGURE 17.10
a = [22, 36, 53] b=20, 32, 52].

17.5 IBITS

Case I: This intrinsic takes three integers as arguments: the first one is an integer variable and the second (pos) and third (len) ones are integer variables or constants or expressions. This function returns an integer after extracting len number of bits starting from the position pos of the integer variable. The extracted bits are stored right adjusted in the destination field (Figure 17.11).

```
integer :: i, j
data i / b"10110010"/
j=ibits(i,4,3)
print *, j
end
```

1	0	1	1	0	0	1	0

0	0	0	0	0	0	1	1

FIGURE 17.11
i = b"10110010", j = b"00000011".

The second argument pos satisfies the following inequality: len+pos <= bit_size(i).

Case II: The first argument can be an integer array. In this case, the intrinsic returns an array of the same type and shape as the first argument. The intrinsic performs the same operations mentioned under case I on every element of the array (Figure 17.12).

```
integer, dimension (3) :: a
integer, dimension (3) :: b
data a / b"10110010", b"10011010", b"01111001"/
b=ibits(a,4,3)
print *, b
end
```

a array	b array
1 0 1 1 0 0 1 0	0 0 0 0 0 0 1 1
1 0 0 1 1 0 1 0	0 0 0 0 0 0 0 1
0 1 1 1 1 0 0 1	0 0 0 0 0 1 1 1

FIGURE 17.12
Arrays a and b.

The elements of the b array become 3, 1 and 7, respectively.

Case III: When the first argument is an integer array, the second argument can also be an array of the same shape and size. In this case, the bits are extracted from each element of the array taking the starting position from the corresponding element of the pos array.

```
integer, dimension (3) :: a, b
integer, dimension (3) :: pos
data a / b"10110010", b"10011010", b"01111001"/
data pos/ 4,1,3/
b=ibits(a,pos,3)
print *, b
end
```

a array	b array
1 0 1 1 0 0 1 0	0 0 0 0 0 0 1 1
1 0 0 1 1 0 1 0	0 0 0 0 0 1 0 1
0 1 1 1 1 0 0 1	0 0 0 0 0 1 1 1

FIGURE 17.13
a and b arrays—case III.

It may be observed from Figure 17.13 that the elements of the b array become 3, 5 and 7, respectively.

Case IV: Like the second argument pos, the third argument len may also be an integer array of shape and size similar to the first argument. In this case, the number of bits extracted depends on the content of the corresponding elements of the len array (Figure 17.14).

```
integer, dimension (3) :: a, b
integer, dimension (3) :: pos, len
data a / b"10110010", &
 b"10011010", b"01111001"/
```

a array
1 0 1 1 0 0 1 0
1 0 0 1 1 0 1 0
0 1 1 1 1 0 0 1
b array
0 0 0 0 0 0 1 1
0 0 0 0 0 0 0 0
0 0 0 0 0 1 1 1

FIGURE 17.14
a and b arrays—case IV.

```
data pos/4,2,3/,len /2,1,3/
b=ibits(a,pos,len)
print *, b
end
```

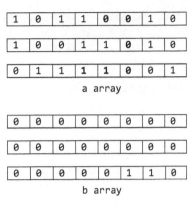

The elements of the b array become 3, 0 and 7, respectively.

Case V: It is perhaps apparent that pos may be a scalar and len may be an array of the same shape and size as the first argument (Figure 17.15).

```
integer, dimension (3) :: a, b
integer, dimension (3) :: len
integer :: pos=2
data a / b"10110010", b"10011010", &
      b"01111001"/, len /2,1,3/
b=ibits(a,pos,len)
print *, b
end
```

FIGURE 17.15

a and b arrays—case V.

The elements of the b array are 0, 0 and 6, respectively.

17.6 LEADZ and TRAILZ

Both these intrinsics take an integer as its argument. The intrinsic `leadz` returns the number of leading zeros, and the intrinsic `trailz` returns the number of trailing zeros. The argument may be a scalar integer or an integer array. In case of the integer array, the intrinsics act on every element of the array and return an integer array of the same type and size as the input array.

```
integer:: ia=0, ib=16
print *, leadz(ia), trailz(ia), leadz(ib), trailz(ib)
end
```

Assuming the default integer takes 32 bits to store an integer, the outputs of the preceding program are 32 32 27 4. When the integer is 0, all the bits are switched off. This explains the first two numbers of the output. However, when the value of the integer is 16, the bit pattern is as shown in Figure 17.16.

FIGURE 17.16

Bit pattern of ib.

There are 27 zeros before 1 and 4 zeros after 1. The readers may verify the output of the following program:

```
integer, dimension(3):: ic=[2, 8, 18]
print *, leadz(ic)
print *, trailz(ic)
end
```

The output is as follows:

```
30  28  27
 1   3   1
```

If the integer argument is not of default kind, the number of leading zeros is calculated based on the number of bits required to represent the integer of that particular kind.

```
integer, parameter::s1=selected_int_kind(2)
integer(kind=s1):: k1=16
print *, leadz(k1), trailz(k1)
end
```

0	0	0	1	0	0	0	0

FIGURE 17.17
Bit pattern of k1.

The output is as follows:

```
3 4
```

In the preceding program, k1 requires 8 bits to store its value. When k1 is 16, there are 3 zero bits at the front and 4 zero bits at the rear (Figure 17.17).

17.7 POPCNT

This function takes one integer (or integer array) as its argument. It returns the number of 1 bit: popcnt(2) returns 1 and popcnt([1,2,3,4,5]) returns [1,1,2,1,2] whose corresponding binary representations are, respectively, $(1)_2$, $(10)_2$, $(11)_2$, $(100)_2$ and $(101)_2$.

17.8 POPPAR

This function takes one integer (or integer array) as its argument. It returns either 1 or 0 (default integer) depending on the number of 1 bit. If the number of 1 bit is odd, it returns 1; otherwise, it returns 0. Here, poppar(5) returns 0 (the number of 1 bit is 2); poppar([1,2,3,4,5]) returns [1,1,0,1,0].

17.9 MASKL

This function is a left-justified mask. This takes two integers as arguments. The second argument (kind) is optional. The first argument, i, must be non-negative and should be less than or equal to bit_size(i). The function returns default integer if kind is absent; otherwise, the returned value is of the kind specified by the second argument. The returned value is an integer having the left-most i bits set to 1 and the rest set to 0. `maskl(3, kind=selected_int_kind(2))` returns the bit pattern 11100000; `maskl(3)` is the same as `shiftl(7_1,bit_size(0_1)-3_1)`. If the parameter `kind` is absent, the returned value is a 32-bit integer (NAG compiler), which is the default integer for this processor.

17.10 MASKR

This function is a right-justified mask. This takes two integers as arguments. The second argument (`kind`) is optional. The first argument, i, must be non-negative and should be less than or equal to `bit_size(i)`. The function returns default integer if `kind` is absent; otherwise, the returned value is of the kind specified by the second argument. The returned value is an integer having the right-most i bits set to 1 and the rest set to 0. `maskr(3, kind=selected_int_kind(2))` returns the bit pattern 00000111, which is 7.

17.11 IAND

Case I: This function takes two integers as arguments: the arguments must be of the same kind. It returns an integer of the same kind as its arguments after logical 'anding' of the bits of the first argument with the corresponding bits of the second argument. The truth table for and is shown in Figure 17.18.

```
integer :: i, j, k
i=13
j=25
k=iand(i,j)
print *, k    ! k=9
end
```

i		0	0	1	1
j		0	1	0	1
iand(i, j)		0	0	0	1

Truth table – iand

0	0	0	0	1	1	0	1

i

0	0	0	1	1	0	0	1

j

0	0	0	0	1	0	0	1

k

FIGURE 17.18
Truth table of iand and i, j and k.

Iand intrinsic may be used to switch off a particular bit (or bits) of an integer without disturbing other bits by using a suitable mask (second argument). The bits of the mask corresponding to the bits of the source that are not to be disturbed are set to 1, and the bits of the mask corresponding to the bits of the source that are to be switched off are set to 0.

```
integer, parameter :: s=selected_int_kind(2)
integer (kind=s) :: i, mask, k
data i/b"10110110"/, mask /b"11111001"/
k=iand (i, mask) ! bits 1 and 2 are switched off
```

Case II: The first argument is an integer array, and the second argument is a scalar (integer). All the elements of the array are 'anded' with the second argument, and the function returns an array of the same shape and size as the first argument.

```
integer, dimension(4)::a
a=[1,2,3,4]
print *, iand(a,2)
end
```

The output is as follows:

```
0 2 2 0
```

`iand(a,2)` is equivalent to `iand(a(1),2)`, `iand(a(2),2)`, `iand(a(3),2)` and `iand(a(4),2)`.

Case III: Both the arguments of `iand` may be an array of the same shape and size. Each element of the first argument is 'anded' with the corresponding element of the second array. The function returns an array of the same type as the first argument.

```
integer, dimension(4)::a,b,c
a=[1,2,3,4]
b=[7,8,9,10]
c= iand(b,a)
print *, c
end
```

The output is as follows:

```
1 0 1 0
```

`a(1)` is 'anded' with `b(1)` and the result is stored in `c(1)`, `a(2)` is 'anded' with `b(2)` and so on. Bitwise 'and' may conveniently be used to convert lowercase letters to the corresponding uppercase letters using the fact that by switching the bit 5 off the lowercase letter, it becomes the corresponding uppercase letter. For example, the internal representation of 'a' is '01100001'; and the internal representation of 'A' is '01000001'. In the following program, the mask is so chosen that the bitwise 'and' operation switches off the 5th bit, and other bits remain unaffected.

```
integer, parameter:: sz=selected_int_kind(2)
integer(kind=sz):: ic
integer::i                              ! gfortran gives overflow
integer(kind=sz)::mask=b"11011111"  ! NAG compiler used
character(len=26):: cc='abcdefghijklmnopqrstuvwxyz'
print *,cc
do i=1,26
ic=iachar(cc(i:i),kind=sz)              ! convert char to 8 bit integer
  ic=iand(ic,mask)                      ! switch off 5th bit
  cc(i:i)=achar(ic)                     ! convert to char
end do
print *,cc
end
```

The outputs are as follows:

```
abcdefghijklmnopqrstuvwxyz
ABCDEFGHIJKLMNOPQRSTUVWXYZ
```

17.12 IOR

Case I: This intrinsic is the same as iand in every respect, but the truth table is different (Figure 17.19). First, we consider a case where both the first and second arguments are scalar.

```
integer :: i, j, k
i=12
j=3
k=ior(i,j)   ! k=15
print *, k
end
```

Inclusive or (ior) may be used to switch on a particular bit (bits) of an integer, keeping the other bits untouched under the influence of a mask. The bits of the mask corresponding to the bits of the source that would remain unaffected are set to 0, and the bits of the mask corresponding to the bits of the source that are to be switched on are set to 1.

```
integer :: i1, mask, k1
data i1/b"10101001"/, mask /b"00010010"/
k1=ior(i1, mask) !bits 1,4 switched on
```

i	0	0	1	1
j	0	1	0	1
ior(i, j)	0	1	1	1

Truth table – ior

0	0	0	0	1	1	0	0

i

0	0	0	0	0	0	1	1

j

0	0	0	0	1	1	1	1

k

1	0	1	0	1	0	0	1

i1

0	0	0	1	0	0	1	0

mask

1	0	1	1	1	0	1	1

k1

FIGURE 17.19
Bit patterns of i, j, k, i1, j1 and k1.

Case II: Like iand, the first argument is an array and the second argument is a scalar. All the elements of the array are 'ored' with the second argument, and the function returns an array of the same shape and size as the first argument.

```
integer, dimension(4)::a
a=[1,2,3,4]
print *, ior(a,2)
end
```

The output is as follows:

```
3 2 3 6
```

Case III: Both the arguments of ior may be the arrays of the same shape and size. Each element of the first argument is 'ored' with the corresponding element of the second array.

```
integer, dimension(4)::a,b,c
a=[1,2,3,4]
b=[7,8,9,10]
c= ior(b,a)
print *, c
end
```

The output is as follows:

```
7 10 11 14
```

a(1) is 'ored' with b(1) and the result is stored in c(1), a(2) is 'ored' with b(2) and so on. Bitwise or may conveniently be used to convert the uppercase letters to the corresponding lowercase letters using the fact that by switching the bit 5 on the uppercase letter, it becomes the corresponding lowercase letter. For example, the internal representation of 'a' is '01100001', and the internal representation of 'A' is '01000001'. In the following program, the mask is so chosen that the bitwise or operation switches on the 5th bit, and other bits remain unaffected.

```fortran
integer, parameter:: sz=selected_int_kind(2)
integer(kind=sz):: ic
integer::i
integer(kind=sz)::mask=b"00100000"
character(len=26):: cc='ABCDEFGHIJKLMNOPQRSTUVWXYZ'
print *,cc
do i=1,26
   ic=iachar(cc(i:i),kind=sz) ! convert char to 8 bit integer
   ic=ior(ic,mask) ! switch on 5th bit
   cc(i:i)=achar(ic) ! convert to char
end do
print *,cc
end
```

The outputs are as follows:

```
ABCDEFGHIJKLMNOPQRSTUVWXYZ
abcdefghijklmnopqrstuvwxyz
```

17.13 IEOR

Case I: Like iand, ieor also takes two integers of the same kind as its arguments and returns an integer of the same kind after exclusive 'oring' the first argument with the second argument according to its truth table shown on the right (Figure 17.20).

```fortran
integer :: i1, j1, k1
data i1 /b"10101011"/, j1 /b"10110000"/
k1=ieor(i1,j1)
```

i	0	0	1	1
j	0	1	0	1
ieor(i, j)	0	1	1	0

Truth table – ieor

1	0	1	0	1	0	1	1
			i1				

1	0	1	1	0	0	0	0
			j1				

0	0	0	1	1	0	1	1
			k1				

FIGURE 17.20
Truth table of ieor and bit patterns of i1, j1 and k1.

The exclusive or (ieor) is often used for bit flipping; that is, switching on a particular bit (or bits) if it is off or switching off a particular bit (or bits) if it is on. Again a suitable mask must be chosen—the bits of the mask corresponding to the bits of the source that would remain unaffected are set to 0, and the bits of the mask corresponding to the bits of the source that are to be flipped are set to 1. In the example (Figure 17.20), bits 4, 5 and 7 of the variable i1 are flipped.

Case II: Like i and, the first argument is an integer array and the second argument is a scalar (integer). All the elements of the array are exclusive 'ored' with the second argument, and the function returns an array of the same shape and size as the first argument.

```
integer, dimension(4)::a
a=[1,2,3,4]
print *, ieor(a,2)
end
```

The output is as follows:

```
3 0 1 6
```

Case III: Both the arguments of ieor may be the arrays of the same shape and size. Each element of the first argument is exclusive 'ored' with the corresponding element of the second array. The function returns an array of the same type as the first argument.

```
integer, dimension(4)::a,b,c
a=[1,2,3,4]
b=[7,8,9,10]
c= ieor(b,a)
print *, c
end
```

The output is as follows:

```
6 10 10 14
```

a(1) is exclusive 'ored' with b(1) and the result is stored in c(1), a(2) is exclusive 'ored' with b(2) and so on. Using the property of exclusive or that a ieor b ieor b = a, it is possible to encrypt a string of characters using a key. It can be subsequently decrypted using the same key.

```
implicit none
character(len=100)::str= &
  "This string will be encrypted using exclusive or"
integer,parameter::k=selected_int_kind(2)
integer::i, l
integer(k),dimension(:),allocatable::buf
integer(k)::t,code=83 ! code is the encryption key
l=len_trim(str) ! length of the string
allocate(buf(l))
print *, "Original String:    ",str
do i=1,l
   t=iachar(str(i:i)) ! convert to integer
   buf(i)=ieor(t,code) ! encrypt using exclusive or
   str(i:i)=achar(buf(i)) ! back to character
enddo
print *, "Encrypted String:   ",str
```

```
do i=1,l
  t=iachar(str(i:i))
  buf(i)=ieor(t,code) ! decrypt using exclusive or
  str(i:i)=achar(buf(i))
enddo
print *,"Decrypted String:   ",str
end
```

The output is as follows:

```
Original String: This string will be encrypted using exclusive or
Encrypted String:;: s '!:=4s$:??s16s6=0!*#'67s& :=4s6+0?& :%6s__
Decrypted String:This string will be encrypted using exclusive or
```

The encrypted string contains some unprintable characters, which are not displayed.

17.14 NOT

Case I: Not takes one integer as its argument and returns an integer of the same kind as the argument after changing 0 bit to 1 and 1 bit to 0 (logical complement) (Figure 17.21).

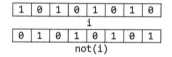

FIGURE 17.21
i and not(i).

```
integer, parameter:: s1=selected_int_kind(2)
integer (kind=s1) :: i, j
data i/b"10101010"/
j=not(i)
```

Case II: The arguments of 'not' may be an integer array. In that case, the returned array is of the same type and size as the input array.

```
integer, dimension (3) :: a, b
b=not(a)
```

'Not' operates on every element of the a array, and the result is stored in the corresponding element of b array; the previous statement is equivalent to b(1)=not (a(1)), b(2)=not(a(2)), b(3)=not(a(3)).

17.15 Bit Sequence Comparison

Four library functions have been provided for bit sequence comparison: bge (greater than or equal), bgt (greater than), ble (less than or equal) and blt (less than). All these functions take two integers or boz constants as arguments. If the size of the bit sequence of the arguments is not equal, the smaller one is made equal to the other one by adding zeros to the left.

Comparison takes place from left to right. If all the bits of the first argument are the same as those of the second argument, they are considered to be equal so far as the bit sequence comparison is concerned. Otherwise, the first unmatched bit decides the issue. The "1 bit" is considered to be greater than the "0 bit." Therefore, the argument, which has "1 bit" at the first unmatched position, is treated as greater than the other argument. All these functions return default logical (`true` or `false`) depending on the result of the comparison. Functions `bge(0,-1)`, `bgt(-1,0)`, `ble(-1,0)` and `blt(0,-1)` return, `false`, `true`, `false` and `true`, respectively. The explanation of the last one is sufficient to understand the remaining three. The bit patterns of `0` and `-1` are shown in Figure 17.22.

0	0		.	.																										.	.	.	0	0
1	1		.	,																										.	.	.	1	1

FIGURE 17.22
Bit patterns of `0` and `-1`.

Although the numerical value of `0` (all the bits are `0`) is greater than the numerical value of `-1` (all the bits are `1`), the bit comparison function `blt(0, -1)` returns `true` as the first unmatched bit (bit `31`) of `0` is `0` and that of `-1` is `1`.

17.16 Programming Example

This program prints an integer in a binary format without using the `'b'` edit descriptor. The logic of the program is to check each bit of the integer; if it is on, (one) `1` is displayed, and if it is off, `0` is displayed. To print the zeros and ones in one line, an option of the `write` statement is used—advance= `'no'`—to prevent carriage movement; that is, the zeros and ones will be displayed in one line. Without this option, the zeros and ones would be displayed in different lines (one character per line).

```
      implicit none
      integer :: num, sz, j, k
      read *, num ! read the integer
      sz=bit_size(num)
      do j=1, sz   ! no of bits
       if(btest(num,sz-j)) then
         k=1 ! tests from the left (bit 31)
       else
         k=0
      endif
      write (6, 20, advance='no')k
20    format(i1)
      enddo
      end
```

17.17 ISHFT

Case I: This function takes two arguments: the first one is an integer variable and the second one is an integer expression (constant, variable, arithmetic expression) called shift count. The function returns an integer of the same kind as the first argument after shifting the bits either to the left or to the right depending on the sign of the shift count. A positive shift count indicates a left shift, and a negative shift count indicates a right shift. In either case, the vacancy created by the shift at either end is filled with zeros (Figure 17.23).

s1=15

| 0 | 0 | 0 | 0 | 1 | 1 | 1 | 1 |

s2=30

| 0 | 0 | 0 | 1 | 1 | 1 | 1 | 0 |

s3=60

| 0 | 0 | 1 | 1 | 1 | 1 | 0 | 0 |

s4=7

| 0 | 0 | 0 | 0 | 0 | 1 | 1 | 1 |

s5=3

| 0 | 0 | 0 | 0 | 0 | 0 | 1 | 1 |

FIGURE 17.23
Bit patterns of s1 s2, s3, s4 and s5.

```
integer, parameter :: s=selected_int_kind(2)
integer(kind=s) :: s1, s2, s3, s4, s5
s1=15; print *, s1                  ! 15
s2=ishft(s1,1) ; print *, s2        ! 30
s3=ishft(s1,2) ; print *, s3        ! 60
s4=ishft(s1,-1) ; print *, s4       ! 7
s5=ishft(s1,-2) ; print *, s5       ! 3
end
```

Case II: The first argument is an array, and the shift count is a scalar. The shift count acts on each element of the array. The intrinsic returns an array of the same type as the first argument (Figure 17.24).

```
integer, parameter :: s=selected_int_kind(2)
integer(kind=s), dimension (3) :: &
  s1=[15,30,50]
integer(kind=s), dimension (3) :: s2, s3
s2=ishft(s1,1) ! s2: 30, 60, 100
print *, s2
s3=ishft(s1,-1) ! s3: 7, 15, 25
print *, s3
end
```

s1=[15,30,50]

0	0	0	0	1	1	1	1
0	0	0	1	1	1	1	0
0	0	1	1	0	0	1	0

s2=[30,60,100]

0	0	0	1	1	1	1	0
0	0	1	1	1	1	0	0
0	1	1	0	0	1	0	0

s3=[7,15,25]

0	0	0	0	0	1	1	1
0	0	0	0	1	1	1	1
0	0	0	0	1	1	0	1

FIGURE 17.24
s1, s2 and s3.

Case III: The shift count may be an array of the same type as the first argument. Each element of the first argument is shifted by an amount specified in the corresponding element of the shift count array. The function returns an array of the same type as the first argument (Figure 17.25).

```
integer, parameter ::s=selected_int_kind(2)
integer(kind=s), dimension (3)::s3=[15,30,22]
integer(kind=s), dimension (3) :: sh=[1, -1, 2]
integer(kind=s), dimension (3) :: s4
s4=ishft(s3,sh) ! s4: 30, 15, 88
print *, s4
end
```

s3=[15,30,22]

0	0	0	0	1	1	1	1
0	0	0	1	1	1	1	0
0	0	0	1	0	1	1	0

s4=[30,15,88]

0	0	0	1	1	1	1	0
0	0	0	0	1	1	1	1
0	1	0	1	1	0	0	0

FIGURE 17.25
s3 and s4.

It may be noted that in this case, the shift counts for the array elements are different—one left shift for s(1), one right shift for s(2) and two left shifts for s(3).

17.18 SHIFTL

This intrinsic takes two integers as arguments. It is essentially the same as `ishft` with a positive shift count (left shift). Both `shiftl(4,1)` and `ishft(4,1)` return 8. The result is an integer of the same kind as the first argument.

17.19 SHIFTR

This intrinsic takes two integers as arguments. It is essentially the same as `ishft` with a negative shift count (right shift). Both `shiftr(16,1)` and `ishft(16,-1)` return 8. The result is an integer of the same kind as the first argument. Both `shiftl` and `shiftr` may take arrays as the first and second arguments like `ishft`.

17.20 SHIFTA

This intrinsic is used to perform a right shift with fill. It takes two arguments: the first one, `i`, is an integer and the second one, `shift`, must be non-negative integer less than or equal to `bit_size(i)`. The result is an integer obtained by shifting `shift` number of bits of `i` to the right and replicating the left-most bit to fill in the gap. The bits shifted out are lost. The result is an integer of the same kind as `i` (assuming `i` as an 8-bit integer): `shifta(i,2)` with `i=b'10000100'` returns `b'11100001'` (Figure 17.26a).

1	0	0	0	0	1	0	0

i

1	1	1	0	0	0	0	1

(a) shifta(i,2)

FIGURE 17.26a
(a) `shifta(i,2)`.

17.21 MERGE_BITS

This function is used to merge bits under mask (Figure 17.26b). It takes three integers as arguments: `i`, `j` and `mask`. It returns an integer formed from `i` and `j`. Bits corresponding to the 1 bits of the mask are taken from `i`, and bits corresponding to the 0 bits of the mask are taken from `j`. This is equivalent to

```
ior(iand(i,mask),iand(j,not(mask)))
merge_bits (13_1, 18_1, 6_1) = 20.
  (NAG compiler)
```

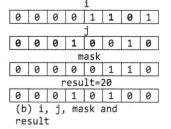

i

0	0	0	0	1	1	0	1

j

0	0	0	1	0	0	1	0

mask

0	0	0	0	0	1	1	0

result=20

0	0	0	1	0	1	0	0

(b) i, j, mask and result

FIGURE 17.26b
(b) `i`, `j`, `mask` and `result`.

17.22 ISHFTC

There are two forms of `ishftc`: the first one takes two arguments like `ishft` and works almost similar to it—the only difference is that the bits shifted out are added to the other end. This shift is also known as a circular shift.

First Form

Case I: Both arguments are scalar (Figure 17.27).

```
integer, parameter :: s=selected_int_kind(2)
integer (kind=s) :: i,sh,j
data i/ b"10011101"/ ! ifort compiler
sh=1
j=ishftc(i,sh) ! one left shift
print 10, i; print 10, j
10    format (b8.8)
end
```

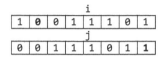

FIGURE 17.27
i and j arrays of case I.

Case II: The first argument is an array (say, rank 1), and the second argument is scalar. All the elements of the array are circularly shifted by the same amount (second argument). The function returns an array of the same type as the first argument. The bits are shown in bold letters in Figure 17.28.

```
integer, parameter :: s=selected_int_kind(2)
integer (kind=s),dimension(3) :: i,j
integer :: sh
data i/ b"10011101",b"01010101",b"10101010"/
sh=-1
j=ishftc(i,sh)
print 10, i; print 10; print 10, j
10    format (b8.8)
end
```

				i			
1	0	0	1	1	1	0	1
0	1	0	1	0	1	0	1
1	0	1	0	1	0	1	0

				j			
1	1	0	0	1	1	1	0
1	0	1	0	1	0	1	0
0	1	0	1	0	1	0	1

FIGURE 17.28
i and j arrays of case II.

Case III: Both the first and second arguments are arrays of the same shape and size. All elements of the first array are shifted according to the values of the corresponding elements of the second array. The function returns an array of the same type as the first argument (Figure 17.29).

```
integer, parameter :: &
  s=selected_int_kind(2)
integer (kind=s),dimension(3) :: i,j
integer (kind=s), dimension(3) :: &
  sh= [1, -1, 2]
data i/ b"10011101", b"01010101", b"10101010"/
j=ishftc(i,sh)
print 10, i; ; print 10; print 10, j
10    format (b8.8)
end
```

				i			
1	0	0	1	1	1	0	1
0	1	0	1	0	1	0	1
1	0	1	0	1	0	1	0

				j			
0	0	1	1	1	0	1	1
1	0	1	0	1	0	1	0
1	0	1	0	1	0	1	0

FIGURE 17.29
i and j arrays of case III.

Discussions related to shifting with `ishft` are all valid in these cases too.

Second Form

The second form of ishftc takes three arguments. The third argument (size) determines the operational zones of the shift operation. When this argument is absent, that is, the first form of ishftc, all the bits participate in the shift operation as shown earlier. If the third argument size is present, only the right-most number of bits as specified in size participates in the shift—other bits are not affected (Figure 17.30).

	i							
1	0	0	1	1	1	0	1

	j							
1	0	0	1	1	0	1	1

FIGURE 17.30
i and j arrays (second form).

```
      integer, parameter :: s=selected_int_kind(2)
      integer (kind=s) :: i,sh,j
      data i/ b"10011101"/
      sh=1 ! left shift
      j=ishftc(i,sh,4)
      print 10, i; print 10, j
10    format (b8.8)
      end
```

The print statements display 10011101 and 10011011, respectively. Let us try to understand the second number from the diagram. As the third argument size is 4, only 4 right-most bits participate in the shift operation, and the other bits remain unaffected. Thus, 1101, after a circular left shift, becomes 1011.

Note that several combinations of the arguments are possible. For example, the various forms of ishftc discussed under cases I–III may also have this size parameter. In addition to this, the size parameter may itself be an array of the same shape and size as the first argument. We illustrate two such cases:

```
      integer, parameter :: s=selected_int_kind(2)
      integer (kind=s), dimension(3) :: i,j
      integer (kind=s), dimension(3) :: sh= [1, -1, 2]
      data i/ b"10011101", b"01010101", b"10101000" /
      j=ishftc(i,sh,4)
      print 10, i; print 10 ; print 10, j
10    format (b8.8)
      end
```

Here, the first and second arguments are rank 1 arrays. The shift operations are restricted to the right-most 4 bits of all elements of the array. We mention the results from the program: 10011101, 01010101, 10101000 from the print *, i and 10011011, 01011010, 10100010 from the print *, j. The next example is a modification of the last example with the third argument as a rank 1 array as well. Depending on the value of different elements of this array, the number of participating bits in the shift operation of the first argument will come into play.

```
      integer, parameter :: s=selected_int_kind(2)
      integer (kind=s), dimension(3) :: i,j, siz(3)
      integer (kind=s), dimension(3) :: sh= [1, -1, 2]
      data i/ b"10011101", b"01010101", b"10101000" /
      data siz /4,3,5/
      j=ishftc(i,sh,siz)
      print 10, i; print 10 ; print 10, j
10    format (b8.8)
      end
```

Again we mention the output from this program: `print *, i` will display `10011101`, `01010101`, `10101000` and `print *, j` will display `10011011`, `01010110`, `10100001`. It may be verified that one left shift, one right shift and two left shifts are performed on `i(1)`, `i(2)` and `i(3)`, where `4`, `3` and `5` right-most bits have participated in the shift operations for the three array elements, respectively, to produce these results.

17.23 DSHIFTL

This intrinsic is used to perform a combined left shift. This function takes three integers as arguments: `i`, `j` and `shift`. Either `i` or `j` may be a `boz` constant. Both of them cannot be a `boz` constant; `shift` must be a non-negative integer less than or equal to `bit_size(i)` if `i` is an (normal) integer or `bit_size(j)`. If one of the arguments is a `boz` constant, it is converted to an integer by an `int` function of the same kind as the kind of the other one. If `i` and `j` are integers, they must be of the same kind. The result is of the same kind as `i` if `i` is an (normal) integer; otherwise, it is of the same kind as `j`. The right-most `shift` bits of the result are the same as the left-most `shift` bits of `j`, and the rest are the right-most bits of `i`. Dshiftl is illustrated with an 8-bit integer (kind=1). If `i=b"1010`1111" and `j=b"1100`0000", then `k=dshiftl(i, j, 4_1)` returns `k=b"11111100"`.

17.24 DSHIFTR

This intrinsic is used to perform a combined right shift. This function takes three integers as arguments: `i`, `j` and `shift`. Either `i` or `j` may be a `boz` constant. Both of them cannot be a `boz` constant; `shift` must be a non-negative integer less than or equal to `bit_size(i)` if `i` is an (normal) integer or `bit_size(j)`. If one of the arguments is a `boz` constant, it is converted to an integer by an `int` function of the same kind as the kind of the other one. If `i` and `j` are integers, they must be of the same kind. The result is of the same kind as `i` if `i` is an (normal) integer; otherwise, it is of the same kind as `j`. The left-most `shift` bits of the result are the same as the right-most `shift` bits of `i`, and the rest are the left-most bits of `j`. Again, dshiftr is illustrated with an 8-bit integer. If `i=b"10101`111" and `j=b"11011`010", then `k=dshiftr(i, j, 3_1)` returns `k=b"11111011"`.

17.25 Logical Operations with Array Elements

In this category, there are three intrinsics: `iall`, `iany` and `iparity`. All these intrinsics take identical arguments and behave in an identical fashion with different truth tables. Intrinsics `iall`, `iany` and `iparity` perform, respectively, `and`, `or` and `exclusive or` operations among the array elements. We discuss `iall` in detail and do not repeat the same discussions for `iany` and `iparity`. These two are discussed with the help of examples.

This intrinsic (iall) performs bitwise 'and' of all array elements of the array. It takes three arguments: array, dim and mask. The last two are optional. Array is an integer array having a rank n, dim is an integer scalar with 1<=dim<=n and mask, if present, must be a logical array conformable with array.

Case I: Mask is absent.

```
integer, dimension(3)::a=[12, 13, 10]
print *, iall(a)
```

prints 8. And operations are performed bit by bit among 12 $(1100)_2$, 13 $(1101)_2$ and 10 $(1010)_2$ according to the truth table of and.

Case II: If mask is present, array elements corresponding to the true value of the mask participate in the 'and' operation. Thus, the print statement, shown in the following program, displays 10 as only the first element 14 $(1110)_2$ and the third element 11 $(1011)_2$ participate in the and operation corresponding to the true value of the mask.

```
integer, dimension(3)::a=[14, 13, 11]
logical, dimension(3)::m=[.true., .false., .true.]
print *, iall(a, mask=m)
```

Case III: The first argument is a two-dimensional array.

```
integer,dimension(3,3)::b=reshape([1,3,5,4,5,6,15,11,9], [3,3])
print *, iall(b)
```

returns 0. If dim is present, the intrinsic returns an array of rank (n-1); in this case rank 1 array of size 3; print *, iall(b, dim=1) returns a rank 1 array containing [1 4 9]. Similarly, print *, iall(b, dim=2) returns a rank 1 array containing [0 1 0].

The readers may verify that the preceding operations are the same as the following:

```
print *, iand(iand(b(1,1),b(2,1)),b(3,1)) ! dim=1
print *, iand(iand(b(1,2),b(2,2)),b(3,2))
print *, iand(iand(b(1,3),b(2,3)),b(3,3))
print *, iand(iand(b(1,1),b(1,2)),b(1,3)) ! dim=2
print *, iand(iand(b(2,1),b(2,2)),b(2,3))
print *, iand(iand(b(3,1),b(3,2)),b(3,3))
```

Case IV: One can use both the dim and the mask arguments. In this case array elements corresponding to the true value of the mask participate in the and operation.

The next intrinsic in this series is iany that uses the truth table for ior.

```
integer, dimension(3)::a=[10, 8, 1]
print *, iany(a)  ! returns 11
integer, dimension(3)::a=[12, 13, 11]
logical, dimension(3)::m=[.true., .false., .true.]
```

```
print *, iany(a, mask=m)      ! displays 15
integer,dimension(3,3)::b=reshape([1,3,5,4,5,6,15,11,9], [3,3])
print *, iany(b)              ! displays 15
```

If `dim` is present, the intrinsic returns a rank `(n-1)`; in this case rank 1 array of size 3.

```
print *, iany(b, dim=1) ! returns a rank 1 array containing [7 7 15].
print *, iany(b, dim=2) ! returns a rank 1 array [15 15 15].
```

In case of `iany`, one can also use both the `dim` and the `mask` arguments. Array elements corresponding to the `true` value of the `mask` participate in the `or` operation.

Finally, the intrinsic `iparity` uses the truth table of `ieor`.

```
integer, dimension(3)::a=[12, 15, 10]
print *, iparity(a)    ! prints 9
integer, dimension(3)::a=[12, 13, 11]
logical, dimension(3)::m=[.true., .false., .true.]
print *, iparity(a, mask=m)   ! displays 7
integer,dimension(3,3)::b=reshape([1,3,5,4,5,6,15,11,9],[3,3])
print *, iparity(b)    ! displays 13
```

If `dim` is present, the intrinsic returns a rank `(n-1)`; in this case rank 1 array of size 3.

```
print *, iparity(b, dim=1) !returns a rank 1 array containing [7 7 13].
print *, iparity(b, dim=2) !returns a rank 1 array containing [10 13 10].
```

In case of `iparity`, one can use both the `dim` and the `mask` arguments. Array elements corresponding to the `true` value of the `mask` participate in the `exclusive or` operation.

17.26 MVBITS

Case I: This intrinsic (subroutine) takes five integers as arguments (Figure 17.31).

```
call mvbits (source, spos, len, dest, dpos)
```

`source` and `dest` are integer variables of the same kind; other arguments are integers or integer expressions. `source`, `spos`, `len` and `dpos` are input to the subroutine, and `dest` is both an input and an output parameter to the subroutine. Bits from the `source` starting from `spos` of length `len` are transferred to `dest` starting from bit position `dpos` of the destination.

```
call mvbits (s,1,4,d,4)
```

source	destination
bit 1	bit 4
bit 2	bit 5
bit 3	bit 6
bit 4	bit 7

FIGURE 17.31
Source and destination of case I.

will transfer 4 bits starting from the bit position 1 of s (bits 4 , 3 , 2 and 1). The bits are transferred to d starting from the bit position 4.

Other bits of the destination are not affected. The following example shows how four 8-bit integers are packed into a 32-bit integer (Figure 17.32):

```
implicit none
integer, parameter :: s1=selected_int_kind(2)
integer, parameter :: s2=selected_int_kind(9)
integer (kind=s1) :: i, j, k, l, sz, dz ! 8 bit
integer (kind=s2) :: t, ipack         ! 32 bit
i=10; j=64; k=31; l=11
sz=bit_size(i)   ! sz=8
dz=bit_size(ipack)      ! dz=32
ipack=0
t=i
call mvbits (t,0,sz,ipack,dz-sz) ! pack, source and destination same kind
t=j ! i goes to bits 31:24 of ipack, j goes to bits 23:16 of ipack
call mvbits(t,0,sz,ipack,dz-2*sz)
t=k
call mvbits(t,0,sz,ipack,dz-3*sz) ! k goes to bits 15:8 of ipack
t=l
call mvbits(t,0,sz,ipack,dz-4*sz) ! l goes to bits 7:0 of ipack
i=0; j=0; k=0; l=0 !  now unpack
l=ibits(ipack,0,sz) ! rightmost 8 bits
ipack=ishft(ipack, -sz)         !right shift
k=ibits(ipack,0,sz) ! rightmost 8 bits
ipack=ishft(ipack, -sz)         !right shift
j=ibits(ipack,0,sz) ! rightmost 8 bits
ipack=ishft(ipack, -sz)         !right shift
i=ibits(ipack,0,sz)
print *, i,j,k,l
end
```

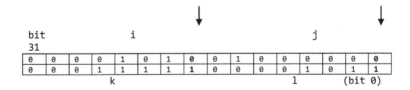

FIGURE 17.32
Destination after packing (shown in two lines).

The following inequalities hold good for mvbits:

```
len >= 0
spos >= 0
dpos >= 0
spos+len <= bit_size (source)
dpos+len <= bit_size (dest)
```

The source and dest may point to the same variable.

Case II: The arguments `source` and `dest` of `mvbits` may be arrays of the same shape and size. Other parameters may be integers or integer arrays. If they are integers, bits from `source(1)` to `dest(1)`, `source(2)` to `dest(2)`, etc. are transferred as discussed earlier. However, if `spos`, `len` and `dpos` are integer arrays, the corresponding array elements are used; that is, while transferring bits from `source(1)` to `dest(1)`, the corresponding parameters `spos(1)`, `len(1)` and `dpos(1)` are used. Readers may convince themselves that the outputs of the following program (Figure 17.33)

```
integer, dimension(2) :: so, sp, len,dest,dpos
sp=[1,2]
dest=0
dpos=[3,4]
so=63
len=[2,3]
call mvbits(so,sp,len,dest,dpos)
print *, dest
end
```
are 24 and 112.

so

| 0 | 0 | 1 | 1 | 1 | 1 | 1 | 1 |
| 0 | 0 | 1 | 1 | 1 | 1 | 1 | 1 |

dest

| 0 | 0 | 0 | 1 | 1 | 0 | 0 | 0 |
| 0 | 1 | 1 | 1 | 0 | 0 | 0 | 0 |

FIGURE 17.33
Source and destination of case II.

17.27 TRANSFER

This intrinsic takes three arguments—the third one, `size`, is optional. The function interprets the bit pattern of the first argument as the type of second argument. The bit pattern of `0.5` is as follows:

```
0 01111110 000 0000 0000 0000 0000 0000
Now,   integer :: i
       i=transfer(0.5,0)
```

interprets the bit pattern of `0.5` as integer, and this bit pattern corresponds to the integer value `1056964608`. This trick may be used to print the bit pattern of a real number with `'b'` format. As the earlier version of Fortran did not allow to use the b edit descriptor with real variable, the `transfer` statement is used to copy the bit pattern to an integer variable so that the same may be displayed in the b format.

Both the first and the second arguments may be array valued. If the second argument is of rank 1 array, the function returns an array of size just sufficient to accommodate the source.

```
complex, dimension(2) :: c1
c1=transfer([1.0,2.0,3.0,4.0], [(0.0, 0.0)])
```

Thus, the complex number `c1(1)` becomes `1+2i` and `c(2)` becomes `3+4i`. It may be noted that the second argument `(0.0, 0.0)` indicates that the first numbers are to be treated as complex numbers. The optional argument `size` must be a scalar and integer. If this is present, the result is an array of rank 1 of size equal to `size`.

```
print *, transfer([1.0,2.0,3.0,4.0],[(0.0, 0.0)],1)
```

will print (1.0, 2.0) because the `transfer` function will create a rank 1 complex array of size 1. Replacing the last parameter by 2, the `print` statement will print (1.0, 2.0) (3.0, 4.0) because now the function will create a rank 1 array of size 2.

18

C–Fortran Interoperability

A Fortran program can call a C function; similarly, a C program can call a Fortran subprogram. Fortran provides a mechanism by which Fortran variables may be interoperable with C. This is achieved by declaring Fortran entries as interoperable with C by an appropriate declaration. We demonstrate this feature by writing short programs in Fortran and C. It will be assumed that the readers have working knowledge in the programming language C.

All these examples given in this chapter have been tested with the NAG Fortran compiler running under Linux. The C compiler that came along with Linux (Ubuntu 18.04) was used to compile C programs.

18.1 Interoperability of Intrinsic Types

To make the entries of Fortran and C interoperable, the Fortran program must use a module called `iso_c_binding`. This module contains several named constants, which are used as `kind` parameters for the entries of Fortran. Table 18.1 contains a list

TABLE 18.1

Interoperability between Fortran and C

Type	Named Constant	Value (NAG)	C Type or Types
Integer	c_int	3	int
	c_short	2	short int
	c_long	3	long int
	c_long_long	4	long long int
	c_signed_char	1	signed char, unsigned char
	c_short_t	3	size_t
	c_int8_t	1	int8_t
	c_int16_t	2	int16_t
	c_int32_t	3	int_32_t
	c_int64_t	4	int64_t
	c_int_least8_t	1	int_least8_t
	c_int_least16_t	2	int_least16_t
	c_int_least32_t	3	int_least32_t
	c_int_least64_t	4	int_least64_t
	c_int_fast8_t	3	int_fast8_t
	c_int_fast16_t	3	int_fast16_t
	c_int_fast32_t	3	int_fast32_t

(Continued)

TABLE 18.1 (*Continued*)

Interoperability between Fortran and C

Type	Named Constant	Value (NAG)	C Type or Types
	c_int_fast64_t	4	int_fast64_t
	c_intmax_t	4	intmax_t
	c_intptr_t	3	intptr_t
Real	c_float	1	float
	c_double	2	double
	c_long_double	−4	long double
Complex	c_float_complex	1	float_complex
	c_double_complex	2	double_complex
	c_long_double_complex	−4	long_double_complex
Logical	c_bool	1	_Bool
Character	c_char	1	char

of such named constants. The numerical values of the named constants are compiler dependent. A compiler may not support every item mentioned in Table 18.1. A negative value indicates that the support for that particular item is not available.

For character-type interoperability, either the length parameter len is omitted or it is set to 1. The reason is that a C character variable can accommodate only one character.

Some other named constants, which are the unprintable characters of C language, are also available. The meaning of these named constants is obvious from their names. Table 18.2 contains a list of these named constants together with their internal values. The table was generated using the following program:

TABLE 18.2

Unprintable Named Constants

Named Constant	Internal Value
c_null_char	0
c_alert	7
c_backspace	8
c_form_feed	12
c_new_line	10
c_carriage_return	13
c_horizontal_tab	9
c_vertical_tab	11

```
use iso_c_binding
integer(c_int)::a
a=iachar(c_null_char); print *, a
a=iachar(c_alert); print *, a
a=iachar(c_backspace); print *, a
a=iachar(c_form_feed); print *, a
a=iachar(c_new_line);  print *, a
a=iachar(c_carriage_return);  print *, a
a=iachar(c_horizontal_tab); print *, a
a=iachar(c_vertical_tab); print *, a
end
```

18.2 C Procedure and Interface Block

In order to make a call to C program, an `interface` block is required in calling a Fortran program. The C procedure is described as a function or subroutine in the block but must contain the `bind(c)` attribute to inform the Fortran compiler that interoperability between C and Fortran is required.

18.3 Function, Subroutine and C Procedure

The concept of subroutine does not exist in C programming language. C program contains only functions. Thus, to treat a C procedure as a Fortran subroutine, C function should not return any value through its name; it should be a `void function`. All other C functions are treated like Fortran functions.

18.4 Interoperability with a C Pointer

C pointers are addresses. To make them interoperable with Fortran, two derived types are available in the module: `c_ptr` and `c_funptr`. These are interoperable with the C object and function pointer types, respectively. The module also provides two more named constants: `c_null_ptr` and `c_null_funptr`.

18.5 Procedures in the Module ISO_C_BINDING

The module `iso_c_binding` contains several procedures.

- c_loc(x) returns the C address of x. The variable x is either an interoperable type or a scalar, nonpolymorphic variable having no length parameter. The argument x must have either a pointer or a target attribute. If it is a pointer, it should be associated. If it is an allocatable, it should have its allocated status `true`. If it is an array, it should not be zero sized, and it should be contiguous. It cannot be a zero-length string and a coindexed object. The result returned from this procedure is of type `c_ptr`.
- c_funloc(x) returns the C address of the procedure x. The argument x is either a pointer associated with an interoperable procedure or an interoperable procedure. It cannot be a coindexed object. The result from this procedure is of type `c_funptr`. The procedure must have a `bind(c)` attribute.
- c_sizeof(x) returns the size of x in bytes. The argument x is an interoperable type and is not an assumed sized array. The procedure returns an integer of `kind`, `c_size_t`.

```
use iso_c_binding
integer(c_int),dimension(10)::a=10
print *, c_sizeof(a)
end
```

The output is 40.

c_associated(c_ptr_1 [,c_ptr_2]) returns true if c_ptr_1 is not c_null_ptr and c_ptr_2 is absent. Both c_ptr_1 and c_ptr2 are scalars of type c_ptr or c_funptr. However, if c_ptr_2 is present, the result is true if both point to the same object. If both of them are set to c_null_ptr, the result is false.

```
use iso_c_binding
integer(c_int),target:: a=10
type(c_ptr)::p1,p2
p1=c_loc(a)
if (c_associated(p1)) then
 print *, "true"              ! this path is chosen
endif
p2=p1
if(c_associated(p1,p2)) then
 print *, "associated"        ! this path is chosen
else
 print *, "not associated"
end if
end
```

The output is as follows:

```
true
associated
```

c_f_pointer(cptr, fptr [,shape]) is a subroutine. It associates a normal Fortran pointer with the target of a C pointer. The first argument, cptr, is the C address of an interoperable variable, or it can be the returned value of the function c_loc with a non-interoperable argument. The second argument, fptr, gets associated with the target of cptr after the subroutine call. The third optional argument is an integer rank 1 array having the same size as the rank of fptr if fptr is an array. It must be present if fptr is an array.

```
use iso_c_binding
integer(c_int), target::x=10
type(c_ptr)::cp
integer, pointer::fp
cp=c_loc(x)
call c_f_pointer(cp,fp)
print *,fp
end
```

It is clear that the output is 10. The following example shows the use of the optional parameter shape:

```
use iso_c_binding
integer(c_int), target::x=10
integer,dimension(1)::sh
```

```
type(c_ptr)::cp
integer, pointer,dimension(:)::fp
cp=c_loc(x)
call c_f_pointer(cp,fp,sh)
print *,fp(1)
end
```

Again the output is 10.

c_f_procpointer(cptr, fptr) is a subroutine that converts cptr of type(c_funptr) to a Fortran procedure pointer fptr.

18.6 Compilation and Linking

Suppose the name of the Fortran source program is myfort.f90 and that of the C source program is myc.c. The compilation and linking using the NAG Fortran compiler (Linux version) are done in the following manner:

Compilation: nagfor -c myfort.f90
              cc -c myc.c

Linking: nagfor -o myout myfort.o myc.o

The name of the executable file is myout, which is executed in a normal way.

18.7 IMPORT Statement

This statement is used within an interface block. This statement allows the entries from the host scoping unit mentioned with the import statement available within the interface body. If no entry is mentioned with the import statement, all the host entries are made available to the interface block by host association.

18.8 Fortran and C Interoperability—Examples

Case I: Dummy parameters of C functions are integers (scalars).

```
! Fortran program
use iso_c_binding
interface
  integer(c_int) function func(a,b,c) bind (c)
```

```
   use iso_c_binding ! this may be replaced by import
   integer(c_int), value :: a,b,c
   end function func
  end interface
  integer(c_int) :: a=10, b=20, c=30
  print *, func(a,b,c)
  end
/* c program */
int func (int a, int b, int c)
{ return (a+b+c);
}
```

In the preceding program, the value attribute is used with the dummy parameter (scalar). It ensures that a copy of the content of the actual parameters is passed to the procedure. The function returns 60, which is displayed by the print statement of the Fortran program. The Fortran program may be written with the import statement.

Case II: Dummy arguments are floating-point variables.

This example is the same as the previous example, except that the dummy arguments are floating-point variables.

```
  ! Fortran program
  use iso_c_binding
    interface
   real(c_float) function func(a,b,c) bind (c)
   use iso_c_binding
   real(c_float), value :: a,b,c
   end function func
    end interface
    real(c_float) :: a=10.0, b=20.0, c=30.0
    print *, func(a,b,c)
    end
/* c program */
float func (float a, float b, float c)
{ return (a+b+c);
}
```

Case III: Dummy arguments are pointers.

Fortran and C pointers are not the same. The former is an alias to a variable, and the latter contains the address of a variable. The Fortran function c_loc(x) returns the address of an object (C address), where x must be of interoperable type with the target attribute. The next program swaps the values of two variables a and b using a C function swap.

```
  ! Fortran program
  use iso_c_binding
  interface
   subroutine swap(a,b) bind(c)
   use iso_c_binding
   type(c_ptr), value :: a,b
   end subroutine swap
  end interface
```

```
      integer(c_int), target:: a=10,b=20
      print *, a, b
      call swap(c_loc(a), c_loc(b))
      print *,a, b
      end
/* c program */
void swap (int *a, int *b)
{ int t;
   t=*a;
   *a=*b;
   *b=t;
}
```

Note that when the C function swap is called from the Fortran routine, the address of the variable obtained through the function c_loc is passed to the swap routine.

If the C argument is, say, int *, it is not necessary that the corresponding Fortran argument has to be of type (c_ptr). It can be just of type (c_int) without the value attribute as follows:

```
! Fortran program
use iso_c_binding
interface
  subroutine  swap(a,b) bind(c)
  import
  integer(c_int) :: a,b
  end subroutine swap
end interface
integer(c_int), target:: a=10,b=20
print *, a, b
call swap(a, b)
print *,a, b
end
```

Case IV: Dummy argument is a dimensioned quantity (rank 1).

In this example, the C function will set every element of a single-dimensional array to some value.

```
! Fortran program
use iso_c_binding
interface
  subroutine init(a,n) bind(c)
  use iso_c_binding
  integer(c_int),dimension(10) ::a
  integer(c_int),value :: n
  end subroutine init
end interface
integer (c_int), dimension(10):: a=10
integer(c_int):: n=10
print *, a
call init (a,n)
print *, a
end
```

```
/* c program */
void init (int a[], int n)
{int i;
 for (i=0; i<n; ++i)
  a[i]=200+i;
}
```

Case V: Accessing the array through pointer.

```
! Fortran program
use iso_c_binding
interface
 subroutine sub(a,n) bind(c)
 use iso_c_binding
 type(c_ptr), value :: a
 integer(c_int), value ::n
 end subroutine sub
end interface
integer(c_int), target, dimension(10)::x=100
integer(c_int):: n=10
print *, x
call sub(c_loc(x(1)),n)
print *, x
end
/* c program */
int sub(int *x, int n)
{int i;
for (i=0; i<n; ++i)
 *(x+i)=200+i;
}
```

Case VI: Dummy argument is an array of rank greater than 1.

There is a major difference between Fortran and C in storing multidimensional arrays. Fortran stores a multidimensional array column-wise, whereas C stores the array row-wise. Therefore, to access a Fortran array having a shape of (2 3 4), that is, a dimension of (2, 3, 4), it must be dimensioned as (4, 3, 2) in the C program. In addition, the Fortran dimension normally specifies the upper bound with the lower bound equal to 1. On the other hand, the C array statement specifies the size with the lower bound equal to 0. Therefore, the upper bound is equal to (size-1). Thus, a Fortran array with a dimension of (2, 3, 4) when accessed within a C program should be dimensioned as (4, 3, 2), with lower bounds equal to 0 and upper bounds being 3, 2 and 1, respectively. The following program illustrates these points. For example, a(1, 2, 3) must be accessed as a[2][1][0] within the C program. The rule of thumb is as follows: invert the Fortran indices, and subtract 1 from each dimension. This argument is valid when the lower bound of the Fortran array in each dimension is 1.

Note that if the Fortran array is declared as a(2:3, 3:5, 4:7), the corresponding C array must be declared as a[4][3][2].

```
! Fortran program
use iso_c_binding
  interface
```

```
      subroutine sub(a) bind(c)
      use iso_c_binding
      integer(c_int), dimension(2,3,4)::a
      end subroutine sub
    end interface
   integer(c_int), dimension(2,3,4)::a
   integer:: i,j,k
   a=reshape([(i, i=1,24)], [2,3,4])
   do k=1,4
    do j=1,3
     do i=1,2
      print *,'a(',i,',',j,',',k,') = ',a(i,j,k)
     enddo
    enddo
   enddo
   print *,'---------- '
   call sub(a)
   do k=1,4
    do j=1,3
     do i=1,2
      print *,'a(',i,',',j,',',k,') = ',a(i,j,k)
     enddo
    enddo
   enddo
   end
/* c program */
void sub(int a[4][3][2])
{ int i,j,k;
  for(i=0; i<4; ++i)
   for(j=0; j<3; ++j)
    for(k=0; k<2; ++k)
     a[i][j][k]=100*i+10*j+k;
}
```

The output of the program is as follows:

```
a( 1 , 1 , 1 ) =   1, a( 2 , 1 , 1 ) =   2, a( 1 , 2 , 1 ) =   3,
a( 2 , 2 , 1 ) =   4, a( 1 , 3 , 1 ) =   5, a( 2 , 3 , 1 ) =   6,
a( 1 , 1 , 2 ) =   7, a( 2 , 1 , 2 ) =   8, a( 1 , 2 , 2 ) =   9,
a( 2 , 2 , 2 ) =  10, a( 1 , 3 , 2 ) =  11, a( 2 , 3 , 2 ) =  12,
a( 1 , 1 , 3 ) =  13, a( 2 , 1 , 3 ) =  14, a( 1 , 2 , 3 ) =  15,
a( 2 , 2 , 3 ) =  16, a( 1 , 3 , 3 ) =  17, a( 2 , 3 , 3 ) =  18,
a( 1 , 1 , 4 ) =  19, a( 2 , 1 , 4 ) =  20, a( 1 , 2 , 4 ) =  21,
a( 2 , 2 , 4 ) =  22, a( 1 , 3 , 4 ) =  23, a( 2 , 3 , 4 ) =  24.
----------
a( 1 , 1 , 1 ) =   0, a( 2 , 1 , 1 ) =   1, a( 1 , 2 , 1 ) =  10,
a( 2 , 2 , 1 ) =  11, a( 1 , 3 , 1 ) =  20, a( 2 , 3 , 1 ) =  21,
a( 1 , 1 , 2 ) = 100, a( 2 , 1 , 2 ) = 101, a( 1 , 2 , 2 ) = 110,
a( 2 , 2 , 2 ) = 111, a( 1 , 3 , 2 ) = 120, a( 2 , 3 , 2 ) = 121,
a( 1 , 1 , 3 ) = 200, a( 2 , 1 , 3 ) = 201, a( 1 , 2 , 3 ) = 210,
a( 2 , 2 , 3 ) = 211, a( 1 , 3 , 3 ) = 220, a( 2 , 3 , 3 ) = 221,
a( 1 , 1 , 4 ) = 300, a( 2 , 1 , 4 ) = 301, a( 1 , 2 , 4 ) = 310,
a( 2 , 2 , 4 ) = 311, a( 1 , 3 , 4 ) = 320, a( 2 , 3 , 4 ) = 321.
```

It may be noted from the preceding output that inside the C function, a[2][1][0] is set to 2*100+1*10+0=210, which is shown as a(1, 2, 3) within the Fortran program following the aforementioned prescription.

Case VII: Dummy argument is a derived type variable.

In this program, the import statement is used. This statement allows us to access the named entries from the host scoping unit within the interface block. The components must be of C type. Pointers and allocatables are not allowed, and also must not be a sequenced type or type-bound procedure. This type cannot be extended. It is obvious that items to be imported are declared in the host scoping unit before the interface block. The bind attribute must be used explicitly to make the user-defined type (derived type) interoperable.

```
! Fortran program
use iso_c_binding
type, bind(c):: mytype
  integer(c_int) :: a,b
  real(c_float) :: s
end type mytype
interface
  subroutine sub(y) bind(c)
  import
  type(c_ptr),value::y
  end subroutine sub
end interface
type(mytype), target:: y
y%a=20
y%b=30
y%s=40.0
print *, y%a, y%b, y%s
call sub(c_loc(y))
print *, y%a, y%b, y%s
end
/* c program */
  typedef struct{
  int a, b;
  float c;
  }mytype;
  int sub ( mytype *x)
  { x->a=130;
    x->b=240;
    x->c=99.0;
  }
```

Note that, in this case, the address of the variable (pointer) is passed to the C function, and through the pointer the elementary items of the derived data type are accessed.

Case VIII: One of the elementary items of the derived type is a rank 1 array.

```
!   Fortran program
use iso_c_binding
type, bind(c):: mytype
```

```
   integer(c_int) :: a,b
   real(c_float) :: s
   integer(c_int), dimension(5)::d
 end type mytype
 interface
   subroutine sub(y) bind(c)
   import
   type(c_ptr),value::y
   end subroutine sub
 end interface
 type(mytype), target:: y
 y%a=20
 y%b=30
 y%s=40.0
 y%d=450
 print *, y%a, y%b, y%s, y%d
 call sub(c_loc(y))
 print *, y%a, y%b, y%s, y%d
 end
/* c program */
 typedef struct{
 int a, b;
 float c;
 int d[];
 }mytype;
 int sub(mytype *x)
 { int i;
   for (i=0; i<5;++i) /* size of the array could be an argument */
   x->d[i]=i*3;
   x->a=130;
   x->b=240;
   x->c=99.0;
 }
```

The C function changes all the elementary items of the derived type, and the changes are reflected in the calling program.

Case IX: Function name as argument.

In the following example, the dummy argument is the address of a function. The Fortran program calls a C function, which, in turn, calls Fortran functions `fun1` and `fun2` depending on the address that is passed through the first argument of `cfunc`.

```
!   Fortran program
program main
use iso_c_binding
interface
   integer(c_int) function fun1(a,b) bind(c)
    use iso_c_binding
    integer(c_int),value:: a,b
   end function fun1
   integer(c_int) function fun2(a,b) bind(c)
    use iso_c_binding
    integer(c_int),value:: a,b
```

```
     end function fun2
     integer(c_int) function cfunc(f,a,b) bind(c)
      use iso_c_binding
      type(c_funptr), value:: f
      integer(c_int), value :: a,b
     end function cfunc
   end interface
   integer(c_int):: x=120, y=150, z
   z=cfunc(c_funloc(fun1),x,y)          ! cfunc will call function fun1
   print *, z                 ! z=x+y
   z=cfunc(c_funloc(fun2),x,y)          ! cfunc will call function fun2
   print *, z         ! z=x-y
   end program main
! Functions begins
   integer(c_int) function fun1(a,b) bind(c)
    use iso_c_binding
    integer(c_int),value:: a,b
    fun1=a+b
   end function fun1
   integer(c_int) function fun2(a,b) bind(c)
    use iso_c_binding
    integer(c_int),value:: a,b
    fun2=a-b
   end function fun2
/* c program */
int cfunc(int (*f)(int, int), int a, int b)
{ return((*f)(a,b));
}
```

Case X: Use of c_f_procpointer (with function).

The function returns value through its name. Therefore, the arguments a and b are passed by value. The procedure pointer was discussed in Chapter 15.

```
   use iso_c_binding
   interface
    integer(c_int) function fun(a,b) bind(c)
    import
    integer(c_int),value::a,b
    end function fun
   end interface
   procedure(fun),pointer::f
   type(c_funptr)::f1
   integer(c_int)::x=10, y=20,z
   f1=c_funloc(fun)
   call c_f_procpointer(f1,f)
   z=f(x,y)
   print *,z
   end
/* c program */
int fun(int x, int y)
{
 return(x+y);
}
```

Case XI: Use of c_f_procpointer (with subroutine).

The subroutine returns the value through its argument. Therefore, the c address of the third argument is passed during the call. Note that the Fortran subroutine is declared as a void function within the C program.

```
  use iso_c_binding
  interface
   subroutine sub(a,b,c) bind(c)
   import
   integer(c_int),value::a,b
   type(c_ptr),value::c
   end subroutine sub
  end interface
  procedure(sub),pointer::f
  type(c_funptr)::f1
  integer(c_int)::x=10, y=20
  integer(c_int),target::z
   f1=c_funloc(sub)
  call c_f_procpointer(f1,f)
  call f(x,y,c_loc(z))
  print *,z
  end
/* c program */
void sub(int a, int b, int *c)
{
 *c=a*b;
}
```

Case XII: Passing a Fortran character string to a C function.

C language supports the character variable of size 1. A character string in C language is a character array terminated by a null character. A Fortran character variable can accommodate more than one character specified by the len attribute. However, a Fortran character variable of kind c_char and of length greater than 1 can be passed to a C function that C can treat as a character array—the last character must be a null character according to the requirement of C language. The following Fortran program passes a Fortran character variable, terminated by a null character, to a C function, and the C function prints the character array.

```
  use iso_c_binding
   interface
    subroutine pprint(ch) bind(c)
     import
     character(kind=c_char), dimension(*)::ch
    end subroutine pprint
   end interface
  character(kind=c_char, len=10)::ch
  ch="I am here" //c_null_char   ! string terminated by null character
  call pprint(ch)
  end
/* c program */
#include <stdio.h>
```

```
void pprint(char ch[])
{ printf("%s\n", ch); /* this will be displayed: I am here */
}
```

18.9 Interoperation with Global Variables

Fortran and C program units can share the location. Two cases may arise:

Case I: In this case the C global variable is accessed by Fortran using the `bind(c)` attribute. It can only be declared within a module. It cannot be a part of a common block.

```
!    Fortran program
module mymod
 use iso_c_binding
 integer (c_int), bind(c):: ftnc
end module mymod
program main
use mymod
interface
  subroutine sub() bind(c)
  end subroutine sub
end interface
ftnc=350
print *, ftnc
call sub
 print *, ftnc
end
/* c program */
int ftnc;
void sub()
{ ftnc=587;
}
```

Case II: The variable is part of a common block. If the common block contains only one variable, the C program unit declares the variable through the name of the common block. The common block is also declared with the `bind` attribute.

```
!    Fortran program
use iso_c_binding
common /blk/a
integer(c_int)::a
bind (c):: /blk/
interface
 subroutine sub() bind(c)
 end subroutine sub
end interface
a=2
print *, a
```

```
   call sub()
   print *, a
   end
/* c program */
int blk;
void sub()
{ blk=30;
}
```

If the common block contains more than one variable, the corresponding C variable is declared as a structure with the name of the common block as the variable name. The elementary items of the structure correspond to the variables of the Fortran program that belong to the common block.

```
!   Fortran program
use iso_c_binding
interface
  subroutine sub() bind(c)
  end subroutine sub
end interface
common /blk/r, s
real (c_float) :: r,s
bind(c) :: /blk/
r=10.0
s=20.0
print *, r,s
call sub
print *, r,s
end
/* c program */
struct {float r,s;} blk;
void sub()
{ blk.r=300.56;
  blk.s=400.45;
}
```

The preceding programs show how a Fortran variable is modified within a C function and the modified values are carried over to the calling program.

Case III: Fortran variable a is used in the C procedure with a different name (myowna).

```
!   Fortran program
module mymod
 use iso_c_binding
 integer(c_int) ::a
 bind(c,name='myowna') ::a
end module mymod
program main
use mymod
interface
 subroutine sub() bind(c)
 end subroutine sub
end interface
```

```
    a=27
    print *, a
    call sub
    print *, a
    end
/* c program */
int myowna;
void sub()
{ myowna=37;
}
```

In the preceding example, the Fortran variable 'a' will be called as 'myowna' within the C function. The bind statement contains name='myowna', which does this job.

Case IV: The function name within Fortran is sub—the name of the C function is mysub.

```
    !    Fortran program
    module mymod
     use iso_c_binding
      integer(c_int) ::a
      bind(c,name='myowna') ::a
    end module mymod
    program main
    use mymod
    interface
      subroutine sub() bind(c,name='mysub')
      end subroutine sub
    end interface
    a=27
    print *, a
    call sub
    print *, a
    end
/* c program */
int myowna;
void mysub()
{ myowna=37;
}
```

In the preceding example, Fortran variable 'a' is known as 'myowna' within the C program. In addition, a call to subroutine 'sub' generates a call to C routine 'mysub'. All these are accomplished through the name parameter of bind. If the C name and the Fortran name are the same, the Fortran name is converted into lower-case. It is possible to choose a different name with the name clause. However, the C routine uses the C name, and the Fortran routine uses the Fortran name.

18.10 C–Fortran Interoperation

In this section we show how a C main program can call a Fortran subprogram.

```
/* c program */
#include <stdio.h>
int   main()
{ int a=10, b=20, c=30, d;
  int func(int, int, int);
  d=func(a,b,c);
  printf("%d %d %d %d \n", a,b,c,d);
}
    !    Fortran program
    function func(a,b,c) result (res) bind(c)
    use iso_c_binding
    integer(c_int), value::a,b,c
    integer(c_int) :: res
    res = a+b+c
    end
```

The Fortran function calculates a+b+c and returns the result to the calling C program. The dummy parameters must be of C type. The arguments cannot be of assumed shape, allocatable, optional or pointer.

The next program calls a Fortran subroutine and gets back the result through one of the arguments of the subroutine.

```
/* c program */
#include <stdio.h>
int   main()
{ int a=10, b=20, c=30, d;
  void func(int, int, int, int *); /* 4th argument is a pointer */
  func(a,b,c,&d);
  printf("%d %d %d %d  \n", a,b,c,d);
}
    !    Fortran program
    subroutine func(a,b,c,d) bind(c)
    use iso_c_binding
    integer(c_int),value::a,b,c
    integer(c_int), intent(out)::d
    d=a+b+c
    end
```

The point to be noted here is that the fourth actual parameter of the 'call' statement is a C pointer, and the corresponding Fortran dummy argument has been declared with intent(out). This is necessary because C parameters are passed by value and Fortran parameters are passed by reference. That is why the first three dummy arguments of the Fortran routine have the value attribute associated with them.

The next program swaps two C variables using a Fortran subroutine.

```
/* c program */
#include <stdio.h>
```

```
int main()
{ int a=10, b=20;
  void swap (int *, int *);
  printf("%d %d\n", a, b);
  swap (&a, &b);
  printf("%d %d\n", a, b);
}
  !    Fortran program
  subroutine swap (a,b) bind(c)
  use iso_c_binding
  integer(c_int), intent(inout)::a,b
  integer :: t
  t=a
  a=b
  b=t
  end subroutine swap
```

The statements that require special attention are shown in bold. The subroutine swap is called with the addresses of a and b (pointers). The corresponding dummy parameters are declared as intent(inout).

In this chapter a few examples of C–Fortran interoperability are shown. Readers are encouraged to do experiments with different kinds of variables.

18.11 ENUMERATOR

A set of named integer constants, interoperable with C are called enumators. The system chooses the kind in such a way that these are interoperable with the corresponding C-type enumeration.

```
use iso_c_binding
implicit none
enum, bind(c)
   enumerator :: red=1, green=3
   enumerator :: blue=5
end enum
integer (c_int) :: a
a=red
print *, a
a=green
print *,a
end
```

The outputs are 1 and 3. If the value is not specified, it is taken as one more than the previous one.

```
enum, bind(c)
   enumerator :: red=1, green=3
   enumerator :: blue
end enum
```

The value of blue, in this case, is 4, one greater than 3 (green). If the first item is not given any value, it is taken as 0.

```
enum, bind(c)
   enumerator :: red, green
   enumerator :: blue
end enum
```

The values of red, green and blue are 0, 1 and 2, respectively. However, in the following declaration:

```
enum, bind(c)
   enumerator :: red, green=5
   enumerator :: blue
end enum
```

the values of red, green and blue are 0, 5 and 6, respectively.

19

Object-Oriented Programming

Strictly speaking, Fortran is not an object-oriented programming language like Java. Fortran supports some features similar to the features of an object-oriented programming language. In the next section, we discuss about object and its properties.

19.1 Object and Its Properties

The world around us contains a number of objects, such as animals, plants, rivers and mountains. Also, around us there are millions of man-made objects like buildings, computers and cars. If we look carefully at these objects, we find that each of these objects is associated with certain characteristics. For example, some characteristics associated with a person could be his or her height or the color of the skin and eyes. The characteristics of a computer may be its clock speed, size of the memory and hard drive, peripheral attached to the computer, etc. Similarly properties associated with a car are its model, color, fuel consumption, maximum speed, etc.

Based on these characteristics, some other information may be derived. For a car, knowing the rate of fuel consumption can help determine the fuel required to traverse 100 km. It is thus observed that each object is associated with certain properties, and the behavior of the object based on these properties can be obtained by proper procedure. These properties are called the data associated with the object, and the behavior of the object is obtained through methods associated with the object.

An object is a software model of a real object consisting of data and procedures. The object is based on a template consisting of data and procedures. An object is treated as a single unit consisting of data and procedure that operate on the data. An object is defined in terms of a template. The data items associated with the object are called instance variables of the object. Similarly, the methods to manipulate the instance variables are called instance methods.

Objects of same type maintain their private copy of instance variables, but they use same instance methods. Normally, the instance variables are accessed through instance methods—they cannot be accessed directly. For example, to set an instance variable to a value, appropriate instance method must be called—it cannot be accessed directly. Similarly, to display an instance variable on the screen, corresponding methods must be used. In other words, instance variables are usually hidden from the programmer. This prevents accidental and undesirable modification of the instance variables. This property of the object is known as data encapsulation and data hiding.

The next important property of objects is inheritance. Consider an object fruit. It has certain general properties. Mango is a kind of fruit that has its own properties that distinguishes it from other fruits in addition to the general properties of the fruit. This may be stated in a different way: the object mango being the derived object of fruit may use all the variables and methods of the object fruit plus its own variables and methods.

Polymorphism is another property of object-oriented programming language. As mentioned in Chapter 1, variables may assume different values during the execution of the program, keeping the type of the values same. Polymorphic variables on the other hand may assume different types during the execution of the program.

In subsequent sections, we show how these properties have been implemented in Fortran.

19.2 Inheritance

A derived type may inherit all components from a previously defined derived type. This may or may not have additional components.

```
type bday
  integer:: dd, mm, yy
end type bday
type person
   character (len=30) :: name
   character (len=50) :: address
   integer :: age
   character (len=2) :: blood_group
   character (len=1) :: rh_factor
   type(bday) :: birth_day
   character (len=15), dimension(3):: phone_no
end type person
type, extends (person) :: active_donor
   integer :: no_donation ! additional component
end type active_donor
type (active_donor) :: donor_list
donor_list%name='SR'
donor_list%no_donation=80
print *, donor_list%name
print *, donor_list%no_donation
end
```

Variable donor_list inherits the entire components from its parent, that is, the derived type **person**. In addition, it has its own component. Variable donor_list of derived type **active_donor** can access all the components of its parent in the usual manner. In addition, all the components defined in the parent can be accessed through the instruction donor_list%person. Thus, print *, donor_list%person will display the values associated with the variables donor_list%name, donor_list%address, ..., donor_list%rh_factor.

The extended type may not have any additional component. In such a case, the extended type is merely another name of the parent type, though it is considered to be of different type. We consider another example.

```
type shape
   real:: weight=10.0
end type shape
type, extends(shape) :: circle
   real:: radius=1.0
   real:: thickness=1.0
end type circle
type, extends(shape) :: rectangle
   integer:: length=2
   integer:: width=1
end type rectangle
type, extends(rectangle) :: square
end type square
type (shape) :: s1
type (circle) :: c1
type (rectangle) :: r1
type (square) :: sq1
print *, "s1%weight = ", s1%weight
c1%weight=23.3
print *, "c1%width = ", c1%weight
r1%length=3
r1%width=2
r1%weight=5.5
print *, "r1%length, r1%width, r1%weight= ", r1%length,&
   r1%width, r1%weight
sq1%weight=12.5
sq1%length=4
sq1%width=4
print *, "sq1%weight, sq1%length, sq1%width= ", &
   sq1%weight,sq1%length,sq1%width
end
```

The outputs are as follows:
```
s1%weight = 10.0000000
c1%width =  23.2999992
r1%length, r1%width, r1%weight=  3 2 5.5000000
sq1%weight,sq1%length,sq1%width=  12.5000000 4 4
```

Objects c1, r1 and sq1 can access weight through inheritance. The extended type square does not have any additional component. As mentioned earlier, it is just another name of its parent. Note that rectangle is an extension of shape and square is an extension of rectangle. Therefore, shape may be considered as the grandparent of square.

19.3 ASSOCIATE

This statement is used to associate name entries with expressions or variables within the associate block. The so-called associate name exists only within the associate block. The statement is very suitable especially when the derived type variable contains embedded derived type objects like a%b%c. A name can be associated with this object to avoid repeating it. The `associate` statement is demonstrated with the help of an example.

```
real ::  x, theta,  ym
real :: r=1.5
x = 10.0
theta = 0.9
ym = 0.7
associate ( z => r * cos(theta)+r*sin(theta), t => ym)
print *, t, x+z, x-z, z
t = 2.5*t
end associate
print *, ym
end
```

Within the `associate` block x+z and x-z are, respectively, x+(r*cos(theta)+r*sin(theta)) and x-(r*cos(theta)+r*sin(theta));t is set to 2.5*t. This will be reflected in the `print` statement outside the `associate` block (print *, ym). Note that both z and t do not exist outside the `associate` block. The readers may verify that the outputs of this program are as follows:

```
0.70 12.107 7.892 2.107
1.750
```

The `associate` statement may have a label. If a label is used, the end `associate` statement must have the same label (it is optional).

```
l1: associate (z=>.....)
    .
    end associate l1
```

Associate statement may be nested.

```
associate (z=>sqrt(x**2+y**2))
  .
  associate (w=>z**3)
    .
    print *, w, z                  ! z is visible in the inner block
    .
  end associate
  .
print *, z                         ! w defined in the inner block is
  .                                ! not available in the outer block
end associate
```

In case of a nested `associate`, if labels are used, they must be different. The variable cannot be a coindexed object.

There may be a branch statement within the `associate` statement that may transfer control to the corresponding end `associate` statement.

```
associate (z=>sqrt(x**2+y**2))
  if(z>10.0) then
```

```
      go to 30
    endif
    .
    print *,z
    .
30  end associate
```

Care should be taken for a nested `associate` statement—a branch statement from the outer `associate` to inner `associate` is not allowed.

```
    associate (z=>sqrt(x**2+y**2))
    .
    if(z>10) then
     go to 40   ! allowed
    end if
    associate (w=>z**3)
    .
    print *, w
    .
30  end associate
    .
    print *,z
    .
40  end associate
```

However, if the statement `go to 40` is replaced by `go to 30`, the compiler should issue a warning (or flag an error).

19.4 Rules of ASSOCIATE

1. Only a branch to the `end associate` is allowed.
2. An associated entry assumes the same rank as its selector.

```
integer, dimension(5)::a=10
associate (x=>a*10)
 print *, x(1), x(5), size(x)       ! x(1)=a(1)*10, x(5)=a(5)*10,
end associate                        ! size(x)=5
end
```

The `print` statement will print `100 100 5`; `x` within the `associate` block is a rank 1 array of size 5.

3. If the selector has an allocatable attribute, it must be allocated.
4. The associated entry is local to the `associate` block. There is no conflict if there is any variable having the same name outside the `associate` block.
5. The associated entry need not be a pointer even if it is associated with an entry having `target` attribute.
6. The associated entry acquires `target`, `volatile` or `asynchronous` attribute should the selector have these attributes.

7. If a label is used with the `associate`, the end `associate` must have the same label. The label is optional.

8. For a nested `associate`, the inner one must terminate before the outer one.

9. In case of a nested `associate` block, if same associated names are used always, the local name prevails over the global name.

```
integer, dimension(5)::a=5, b=10
associate (x=>a)
 print *, x              ! prints a(1) … a(5)
 associate(x=>b)
  print *, x             ! prints b(1) … b(5)
 end associate
end associate
end
```

19.5 Polymorphic Variables

A variable is called polymorphic if its data type may vary during the execution of the program. It is a pointer or an allocated variable or a dummy data object. A polymorphic variable is defined by the `class` keyword. The rules for `type` and `class` are similar, but there is a major difference. Normally, during a procedure call, the formal arguments and dummy parameters must match. In addition, a pointer must point to a target of correct type, and it should match exactly. Similarly, there should be a match between an allocatable variable and the corresponding data. If a pointer or an allocatable variable or a dummy argument to a procedure is declared with `class` (user-defined type), this variable will match with this user-defined type and all its descendants (extensions). The rules for `class` are summarized next:

- It is used to declare a polymorphic object.
- The expression `class(*)` refers to an unlimited polymorphic object (to be discussed shortly).
- The `class` keyword is used to declare an entity, which must either be a dummy argument or have the pointer/allocatable attribute. The derived type must not be a sequenced type, a bind type or an intrinsic type.

```
type twod
   real:: x,y
end type twod
class(twod), pointer:: ptr
```

Pointer `ptr` may point to any object of `type(twod)` or its extensions. This pointer can access the components of the declared elements in the `type` declaration directly. It cannot access the components of the extensions directly, though it can point to the extension during the execution of the program. During the compilation time, the compiler knows the declaration of the pointer and the type of the object the pointer is supposed to point. Therefore, during execution, this pointer `ptr` can point to an object of type:

```
type, extends(twod)::threed
   real::z
end type threed
```

but it cannot access the component z directly, though it can access components x and y. The next program demonstrates this feature:

```
type:: two_d
   real:: x=10.0
   real:: y=20.0
end type two_d
type, extends(two_d):: three_d
   real::z=30.0
end type three_d
type(two_d), target:: coordi
type(three_d), target:: vector
class(two_d), pointer:: class_p
class_p=>coordi
print *, class_p%x
print *, class_p%y
class_p=>vector
print *, class_p%x
print *, class_p%y
!  print *, class_p%z
end
```

Pointer class_p has been declared having type two_d. It is true that it is polymorphic, and as such, it can point to vector (extension of the derived type two_d). However, it cannot access class_p%z. There will be compilation error if the comment ("!" character) is removed from the statement print *, class_p%z.

Unlimited polymorphic entities are defined using (*) in place of class specifier.

```
class (*), pointer ::p
```

This allows a pointer to refer to any object. This type of object can be used as an actual argument of a subprogram or as a selector of a select type statement (Section 19.6). This object cannot appear on the left-hand side of the assignment sign.

```
class(*), pointer :: x
character(len=20),target :: ch="abcd"
real, target::r=25.0
integer,target::ia=10
logical, target::l=.true.
complex, target:: z=cmplx(2.0,3.0)
real(kind=kind(0.0d0)),target::dp=3.14159265d0
!------------------------------------
!    x => ch
!    x=> r
!    x=> ia
!    x=> l
!    x=> z
!------------------------------------
```

```
      x=> dp
      select type(x) ! this is discussed in detail in the next section
      type is (complex)
         print *,"complex : ",z
      type is (character(len=*))
         print *,"character : ",ch
      type is (real)
         print *, "real : " , r
      type is (integer)
         print *, "integer : ", ia
      type is (logical)
         print *, "logical : ", l
      type is (real(kind=kind(0.0d0))) ! this path is chosen
         print *, "double precision : ", dp
      end select
      end
```

In the preceding program, pointer x can point to any object. Readers are advised to remove the comments one at a time from the program, compile and run the program. It is clear that depending upon the pointer assignment, different blocks of the select construct will be chosen. The next example shows that an unlimited polymorphic pointer class_p can point to derived types coordi and vector, but it cannot access directly any component of the dynamic data type.

```
      type:: two_d
       real:: x=10.0
       real:: y=20.0
      end type two_d
      type, extends(two_d):: three_d
       real::z=30.0
      end type three_d
      type(two_d), target:: coordi
      type(three_d), target:: vector
      class(*), pointer:: class_p
      class_p=>coordi
 !    print *, class_p%x
 !    print *, class_p%y
      class_p=>vector
 !    print *, class_p%x
 !    print *, class_p%y
 !    print *, class_p%z
      end
```

In fact, there will be compilation error if the comment (the character "!") is removed from any print statement.

19.6 SELECT TYPE Construct

Using a super class pointer, the select type construct allows finding the associated subclass of an object. This construct allows the execution of one of its constituent blocks depending upon the dynamic type of a variable or an expression. It is possible to access the components of the extensions. The syntax of the `select type` construct is as follows:

```
select type(associate-name=>selector)
type-guard-block
...
end select
```

Note that `associate-name` is similar to the `associate` construct discussed in section 19.3.

The `type-guard-block` may be one of the following:

```
type is (derive-type-spec)
class is (derive-type-spec)
class default
```

The block attached to a `type-guard-block` is executed if the dynamic type selector matches exactly with the derived type. The block is selected according to the following prescription:

- If the `type is` block matches, it is executed, otherwise
- If a single `class is` matches, it is executed, otherwise
- If there is a match with several `class is` blocks, the one which is the extension of all others is taken. The simple rule of thumb is to select the nearest `class is` containing the nearest ancestor.

There may be a `class default` block. If there is no match with any of the blocks, this is executed. If `class default` is not present, `select type` is ignored.

All these are explained with the help of examples:

```
type :: twod
   real :: x=1.0
   real :: y=2.0
end type twod
type, extends(twod) :: threed
   real:: z=4.0
end type threed
type, extends(twod) :: fourd
   real:: t=20.0
end type fourd
type(twod), target:: p2
type(threed), target:: p3
type(fourd), target:: p4
class(twod), pointer:: cp ! superclass pointer
cp=>p3
select type(cp)
   type is (twod)
    print *, "Enter Twod ..."
    print *, cp%x, cp%y
   type is (threed)
    print *, "Enter Threed ..."
    print *, cp%x, cp%y, cp%z
```

```
      type is (fourd)
        print *, "Enter Fourd ..."
        print *, cp%x, cp%y, cp%t
      end select
      end
```

The output is as follows:

```
Enter Threed ...
1.0000000    2.0000000   4.0000000
```

The program treats the cp pointer as if it is p3 pointer, thus allowing access to the instance variable z. If statement cp=>p3 is replaced by cp=>p4, the output will be as follows:

```
Enter Fourd ...
1.0000000    2.0000000 20.0000000
```

This is due to the fact that, in this case, type is (fourd) is chosen. The next program gives a demonstration of the default path.

```
      type :: twod
        real :: x=1.0
        real :: y=2.0
      end type twod
        .
      class(twod), pointer:: cp
      cp=>p4
      select type(cp)
        type is (twod)
          print *, "enter twod ..."
          print *, cp%x, cp%y
        type is (threed)
          print *, "enter threed ..."
          print *, cp%x, cp%y, cp%z
 !      type is (fourd)
 !        print *, "enter fourd ..."
 !        print *, cp%x, cp%y, cp%
      class default
          print *, "Default.."
      end select
      end
```

As the path type is (fourd) has been commented, there will not be any match. As a result, the default path (class default) is selected. The output in this case will be Default.....

Now we discuss the difference between type is and class is. Consider the previous program.

```
      cp=>p4 !fourd
      select type(cp)
        type is (twod)
          print *, "Enter Twod ..."
          print *, cp%x, cp%y
        type is (threed)
```

```
      print *, "Enter Threed ..."
      print *, cp%x, cp%y, cp%z
!     type is (fourd)
!       print *, "Enter Fourd ..."
!       print *, cp%x, cp%y, cp%t
      class is (fourd)
      print *, "Enter Fourd (class) ..."
      print *, cp%x, cp%y, cp%t
      class default
       print *, "No Match"
   end select
   end
```

There is no type is (fourd) block. However, there is one class is (fourd) block. In the absence of type is (fourd) block, the program will choose the class is (fourd) path, and the output will be as follows:

```
Enter Fourd (class) ...
1.0000000    2.0000000 20.0000000
```

If the comments from the type is (fourd) block are removed, the program will choose this block as type is gets preference over class is. The output will be as follows:

```
Enter Type Fourd...
1.0000000    2.0000000    20.0000000
```

Consider the following program:

```
type :: two
   real :: x=1.0
   real :: y=2.0
end type two
type, extends(two):: three
   real:: z=4.0
end type three
type, extends(three):: four
   real:: t=20.0
end type four
type, extends(four):: five
   character(len=10)::c="avbdwb"
end type five
type (two), target:: p2
type (three), target:: p3
type(four), target:: p4
type (five), target:: p5
class(two),pointer::cp
cp=>p5
select type(cp)
   type is (two)
     print *, "Enter Two ..."
   print *, cp%x, cp%y
 class is (three)
  print *, "Enter Three ..."
  print *, cp%x, cp%y, cp%z
!  class is (four)
```

```
!      print *, "Enter Four ..."
!      print *, cp%x, cp%y, cp%z,cp%t
!    class is (five)
!      print *, "Enter Five ..."
!      print *, cp%x, cp%y,cp%z, cp%t, cp%c
    class default
      print *, "No Match"
    end select
    end
```

This program does not contain type is (five) or type is (four) or class is (five) or class is (four). Therefore, class is (three) block (Grand Father) will be selected, and the output is as follows:

```
Enter Three ...
1.0000000    2.0000000   4.0000000
```

If the comments from the line class is (four) and the print statements are removed, following the same prescription it is easy to understand that the program will choose class is (four) block (Father), and the output is as follows:

```
Enter Four ...
1.0000000    2.0000000   4.0000000   20.0000000
```

We give another example. Here four is a child of three and three is a child of two and five is a child of two. Readers may verify that the output of the following program

```
    type :: two
      real :: x=1.0
      real :: y=2.0
    end type two
    type, extends(two):: three
      real:: z=4.0
    end type three
    type, extends(three):: four
     real:: t=20.0
    end type four
    type, extends(two):: five
      character(len=10)::c="avbdwb"
    end type five
    type (two), target:: p2
    type (three), target:: p3
    type(four), target:: p4
    type (five), target:: p5
    class(two),pointer::cp
    cp=>p5
    select type(cp)
    class is (two)
      print *, "Enter Two ..."
      print *, cp%x, cp%y
    class is (three)
      print *, "Enter Three ..."
      print *, cp%x, cp%y, cp%z
!    class is (four)
```

```
!       print *, "Enter Four ..."
!       print *, cp%x, cp%y, cp%z,cp%t
!   class is (five)
!       print *, "Enter Five ..."
!       print *, cp%x, cp%y, cp%c
    class default
        print *, "No Match"
    end select
  end
```

displays (Father):

```
Enter Two ...
1.0000000    2.0000000
```

Finally, we give another example, and the outputs obtained from this program are shown. Here the polymorphic variable has been declared as allocatable.

```
type point2d
  real:: x=1.0
  real:: y=2.0
  end type point2d
  type, extends(point2d):: point3d
   real:: z=3.0
  end type point3d
  type, extends(point3d):: point4d
   real:: t=4.0
  end type point4d
  class(point2d), allocatable:: z
  integer:: index
  do index=1,3
   select case(index)
    case(1)
     allocate(z,source=point2d())
    case(2)
     allocate(z,source=point3d())
    case(3)
     allocate(z,source=point4d())
    end select
   select type(z)
    type is (point2d)
     print *, "inside point2d ", z%x,z%y
    type is(point3d)
     print *, "inside point3d ", z%x,z%y,z%z
    type is (point4d)
     print *, "inside point4d ", z%x,z%y,z%z,z%t
    end select
   deallocate(z)
   end do
  end
```

The outputs are as follows:

```
  inside point2d  1.0000000 2.0000000
  inside point3d  1.0000000 2.0000000 3.0000000
  inside point4d  1.0000000 2.0000000 3.0000000 4.0000000
```

19.7 Allocation and Polymorphic Variables

An unlimited polymorphic item may be allocated to any intrinsic type.

```fortran
class(*), pointer:: s
allocate(character(len=10)::s) ! gfortran gives error, ifort ok
select type(s)
type is(integer)
   s=10
   print*, s
type is (double precision)
   s=3.14169265d0
   print *, s
type is(complex)
   s=cmplx(2.0,3.0)
   print *, s
type is (logical)
   s=.true.
   print *, s
type is (real)
   s=1.25
   print *, s
type is (character(len=*))
   s="abcd"
   print *, s
end select
end
```

Readers are advised to change the statement `allocate(character(len=10)::s)` to `allocate(integer::s)`, `allocate(real::s)`, `allocate(double precision:: s)`, `allocate(complex:: s)`, `allocate(logical:: s)` and see the result. The next example shows how unlimited and standard polymorphic items are allocated.

```fortran
type:: aloc
   real :: x=1.0
   real :: y=2.0
end type aloc
type, extends(aloc):: aloc1
   real:: z=3.0
end type aloc1
class(aloc), pointer::p(:),q, r
class(*), pointer::pa(:), s
allocate(aloc::p(5))   ! allocates p having type aloc of size 5
p(1)%x=10.0
p(1)%y=20.0
print *, p(1)%x, p(1)%y
allocate(aloc1::pa(10)) ! allocates pa having type aloc1 of size 10
allocate(aloc1::q)     ! allocates q having type aloc1
allocate(aloc::r)      ! allocates r having type aloc
allocate(aloc1::s)     ! automatic deallocation before fresh allocation
allocate(aloc::p(2))
print *, p(1)%x, p(1)%y ! prints the default values
end
```

19.8 Type Bound Procedure

It is possible to associate a procedure with a derived type. The method of accessing such a procedure is similar to accessing any other data element of the derived type. Several attributes are available for binding. They are pass, nopass, public, private, non_overridable and deferred. We first discuss pass and nopass. If pass or nopass is not specified or pass without argument is specified, the scalar through which the type bound procedure is accessed becomes the first argument of the procedure. For example, p%dist(q) essentially means calling the type bound procedure dist with two arguments p and q.

If the type bound procedure is declared with the attribute nopass, the name of the variable that has invoked the procedure is not passed as the first argument, and in that case, all required arguments are to be passed explicitly.

Let us suppose that we have a type bound procedure addlist that takes two arguments d1 and d2 of type dist. If the pass attribute has been used, d1%addlist(d2) essentially invokes the type bound procedure addlist of d1 with d1 as the first argument and d2 as the second argument. However, if nopass attribute has been chosen, d1%addlist(d2) invokes addlist with the sole argument d2. In fact, if the type bound procedure requires two arguments d1 and d2 of type dist, it must be, in this case, invoked (as nopass is used): d1%addlist(d1, d2) or d2%addlist(d1, d2). Therefore, if pass attribute is used, a procedure requiring only one argument is to be invoked without any argument as the variable invoking the procedure is automatically passed as the first argument. In other words, a type bound procedure declared with pass attribute requiring n number of arguments must be invoked with n-1 arguments.

We illustrate type bound procedure with a few examples. First, a very simple example that perhaps needs no explanation uses nopass attribute.

```
module m
type, public:: t
   integer, private::x,y
contains
   procedure, nopass:: set_val
   procedure, nopass:: get_val
end type t
contains
   subroutine set_val(a,b)
   class(t), intent(inout)::a
   integer, intent(in)::b
   a%x=b
   a%y=b*10
   end subroutine set_val
   subroutine get_val(a,b,c)
   class(t), intent(in)::a
   integer, intent(inout)::b,c
   b=a%x
   c=a%y
   end subroutine get_val
end module m
use m
type(t)::m1
integer::p,q,r=10
```

```
call m1%set_val(m1,r)
call m1%get_val(m1,p,q)
print *, p,q
end
```

The following program uses pass attribute:

```
module m
implicit none
type, public:: t
   integer, private::x,y
contains
   procedure, pass:: set_val
   procedure, pass:: get_val
end type t
contains
 subroutine set_val(a,b)
  .
 end subroutine set_val
 subroutine get_val(a,b,c)
  .
 end subroutine get_val
end module m
use m
type(t)::m1
integer::p,q,r=10
call m1%set_val(r)
call m1%get_val(p,q)
print *, p,q
end
```

Now we use our old example—addition of English distance.

```
module mymod
implicit none
type dist
  integer, private :: mile
  integer, private :: yds
contains
  procedure, public, pass :: adddist
  procedure, public, pass :: set_val
  procedure, public, pass :: get_val
end type dist
contains
function adddist(d1,d2) result(d3)
class(dist), intent(in)::d1,d2
type(dist)::d3
d3%mile=0
d3%yds=d1%yds+d2%yds
if (d3%yds >=1760) then        ! 1760 yds = 1 mile
  d3%mile=d3%mile+1
  d3%yds=d3%yds-1760
endif
d3%mile=d3%mile+d1%mile+d2%mile      ! d1+d2 is returned
```

```
end function adddist
subroutine set_val(d,m,y)
class(dist), intent(out)::d
integer, intent(in)::m,y
d%mile=m
d%yds=y
end subroutine set_val
subroutine get_val(d,m,y)
class(dist), intent(in)::d
integer, intent(out)::m,y
m=d%mile
y=d%yds
end subroutine get_val
end module mymod
program distance_add
use mymod
type(dist):: d1,d2,d3               ! add "english" distance
integer:: m1, y1, m2, y2,m3, y3     ! read elementary items of d1 and d2
read *,m1,y1,m2,y2                  ! data validation is not done for
                                    ! simplicity
call d1%set_val(m1,y1)             ! yds < 1760
call d2%set_val(m2,y2)
d3=d1%adddist(d2)
print *," d1.miles = ", m1, " d1.yds = ",y1
print *," d2.miles = ", m2, " d2.yds = ",y2
call d3%get_val(m3,y3)
print *," d3.miles = ", m3, " d3.yds = ",y3
end
```

Note that derived type arguments have been declared with `class` keyword instead of `type` keyword.

If neither pass nor nopass is present, the default attribute is pass. Thus, `procedure, pass :: adddist` and `procedure :: adddist` are equivalent.

It has been already mentioned that if pass is replaced by nopass, the type bound procedure is to be invoked in a different way: `d3=d1%adddist(d1,d2)`. In this case, the statement can be replaced by `d3=d2%adddist(d1,d2)` as d1+d2 is the same as d2+d1.

The type bound procedure adddist may be renamed; that is, a call to adddist can initiate a call to some other procedure `mileadd`.

```
module mymod
implicit none
type dist
  integer, private :: mile
  integer, private :: yds
contains
procedure, public, pass :: adddist=>mileadd
  procedure, public, pass :: set_val
  procedure, public, pass :: get_val
  procedure, public, pass :: mileadd
end type dist
contains
function mileadd(d1,d2) result(d3)
class(dist), intent(in)::d1,d2
```

```
type(dist)::d3
d3%mile=0
d3%yds=d1%yds+d2%yds
if (d3%yds >=1760) then        ! 1760 yds = 1 mile
 d3%mile=d3%mile+1
 d3%yds=d3%yds-1760
endif
d3%mile=d3%mile+d1%mile+d2%mile      ! d1+d2 is returned
end function mileadd
subroutine set_val(d,m,y)
class(dist), intent(out)::d
integer, intent(in)::m,y
d%mile=m
d%yds=y
end subroutine set_val
subroutine get_val(d,m,y)
class(dist), intent(in)::d
integer, intent(out)::m,y
m=d%mile
y=d%yds
end subroutine get_val
end module mymod
program distance_add
use mymod
type(dist):: d1,d2,d3  ! add "english" distance
integer:: m1, y1, m2, y2,m3, y3 ! read elementary items of d1 and d2
read *,m1,y1,m2,y2      !data validation is not done for simplicity
call d1%set_val(m1,y1) ! yds < 1760
call d2%set_val(m2,y2)
d3=d1%adddist(d2)
print *," d1.miles = ", m1, " d1.yds = ",y1
print *," d2.miles = ", m2, " d2.yds = ",y2
call d3%get_val(m3,y3)
print *," d3.miles = ", m3, " d3.yds = ",y3
end
```

The call to adddist [d3=d1%adddist(d2)] actually generates a call to the procedure mileadd because of the presence of the statement procedure, pass:: adddist=>mileadd in the type declaration.

A few more examples of type bound procedures are shown next:

```
module mymod
type :: point
real, private :: x, y
contains
  procedure, pass :: dist => p2d
  procedure, pass :: p2d1
  procedure, pass :: set_val
  procedure, pass :: get_val
end type point
private p2d, p2d1
contains
  real function p2d(a, b)
  class (point), intent (in) :: a, b
```

```
        p2d = sqrt ( (a%x - b%x)**2 + (a%y - b%y)**2 )
        end function p2d
        real function p2d1(a)
        class(point), intent(in)::a
        p2d1=sqrt((a%x+a%y))
        end function p2d1
        subroutine set_val(p,a,b)
        class(point),intent(out)::p
        real,intent(in)::a,b
        p%x=a; p%y=b
        end subroutine set_val
        subroutine get_val(p,a,b)
        class(point),intent(in)::p
        real,intent(out)::a,b
        a=p%x; b=p%y
        end subroutine get_val
end module
program main
use mymod
type (point):: p,q
real:: c1,c2, l
c1=7.0; c2=4.0
call p%set_val(c1,c2)
c1=2.0; c2=5.0
call q%set_val(c1,c2)
l=p%dist(q)
print *, l
l=p%p2d1()
print *, l
end
```

The type bound procedures dist and p2d1 are public. Note that module procedures p2d and p2d1 are private. Also, observe that in the type bound procedure declaration (shown in bold letters), the explicit binding is shown (dist=>p2d). This is not necessary if the name of the actual procedure is the same as the type bound procedure (p2d1); that is, the name of the type bound procedure and the actual procedure is the same (the public declaration is applicable to the actual procedure not to the type bound procedure).

Note the statements l=p%dist(q) and l=p%p2d1(). The procedures are invoked by a scalar variable of type point. Because of the presence of pass, these are passed as the first argument to the procedures, and for the first call, q is passed as the second argument to the procedure.

However, if in the procedure declaration the pass of p2d1 is replaced by nopass, the corresponding call to p2d1 would be p%p2d1(p) and not p%p2d1() because of the presence of nopass.

PASS(arg-name): If pass(arg-name) is specified for a type bound procedure and the type bound procedure has a dummy argument with the same name, the variable invoking the procedure takes the corresponding position during the actual call. The relevant lines are shown in bold in the following program:

```
module m
type, public:: t
   integer, private::x,y
contains
```

```
         procedure, pass(a):: set_val
         procedure, pass(a):: get_val
      end type t
      contains
        subroutine set_val(b,a)
        class(t), intent(inout)::a
        integer, intent(in)::b
        a%x=b; a%y=b*10
        end subroutine set_val
        subroutine get_val(b,a,c)
        class(t), intent(in)::a
        integer, intent(inout)::b,c
        b=a%x; c=a%y
        end subroutine get_val
      end module m
      use m
      type(t)::m1
      integer::p,q,r=10
      call m1%set_val(r)        ! equivalent to call set_val(r,m1)
      call m1%get_val(p,q)      ! equivalent to call get_val(p,m1,q)
      print *, p,q
      end
```

First, consider the declarations `procedure, pass(a):: set_val` and `subroutine set_val(b,a)`. The argument of `pass` is `'a'` and the second dummy argument of subroutine `set_val` is `a`. When the subroutine is invoked with `call m1%set_val(r)`, `m1` becomes the second actual argument to the subroutine. Similarly, `call m1%get_val(p,q)` is translated to `call get_val(p,m1,q)`.

It is also possible to call a specific procedure depending upon the dynamic type and type parameter of a polymorphic object.

19.9 Generic Binding

This section contains a discussion on type bound generic procedures. Two or more type bound procedures may have a generic name, and depending upon the type of arguments, the appropriate procedure is invoked. The program that follows calls the generic name twice: once with a real argument and the next time with a complex argument. Depending upon the argument, either p1 or p2 is invoked.

```
      module genmod
      private
      type, public :: gent
        real:: x,y
      contains
        private
        procedure :: p1, p2
        generic, public ::gproc=>p1,p2
      end type
      contains
```

```
         subroutine p1(a,b)
         class(gent), intent(in)::a
         real, intent(out):: b
         print *, 'Entering .... p1'
         b=sqrt(a%x**2+a%y**2)
         end subroutine p1
         subroutine p2(a,b)
         class(gent), intent(in)::a
         complex,intent(out)::b
         print *, 'Entering .... p2'
         b=cmplx(a%x,a%y)
         end subroutine p2
      end module genmod
      use genmod
      type(gent)::x=gent(2.0,3.0)
      complex :: c
      real :: d
      call x%gproc(d)              ! calls p1 (x,d)
      print *, d
      call x%gproc(c)              ! calls p2 (x,c)
      print *,c
      end
```

Note that the generic name is `public`, but the individual procedure names p1 and p2 are `private`. Therefore, procedures p1 and p2 cannot be called [like `call x%p1 (d)`] from the main program directly. They must be called through the generic name.

A generic type bound procedure may be an assignment as shown next:

```
module m
   type mydata
      real:: p,q
   end type mydata
   type point2d
      real:: x,y
   contains
      procedure, private:: p1=>setpoint
      generic:: assignment(=) => p1
   end type point2d
   contains
      subroutine setpoint(a,b)
      class(point2d), intent(out):: a
      type(mydata), intent(in)::b
      a%x= b%p; a%y= b%q
      end subroutine setpoint
end module m
   use m
   type(point2d):: x=point2d(3.0,4.0)
   type(mydata):: z=mydata(100.0,200.0)
   print *, x        ! prints 3 and 4 (before)
   x=z               ! print 100 and 200 (after)
   print *,x
   end
```

A generic type bound procedure may be an operator also:

```
module m
type mydata
  real:: p,q
end type mydata
type point2d
  real:: x,y
contains
  procedure, private:: p1=>typemul
  generic:: operator(*) => p1
end type point2d
contains
  type(point2d) function typemul(b,c) result(res)
  class(point2d), intent(in)::b
  type(mydata), intent(in)::c
  res%x=b%x*c%p ! arbitrary multiplication rule defined
  res%y=b%y*c%q
  end function typemul
end module m
use m
type(point2d):: y=point2d(3.0,4.0)
type(mydata):: z=mydata(100.0,200.0)
print *, y
y=y*z               ! calls typemul
print *,y           ! result 300 800
end
```

19.10 Overriding Type Bound Procedures

If the binding name specified in the type definition is identical to the binding name inherited from the parent, the binding obtained from the type definition overrides the binding inherited from the parent.

```
module mymod
type :: p2d
  real :: x, y
contains
  procedure, pass:: dist => dist2d
end type p2d
type, extends (p2d) :: p3d
  real :: z
contains
  procedure, pass :: dist => dist3d
end type p3d
contains
real function dist2d (a, b)
  class (p2d), intent (in) :: a, b
  dist2d = sqrt ((a%x - b%x)**2 + (a%y - b%y)**2)
```

```
      end function dist2d
   real function dist3d ( a, b )
      class (p3d), intent (in) :: a
      class (p2d), intent (in) :: b
   select type(b)
      class is (p3d)
      dist3d = sqrt((a%x-b%x)**2 + (a%y-b%y)**2 + (a%z-b%z)**2)
      print *, 'return from dist3d'
      return
   end select
   print *, " p3d, dynamic type: incorrect argument"
   stop
   end function dist3d
   end module mymod
   program main
   use mymod
   type (p3d):: x=p3d(4.0,5.0,6.0)
   type (p3d):: y=p3d(5.0,6.0,8.0)
   real:: r
   r=x%dist(y)
   print *, r
   end
```

In this case, dist3d will be called as type definition; (dist3d) overrides the inherited binding (dist2d). This overriding may be prevented by the following declaration in the parent type:

```
type :: p2d
   real :: x, y
contains
   procedure, pass, non_overridable:: dist => dist2d
end type p2d
```

The program segment is shown next:

```
module mymod
type :: p2d
   real :: x, y
contains
   procedure, pass, non_overridable:: dist => dist2d
end type p2d
type, extends (p2d) :: p3d
   real :: z
! the next lines are to be removed otherwise there will be
! compilation error. here lines have been commented.
!   contains
!   procedure, pass :: dist => dist3d
   end type p3d
contains
   .
   r=x%dist(y)
   print *, r
   end
```

19.11 Deferred Binding

Sometimes a template of a type is created that is used to create an extension. The base type
is not used to create any object. In such a case, type bound procedures have a `deferred`
attribute and the definition of the type contains an `abstract` keyword. The following
program demonstrates these features.

```
module mymod
implicit none
type, abstract :: abstype
contains
   procedure(absint), deferred :: p1
   procedure(absint), deferred :: p2
end type abstype
abstract interface
   subroutine absint(a)
   import :: abstype
   class (abstype), intent(inout):: a
   end subroutine absint
end interface
type, extends(abstype):: newabstype
   real, private :: i,j
contains
   procedure :: p1=>newp1
   procedure :: p2=>newp2
   procedure :: setvalue
   procedure :: getvalue
endtype newabstype
contains
   subroutine newp1(a)
   class(newabstype), intent(inout)::a
   a%i=a%i+a%j
   end subroutine newp1
   subroutine newp2(a)
   class(newabstype), intent(inout)::a
   a%i=a%i-a%j
   end subroutine newp2
   subroutine setvalue(a,l,m)
   class(newabstype),intent(out)::a
   real, intent(in)::l,m
   a%i=l; a%j=m
   end subroutine setvalue
   subroutine getvalue(a)
   class(newabstype),intent(in)::a
   print *, a%i, a%j
   end subroutine getvalue
   end module mymod
   program main
   use mymod
   type(newabstype):: a, b
   call a%setvalue(4.0, 5.0); call b%setvalue(6.0, 7.0)
   call a%getvalue(); call a%p1()
```

```
call a%getvalue(); call b%getvalue()
call b%p2(); call b%getvalue()
end
```

The outputs are as follows:

```
4.0    5.0
9.0    5.0
6.0    7.0
-1.0   7.0
```

19.12 Finalization

A derived type may be made finalizable through the `final` statement. The `final` statement specifies the name of the subroutine to be called automatically by the compiler when local variables are destroyed at the end of procedure call. It is somewhat similar to destructor of C++ language. Essentially, the finalization procedure is used for the cleaning operation. This facility is available only for derived types. The conditions are that there are no `sequence` or `bind` attributes.

The `final` subroutine name, a module procedure, is defined with exactly one dummy argument. The argument must not be a pointer or an allocatable variable or a polymorphic variable. The dummy argument must not have `intent(out)`. Moreover, two final subroutines must not have the same type of dummy argument. The following is an example of `final` statement.

```
module m
type:: one   ! ifort compiler used
 integer::x=10
contains
 final:: onefinal
end type one
type:: two
integer:: y=20
contains
 final:: twofinal
end type two
contains
  subroutine onefinal(a)
  type(one):: a
  print *, "enter subroutine onefinal"
  print *, "exit subroutine onefinal"
  end subroutine onefinal
  subroutine twofinal(b)
  type(two):: b
  print *, "enter subroutine twofinal"
  print *, "exit subroutine twofinal"
  end subroutine twofinal
end module m
program main
call sub
end program main
```

```
subroutine sub
use m
type(one):: p
type(two):: q
end subroutine sub          :
```

The output is as follows:

```
enter subroutine onefinal
exit subroutine onefinal
enter subroutine twofinal
exit subroutine twofinal
```

This shows that when variables p and q are being destroyed, the corresponding final routines are automatically called. The next example shows when one derived type is an extension of another derived type, the final routine of both the child and parent is executed.

```
module m
type:: one
integer::x=10
contains
  final:: onefinal
end type one
type, extends(one):: two
integer:: y=20
contains
  final:: twofinal
end type two
contains
 subroutine onefinal(a)
 type(one):: a
 print *, "enter subroutine onefinal"
 print *, "exit subroutine onefinal"
end subroutine onefinal
subroutine twofinal(b)
 type(two):: b
 print *, "enter subroutine twofinal"
 print *, "exit subroutine twofinal"
 end subroutine twofinal
 end module m
 program main
 call sub
 end program main
 subroutine sub
 use m
 type(one):: p
 type(two):: q
 end subroutine sub
```

The output is as follows:

```
enter subroutine onefinal
exit subroutine onefinal
```

```
enter subroutine twofinal
exit subroutine twofinal
enter subroutine onefinal
exit subroutine onefinal
```

When variable q is being destroyed first, its own final routine is called and then the final routine of its parent is invoked by the system. Note that there is no explicit call to the final routines within the program. The last example shows how allocatable variables are deallocated with the final routines.

```
module mymod
type :: first
 real a,b
end type first
type, extends(first) :: second
 real, pointer :: c(:),d(:)
contains
 final :: fsecond
end type second
type, extends(second) :: third
 real, pointer :: e
contains
 final :: fthird
end type third
contains
 subroutine fsecond(x) ! type (second) finalizer
 type(second) :: x
 print *, 'enter fsecond'
 if (associated(x%c)) then
    print *, ' c was allocated and is now being deallocated'
    deallocate(x%c)
 end if
 if (associated(x%d)) then
  print *, ' d was allocated and is now being deallocated'
  deallocate(x%d)
 end if
 print *, "exit fsecond"
end subroutine
subroutine fthird(y) ! type(third) finalizer
type(third) :: y
print *, 'enter fthird'
if (associated(y%e)) then
 print *, ' e was allocated and is now being deallocated'
 deallocate(y%e)
end if
print *, "exit fthird"
end subroutine
end module
program main
call sub
end program
subroutine sub
use mymod
type(first):: x1
```

```
      type(second):: x2
      type(third):: x3
      allocate(x2%c(5))
      x2%c=[1.0, 2.0, 3.0, 4.0, 5.0]
      allocate(x2%d(5))
      x2%d=[10.0,20.0,30.0,40.0,50.0]
      allocate(x3%e)
      x3%e=10.0
!     x1 is not finalizable. so no routine is called
!     compiler generated call for finalization
!     call fsecond(x2)
!     call fthird(x3)
!     call fsecond(x3%second)
      end subroutine
```

The output is as follows:

```
enter fsecond
  c was allocated and is now being deallocated
  d was allocated and is now being deallocated
exit fsecond
enter fthird
  e was allocated and is now being deallocated
exit fthird
enter fsecond
exit fsecond
```

Just before subroutine sub is exited, the compiler generated calls to the final routines fsecond and fthird. In this case, the final routines deallocate the allocated variables.

Before we close this section, we point out that if a derived type c is an extension of the derived type b, which is again an extension of the derived type a, when a variable x having type c is being destroyed, 'final' routines corresponding to c, b and a will be called in that order.

19.13 SAME_TYPE_AS

This inquiry function takes two arguments a and b, both of which are objects of extensible type. If they are pointers, their states should not be of undefined association.

The function returns true if the dynamic type of a and the dynamic type of b are same; it returns false otherwise.

```
      type:: x
        integer:: a=10
      end type x
      type, extends(x):: y
        integer:: b=20
      end type y
      type(x):: p,r
      type(y):: q,s
        if(same_type_as(p,r)) then
          print *, "same"
```

```
else
    print *, "not same"
endif
  if(same_type_as(q,s)) then
    print *, "same"
else
    print *, "not same"
endif
    if(same_type_as(p,q)) then
      print *, "same"
else
    print *, "not same"
endif
    if(same_type_as(r,s)) then
    print *, "same"
else
    print *, "not Same"
endif
end
```

In the first case, the intrinsic returns "true" (same). The result is also true (same) when the arguments to the intrinsic are q, s, namely, same_type_as(q,s). However, the intrinsic returns false (not same) for the following cases: same_type_as(p,q) and same_type_as(r,s).

19.14 EXTENDS_TYPE_OF

This inquiry function takes two arguments: a and mold. It tests whether the dynamic type a is an extension of dynamic type mold and returns true or false accordingly. Both a and mold must be objects of extensible type, and if they are pointers, they should not have undefined association. The result is also true if mold is an unallocated allocated variable or a disassociated pointer or an unlimited polymorphic variable.

```
type:: x
  integer:: a=10
end type x
type, extends(x):: y
  integer:: b=20
end type y
type, extends(y):: z
  integer:: c=30
end type z
type(x):: p
type(y):: q
type(z):: t
if(extends_type_of(t,q)) then
 print *, "same"
else
 print *, "not same"
endif
```

```
 if(extends_type_of(t,p)) then
   print *, "same"
else
   print *, "not same"
endif
 if(extends_type_of(q,t)) then
   print *, "same"
else
   print *, "not same"
endif
 if(extends_type_of(p,t)) then
   print *, "Same"
else
   print *, "Not Same"
endif
end
```

The first result is true (same). Similarly, extends_type_of(t, p) returns true (same). However, extends_type_of(q, t) and extends_type_of(p, t) return false (not same).

19.15 Derived Type Input and Output

User-defined input/output can be used to specify how the derived type input is available on the input device or how the derived type data will be displayed on the output device. It is not necessary to know the individual items of the data; even some of the components may be hidden (private) from the programmer. This section could not be accommodated with the previous chapters, as this requires the knowledge of class.

Input/output operations involving derived type object can be controlled by subroutines, which are automatically called by the processor, and therefore the implementation detail may not be necessary.

There are four different subroutines for performing input/output operations— read (formatted), write (formatted), read (unformatted) and write (unformatted). These are, respectively, used for formatted input, formatted output, unformatted input and unformatted output. It is not necessary to provide with all the sub-routines—only the required routines must be present. The subroutines are specified either by interface block or by generic binding. The actual name of the subroutines does not matter.

Suppose names of the subroutines are as follows:

```
read_der_for, write_der_for, read_der_unfor and write_der_unfor
```

The subroutines must have arguments as shown next:

```
subroutine read_der_for(dtv,unit,iotype,v_list,iostat,iomsg)
subroutine write_der_for(dtv,unit,iotype,v_list,iostat,iomsg)
subroutine read_der_unfor(dtv,unit,iostat,iomsg)
subroutine write_der_unfor(dtv,unit,iostat,iomsg)
```

The first two are for formatted read/write and last two are for unformatted read/write operations.

The arguments are declared as follows:

```
read(formatted)
class(derive-type-spec), intent(inout):: dtv
integer, intent(in):: unit
character(len=*), intent(in):: iotype
integer, intent(in), dimension(:):: v_list
integer, intent(out):: iostat
integer, intent(inout):: iomsg
write(formatted)
class(derive-type-spec), intent(in):: dtv
integer, intent(in):: unit
character(len=*), intent(in):: iotype
integer, intent(in), dimension(:):: v_list
integer, intent(out):: iostat
integer, intent(inout):: iomsg
read(unformatted)
class(derive-type-spec), intent(inout):: dtv
integer, intent(in):: unit
integer, intent(out):: iostat
integer, intent(inout):: iomsg
write(unformatted)
class(derive-type-spec), intent(in):: dtv
integer, intent(in):: unit
integer, intent(out):: iostat
integer, intent(inout):: iomsg
```

Needless to say, the names of the dummy arguments are arbitrary. We now discuss the properties of the argument. The first argument dtv is the user-defined object, which is supposed to take part in input/output operation. Its intent is inout for input operation and in for output operation. For extendable type, class must be used. If it is not extendable, type may be used. The next argument is the unit number of the device to be used by the subroutine. It is same as the unit number used by the parent. For list-directed input/output, it is obtained from the module iso_fortran_env. The processor passes this to the subroutine. The third argument (not required for unformatted input/output) iotype contains the character string 'LISTDIRECTED' or 'NAMELIST' or 'DT' depending on the mode of data transfer (processor passes this using capital letters); with dt the character string, if any, goes as it is—small or capital. This is passed to the subroutine by the processor. The next argument v_list gets the value from the argument of the edit descriptor dt. The size of the array becomes the number of items with the dt edit descriptor. If the dt descriptor does not contain any argument list, the processor passes a zero-sized array to the subroutine. The next item, iostat, reports error, if any. The subroutine returns a positive integer when an error occurs. The last argument iomsg returns additional description in case there is any error. It is not modified if there is no error. Intel ifort compiler has been used to test the programs given in this section.

```
module m
type:: mytype
 integer, dimension(4):: a
end type mytype
interface read(formatted)
 procedure myr
end interface
```

```
   interface write(formatted)
    procedure myw
   end interface
   contains
    subroutine myr(dtv, unit, iotype, v_list, iostat,iomsg)
    class(mytype), intent(inout):: dtv
    integer, intent(in):: unit
    character(*), intent(in):: iotype
    integer, intent(in), dimension(:):: v_list
    integer, intent(out):: iostat
    character(*), intent(inout)::iomsg
    read(unit,fmt=*,iomsg=iomsg, iostat=iostat) dtv%a
    print *, "myr --",iotype
    print *
    end subroutine myr
    subroutine myw(dtv,unit, iotype, v_list, iostat,iomsg)
    class(mytype), intent(in):: dtv
    integer, intent(in):: unit
    character(*), intent(in):: iotype
    integer, intent(in), dimension(:):: v_list
    integer, intent(out):: iostat
    character(*), intent(inout)::iomsg
    print *, "myw --", iotype
    write(unit, fmt=*, iomsg=iomsg, iostat=iostat) dtv%a
    end subroutine myw
   end module m
   program test_dt
   use m
   type(mytype):: x
   read(5,fmt=*)x ! control is transferred to myr,the input: 100 200 300 400
   write(6,fmt=*)x ! control is transferred to myw
   end
```

The input data are as follows:

```
100 200 300 400
```

The output are as follows:

```
myr --LISTDIRECTED
myw --LISTDIRECTED
    100    200    300    400
```

From the output, it is clear that input/output operations are performed through user-defined routines myr and myw. In addition, it is to be noted that iotype in both cases are LISTDIRECTED. Instead of using an interface block, generic binding can be used as shown next:

```
   module m
    type:: mytype
    integer, dimension(4)::a
    contains
     procedure myr
     generic:: read(formatted) => myr
    procedure myw
```

```
generic:: write(formatted)=> myw
end type mytype
contains
 subroutine myr(dtv, unit, iotype, v_list, iostat,iomsg)
   .
 end subroutine myr
 subroutine myw(dtv,unit, iotype, v_list, iostat,iomsg)
   .
 end subroutine myw
end module m
program test_dt
use m
type(mytype):: x
read(5,fmt=*)x ! control is transferred to myr, the input: 100 200 300 400
write(6,fmt=*)x ! control is transferred to myw
end
```

The edit descriptor dt has the form dt 'string'(arg-list). Now consider the
edit descriptor dt'mystring'(5, 4, 7). This will call appropriate subroutine with
iotype='dtmystring', v_list(1)=5, v_list(2)=4 and v_list(3)=7. If the
character string is absent, a null string is passed. DT(7, 3, 5) will call the subroutine with
iotype='dt', v_list(1)=7, v_list(2)=3 and v_list(3)=5. The next program
segment demonstrates use of dt edit descriptor.

```
module mdt
 type:: dob
   integer:: dd
   integer:: mm
   integer:: yy
 end type dob
 type:: mytype
   character (len=15):: name
   integer:: age
   character:: sex
   type(dob):: birthday
 end type mytype
 interface write(formatted)
  procedure myr
 end interface
 contains
 subroutine myr(dtv,unit, iotype, v_list, iostat,iomsg)
   class(mytype), intent(in):: dtv
   integer, intent(in):: unit
   character(*), intent(in):: iotype
   integer, intent(in), dimension(:):: v_list
   integer, intent(out):: iostat
   character(*), intent(inout)::iomsg
   character(len=50):: temp=" "
   write(temp, '(a,i2,a,i1,a,i1,a,i1,a,i1,a,i1,a)') '(a', v_list(1), &
     ',i',v_list(2), ',a', v_list(3), ',i',v_list(4),',i',v_list(5), &
     ',i',v_list(6), ')'
   print *, "temp --- ",temp
   print *, "iotype ----     ", iotype
   print *, "v_list ----", v_list
```

```
      write(unit, fmt=temp, iomsg=iomsg, iostat=iostat) dtv%name, &
      dtv%age,dtv%sex, dtv%birthday%dd, dtv%birthday%mm, dtv%birthday%yy
   end subroutine myr
   end module mdt
   program test_dt
   use mdt
   type(mytype):: x
   x%name="subrata ray"
   x%age=70
   x%sex="m"
   x%birthday%dd=19
   x%birthday%mm=4
   x%birthday%yy=47
   write(6,fmt="(dt(15,6,4,3,3,3))")x
   end
```

The output is as follows:

```
temp --- (a15,i6,a4,i3,i3,i3)
iotype ----DT
v_list ---- 15   6 4 3 3   3
subrata ray70 m 19  4 47
```

This program needs some explanation. When the write routine is invoked automatically, the subroutine gets values from the calling program iotype= "dt" (note that there is no string attached to dt in the test_dt, so a null string is passed along with dt). v_list(1) to v_list(6) become 15, 6, 4, 3, 3, 3.

Array temp is constructed as (a15,i6,a4,i3,i3,i3) using the following statement:

```
write(temp, '(a,i2,a,i1,a,i1,a,i1,a,i1,a,i1,a)') '(a', v_list(1), &
     ',i',v_list(2), ',a', v_list(3), ',i',v_list(4),',i',v_list(5), &
     ',i',v_list(6), ')'
```

This temp array is used as edit descriptor to display various items through the statement:

```
write(unit, fmt=temp, iomsg=iomsg, iostat=iostat) dtv%name, &
      dtv%age,dtv%sex, dtv%birthday%dd, dtv%birthday%mm, dtv%birthday%yy
```

In the actual output, the last line is as follows:

```
subrata ray 70   m 19  4 47
```

However, if the write statement in the main program is changed to

```
write(6,fmt="(dt'mysample'(15,6,4,3,3,3))")x
```

the print *, iotype in the subroutine will display DTmysample. The next example shows how to read and write using unformatted read and write routines.

```
   module m
   type:: mytype
      integer, dimension(10)::a
   end type mytype
   interface read(unformatted)
      procedure myr
   end interface
```

```
interface write(unformatted)
   procedure myw
end interface
interface write(formatted)
   procedure myfw
end interface
contains
 subroutine myr(dtv,unit, iostat,iomsg)
 class(mytype), intent(inout):: dtv
 integer, intent(in):: unit
 integer, intent(out):: iostat
 character(*), intent(inout)::iomsg
 read(unit,iomsg=iomsg, iostat=iostat) dtv%a
 end subroutine myr
 subroutine myw(dtv,unit, iostat,iomsg)
 class(mytype), intent(in):: dtv
 integer, intent(in):: unit
 integer, intent(out):: iostat
 character(*), intent(inout)::iomsg
 write(unit, iomsg=iomsg, iostat=iostat) dtv%a
 end subroutine myw
 subroutine myfw(dtv,unit, iotype, v_list, iostat,iomsg)
class(mytype), intent(in):: dtv
integer, intent(in):: unit
character(*), intent(in):: iotype
integer, intent(in), dimension(:):: v_list
integer, intent(out):: iostat
character(*), intent(inout)::iomsg
print *, "myfw --", iotype
write(unit, fmt=*, iomsg=iomsg, iostat=iostat) dtv%a
end subroutine myfw
end module m
program test_dt
use m
type(mytype):: x
x%a=[10,20,30,40,50,60,70,80,90,100]
write(10)x
close(10)     ! file is to be closed before reading
read(10)x     ! otherwise end of file error will be signaled
write(6, fmt=*)x
end
```

The next program uses generic binding instead of interface block.

```
module m
 type:: mytype
  integer, dimension(10)::a
 contains
  procedure myr
  generic:: read(unformatted) =>myr
  procedure myw
  generic:: write(unformatted) => myw
  procedure myfw
  generic:: write(formatted) => myfw
 end type mytype
```

```
contains
 subroutine myr(dtv,unit, iostat,iomsg)
   .
 end subroutine myr
 subroutine myw(dtv,unit, iostat,iomsg)
   .
 end subroutine myw
 subroutine myfw(dtv,unit, iotype, v_list, iostat,iomsg)
   .
 end subroutine myfw
end module m
program test_dt
use m
type(mytype):: x
  .
end
```

The final example demonstrates the use of namelist. In addition, the program shows that the string associated with dt may be used to control the flow within the subroutine (in this case myr).

```
module m
   integer, parameter:: sz=selected_real_kind(16,307)
   character (len=25):: buf=" "
type:: nm
   integer, private:: n = 2
   integer, private:: m = 3
end type nm
interface write(formatted)
   procedure myr
end interface
contains
 subroutine myr(dtv,unit, iotype, v_list, iostat,iomsg)
 class(nm), intent(in):: dtv
 integer, intent(in):: unit
 character(*), intent(in):: iotype
 integer, intent(in), dimension(:):: v_list
 integer, intent(out):: iostat
 character(*), intent(inout)::iomsg
 integer :: id1, id2
 namelist /subnml/d1, d2, t
 id1=dtv%n
 id2=dtv%m
 print *, "Iotype = ",iotype ! iotype - different paths will be chosen
 if(iotype .eq."LISTDIRECTED") then
   write(unit, fmt=*, iomsg=iomsg, iostat=iostat) dtv%n, dtv%m
 else if (iotype .eq. "DTiform") then
   write(unit, "(i3,i3)", iomsg=iomsg, iostat=iostat) dtv%n, dtv%m
 else if (iotype .eq. "DTfdivform") then
   write(unit, "(f15.6)", iomsg=iomsg, iostat=iostat) &
     real(dtv%n)/ real(dtv%m)
 else if (iotype .eq. "DTfdivdoubleform") then
   write(unit, "(f25.16)", iomsg=iomsg, iostat=iostat) &
     real(dtv%n, kind=sz)/ real(dtv%m, kind=sz)
 else if (iotype .eq. "DTiformwidthgiven") then
   buf= " "
```

```
          write(buf,'(a,i1,a,i1,a)') "(i",v_list(1),",i",v_list(2), ")"
          write(unit, fmt=buf, iomsg=iomsg, iostat=iostat) dtv%n, dtv%m
        else if (iotype .eq. "DTfdouble") then
          buf=" "
          write(buf,'(a,i2,a,i2,a)') "(f",v_list(1),".",v_list(2),")"
          write(unit, fmt=buf, iomsg=iomsg, iostat=iostat) &
            real(dtv%n, kind=sz)/ real(dtv%m, kind=sz)
        else if (iotype .eq. "NAMELIST") then
          d1= real(dtv%n)
          d2= real(dtv%m)
          t= d1/d2
          write(6, nml=subnml)
        endif
      end subroutine myr
    end module m
    program test_dt
    use m
    type(nm):: x
    namelist /mainnml/x
    write(6,*)x
    write(6,"(dt'iform')")x
    write(6,"(dt'fdivform')")x
    write(6,"(dt'fdivdoubleform')")x
    write(6,"(dt'iformwidthgiven'(4,6))")x
    write(6,"(dt'fdouble'(20,16))")x
    write(6, nml=mainnml)
    end
```

The output is as follows:

```
Iotype = LISTDIRECTED
            2           3
Iotype = DTiform
  2 3
Iotype = DTfdivform
      0.666667
  Iotype = DTfdivdoubleform
      0.6666666666666667
  Iotype = DTiformwidthgiven
  2     3
  Iotype = DTfdouble
  0.6666666666666667
  &MAINNML
 Iotype = NAMELIST
 X=
&SUBNML
 D1  =    2.000000  ,
 D2  =    3.000000  ,
 T   =    0.6666667
 //
```

The readers are advised to trace this program and interpret the results. When the input/output operation is active, `open`, `close`, `backspace`, `endfile`, `rewind`, `flash` and `wait` statements cannot be executed. Moreover, the edit descriptor `dt` can be used only with synchronous data transfer.

20

Parallel Programming Using Coarray

One of the principal characters, King Ravana, of the Indian epic *Ramayana* had 10 heads. It is not clear how the king used to eat or whether he could think 10 different war strategies simultaneously using his 10 heads. The sage-poet Valmiki who authored this famous epic probably wanted to convey that the king was 10 times more intelligent than an ordinary intelligent person was.

Since the dawn of the computer era, there had been a constant demand for a faster machine. The word faster implies that a machine that could compute at a much faster speed than the existing machines. During the past 70 years, there had been a tremendous development in the field of both computer hardware and software—vacuum tube to VLSI (very large-scale integration) and no operating system to very sophisticated time-sharing operating system. Still there are three hurdles from the point of view of computer hardware. First, the density of the active components cannot be increased arbitrarily. Second, as the density of the active components of a VLSI chip increases, heat dissipation becomes a major problem. Finally, the speed of any signal cannot exceed the speed of light according to Einstein's special theory of relativity. Thus, it becomes necessary to think in a different line to build a faster computer.

20.1 Parallel Computing

Let us consider a very simple problem: sum the series of the first 1000 natural numbers. Assuming no standard formula, a typical Fortran program would be as follows:

```
integer:: sum=0, i
do i=1, 1000
 sum=sum+i
end do
```

If one addition takes t units of time, the total time required to perform this task would be 1000t units of time, neglecting the time involved to perform the loop operations.

Let us now break this problem into two—sum over the odd series and the even series.

```
integer:: s1=0, s2=0, i, s
do i=1, 1000, 2
 s1=s1+i            ! sum over odd numbers
end do
do i=2, 1000, 2
 s2=s2+i            ! sum over even numbers
end do
s=s1+s2
```

Here, s1 is the sum of the odd series, s2 is the sum of the even series and s is the total (even+odd series). Therefore, each of these loops will take 500t units of time to complete its job. If, somehow by the combined operations of hardware and software, these two loops are handed over to two different processors (assuming identical processors for simplicity) and if these processors run simultaneously (in parallel), each will take 500t units of time and the whole process will take 501t units of time (s=s1+s2 will take one unit of time). It is thus seen that by using two processors, the execution speed can be increased by a factor of 2. The execution speed can be increased further by adding a few more processors and sharing the computing load among them. In fact, individual processors may not be very fast—they can be inexpensive. The combined effect of hundreds of such processors, when running in parallel, produces a computer system having a very high computing speed.

20.2 Coarray

In a coarray environment, a single program is replicated in the processors participating in a particular computation. These programs are called images. The images are numbered from 1 and go up to n, where n is the total number of processors. Although the same code is replicated in all the images, obviously the same path will normally not be followed in every image. For example, if the same code is replicated in two processors where one processor is entrusted to sum the odd numbers between 1 and 1000 and the other is supposed to sum the even numbers, there should be some mechanism to sum two different series using the same code. In addition, there should be special variables in the code so that these variables can be shared among the images. If the image number 1 calculates the odd series (say, s1) and the image number 2 calculates the even series (say, s2), then to get the complete series, the partial sum of one series should be made available (or visible) to the other image (s=s1+s2). These special variables are called coarray. The coarray may be a scalar or a dimensioned quantity. This extension, coarray, uses a programming model called "single program, multiple data" (SPMD). The number of images participating in a given computation may be decided during the compilation time, at the link time or at the run time. It may or may not be the number of CPUs available for computation.

20.3 Compilation of Fortran Program with Coarray

The compilation of Fortran program with coarray is performed in the following manner.

Intel ifort compiler (Windows): ifort/Qcoarray/Qcoarray-num-images=4 x.f90 (four processors)

GNU gfortran compiler (under Ubuntu 18.4): Installation

sudo apt update
sudo apt install gfortran
sudo apt install open-coarrays-bin
sudo apt install libmpich-dev
sudo apt install libopenmpi-dev
Compilation and Execution:
caf a.f90 –oa
cafrun –n 4 a [4 the number of processors]

20.4 Declaration

A coarray is declared with the codimension and is given within a square bracket.

```
integer :: a [*]
real :: b [*]
double precision :: c [*]
complex:: d [*]
character (len=10) :: e [*]
logical:: l[*]
```

The variables a, b, c, d and e are coarrays (scalar in this case). The last dimension (actually called codimension) must be of assumed size (*). When the same code is replicated across the images, the coarray without the square bracket is a variable corresponding to the same image. To access the coarray of another image, the coarray with the image number as the subscript must be used. Assuming that at this moment there are two images participating, a [1] refers to the coarray corresponding to image 1 and a [2] refers to the coarray corresponding to image 2. The image 1 can access the coarray of image 2 by writing a [2], and similarly image 2 can access the coarray of image 1 as a [1]. Note that 'a' without any subscript refers to the variable (coarray) corresponding to that particular image. If the code contains a statement, say, a=0, this statement is executed in all the images. When it is necessary to execute a selected portion of the code in a particular image, if statement with the image number as argument may be used. This will be discussed at the appropriate place.

The previous declarations may also be made in the following manner:

```
integer, codimension[*]:: a
real, codimension[*]:: b
double precision, codimension [*]:: c
complex, codimension[*]:: d
character(len=10), codimension[*]:: e
logical, codimension[*]:: l
```

On every image, a particular coarray is of the same type and has the same set of bounds.

20.5 Initialization

A coarray can be initialized either by `data` statement or along with its declaration.

```
integer, codimension[*]::  i=2
real, codimension[*]:: r=1.5
double precision, codimension[*]:: d=3.14159265d0
complex, codimension[*]:: e=cmplx(2.0,3.0)
character(len=4), codimension[*]:: c="avbd"
logical, codimension[*]:: l=.true.
```

Suppose that we are using two processors. In each of the processor, the code file will be replicated, and when the execution begins, the variables of each image (coarray) will be initialized properly. Variable `i` will be initialized to 2, and similarly other variables will be initialized with their corresponding values.

A coarray initialized with the `data` statement is as follows:

```
integer, codimension[*]:: x
data x/100/
```

This will initialize the coarray x to 100 in every image. Note that the `data` statement cannot contain the reference to any image.

Hence,

```
integer, codimension[*]:: x
data x[2]/100/                    ! not allowed
```

is not allowed.

20.6 Input and Output with Coarray

When a coarray is used, only image 1 can read from the standard input device. However, any image can write on the standard output device.

20.7 THIS_IMAGE

The intrinsic `this_image()` without any argument returns the image number of the code executing the instruction. For the first image, it is 1; for the second image, it is 2 and so on.

Let us add the following two lines to the previous program segment and execute the same with two processors:

```
print *, this_image(), i,r,d,e,c,l
end
```

The program will produce two lines of output: one from processor 1 and the other from processor 2. The processor number will be printed as the first item in each line. Note that in this case, an identical code will be executed in both the processors. The ordering of the outputs cannot be predicted—it may vary from one run to another. Let us replace the print statement as follows:

```
if (this_image()==1) then
  print *, this_image(), i,r,d,e,c,l
endif
end
```

The print statement for image 2 will be skipped as this_image()==1 is false. Note that the paths followed by image 1 and image 2 are different although the same code is being executed by both image 1 and image 2.

Consider the following program:

```
program demo_parallel
integer :: a=30,b=10,c
integer:: n                   ! executed with 4 images
n=this_image()
select case (n)
 case(1)
  c=a+b
  print *, "image no : ", n, "  a+b=c  ", a, b, c
 case(2)
  c=a-b
   print *, "image no : ", n, "  a-b=c  ", a, b, c
 case(3)
  c=a*b
   print *, "image no : ", n, "  a*b=c  ", a, b, c
 case(4)
   c=a/b
   print *, "image no : ", n, "  a/b=c  ", a, b, c
end select
end program demo_parallel
```

The outputs are as follows:

image no :	4	a/b=c	30	10	3
image no :	1	a+b=c	30	10	40
image no :	2	a-b=c	30	10	20
image no :	3	a*b=c	30	10	300

This program was executed with four images. A specific task was assigned to each of the images. This was controlled by the image number (1 to 4) and the case statement. It is not possible to predict which image will first reach the end of task. The ordering of images in the output may vary from one run to another. The compiler options (for ifort) /Qcoarray /Qcoarray-num-images=4 are used to compile this program.

20.8 NUM_IMAGES

This intrinsic returns the number of images participating in the computation process during the execution of the program. This intrinsic takes several optional arguments. For our purpose, the function is called without any optional arguments.

```
print *, num_images()
```

will display the number of processors being used during the execution of the code. If the number of processors is 25, it would display 25 (of course, 25 times unless there is an if statement like Section 20.7).

20.9 SYNC ALL

Consider the following program:

```
integer, codimension[*]:: a
integer:: tot
if(this_image()==1) then
  a=10
endif
if(this_image()==2) then
  a=20
endif
sync all                      ! synchronize
if(this_image()==1) then
  tot=a+a[2]
  print *, tot
endif
end
```

When the replicated code is being executed by the processors simultaneously, there may be (actually will be) a time difference in arriving at a particular point of the code by different images. Moreover, normally all the images are not supposed to follow the same path. Under this circumstance if the value of a particular variable (say, a[2]) of image 2 is required by another image, say, image 1, it is necessary that the computation required to calculate a[2] must be over before it is used. Therefore, there should be synchronization among different images. The sync all statement ensures that no image can proceed further until all the images reach the sync all statement. This is a barrier that can only be crossed by the images when all of them reach at that point. If any image reaches earlier than the other does, it must wait at that point until all other images reach there. In the present case when the images cross the barrier, the coarray 'a' has been set to the correct value in different images. This guarantees that when the computation of tot is performed, the calculations (in this case there is actually no calculation!) of a[1] and a[2] are over—tot is being calculated by image 1; therefore, the calculation related to a[2] must be over before it is used.

20.10 Array of Coarray

A coarray may also be an array itself.

```
integer, dimension(5), codimension[*]:: a
```

The size of the coarray in every image is the same; here, it is 5. The dimensioned coarray is accessed in the usual manner. We noted earlier that inside a particular image, the subscript of the coarray (cosubscript) is not required; a and a[1] are the same inside image 1. The first element of coarray within image 1 is referred to as a(1) or a(1)[1]. However, if image 1 wants to access the third element of the coarray 'a' belonging to image 2, it must be written as a(3)[2].

The next program illustrates this fact. We assume that there are only two processors.

```
integer, dimension(5), codimension[*]:: a
integer:: i, t, tot
t=this_image()            ! executed with 2 processors
do i=1, 5
 a(i)=t*i                 ! store some value depending on
end do                    ! the image number
sync all                  ! wait till a's are generated
if(t==1) then
 tot=0                    ! final addition will be done within image 1
 do i=1,5
  tot=tot+a(i)+ a(i)[2] ! a(i) is same as a(i)[1]
 end do
print *, tot
endif
end
```

20.11 Multidimensional Coarray

The coarray may have more than one codimension. The upper bound of the last codimension is never specified. In fact, the compiler will flag error if the upper bound of the last codimension is specified.

```
integer, codimension[2, 3, *]:: z
real, dimension(10), codimension[2, 2, *]::x
real, dimension(10):: x[2, 2, *]
real:: x(10)[2, 2, *]
```

The last three statements, shown in bold letters, are equivalent.

20.12 Upper Bound of the Last CODIMENSION

During the execution, the number of participating processors decides the actual value of the last codimension (actually upper bound). We illustrate this point with examples.

```
integer, codimension[*]::x
```

If this is executed with 25 processors, the 'value of the asterisk (*)' will be 25. If the declaration is

```
integer, codimension[0: *]::x
```

the 'value of the asterisk (*)' is 24. The coindices (in this case coindex) are mapped onto a single dimension starting from 1 goes up to the total number of processors. When the codimension is declared as

```
integer, codimension[*]::x
```

there is no difference between the coindex and the processor number. Index 1 corresponds to processor 1, index 2 corresponds to processor 2 and so on. When the codimension is defined as

```
integer, codimension[0: *]::x
```

index 0 corresponds to image 1, index 1 corresponds to image 2 and so on. Similarly, for a multidimensional coarray, the multi-codimension is mapped onto a single dimension corresponding to the image number.

```
integer, codimension[2, 3, *]:: z
```

With 25 processors, the upper bound of the last dimension is 5. The relations between the coindex and the images are given below. The first index varies faster than the second does, and the second index varies faster than the third does.

```
1, 1, 1   -> 1
2, 1, 1   -> 2
1, 2, 1   -> 3
2, 2, 1   -> 4
1, 3, 1   -> 5
   .
2, 3, 4   -> 24
1, 1 ,5   -> 25
2, 1, 5   -> 26
1, 2, 5   -> 27
2, 2, 5   -> 28
1, 3, 5   -> 29
2, 3, 5   -> 30
```

The upper bound of the last codimension of the coarray is the smallest value that is required to represent the largest processor number in a particular run. The processor numbers shown in bold are non-existent.

Note that to accommodate image number 25, the last codimension has to be, in this case, 5 during the run time. As the maximum number of images is 25, images 26, 27, 28, 29 and 30 do not have any meaning. If the lower bound is not 1 like

```
integer, codimension [-1:0, -1:1, 0:*] :: w
```

the mapping will be as follows:

```
-1, -1,  0      -> 1
 0, -1,  0      -> 2
-1,  0,  0      -> 3
 0,  0,  0      -> 4
-1,  1,  0      -> 5
 .
 0,  1,  3      -> 24
-1, -1,  4      -> 25
```

20.13 Properties of Coarray

As already mentioned, a coarray might have both the dimension and the codimension attributes.

```
real, dimension(10), codimension[*] :: x
integer, dimension(3,4,5), codimension[2,3,*] :: y
```

We discuss in detail the second declaration. Let us assume that during execution the number of images is 25; y has a rank 3 of shape [3 4 5] with lower bounds 1, 1, 1 and upper bounds 3, 4, 5. In addition, it has a corank 3 of coshape [2 3 5] with lower cobounds 1, 1, 1 and upper cobounds 2, 3, 5 (to accommodate 25 images). Note that some of the cosubcripts (like 2 1 5) are non-existent as they correspond to non-existent images (like 26). The sum of the ranks and coranks cannot exceed 15. In the present case, the sum of the ranks and coranks is 6 (3+3). Needless to say that the rank and corank can be anything but need not be the same; the only restriction is that the sum cannot exceed 15. The following is a valid declaration:

```
real, dimension(2,3,4,4,3,2,5,6), codimension[5,4,5,2,3,5,*] :: z
```

Here, the rank of z is 8 and the corank is 7.

20.14 LCOBOUND

This intrinsic takes three arguments; two of them are optional. It is identical to the intrinsic lbound (Chapter 6) except that the first argument is a coarray. The second and third arguments, dim and kind, are integers. Using the declaration of y from Section 20.13, lcobound(y) returns a rank 1 array of size 3 containing the lower bounds of the array [1 1 1].

If the argument `dim` is present, the intrinsic returns a scalar corresponding to the lower bounds for that dimension.

```
print *, lcobound(y, dim=1); print *, lcobound(y, dim=2)
print *, lcobound(y, dim=3)
```

return an identical scalar having value 1. If the third argument `kind` is used, the returned values are of kind type `kind`; if `kind` is absent, the returned values are of default `kind`.

20.15 UCOBOUND

This intrinsic takes three arguments; two of them are optional. It is identical to the intrinsic ubound (Chapter 6) except that the first argument is a coarray. The second and third arguments, `dim` and `kind`, are integers. Using the declaration of `y` from Section 20.13, ucobound(y) returns a rank 1 array of size 3 containing the upper bounds of the array [2 3 5] if the actual code is executed with 25 processors. If the argument `dim` is present, the intrinsic returns a scalar corresponding to the upper bounds for that dimension.

```
print *, ucobound(y, dim=1)
print *, ucobound(y, dim=2)
print *, ucobound(y, dim=3)
```

return scalars 2, 3 and 5 (number of processors is 25), respectively. If the third argument `kind` is used, the returned values are of kind type `kind`; if `kind` is absent, the returned values are of default `kind`.

20.16 COSHAPE

The intrinsic takes two arguments: coarray (must not be an unallocatable coarray) and an optional variable integer variable `kind`. The intrinsic returns an integer array of rank 1 of size equal to the corank of the coarray. If `kind` is present, the returned array is of type `kind`; otherwise, the returned array is of default `kind`. This intrinsic is similar to the shape intrinsic discussed in Chapter 6.

```
print *, coshape(y)
```

returns a rank 1 integer array containing [2 3 5] (defined in Section 20.13).

20.17 THIS_IMAGE with Argument

This intrinsic without any argument has already been introduced. It was noted that this intrinsic without any argument returns the image number of the image executing the intrinsic. This intrinsic may take three optional arguments: coarray, team and dim; here we discuss two: coarray and dim. this_image(coarray=y) or this_image(y) returns a rank 1 array of size equal to the codimension of y (in this case 3) containing the cosubscript executing the image. Again, assuming the number of processors participating during the execution of the program to be 25, if image 7 executes

```
print *, this_image(y)
```

the corresponding output will be [1 1 2], the cosubscript corresponding to image 7. This may be easily understood following the mapping of multidimensional coindex onto a single dimension.

If the second optional parameter dim is present, the cosubscript for that dimension of the coarray (scalar) is returned.

```
print *,this_image(y,dim=1)
print *,this_image(y,dim=2)
print *,this_image(y,dim=3)
```

display 1, 1 and 2, respectively.

The next example illustrates the use of lcobound, ucobound and this_image with argument.

```
real, codimension[2, 2, *]:: a=1.25    ! executed with 5 processors
print *, "Image No: ", this_image(),  " Lcobound: ", lcobound(a),  &
  " Ucobound: ", ucobound(a), " Coindex: ", this_image(a)
end
```

The outputs are as follows:

```
Image No: 1 Lcobound: 1 1 1 Ucobound: 2 2 2 Coindex: 1 1 1
Image No: 2 Lcobound: 1 1 1 Ucobound: 2 2 2 Coindex: 2 1 1
Image No: 3 Lcobound: 1 1 1 Ucobound: 2 2 2 Coindex: 1 2 1
Image No: 4 Lcobound: 1 1 1 Ucobound: 2 2 2 Coindex: 2 2 1
Image No: 5 Lcobound: 1 1 1 Ucobound: 2 2 2 Coindex: 1 1 2
```

As there are five processors, five lines of outputs will be generated corresponding to each processor.

20.18 IMAGE_INDEX

This intrinsic takes three arguments. Here we discuss two arguments. The first one (coarray) is a coarray, and the second one (sub) is a rank 1 integer array of size equal to the corank of the coarray. If the array sub contains valid coindices, the intrinsic returns the corresponding image index.

```
integer, dimension(3,4,5), codimension[2,3,*]:: y
print *, image_index(coarray=y, sub=[2, 3, 1])
end
```

Executing the program for 30 processors, the print statement returns 6. If the array contains invalid coindices, the intrinsic returns 0.

```
print *, image_index(coarray=y, sub=[2, 3, 6])
```

returns 0.

20.19 Synchronization

When the execution of a code containing a coarray begins, all the images start executing the code at the same time. Even when all the processors are identical, not all the images may proceed with the same speed. In addition, the execution time of different images may vary from run to run. It may sometimes be necessary to synchronize the images at a particular point before further execution of code may take place. Already we have seen the sync all statement where no image can proceed further until all the images reach the sync all statement. In this section, we consider two more synchronization statements: sync images and sync memory.

SYNC IMAGES: This is used to synchronize pairs of images. The simplest case is when image 1 is asked to wait till image 2 executes sync images(1). Similarly, image 2 waits till image 1 executes sync images(2). Actually, the image that reaches the sync images statement earlier waits for the other to reach the sync images statement. Once this is resolved, both can continue.

```
integer, codimension[*]:: x
 .
if(this_image()==1) then
  sync images(2)
else
  sync images(1)
endif
```

The image specification within the sync images may be a single-dimensional integer array.

```
sync images([2, 3, 4])
```

If the sync images(2) statement is replaced by the preceding statement, image 1 can proceed further when images 2, 3 and 4 execute the sync images(1) statement. It is obvious that the following statements may replace the preceding statement:

```
     integer, dimension(3):: z=[2, 3, 4]
     sync images(z)
```

We illustrate this with a very simple example:

```
     integer :: a[*]
     integer:: t
     t=this_image()                        !  4 processors are used
     select case(t)
      case (1)
        a[1]=100
        sync images([2, 3, 4])
!     wait till images 2, 3 and 4 execute sync images(1) statement
      case (2)
        a[2]=200
        sync images(1)
      case (3)
        a[3]=300
        sync images(1)
      case(4)
          a[4]=400
        sync images(1)
     end select
     if (t==1) then
      a[1]=a[1]+a[2]+a[3]+a[4]
      print *, a[1]
     end if
     end
```

There is another form of the sync images statement, named sync images(*). This refers to all other images except the image executing sync images(*). This is different from sync all where all the images wait for their completion. In the preceding program, the result will be same if sync images([2, 3, 4]) is replaced by sync images(*).

SYNC MEMORY: This statement is used to force the pending coarray write to take place. This also flushes the coarray read. The latest values are used before further processing can take place.

```
     integer, codimension[*]:: x=100
     integer:: t
     t=this_image()
     if(t==1) then
      x[2]=27         !  at this moment x[2] in image 2 may not be 27
      sync memory     !  this forces all preceding pending write to
                      !  take place
     endif
     end
```

The sync all statement also implies an implicit sync memory statement.

20.20 CRITICAL Section

Let us consider a simple case. We want to sum the output of this_image() of each image
and store in a coarray of image 1. All the images will use a single variable in image 1. It is
thus necessary that not all the images should access this variable simultaneously—the
addition should be performed one at a time. Here the critical section comes into play.
The codes surrounded by the critical-end critical block are not executed simultaneously by all the images; rather when one image is executing the critical block, all
other images wait till the completion of the critical block is over. Then the critical
block of another image starts. The point to be noted is that at any time only one image
can execute the codes surrounded by the critical-end critical block. It is not
permitted to enter into or to come out of the critical block using the branch statement.

```
integer, codimension[*] :: total=0
integer:: n
n=this_image()
critical
  total[1]=total[1]+n
end critical
sync all
if(n==1) then
  print *, "sum of the image numbers : ", total
end if
end
```

If the statement total[1]=total[1]+n is allowed to be executed in different images
simultaneously, it is likely that before the contribution from one of the images is stored
in total[1], it is fetched by another image, and thus, the result of computation will be
wrong. Therefore, it should be executed one at a time. The sync all statement ensures
that the contributions from all the images are stored in total[1] before image 1 can
proceed further to print the result. Of course, this is a very simplified demonstration of the
critical block. Perhaps a better and an easier way to the same task is to find the number
of the participating processors through num_images() as follows:

```
if (this_image() == 1) then
  nt=num_images()
  total=0
  do i=1, nt
    total=total+i
  end do
endif
```

The critical-end critical block can also be simulated by using sync images. No
image control statement (sync all, sync images, sync memory, lock, unlock,
allocate, deallocate, critical, end critical, end, return, block, stop
and end program) is allowed within the critical block.

```
integer, codimension[*] :: total=0
integer:: t, n
t=this_image()        ! executed with 3 processors
n=num_images()
```

```
if (t>1) then
 sync images(t-1)
endif
total[1]=total[1]+t
if(t<n) then
 sync images(t+1)
endif
sync all
if (t==1) then
 print *, total
endif
end
```

We now analyze the flow of the program in different images, assuming that there are three processors participating in the computing process. The simplified picture of the code files of the three images is as follows:

Image 1:

```
t=1
n=3
total[1]=total[1]+1
sync images(2)
 .
```

Image 2:

```
t=2
n=3
sync images(1)
total[1]=total[1]+2
sync images(3)
 .
```

Image 3:

```
t=3
n=3
sync images(2)
total[1]=total[1]+3
 .
```

At t=0, all the images begin their execution. Image 1 executes the first three lines, and then sync images(2) is executed. Image 2 after executing the first two lines waits till sync images(2) is executed by image 1. Image 3 after executing the first two statements waits till sync images(3) is executed by image 2. Note that after executing the first two lines, both images 2 and 3 wait. So when image 1 executes sync images(2), image 2 executes the instruction following sync images(1). It then executes sync images(3), which, in turn, allows image 3 to proceed further. In other words the statement following sync images(2) in image 3 is executed. It is thus clear that the original statement total[1]=total[1]+t is executed one at a time sequentially in different images according to the image number. This construction is, in effect, similar to a critical block.

20.21 ALLOCATABLE Coarray

A coarray may be an allocatable array. The upper and lower bounds must be the same in every image.

```
real, allocatable, dimension(:), codimension[:]:: y
integer, allocatable, dimension(:, :), codimension[:,:]:: z
real, allocatable, dimension(:, :, :), codimension[:,:,:]:: x
allocate(y(100)[*])
allocate(z(5, 6)[3, *])
allocate(x(3, 4, 5)[2, 3,*])
```

From the preceding examples, it should be clear that the cobounds must be present in the `allocate` statement. In addition, the last cobound must be an asterisk. There is an implicit barrier associated with the `allocate` statement. Unless `allocate` statements of all the images are executed, computation (execution of the next statement) cannot proceed. The reason is not difficult to guess. Suppose a coarray is allocated in image n and the next instruction tries to access the same coarray of image m, which has not yet been allocated. In that case, image n would be trying to access a non-existent array. For this reason, no image can proceed further until all the images finish their allocation. It is, as if, there is an implicit barrier associated with the `allocate` statement. More than one coarray may be allocated by a single `allocate` statement.

The `deallocate` statement deallocates a coarray, and like the `allocate` statement, no image can proceed further until all the images finish the same `deallocate` statement.

```
integer, allocatable, dimension(:), codimension[:]::a
allocate(a(10)[*])
.
deallocate(a)
```

will deallocate the coarray a. More than one coarray may be deallocated by a single `deallocate` statement.

If a coarray is allocated inside a procedure without the `save` attribute, the coarray is deallocated; when the procedure is exited, automatic synchronization takes place among the different images.

20.22 CO Routines

There are five intrinsics in this category: `co_max`, `co_min`, `co_sum`, `co_reduce` and `co_broadcast`. All of them have two common optional parameters. The optional parameters `stat` and `errmsg` have been discussed in detail in Chapter 15. The other optional argument of `co_max`, `co_min`, `co_sum` and `co_reduce` is `result_image=image number` (integer). If it is present, the result is returned from routines to that particular image. The variables in other images (`result_image`) become undefined. If `result_image` is absent, the result is broadcast to all the images.

In subsequent discussions, random numbers are used to fill in the variables, and as random_seed is called, the outputs will change from one run to another. In addition, four images are utilized for each run.

20.23 CO_MAX

This routine returns the element-wise (for array) maximum value of the first argument of all the images of the team. The first argument is an integer or a real or a character variable (scalar or array). It is of the same type and shape having the same type parameters in all the images of the team. The second argument is an integer that, if present, indicates the image number where the result will be returned by the subroutine; otherwise, the result is returned to all the images. This parameter must have the same value in all the images within the team. The result will be stored in the first argument. Two other optional arguments of the subroutines are stat and errmsg.

```
implicit none
real :: val
integer:: myid          ! tested with gfortran (Linux)
myid=this_image()       ! 4 images used
call random_seed()
call random_number(val)
sync all
select case (myid)
case (1)
  print *, "image No ", myid, val
case (2)
 print *,  "image No ", myid, val
case (3)
 print *,  "image No ", myid, val
 case (4)
 print *,  "image No ", myid, val
end select
call co_max(val, result_image=1)
if (myid==1) then
 print *, "The Maximum is :", val
end if
end
```

The outputs are as follows:

```
image No           2  0.179055393
image No           4  0.982329667
image No           1  0.511657298
image No           3  0.588915586
The Maximum is :  0.982329667
```

Each image will call a random number, and different values will be stored in the variable val. Subsequently, a call to co_max will return the maximum among the val variable in different images to the val variable of image 1. Readers may try to explain the results by changing the first instruction to real:: val(3).

20.24 CO_MIN

This subroutine is identical to co_max, except that it returns the minimum value.

```
implicit none
real :: val
integer:: myid
myid=this_image() ! 4 images used
call random_seed()
call random_number(val)
sync all
select case (myid)
case (1)
  print *, "image No ", myid, val
case (2)
 print *,  "image No ", myid, val
case (3)
 print *,  "image No ", myid, val
case (4)
 print *,  "image No ", myid, val
end select
call co_min(val, result_image=1)
if (myid==1) then
 print *, "The Minimum is :", val
end if
end
```

The outputs are as follows:

```
image No             3   0.416651547
image No             4   0.351018369
image No             1   0.102243423
image No             2   0.182368159
The minimum is :  0.102243423
```

20.25 CO_SUM

The first argument of this subroutine is of numeric type (integer, real or complex variable or array). This subroutine adds the first argument of all the images (or element by element if it is an array). This argument must be of the same type having the same type parameter within the team. The following program demonstrates this subroutine using a complex variable (real and imaginary parts are added separately).

```
implicit none
real :: x,y
integer:: myid
complex:: cmp
myid=this_image()
call random_seed()
```

```
call random_number(x)
call random_number(y)
cmp=cmplx(x,y)          ! generate the complex number
sync all
select case (myid)
case (1)
  print *, "image No ", myid, cmp
case (2)
  print *, "image No ", myid, cmp
case (3)
  print *, "image No ", myid, cmp
case (4)
  print *, "image No ", myid, cmp
end select
call co_sum(cmp, result_image=1) ! adds the real and imaginary part
if (myid==1) then       ! separately. result is available in image 1
  print *, "The sum is :", cmp
end if
end
```

The outputs are as follows:

```
image No            2       (0.134604454,6.049692631E-02)
image No            4       (7.629013062E-02,3.661173582E-02)
image No            1       (0.825442970,2.105647326E-02)
image No            3       (0.641139805,0.919207215)
The sum is :                (1.67747736,1.03737235)
```

For complex arguments, the real and imaginary parts are separately added.

20.26 CO_REDUCE

This routine reduces the values of the first argument (' a ' is a variable or array of the same type and shape in all the images) according to the pure function passed as an operator to the routine to pair-wise reduce the values of the first argument. The first argument must not be polymorphic. If it is allocatable, it should be allocated; if it is a pointer, it should be in an associated state. Either the variable ' a ' from different images or the result of previous reduction is passed to the function. This is demonstrated with the help of integer arrays. The reduction in this case involves multiplication of the variables passed to the function; it takes place element by element of the array.

```
implicit none
integer:: myid
integer, dimension(3)::intval
real, dimension(3)::val
myid=this_image()
call random_seed()
call random_number(val)
intval(1)=val(1)*20.0          ! truncated to integer after
                               ! multiplication
intval(2)=val(2)*22.0
```

```
intval(3)=val(3)*24.0
sync all
select case (myid)
  case (1)
    print *, "image No ", myid, "intval =",intval
  case (2)
    print *, "image No ", myid, "intval =",intval
  case (3)
    print *, "image No ", myid, "intval =",intval
  case (4)
    print *, "image No ", myid, "intval =",intval
end select
call co_reduce(intval, result_image=1, operator=ownprod)
if (myid==1) then
  print *, "The product - Element-wise are:", intval
end if
contains
  pure integer function ownprod(a,b)
  integer, value:: a,b
  ownprod=a*b
  end function ownprod
end
```

The outputs are as follows:

```
image No           2 intval =          15          2          11
image No           4 intval =           8         14           6
image No           1 intval =          16         13          15
image No           3 intval =          11          8           4
The product - Element-wise are:      21120       2912        3960
```

20.27 CO_BROADCAST

This subroutine copies the value of the first argument (a) from the image specified by the second argument (source_image) to all other images. The first argument should be of the same dynamic type and type parameters in all the images of the team. If it is an array, it must have the same shape in all the images of the team. In the following example, first the val array is filled with some random numbers. Subsequently these are copied onto the val variable of the other images through the routine co_broadcast.

```
implicit none
integer:: myid
real, dimension(3)::val
myid=this_image()
if (myid==1) then
  call random_seed()
  call random_number(val)
  val(1)=val(1)*105.0 ! val's of the first images are filled with random
  val(2)=val(2)*217.0 ! numbers multiplied by some arbitrary numbers and
  val(3)=val(3)*305.0
```

```
      endif
      sync all
      call co_broadcast(val, source_image=1)
      print *, "Image no: ", myid, "val :",val
      end
```

The outputs are as follows:

```
Image no:          1 val :   43.3511543     213.145920      265.393494
Image no:          3 val :   43.3511543     213.145920      265.393494
Image no:          2 val :   43.3511543     213.145920      265.393494
Image no:          4 val :   43.3511543     213.145920      265.393494
```

20.28 Coarray and Subprogram

The dummy argument of a subprogram may be a coarray. In that case, the corresponding actual argument must be a coarray.

1. Coarray as argument:

```
program cosub1
interface
  subroutine sub(a,b,c)
  integer, intent(in):: a[*], b[*]
  integer, intent(out):: c[*]
  end subroutine sub
end interface
integer:: x[*], y[*], z[*]
integer:: i
x=100
y=200
call sub(x,y,z)
sync all
if(this_image()==1) then
  do i=1, num_images()
    print *, x[i], y[i], z[i]
  end do
endif
end program cosub1
subroutine sub(a,b,c)
integer, intent(in):: a[*], b[*]
integer, intent(out):: c[*]
integer:: n
n=this_image()
c=n*(a+b)
return
end
```

2. Array of coarray as argument (assumed shape):

```
program cosub2
 interface
  subroutine sub(a,b,c)
  integer,intent(in),dimension(:),codimension[*]::a,b
  integer,intent(out),dimension(:),codimension[*]::c
  end subroutine sub
 end interface
 integer,dimension(5),codimension[*]::a,b,c
 integer::t,i
 t=this_image()
 do i=1,5
  a(i)=t*(t+1)+i
  b(i)=t*(t+2)+i
 end do
 call sub(a,b,c)
 sync all
 if(this_image()==1) then
  print *, a; print *,b; print *,c
 endif
 end
subroutine sub(a,b,c)
 integer, intent(in),dimension(:),codimension[*]::a,b
 integer, intent(out),dimension(:),codimension[*]::c
 c=a+b
end subroutine sub
```

3. Array of coarray as argument (assumed size):

```
program cosub3
 interface
  subroutine sub(n,m,a)
  integer,intent(in)::n,m
  integer,intent(out),dimension(n,*),codimension[*]::a
  end subroutine sub
 end interface
 integer,dimension(5,5),codimension[*]::a
 integer::n=5,m
 m=size(a)/n
 call sub(n,m,a)
 if (this_image()==2) then! Only output of image 2 will be displayed
  print *, a
 endif
 end
subroutine sub(n,m,a)
  integer,intent(in)::n,m
  integer,intent(out),dimension(n,*),codimension[*]::a
  integer::i,j,t
  t=this_image()
  do i=1,n
   do j=1,m
```

```
   a(i,j)=i*t+j*t
    end do
  enddo
  end subroutine sub
```

4. Array of coarray as argument (allocatable):

```
program cosub4
interface
  subroutine sub(t, a)
  integer,intent(in),allocatable, dimension(:),codimension[:]::a
  integer, intent(in)::t
  end subroutine sub
end interface
integer, allocatable, dimension(:),codimension[:]::x
integer::i, t
allocate(x(5)[*])
t=this_image()
do i=1, 5
 x(i)=t*(t+1)+i
enddo
call sub(t,x)
end
subroutine sub(t,a)
integer, intent(in), allocatable, dimension(:),codimension[:]::a
integer, intent(in)::t
print *, t, a
end subroutine sub
```

If the dummy argument is an allocatable coarray, the corresponding actual argument must be an allocatable coarray having the same rank and corank. Moreover, the intent of the allocatable coarray must be in or inout.

5. Multidimensioned coarray as argument:

```
program cosub5
 interface
   subroutine sub(n,a)
   integer,intent(in)::n
   integer,intent(out),dimension(n),codimension[n,*]::a
   end subroutine sub
 end interface
 integer,dimension(3),codimension[3,*]::a
 integer::n=3
 call sub(n,a)
 print *, a(this_image())
 end
subroutine sub(n,a)
   integer,intent(in)::n
   integer,intent(out),dimension(n),codimension[n,*]::a
   integer::i,j,t
   t=this_image()
   a(t)=this_image(a,dim=1)+this_image(a,dim=2)
end subroutine sub
```

An automatic array cannot be a coarray. The following will give rise to a Fortran error:

```
subroutine sub(n)
integer:: n
integer, dimension(n), codimension[*]::a
```

6. Coarray and module:

Two examples of using coarray and module are shown.

```
module m
  integer:: x[*]
end module m
use m
x=this_image()*100
print *, x
end
```

The next example uses a subroutine with coarray as arguments:

```
module m
  interface
    subroutine sub(x,y,z)
    integer,intent(in)::x[*],y[*]
    integer,intent(out)::z[*]
    end subroutine sub
  end interface
end module m
use m
integer:: a[*],b[*],c[*]
integer:: t
t=this_image()
a=100*t
b=200*t
call sub(a,b,c)
print *, a,b,c
end
subroutine sub(x,y,z)
integer,intent(in)::x[*],y[*]
integer,intent(out)::z[*]
z=x+y
end subroutine sub
```

20.29 Coarray and Function

A function name (or if there is a `result` clause) cannot return a coarray.

20.30 Coarray and Floating Point Status

In the coarray environment, each image will maintain its own floating point status.

20.31 User Defined Type and Coarray

1. User-defined type as coarray:

```
type:: mytype
 integer::a
 integer::b
end type mytype
integer:: t
type (mytype):: x[*]
t=this_image()
x%a=100*t
x%b=200*t
print *, t,x%a,x%b
end
```

2. User-defined type as argument to a subroutine:

```
type:: mytype
 sequence
 integer::a
 integer::b
end type mytype
type(mytype):: y[*]
interface                ! interface is required
 subroutine sub(x)
 type:: mytype
 sequence
 integer:: a
 integer:: b
 end type mytype
 type(mytype):: x[*]
 end subroutine sub
end interface
integer::t
t=this_image()
call sub(y)
print *, t,y%a, y%b
end
subroutine sub(x)
type:: mytype
 sequence
 integer:: a
 integer:: b
```

```
end type mytype
type(mytype):: x[*]
integer:: t
t=this_image ()
x%a=100*t
x%b=200*t
end subroutine sub
```

When a user-defined type is passed as an argument, sequence must be present in the type definition.

3. User-defined type, module and coarray:

```
module m
 type:: mytype
  sequence
  integer:: a
  integer:: b
 end type mytype
end module m
program test
use m
type(mytype)::y[*]
integer:: t
interface
 subroutine sub(x)
  type:: mytype
  sequence
  integer:: a
  integer:: b
  end type mytype
  type(mytype):: x[*]
 end subroutine sub
end interface
t=this_image()
call sub(y)
print *, t,y%a, y%b
end
subroutine sub(x)
 type:: mytype
  sequence
  integer:: a
  integer::b
end type mytype
type(mytype):: x[*]
integer:: t
t=this_image()
x%a=100*t
x%b=200*t
end subroutine sub
```

4. User-defined type, coarray and internal procedure:

```
type mytype
 integer:: a
 integer::b
```

```
 end type mytype              !  contains internal procedure
 type(mytype):: y[*]
 integer::t
 t=this_image()
 call sub(y)
 print *, t, y%a, y%b
contains
 subroutine sub(x)
  type(mytype):: x[*]
  integer:: t
  t=this_image()
  x%a=100*t
  x%b=200*t
  end subroutine sub
  end
```

5. Use of interface using the `import` statement:

```
type:: mytype
 sequence
 integer:: a
 integer::b
end type mytype
interface
 subroutine sub(x)
 import mytype              ! discussed in chapter 19
 type(mytype):: x[*]
 end subroutine sub
end interface
type(mytype):: y[*]
integer::t
t=this_image()
call sub(y)
print *, t,y%a, y%b
end
subroutine sub(x)
type:: mytype
 sequence
 integer::a
 integer::b
end type mytype
type(mytype):: x[*]
integer:: t
t=this_image()
x%a=100*t
x%b=200*t
end subroutine sub
```

6. User-defined type with allocatable component (scalar):

```
type mytype
 integer, allocatable::a[:]
end type mytype
integer::t
type(mytype):: x
allocate(x%a[*])
```

```
t=this_image()
x%a=t**3
print *, this_image(),x%a
deallocate(x%a)
end
```

7. User-defined type with allocatable component (dimensioned quantity):

```
type mytype
  integer, dimension(:), allocatable::a[:]
end type mytype
integer::t
type(mytype):: x
allocate(x%a(1)[*])
t=this_image()
x%a(1)=t**3
print *, this_image(), x%a(1)
deallocate(x%a)
end
```

8. User-defined type with allocatable component (multidimensioned quantity):

```
type mytype
  integer, dimension(:,:), allocatable::a[:,:,:]
  end type mytype
  integer::t
  type(mytype):: x
  allocate(x%a(3,4)[2,3,*])
  t=this_image()
  x%a(1,1)=t*this_image(x%a,dim=1)* this_image(x%a,dim=2)* &
      this_image(x%a,dim=3)
  print *,t, x%a(1,1)
  deallocate(x%a)
end
```

Readers may verify if the program is executed with six processors (with six processors the codimension would be [2, 3, 1]). The outputs are as follows:

```
1    1
2    4
3    6
4    16
5    15
6    36
```

9. User-defined type with structure inside:

```
type m1
  integer, allocatable, dimension(:), codimension[:]::dd
end type m1
type m2
  type(m1)::d
end type m2
```

```
integer::t
type(m2)::day
t=this_image()
allocate(day%d%dd(1)[*])
day%d%dd(1)=t**3
print *, day%d%dd(1)
deallocate(day%d%dd)
end
```

If any object of a user-defined type is a coarray, its parent (ancestor) cannot be allocatable/pointer/coarray. Moreover, it must be a scalar.

Program segments of this section and Section 20.28 are for demonstration purposes only. These program segments show how one can use a coarray in various situations. In real life, the programming logic is much more complicated.

20.32 COARRAY and POINTER

A coarray cannot be a pointer (also `c_ptr` or `c_funptr`). However, a pointer may be a component or allocatable component of a user-defined type coarray. In this case, the target must be local.

```
type mytype
 integer, pointer:: p
end type mytype
type(mytype)::x[*]
integer, target::y=8    ! local
y=y*this_image()
x%p=>y; print *, x%p    ! prints 8, 16, 24, 32
end
```

The user-defined type coarray may be an allocatable variable:

```
type mytype
  integer, pointer::x
end type mytype
type(mytype), allocatable::y[:]
integer,target::z
allocate(y[*])
z=10*this_image()
y%x=>z; print *, y%x ! executed with 2 processors, output: 10 20
end
```

The pointer itself may be allocatable:

```
type mytype
 integer, pointer::x(:)
end type mytype
type(mytype),allocatable::y[:]
integer, target::z
allocate(y[*])
```

```
allocate(y%x(5))
z=10*this_image()
y%x(1)=z
print *, y%x(1); deallocate(y%x); deallocate(y)
end
```

This example shows the method of accessing the coarray using a local pointer:

```
type mytype
 integer, pointer::x
end type mytype
type(mytype), allocatable::y[:]
integer, target::z
integer, pointer::ptr        ! use of local pointer
allocate(y[*]); z=10*this_image(); y%x=>z
ptr=>y%x
print *, ptr
end
```

20.33 Operator Overloading and Coarray

We again use our old example of adding English distance (Chapter 12). The addition operator (+) will be overloaded. We assume that there are two processors.

```
program distance
 type dist
  sequence
  integer :: mile !  add english distance with coarray
  integer :: yds
end type dist
interface operator (+)
 function adddist(d1,d2) result(res)
 import dist
 type(dist), intent(in)::d1,d2
 type(dist)::res
 end function adddist
end interface
type(dist):: d1[*],d2[*],d3[*]
integer::t,i,n, m1,y1,m2,y2
t=this_image()
n=num_images()
if(t==1) then
 do i= 1,n ! the input depends upon on the number of images
  print *, "input? "
 read *, m1,y1,m2,y2
 d1[i]%mile=m1; d1[i]%yds=y1; d2[i]%mile=m2; d2[i]%yds=y2
 !  read elementary items of d1 and d2 (for each image)
 !  data validation is not done for simplicity, yds must be less than 1760
 enddo
```

```
  endif
  sync all        ! wait till inputs are over
  d3=d1+d2    ! + operator has been overloaded
  sync all
  print *, "Image No : ", t," d1.miles = ", d1%mile, " d1.yds = ",d1%yds
  print *, "Image No : ", t," d2.miles = ", d2%mile, " d2.yds = ",d2%yds
  print *, "Image No : ", t," d3.miles = ", d3%mile, " d3.yds = ",d3%yds
  end
function adddist(d1,d2) result(res)
  type dist
    sequence
    integer :: mile
    integer :: yds
  end type dist
type(dist), intent(in)::d1,d2
type(dist)::res,d3
d3%mile=0
d3%yds=d1%yds+d2%yds
if (d3%yds >=1760) then  ! 1760 yds = 1 mile
  d3%mile=d3%mile+1
  d3%yds=d3%yds-1760
endif
d3%mile=d3%mile+d1%mile+d2%mile
res=d3   ! d1+d2 is returned
end function adddist
```

The read statement will read data for each image. Other images wait until the reading of data for all the images is over. Subsequently, each image will add two English distances.

20.34 Atomic Variables and Subroutines

Fortran allows a special type of variables called atomic variables, which may be modified only through special atomic subroutines. The variable must be a scalar integer or logical coarray or coindexed object of either atomic_int_kind or atomic_logical_kind having intent inout. These atomic kinds are defined in the module iso_fortran_env. Naturally, to use atomic variable, this module must be used.

There are 11 atomic subroutines in Fortran. These subroutines act on the atomic variable instantaneously—as if exclusive access is given to the atomic variable. In other words, atomic action never overlaps with other atomic actions that may take place asynchronously.

In the subsequent descriptions of the atomic subroutines, atom stands for an atomic variable of either integer of atomic_int_kind or logical atomic_logical_kind, and stat, an integer, having intent inout returns the status of the operations. It returns a processor-dependent positive number if the subroutine encounters any error; otherwise, it returns 0—stat is an optional argument. If stat is absent and an error occurs, the job is terminated. In all subsequent discussions related to atomic subroutines, this optional parameter would be omitted.

20.35 ATOMIC_DEFINE (ATOM, VALUE)

`Atom`: Integer, scalar coarray or coindexed object

`Value`: Integer scalar, same type as `atom`, intent `in`

Result: `atom=value`

Example: `call atomic_define (i[2], 10)`

The variable `i` of image 2 becomes `10` after the atomic operation.

20.36 ATOMIC_REF (VALUE, ATOM)

`Atom`: Integer, scalar coarray or coindexed object

`Value`: Integer scalar, same type and kind as `atom`, intent `in`

Result: `value=atom`

Example: `call_ref (value, i[2])`

If the variable `i` of image 2 is `13`, `value` is set to `13` after the atomic operation.

20.37 ATOMIC_ADD (ATOM, VALUE)

`Atom`: Integer, scalar coarray or coindexed object

`Value`: Integer scalar, intent `in`

Result: `atom+value`

Example: `call atomic_add (i[2], 10)`

If the variable `i` of image 2 is `20`, it becomes `30 (20+10)` after the atomic operation.

20.38 ATOMIC_FETCH_ADD (ATOM, VALUE, OLD)

`Atom`: Integer, scalar coarray or coindexed object

`value`: Integer scalar, intent `in`

Result: `old=atom, atom=atom+value`

Example: `call atomic_fetch_add(i[2], 10, old)`

If the variable `i` of image 2 is `20`, it becomes `30 (20+10)` and `old` becomes `20` after the atomic operation.

20.39 ATOMIC_AND (ATOM, VALUE)

`Atom:` Integer, scalar coarray or coindexed object

`Value:` Integer scalar, intent in

Result: `iand(atom,value)`

Example: `call atomic_and (i[2], 5)`

If the variable i of image 2 is 6, it becomes 4 after the atomic operation.

20.40 ATOMIC_FETCH_AND (ATOM, VALUE, OLD)

`Atom:` Integer, scalar coarray or coindexed object

`Value:` Integer scalar, intent in

`Old:` Scalar, same kind and type as `atom`, intent out

Result: `old=atom, iand(atom,value)`

Example: `call atomic_fetch_and(i[2], 5, old)`

If the variable i of image 2 is 6, it becomes 4 and `old=6` after the atomic operation.

20.41 ATOMIC_OR (ATOM, VALUE)

`Atom:` Integer, scalar coarray or coindexed object

`Value:` Integer scalar, intent in

Result: `ior(atom,value)`

Example: `call atomic_or (i[2], 5)`

If the variable i of image 2 is 6, it becomes 7 after the atomic operation.

20.42 ATOMIC_FETCH_OR (ATOM, VALUE, OLD)

`Atom:` Integer, scalar coarray or coindexed object

`Value:` Integer scalar, intent in

`Old:` Scalar, same kind and type as `atom`, intent out

Result: `old=atom, ior(atom,value)`

Example: `call atomic_fetch_and(i[2], 5, old)`

If the variable i of image 2 is 6, it becomes 7 and `old=6` after the atomic operation.

20.43 ATOMIC_XOR (ATOM, VALUE)

Atom: Integer, scalar coarray or coindexed object

Value: Integer scalar, intent in

Result: ieor(atom,value)

Example: call atomic_xor(i[2], 1)

If the variable i of image 2 is 3, it becomes 2 after the atomic operation.

20.44 ATOMIC_FETCH_XOR (ATOM, VALUE, OLD)

Atom: Integer, scalar coarray or coindexed object

Value: Integer scalar, intent in

Old: Scalar, same kind and type as atom, intent out

Result: old=atom, ieor(atom,value)

Example: call atomic_fetch_xor(i[2], 1, old)

If the variable i of image 2 is 3, it becomes 2 and old=3 after the atomic operation.

20.45 ATOMIC_CAS (ATOM, OLD, COMPARE, NEW)

Atom: Integer, scalar coarray or coindexed object

Value: Scalar, same type and kind as atom, intent inout

Old: Scalar, same type and kind as atom, intent out

Compare: Scalar, same type and kind as atom, intent in

New: Scalar, same type and kind as atom, intent in

Result: old=current value of atom. If atom is an integer and equal to compare or a logical and equivalent to compare, it takes the value new; otherwise, it retains its current value.

Example: call atomic_cas(i[2], old, 5, 7)

If the variable i of image 2 is 10, it remains 10 and old becomes 10 after the atomic operation. However, if i[2] is 12, call (i[2], old, 12, 24) will set i[2] to 24 and old to 12.

The following program shows how atomic subroutines can be used:

```
use iso_fortran_env
logical(atomic_logical_kind):: x[*]=.true.
logical :: l
integer:: t
t=this_image()
if(t==1) then
 sync memory
 call atomic_define(x[2], .false.)
else if(t==2) then
   l=.true.
   do while(l)
     .
    call atomic_ref(l,x)
   enddo
   sync memory
end if
end
```

The do loop of image 2 continues until image 1 becomes false, and in each cycle the value of image 1 is refreshed through the atomic subroutine.

20.46 LOCK and UNLOCK

The lock statement is used to lock a coarray or a sub-object of a coarray. The concerned variable is of derived type lock_type defined in iso_fortran_env.

```
use iso_fortran _env
type(lock_type):: lck[*]
```

It can exist in two states: lock and unlock, with unlock being the default value. The states may be changed with lock and unlock statements.

```
lock(lck)
unlock(lck)
```

If an image has executed a lock statement, only the same image can unlock the lock variable. If an image, say, I1, tries to lock a lock variable and if it is locked by another image, say, I2, image I1 waits till the lock variable is unlocked by image I2.

There is another version of lock, which does not cause an image to wait if it is locked by another image.

```
logical:: sta
lock(lck, acquired_lock=sta)
```

The variable `sta` is set to `true` if `lck` is not locked (unlocked) when the `lock` statement is executed, but it returns `false` if the variable is already locked and the program continues (i.e., the concerned image does not wait). It may be noted that the `critical-end critical` block is similar to the `lock-unlock` pair with its own `lock` variable.

20.47 Status Specifiers

In Chapter 15, we discussed the status specifiers `stat=` and `errmsg=` in detail in connection with `allocate` and `deallocate` statements. Statements `syncall`, `sync images`, `sync memory`, `lock` and `unlock` may also use `stat` and `errmsg` specifiers.

20.48 ERROR STOP

Execution of the `error stop` statement causes all the images to be terminated abnormally as soon as possible. The statement `error stop` may take an optional integer or character as stop code. This is, if present, displayed on the screen.

20.49 Coarray and Interoperability

As C programming language does not have the concept of coarray, the interoperability between the C program and the coarray is not possible.

20.50 COMMON, EQUIVALENCE and Coarray

A coarray cannot be used in `common` and `equivalence` statements.

20.51 VOLATILE Variable

If a variable is expected to be modified by another program (not necessarily a Fortran program) being executed in parallel, it is to be declared as `volatile`. In other words, the variable is available outside the scope of the present program.

```
     integer, volatile :: a
or   integer :: a
     volatile :: a
```

For a pointer this refers to its association with the target and not with the content of the target. If an object is declared as volatile, all its sub-objects become volatile too.

As the volatile variable can be changed by other program(s), every reference to the variable should fetch its value from the main memory, rather than obtaining from the registers that may not contain the current value.

Consider a program waiting for a signal from another program (say, from an instrument). So long as the signal is absent, the present program is in an infinite loop. When the signal comes (actually the instrument somehow can access a variable of this program) from the instrument, the instrument modifies a variable and the program proceeds.

```
integer, volatile::signal
signal=0
do while (signal .eq. 0)      ! infinite loop
.
end do                        ! can come out when the signal becomes nonzero
.
end
```

This is an infinite loop. The control can come out of the loop when the signal becomes nonzero. The value of the variable signal is changed by some other program that has access to this variable.

20.52 EVENT

Using this facility one image can suspend the execution of another image temporarily. The second image waits until a suitable signal from the other images is received. This is another way of synchronizing different images. To use this facility, the module iso_fortran_env must be included in the program (use iso_fortran_env). The module contains a user-defined type event_type:

```
type (event_type) :: ev[*]
```

ev, event_variable, is a scalar coarray; all its components are private. The variable event_variable includes a variable event count. This variable has a value 0 at the beginning. This variable is incremented by 1 when an event post statement is executed for a particular image, and it is decremented by 1 when an event wait statement is executed for this image.

20.53 EVENT POST

This statement takes one argument of type event_type, for example, event post (ev[2]). This increases the event counter corresponding to image 2 by 1.

20.54 EVENT WAIT

An image may wait using an event wait statement till an appropriate signal is received from other images. The syntax is event wait (event variable, until_count=scalar expression, stat, errmsg). The last three arguments are optional. The event variable must not be coindexed. If until_count is absent, it is assumed to be 1. This value is called the threshold. Until the threshold is reached (or greater than the threshold), the image waits at the event wait statement. Execution of the corresponding image resumes as soon as the threshold is reached. The counter is decremented atomically.

```
use iso_fortran_env
integer:: a[*]=10
integer:: count
type(event_type):: ev[*]
if(this_image()==1) then
  print *, "Image 1"
  print *, "Input ?"
  read *, a
  event post (ev[2])
else if (this_image()==2) then
  print *, "i am here"
  event wait(ev)
  print *, a[1]**2
  end if
end
```

The outputs on the screen are as follows:

```
Image 1
I am here
Input ?
20 [input given]
    400
```

Let us try to understand the output. The first line comes from image 1. The second line comes from image 2. Image 1 now waits for the input, and image 2 waits for an event post from image 1. The event post command is executed by image 1 when the input is given (in this case it is 20). Having received the event post with ev[2] from image 1, now image 2 can proceed, and it calculates and prints the square of the input (in this case it is $20^2=400$).

The starting value of the event variable is 0. Each time an event post is executed, it is incremented by 1.

```
    use iso_fortran_env
    integer:: count
    type(event_type):: ev[*]
    if(this_image()==1) then
      print *, "i am inside image 1" ! executed with 3 processors
      event post (ev[2])
!     event post(ev[2])
      event post(ev[3])
    else if (this_image()==2) then
```

```
     event wait(ev, until_count=2)
     print *, "I am inside image 2"
   else if (this_image()==3) then
     event wait(ev)
     print *, "I am inside image 3"
   end if
   end
```

In the preceding program, the output will be I am inside image 1 in the first line and I am inside image 3 in the second line. The program then waits forever because image 2 has received only one event post. To continue processing, image 2 needs two event post signals because of until_count=2. If the comment is removed from the event post, the job will be terminated normally with three lines of output, I am inside image 1, I am inside image 2 and I am inside image 3. It may be noted that the event post to image 2 may come from anywhere. The moot point is that image 2 must receive two event post instructions so that it can go beyond the event wait statement.

20.55 EVENT_QUERY

This subroutine takes three arguments: event, count and stat. The third one is optional. The first argument is of type event_type, and the second argument is of type integer. The output parameter count returns the number posted to ev but is not removed by the event wait.

```
   use iso_fortran_env
   integer:: count, dummy
   type(event_type):: ev[*]
   if(this_image()==1) then
      read *, dummy      ! wait so that image 2 can get time to send
      call event_query(ev,count)
      print *, "Count = ", count
   else if (this_image()==2) then
         event post(ev[1]); event post(ev[1])
   end if
   end
```

The output of the preceding program is Count = 2, as two events have been posted and there are no corresponding event wait.

20.56 Programming Examples Using Coarray

In this example, a function will be integrated with coarray using trapezoidal rule. Integration is essentially addition of values of the function at different points. The program is quite general. It is independent of the number of processors being used. It can be used to perform calculation using any number of processors. To make our life simple, only two processors will be used. The logic is straightforward, and the do loop shown in bold in the program essentially distributes the task among the processors. In the present case, since nt=2, the do loop index for image 1 will be 1, 3, 5, ..., and for image 2 it will be 2, 4, 6, ... The partial sum is stored in the coarray sum. The final computation, that is, the addition of these partial sums, is performed in image 1. The result of this integration is known to be the value of π.

```fortran
program trap
implicit none
double precision, codimension[*]::sum   ! sum is a scalar
double precision:: pi,h,x,lima,limb
integer::n,myid,i,nt
lima=0.0d0; limb=1.0d0; sum=0.0d0; n=100000
h=(limb-lima)/n
myid=this_image()  !  id of the image executing this line
nt=num_images()    !  total number of images, given at the run time
if(myid==1) then
   sum=h*(f(lima)+f(limb))*0.5d0 !first and last points
endif
do i=myid,n-1,nt  !  this will control the loop in different images
 x=lima+i*h
 sum=sum+f(x)
enddo
sum=sum*h
sync all
if (myid==1) then
 pi=sum  !  contribution from the first image, it is the same as sum[1]
 if(nt>1) then
  do i=2,nt
   pi=pi+sum[i]   !  contribution from other images if any
  enddo
 endif
endif
if(myid==1) then
 print *,"The value of pi is = ",pi !  print once from the first image
endif
contains
 double precision function f(y)
 double precision:: y
 f=4.0d0/(1.0d0+y*y)
 end function f
end program trap
```

Careful examination of the do statement, shown in bold, reveals that when the number of processors is, say, 4, image 1 will perform the calculation for the do loop index 1, 5, 9, ..., and image 2 will perform the calculation for the do loop index 2, 6, 10, ... Similarly, images 3 and 4 will perform calculations for the do loop index 3, 7, 11, ... and 4, 8, 12, ..., respectively.

The next program uses a similar argument and calculates the value of π using the expression:

$\pi = 4\sum$ (-1)**(n-1)/(2n-1), the sum is 1 to infinity

```
program pi_calc
implicit none
double precision:: pi[*]
integer:: i, limit=100000, im, nm
integer, parameter:: dk=kind(0.0d0)
im=this_image()
nm=num_images()
do i = im, limit, nm
  pi = pi + (-1)**(i-1) / real( 2*i-1, kind=dk )
end do
sync all ! wait, barrier
if (im .eq. 1) then
 do i = 2, nm
 pi = pi + pi[i]
 end do
pi = pi * 4.0_dk
print *, 'pi = ',pi
end if
end program pi_calc
```

The output is as follows:

```
The result is 3.1415…
```

The next program calculates maximum and minimum values using coarray

```
program find_max
implicit none
real:: imgmax[*], imgmin[*]
integer:: noimages
real, dimension(100)::x
!    call random_init(.false., .true.) not yet supported
call random_seed
call random_number(x) ! 100 random number for each image
imgmax=maxval(x) ! maximum value for a particular image
imgmin=minval(x) ! minimum value for a particular image
sync all ! wait till all the images finish their calculation
if(this_image()==1) then
 do noimages=2, num_images()
   immax=max(imgmax,imgmax[noimages]) ! global maximum and minimum
   imgmin=min(imgmin,imgmin[noimages])  !are calculated
 enddo
print *, "The largest number : ", imgmax
print *, "The smallest number : ", imgmin
endif
end program find_max
```

The final program calculates the value of π using monte_carlo method

```
program pi_monte_carlo
implicit none
double precision:: x, y ! algorithm discussed in Section 22.50
```

```
double precision:: tc[*]=0, ts[*]=0
integer(kind=selected_int_kind(18)):: limit=10000000, i
call random_seed
if (this_image()==1) then
  print *,'The number of images for this run: ', num_images()
endif
do i=1, limit
call random_number(x); call random_number(y)
if (x*x+y*y <=1.0) then
 tc=tc+1.0d0; ts=ts+1.0d0
else
 ts=ts+1.0d0
endif
enddo
sync all
if (this_image()==1) then
   do i=2, num_images()
     tc=tc+tc[i]; ts=ts+ts[i]
   end do
   print *, 'The value of pi is : ',4.0d0*tc/ts
endif
end program pi_monte_carlo
```

21

Parallel Programming Using OpenMP

OpenMP stands for open multiprocessing. At the outset, it may be pointed out that Openmp (OpenMP) is neither a programming language nor a part of Fortran 2018 standard. It is actually composed of a few compiler directives to the Fortran or C/C++ compilers associated with runtime library procedures and environment variables to make a portion of the program parallel. Openmp gives full control to the programmer to parallelize a portion of his or her code. It does not create a parallel region on its own. It is the programmer who identifies a region suitable for parallelization and uses the Openmp directive to make the region parallel to gain execution speed.

Three compilers were used to test most of the programs described in this chapter. They are Intel Fortran compiler, ifort (v 19), gfortran (v 7) and NAG compiler (v 6.2). The current version of Openmp is 5.0. This chapter contains a few essential directives, runtime routines and environment variables related to Openmp. Interested readers may consult the Openmp 5.0 manual for detailed discussions on all the directives, runtime routines and environment variables.

21.1 Thread

In this chapter, we frequently use the word thread. First, let us define the term thread. According to the Openmp report 5.0, a thread is an execution entity with a stack and associated static memory called thread private. Stated in simple language, a thread is a runtime entity containing one or more instructions that can be executed independently. The Openmp directive may create multiple threads depending on the environment, and these threads can run in parallel. Usually, the number of threads created by default is the number of processor/cores available in the system. However, the application may increase or decrease the number of threads within the parallel environment by the appropriate statement/declaration. The system administrator may set an upper limit to the number of threads that a particular user may use. The primary condition is that the threads are independent, and generally execution of the threads can be performed in any order and theoretically the final result is normally independent of the order of the execution of threads.

21.2 Structured Block

A structured block is defined as a block of executable statement having a single entry point at the beginning and a single exit point at the end. The end may contain an Openmp construct also.

21.3 Parallelism

Normally, a program executes instructions sequentially. When a serial program enters the parallel region, the main part, called the master thread, creates additional threads depending upon the environment. All these threads run in parallel and when all the threads finish their work, the slave threads disappear and the master thread continues to run normally in sequential fashion. Usually, there is an implicit barrier at the end of the parallel segment, which forces all of them to finish their work. In other words, if a particular thread finishes its task ahead of all other threads, this thread has to wait until all other threads finish their work. The model is known as Fork-Join model.

21.4 Memory Management

All threads may share the same memory locations, and changes made by one thread are visible, that is, accessible by other threads. The threads may have a number of local (private) variables. Only the concerned thread can access these local memories. The local memories of one thread are not visible from another thread. Moreover, local memories are usually given back to the system once the parallel region ends. These local memories are not available outside the parallel region. As all threads share common memory, this concept is known as symmetric multiprocessing (SMP) architecture.

21.5 Application Program Interface (API)

The application program interface for Openmp has three components:

- Compiler directives
- Runtime library
- Environment variables

These are discussed in detail in appropriate sections.

The interfaces for the runtime library and some constants are available within module omp_lib. Therefore, this module will be used in all our programs that use Openmp.

21.6 Compiler Support

Compiler support for Openmp is available for Fortran and C/C++ compilers. As this book is on Fortran 2018, Openmp related to only Fortran will be discussed in this chapter. Moreover, all compiler directives will assume that the source program is in free form, as fixed form is normally not used nowadays. However, a fixed form source program can use Openmp directives by following the rules of fixed form. In addition, it may be noted that all features of Openmp 5.0 are not yet supported by all Fortran compilers. Moreover, as this chapter is a gentle introduction to Openmp, only some of the important features of Openmp are discussed. It is hoped that the readers will be able to get a taste of the parallel programming after reading this chapter and will be able to parallelize their serial codes. The readers, requiring advanced facilities of Openmp, may consult the Openmp manual version 5.0 (see reference).

21.7 Compilation of Program Containing Openmp Directives

To compile a source program with Openmp directives, certain compiler directives must be used. In these examples, a.f90 is the name of the source program.

GCC gfortran:
gfortran –**fopenmp** a.f90

Intel ifort:
ifort **/Qopenmp** a.f90 (under Windows)
ifort **–qopenmp** a.f90 (under linux)

NAG Fortran:
nagfor **–openmp** a.f90

Manuals of the respective compilers may be consulted for more information.

21.8 Structure of Compiler Directives

Although all the programs in the book are written in free form, the Openmp directive will be written starting from the first position of a line. However, it may start from any position of the line provided there are only blanks before the compiler directives. All compiler directives start with !$ so that the compilers that do not support Openmp treat the Openmp directive as a comment because it starts with !. Thus, such programs are automatically converted to a normal serial program. The compilers that support Openmp treat !$ as the Openmp directive.

Continuation of any Openmp directive is possible by having the continuation character & as the last non-blank character of the original line. The continued line should also have !$ as the first two characters.

```
!$ ...... &
!$..........
```

The general rules for the directives are as follows:

- Appropriate compiler directive must be used during the compilation of the program containing the Openmp directive.
- Each line containing a compiler directive cannot contain more than one directive name.
- Although some directives end with optional end directive name, it is recommended to use this optional end directive for the sake of readability.
- All the Openmp directives should start with `!$omp`.
- Both uppercase and lowercase letters may be used.
- Conditional compilation is possible with the starting characters `!$` as the first two characters. This line is treated as a comment by non-Openmp compliant compiler but accepted as a valid statement by the Openmp compliant compiler. The line may be continued by using the ampersand character as the last character and using the characters `!$` as the first two characters of the next line.
- There should be at least one blank after `!$omp` or `!$`; otherwise, the directive sentinels are treated as a comment.

21.9 Parallel Region

As already mentioned, a serial code having one thread (master thread) when entering into the parallel region creates multiple slave threads depending upon the environment. The master thread gets a thread number 0 while slave threads get thread numbers 1, 2, 3 . . . .

Unless otherwise specified a program described in this chapter entering into the parallel region would normally create four threads (one master and three slaves) by default. If the physical number of processors (cores) is less than the number of threads created within the parallel region, system will create logical threads. This may degrade the performance of the system. In addition, one is allowed not to use all the processors (cores) available with the system.

21.10 Parallelization Directives

The directives that are used to control parallelization may be grouped into several categories. It may be mentioned that not every construct available in Openmp 5.0 is discussed in this book.

Parallel region construct: In this category, there is only one construct: `parallel / end parallel`.

Workshare construct: There are five directives in this category: `do / end do`, `sections / end sections`, `workshare / end workshare`, `single / end single` and `master / end master`.

Combined parallel and workshare construct: The parallel region construct and workshare constructs can be combined, and instead of using two separate !$omp constructs, a single !$omp construct may be used as parallel do / end parallel do, parallel sections / end parallel sections and parallel workshare / end parallel workshare.

Synchronization constructs: The synchronization constructs that are discussed in this book are atomic, barrier, critical / end critical and ordered / end ordered.

Other directives: There are two directives in this category: threadprivate and schedule.

21.11 Clauses Associated with the Directives

The environment of the parallel region may be modified by using clauses. The following clauses are discussed in this book: copyin, copyprivate, default, firstprivate, if, lastprivate, nowait, num_threads, reduction, shared, ordered, schedule. It may be noted that not all clauses are available for all constructs. Table 21.1 is summary of clauses/directives. In the table, 'y' stands for available and 'n' stands for not available.

The Openmp directives master, critical, barrier, atomic, flash, ordered and threadprivate do not have any clauses attached to it.

TABLE 21.1

Summary of Clauses/Directives

| Clause | Directive | | | | | |
	Parallel	Do	Sections	Single	Parallel do	Parallel sections
if	y	n	n	n	y	y
private	y	y	y	y	y	y
shared	y	y	n	n	y	y
default	y	n	n	n	y	y
firstprivate	y	y	y	y	y	y
lastprivate	n	y	y	n	y	y
reduction	y	y	y	n	y	y
copyin	y	n	n	n	y	y
copyprivate	n	n	n	y	n	n
schedule	n	y	n	n	y	n
ordered	n	y	n	n	y	n
nowait	n	y	y	y	n	n

21.12 Parallel Directive

This section contains some of the facilities of the parallel region. The parallel directives are shown next:

```
!$omp parallel [clauses ] [clauses]  ....
            if(scalar logical expression)
            private(list)
            shared(list)
            default(private|firstprivate|shared|none)
            firstprivate(list)
            reduction(operator: list)
            num_threads(integer expression)
<code>
!$omp end parallel
```

The clauses are optional. Some of them are discussed in this section. Rest are treated separately.

```
            program omp1
            use omp_lib
            implicit none    ! the author's laptop has 4 threads by default
            print *, 'Entering parallel region ...'
!$omp parallel
            print *, 'Entered into parallel region'
!$omp end parallel
            end
```

The output is as follows:

```
Entering parallel region ...
Entered into parallel region
Entered into parallel region
Entered into parallel region
Entered into parallel region
```

When the directive `!$omp parallel` is encountered, four threads including the master thread are created and each thread executes the same `print` instruction in this case. The same program is replicated within all threads, and all the four threads execute the same set of instructions. This explains why the same output is displayed four times (the author's laptop creates four threads by default).

If the default for this laptop is changed by setting the environment variable of the same program as

```
    set omp_num_threads=2
```

the following output is obtained.

The output is as follows:

```
Entering parallel region ...
Entered into parallel region
Entered into parallel region
```

Note that in this case, only two threads will be created (one master and one slave).
For Unix system, the command will be `export OMP_NUM_THREADS=2`.

21.13 Lexical and Dynamic Region

The parallel region may contain ordinary statements and also a call to function or subroutine. The codes directly under omp `parallel` up to omp end `parallel` are known as the lexical region. The dynamic region is defined as the lexical region plus the code of the subroutine or function that are called directly or indirectly from the lexical region. Thus, the lexical region is a subset of the dynamic region. Consider the following program:

```
        use omp_lib
          .
!$omp parallel
          .
        call sub
          .
!$omp end parallel
        end
        subroutine sub
          .
        end subroutine sub
```

The codes between `!$omp parallel` and `!$omp end parallel` belong to the lexical region. When the codes belonging to the `subroutine sub` are included, the whole region is known as the dynamic region.

21.14 Three Runtime Routines

Three runtime callable subprograms are introduced here. These are used in subsequent programs/program segments. Some of these runtime routines are discussed in detail at appropriate places. The routines are omp_set_num_threads, omp_get_num_threads and omp_get_thread_num. The first one is a subroutine that takes one positive nonzero integer as its argument. This sets the number of threads that the subsequent parallel region will use. The other two are functions without any argument. They return, respectively, the total number of threads in the parallel region and the thread number that has called this routine. The thread numbers have been assigned by the system to the threads.

In the following program, the routine omp_set_num_threads is called with an argument 4 meaning, thereby creating four threads in the parallel region (one master and three slaves).

```
        program omp2
        use omp_lib
        implicit none
        call omp_set_num_threads(4)
```

```
      print *, 'Entering parallel region'
!$omp parallel
      print *, 'Inside parallel region,  Thread no: ', omp_get_thread_num()
!$omp end parallel
      end
```

The output is as follows:

```
Entering parallel region
Inside parallel region Thread no:          0
Inside parallel region Thread no:          3
Inside parallel region Thread no:          2
Inside parallel region Thread no:          1
```

In this case, four threads are created within the parallel region. All these threads are independent of each other. It cannot be predicted whether thread 0 will finish its job before thread 1 or not. In fact, outputs may vary from one run to another. The threads are executed in random order. The threads move asynchronously, and it cannot be predicted which thread will first finish its work.

21.15 Nested Parallel

A parallel region may contain another parallel region. When the threads from the outer parallel region enter into the inner parallel region and if nesting is permitted, each thread becomes a master thread and creates zero or more threads according to the programming environment. The runtime routine omp_set_nested must be called outside the parallel region with the argument .true. to permit nesting of the parallel regions. The default value is false. Consider the following program:

```
      program omp_nested_1
      use omp_lib
      call omp_set_num_threads(2)
      ! no of threads inside parallel region is 2
      call omp_set_nested(.true.) ! nesting is allowed
      print *, "I am in serial region"
!$omp parallel
      print *, "About to enter nested parallel region", &
         omp_get_thread_num()
!$omp parallel
      print *, "I am inside nested parallel region, my thread no is ", &
         omp_get_thread_num()
!$omp end parallel
!$omp end parallel
      print *, "Back to serial region"
      end program omp_nested_1
```

The output is as follows:

```
I am in serial region
About to enter nested parallel region          0
```

```
About to enter nested parallel region           1
I am inside nested parallel region, my thread no is  0
I am inside nested parallel region, my thread no is  1
I am inside nested parallel region, my thread no is  0
I am inside nested parallel region, my thread no is  1
Back to serial region
```

The first call to omp_set_num_threads sets the number of threads to 2. The next call to omp_set_nested with argument .true. allows creation of two threads for each thread entering into the nested parallel region. Now, once the master and the slave threads from the outer parallel region enter into the inner parallel region, both of them become master threads inside the nested region and create one additional thread (number of threads is 2). Thus, inside the inner parallel region, there will be four threads. However, if the routine omp_set_nested is called with the argument .false. or it is not called at all because the default is false, the output will be as shown next:

```
I am in serial region
About to enter nested parallel region           0
I am inside nested parallel region, my thread no is          0
About to enter nested parallel region           1
I am inside nested parallel region, my thread no is          0
Back to serial region
```

It is easy to explain the preceding output. The outer parallel construct creates two threads. As the nesting is switched off when these two threads enter into the inner parallel construct, both of them become master (thread number becomes 0 inside the nested region) but they do not create any new threads within the inner parallel region.

21.16 Clauses Associated with Parallel Construct

Several clauses are associated with the parallel construct. Some of these clauses are discussed in the next few sections.

21.17 IF Clause

This clause is used if it is necessary to enter into the parallel region depending upon the value of a certain variable. The syntax of an if clause is if (scalar-logical-expression). The expression is evaluated. If the result is true, the computation is done in parallel. If the result is false, the computation is done serially. Consider a simple case. We want to add natural numbers. If the number is less than 1000, we want to add serially. If it is more than 1000, we want to perform a parallel calculation.

```
program omp_if
use omp_lib
implicit none
integer:: index=2
```

```
          call omp_set_num_threads(5)
          print *, 'entering parallel region ...'
!$omp parallel if(index.eq.2)
          print *, 'entered into parallel region', omp_get_thread_num()
!$omp end parallel
          end
```

Since the value of the index is 2, five threads will be created and there will be 5 lines of output—one for each thread. If the values of the index are changed to some other values (not equal to 2), only one line of output will be generated corresponding to thread zero (the master thread).

21.18 NUM_THREADS

This clause is used for running a parallel program for a particular number of threads.

```
!$omp parallel num_threads(3)
          .
!$omp end parallel
```

```
          program omp_num_threads
          use omp_lib
          implicit none
          print *, 'entering parallel region ...'
!$omp parallel num_threads(3)
          print *, 'entered into parallel region', omp_get_thread_num()
!$omp end parallel
          end
```

For this parallel region, three threads will be created (one master and two slaves).

21.19 PRIVATE

If it is necessary that some variables will have different values in each thread, this clause is used. The syntax is private (list-of-variables). When a variable is declared as private, separate locations for the variables are assigned to the variable in each thread. Normally such variables are undefined when the threads are entered, and they become undefined when the parallel region is exited. These variables are local to the thread.

```
!$omp parallel private (a,b,c)
          .
!$omp end parallel
```

Variables a, b and c are local to the threads.

21.20 SHARED

If it is necessary to make certain variables available to all the threads, the clause `shared` is used.

```
!$omp parallel shared (p,q,r)
            .
!$omp end parallel
```

Variables p, q and r are global to all the threads. Barring a few variables like the index of do loop, all variables are shared by default unless they are made private. All the threads may read from and write on the shared variable.

21.21 DEFAULT NONE

If this clause is used, all variables inside the parallel region are to be declared explicitly whether private or shared.

```
!$omp parallel default (none) private (a,b,c) shared(p,q,r)
            .
!$omp end parallel
```

This is somewhat similar to `implicit none` of Fortran.

21.22 DEFAULT PRIVATE

If this clause is used, all variables inside the parallel region by default become private variable. This can be overridden by using a shared clause.

```
!$omp parallel default(private) shared(a,b,c)
            .
!$omp end parallel
```

In this case, all variables other than a, b and c inside the parallel region are private variables.

21.23 DEFAULT SHARED

If this clause is used, all variables inside the parallel region by default become shared variable. This can be overridden by using a private clause.

```
!$omp parallel default(shared) private(p,q,r)
            .
!$omp end parallel
```

If no default clause is specified, `default(shared)` is assumed.

21.24 FIRSTPRIVATE

The variables declared as `firstprivate` are private in nature, but they are initialized with the value of the variables before the declaration.

```
          program omp_pvt1
          use omp_lib
          integer::i=100
          call omp_set_num_threads(3)
!$omp parallel firstprivate(i)
          print *, omp_get_thread_num(), i
!$omp end parallel
          end program omp_pvt1
```

Variable `i` is local to the threads, but this has been initialized to `100`. Therefore, this program will generate 3 lines of output with thread numbers `0`, `1` and `2` and with the value of `i` equal to `100`. If the original variable is an allocatable variable, the variable must be allocated before entering into the parallel region.

```
          program omp_pvt2
          use omp_lib
          integer, allocatable, dimension(:)::a
          allocate(a(4))           ! allocated before entering parallel region
          a=[10,20,30,40]
!$omp parallel firstprivate(a) num_threads (2)
          print *, omp_get_thread_num(), a
!$omp end parallel
          end program omp_pvt2
```

The output is as follows:

```
0    10    20    30    40
1    10    20    30    40
```

If the original variable is a pointer, it must point to an actual variable or an unnamed location created by the allocation.

21.25 Rules for OMP PARALLEL Directive

There are a few simple rules to be followed while creating a parallel region.

- Both the directives `omp parallel` and `omp end parallel` must be within the same program unit.
- It is prohibited to jump into the parallel region from outside the parallel block by a conditional or an unconditional branch statement.

- However, it is permitted to use a conditional or unconditional branch statement within the parallel block.
- If the input/output operations are performed within the parallel block using different units, there is no problem. However, if the operations are performed on the same unit, simultaneous access to the unit may cause trouble. Therefore, the programmer should take proper steps to keep to these rules.

Consider the following two programs:

```
        program omp_par1
        use omp_lib
        implicit none
        print *, 'entering parallel region ...'
        go to 10              ! not allowed
!$omp parallel
     10 continue
        print *, 'entered into parallel region'
!$omp end parallel
        end program omp_par1


        program omp_par2
        use omp_lib
        implicit none
        integer:: n=4
  print *, 'entering parallel region ...'    ! 4 (threads)
!$omp parallel
        if(n.eq.4) then
          go to 10                ! not allowed
        end if
        print *, 'entered into parallel region'
!$omp end parallel
     10  continue
        end omp_par_2
```

The preceding programs would generate compilation errors.
   However, the following program

```
        program omp_par3
        use omp_lib
        implicit none
        integer:: n=4
        print *, 'entering parallel region ...'
!$omp parallel
        if(n.eq.4) then
          go to 10                ! allowed
        end if
        print *, 'entered into parallel region'
     10  continue
!$omp end parallel
        end program omp_par3
```

will compile properly as the jump is within the parallel region.

The next two programs show that the directive announcing the beginning and the end of the parallel region should be in the same unit—even the end of the parallel region cannot be inside the inline subprogram.

```
          program omp_par4
          use omp_lib
          implicit none
          print *, 'entering parallel region ...'
!$omp parallel
          call sub          ! not allowed
          contains
           subroutine sub
           print *, 'thread number: ', omp_get_thread_num(), &
          & ' entered into subroutine'
!$omp end parallel
           end subroutine sub
          end program omp_par4


          program omp_par5
          use omp_lib
          implicit none
          print *, 'entering parallel region ...'
!$omp parallel
          call sub    ! allowed
!$omp end parallel
          contains
           subroutine sub
            print *, 'thread number: ', omp_get_thread_num(), &
             ' entered into subroutine'
           end subroutine sub
          end program omp_par5
```

21.26 Workshare Construct

Within the workshare construct, the job is divided among the threads. In the subsequent discussion, we consider four workshare constructs, namely, do, sections, workshare and single.

21.27 OMP DO/OMP END DO

The do loop within the directives !$omp do - !$omp end do are executed in several parallel chunks. Each thread within a parallel region handles a fixed number of iterations. For example, suppose there are four threads and the statement is do i=1, 100. The system will probably allocate the number of iterations 1–25 to thread 0, 26–50 to thread 1, 51–75 to thread 2 and 76–100 to thread 3. The system must know precisely the number of times the do loop is going to be executed before the do loop is entered. For this reason, do-while and just do (infinite do) cannot be parallelized. The syntax is as follows:

```
!$omp parallel
!$omp do
         do i=....
             .
         end do
!$omp end do
!$omp end parallel
```

This may also be written as:

```
!$omp parallel do
         do i=....
             .
         enddo
!$omp end parallel do
```

Like the !$omp parallel, !$omp do also takes several clauses. These are discussed separately. To use parallel do, the primary condition is that the iterations are independent of each other and that exact order of executing different iterations does not matter. There is an implied barrier at the end of the do loop. Unless all threads finish their work, further computation cannot continue; threads reaching early have to wait until other threads finish their work.

21.28 Rules of OMP DO/OMP END DO

Rule 1: The initial value, final value and stride are integers. The trip count should be available to the system before the do loop is entered.

Rule 2: Do-while and just do (infinite do) cannot be parallelized.

Rule 3: There should be one entry point at the top and one exit point at the bottom.

Rule 4: There cannot be any conditional or unconditional branch state that may bring the control outside the do loop.

Rule 5: However, such a jump within the do loop is permitted.

Rule 6: There cannot be an exit statement within the loop.

Rule 7: There can be a cycle statement within the loop.

Rule 8: The index variable of the do loop by default is a private variable.

21.29 OMP SECTIONS/OMP END SECTIONS

This construct allows assigning a block of code to a thread. Each thread would execute a different code. If the number of available threads within a parallel region is less than the number of sections, some threads may have to handle more than one block of codes, and as a result, the program becomes inefficient. For example, if there are six sections with four threads within a parallel region, it is obvious that two threads have to handle more than one section and as a result, two threads will remain idle during this time. This is a wastage of resources. Within omp sections / end omp sections, each section starts with omp section. In the program sec_1, four sections have been created and each section will print a character string. As it is known that the threads will not be executed in a particular fixed order, no doubt, the output will contain all the four strings—I, am, here, ok—but may not in this order. In fact, it is expected to vary from one run to another. Further, it may be noted that !$omp parallel and !$omp sections may be combined as !$omp parallel sections. Similarly, !$omp end sections and !$omp end parallel may be combined as !$omp end parallel sections.

```fortran
        program sec_1
        use omp_lib
        print *, "i am in serial region"
!$omp parallel sections
!$omp section
        print *, "i", omp_get_thread_num()
!$omp section
        print *, "am", omp_get_thread_num()
!$omp section
        print *, "here", omp_get_thread_num()
!$omp section
        print *, "ok", omp_get_thread_num()
!$omp end parallel sections
        end program sec_1
```

The output is as follows:

```
i            0
am           1
ok           3
here         2
```

The ordering of the character strings on the output may be different in another run. In the next program, each section will call different functions. It calculates area/volume of different shapes. The task is divided among the four threads; different sections (threads) will invoke different functions to calculate the area or volume of different shapes.

```fortran
        program sec_2
        use omp_lib
        implicit none
        real:: r=2.0
        real :: b=3.0, h=4.0
        real, parameter::pi=3.1415926
        integer:: myid
!$omp parallel private(myid)
        myid=omp_get_thread_num()
!$omp sections
!$omp section
        print *, "Area of the Circle    ",circle(r), "Thread No ",myid
!$omp section
```

```
          print *, "Volume of the Sphere    ",sphere(r), "Thread No ",myid
!$omp section
          print *, "Area of the Square    ",square(r), "Thread No ",myid
!$omp section
          print *, "Area of the Triangle ",triangle(b,h), "Thread No ",myid
!$omp end sections
!$omp end parallel
          contains
          real function circle(r)
          real::r
          circle=pi*r*r
          end function circle
          real function sphere(r)
           real::r
           sphere=4.0/3.0*pi*r*r*r
          end function sphere
          real function square(r)
           real:: r
           square=r*r
          end function square
          real function triangle(b,h)
           real:: b,h
           triangle=0.5*b*h
          end function triangle
          end program sec_2
```

The output is as follows:

```
Area of the Circle      12.56637    Thread No        0
Area of the Square       4.000000   Thread No        2
Area of the Triangle     6.000000   Thread No        3
Volume of the Sphere    33.51032    Thread No        1
```

21.30 OMP WORKSHARE

This construct is used to parallelize array operations by dividing the task into units of work. Each such unit of work is handled only by one of the available threads present in the parallel region. The block of code within the workshare construct may contain only the following constructs/statements: array assignment statements, `atomic directive`, `critical` construct, `forall` construct/statement, `parallel` construct, `parallel do` construct, `parallel section` construct, `parallel workshare` construct, `scalar assignment` statement, `where` construct/ statement. For array handling, the intrinsic functions `all`, `any`, `count`, `cshift`, `dot_product`, `eoshift`, `matmul`, `maxloc`, `maxval`, `minloc`, `minval`, `pack`, `product`, `reshape`, `spread`, `sum`, `transpose` and `unpack` are available within the workshare construct. If any user-defined function is called with the construct, it must be an elemental function. The unit of work is defined in the following manner for each type of statement.

Array expression: Each element of array represents one unit. The function listed earlier constitutes any number of units of work.

Assignment statement: Each element of assignment statement constitutes one unit of work. Scalar assignment is treated as one unit of work.

Constructs: Each critical construct constitutes one unit of work. Each parallel construct inside the workshare construct constitutes one unit of work. For forall or where statement or construct, the evaluation of masked expression and masked assignment constitutes one unit of work.

Directives: The atomic directive to update a scalar constitutes one unit of work.

Elemental function: When the argument of the elemental function is an array, each element of the array constitutes one unit of work.

Default: If some statement does not fall under the preceding categories, it is considered as one unit of work.

```
          program sec_1
          use omp_lib
          implicit none
          real, dimension(10)::a=200.0, b
          real::mysum
!$omp parallel
!$omp workshare
          b=a
!$omp end workshare
!$omp end parallel
          print *, b
          end program sec_1

          program sec_2
          use omp_lib
          implicit none   ! use of where
          real, dimension(10)::a
          call random_seed()
          call random_number(a)
!$omp parallel
!$omp workshare
          a=1000.0*a
!$omp end workshare
!$omp workshare
          where (a>500.0)
           a=sqrt(a)
          elsewhere
           a=-999.0
          end where
!$omp end workshare
!$omp end parallel
          print *, a
          end program sec_2

          program sec_3
          use omp_lib
          implicit none   ! use of sum intrinsic function
```

```
              real, dimension(10)::a=[1.0,2.0,3.0,4.0,5.0,6.0,7.0,8.0,9.0,10.0]
              real::mysum
!$omp parallel
!$omp workshare
              a=1000.0*a
!$omp end workshare
!$omp workshare
              mysum=sum(a)
!$omp end workshare
!$omp end parallel
              print *, a
              print *, mysum
              end program sec_3

              program sec_4
              use omp_lib
              implicit none    ! use of user function
              real::mysum
              real, dimension(10)::a=[1.0,2.0,3.0,4.0,5.0,6.0,7.0,8.0,9.0,10.0]
              integer::i
!$omp parallel workshare
              mysum=sum(myfunc(a))
!$omp end parallel workshare
              print *, a; print *, mysum
              mysum=0.0
              do i=1,10
                mysum=mysum+a(i)**2
              end do
              print *, mysum
              contains
               elemental  function myfunc(x) result(res)
               real, intent(in)::x
               real:: res
               res=x**2
              end function myfunc
              end program sec_4
```

21.31 OMP SINGLE/OMP END SINGLE

In the single construct, a single thread within the parallel region executes the enclosed codes within omp single and omp end single. The thread that first reaches the omp single executes the codes within omp single and omp end single. All other threads wait until this thread finishes its work. After that, all the threads proceed to execute the codes in parallel. Suppose it is necessary to read a shared variable from external device before all the participating threads can proceed further, omp single / omp end single ensures that unless the reading of the variable is over all other threads will wait. This is shown in the next program. When one thread enters into the omp single / omp end single block, other threads wait until this thread reads variable r and comes out of the single block. After this, all the threads proceed to execute next instructions in parallel.

```
            program sing_1
            use omp_lib
            implicit none
            real:: r
            integer:: myid
            call omp_set_num_threads(3)
!$omp parallel private(myid)
            myid=omp_get_thread_num()
!$omp single
            read *, r; print *, "Thread No ", myid,  " r inside single ",r
!$omp end single    ! threads wait here
            select case (myid)
             case (0)
              print *, "Thread No ", myid, " r= ", r
             case (1)
              print *, "Thread No ", myid, " r= ", r
             case (2)
              print *, "Thread No ", myid, " r= ", r
             end select
!$omp end parallel
            end program sing_1
```

21.32 OMP MASTER/OMP END MASTER

This construct is similar to omp single / omp end single. There are two differences. Only the master thread executes the block of code surrounded by omp master / omp end master, and there is no implied barrier at the omp end master. Therefore, all the threads move in parallel like omp single / omp end single; threads do not wait till the master thread finishes it task within omp master / omp end master. Consider the following program and the output of the program.

```
            program mast_1
            use omp_lib
            implicit none
            real:: r=33.25
            integer:: myid
            call omp_set_num_threads(3)
!$omp parallel private(myid)
            myid=omp_get_thread_num()
!$omp master
            read *, r
!$omp end master
            select case (myid)
             case (0)
              print *, "Thread no ", myid, " r= ", r
             case (1)
              print *, "Thread no ", myid, " r= ", r
             case (2)
              print *, "Thread no ", myid, " r= ", r
             end select
!$omp end parallel
            end program mast_1
```

The output is as follows:

```
Thread No      1   r=      33.25000
Thread No      2   r=      33.25000
Thread No      0   r=      13.00000
```

We shall now try to explain the output. The master thread is entered and the system waits for the input of variable r. Since there is no implicit barrier, the other threads move according to the program logic and thus generate the first two lines of the output having thread numbers 1 and 2. Variable r was initialized to 33.25. Therefore, threads 1 and 2 display this value. However, when data is supplied in response to read *, r (in this case, 13.0 was supplied from the keyboard), this becomes the value of r, so thread zero displays the value of r as 13.0. Before the data is supplied from the keyboard, threads 1 and 2 finish their work, and therefore, they display the value of r as 33.25 (r was initialized to this value).

21.33 REDUCTION

All participating threads within the parallel region can access a shared variable. When updating a variable is required by all the threads, there is no guarantee that this will be done one at a time; it may be performed by the threads simultaneously. Simultaneous updating of the shared variable may return an incorrect value.

Reduction is a mechanism by which such a situation is avoided. In this process, all threads create a private copy of the shared variable and update this private copy, and finally at the end, all the updated private copies are merged with the original shared variable. The reduction clause is available for expressions like variable=variable operator expression or variable=intrinsic_procedure (variable, expression_list), where the available operators, intrinsic procedures and their initial values are shown in Table 21.2.

TABLE 21.2

Available Operators, Intrinsic Procedures and Their Initial Values

Operator/Intrinsic	Initial Value
+	0
-	0
*	1
.and.	.true.
.or.	.false.
.eqv.	.true.
.neqv.	.false.
max	smallest available number
min	largest available number
iand	all bits on
ior	all bits zero
ieor	all bits zero

The variable must be a scalar of the intrinsic type, and the operator must be a binary operator. We illustrate reduction with a few examples. Consider `program red_1`. This is a serial program to add all numbers from 1 to 1000. This program gives a correct result.

```
program red_1
implicit none
integer:: sum=0,i
do i=1,1000              ! serial program
 sum=sum+i
end do
print *, sum
end program red_1
```

The output is as follows:

```
500500
```

When the loop is parallelized, the updating instruction `sum=sum+i` may not be executed one at a time by the threads. There is always a possibility that more than one thread may try to access the location `sum` simultaneously for updating the location. In such a situation, the returned result will be wrong. The same program when converted to a parallel program (`program red_2`) fails to give correct result because all the threads can access the statement `sum=sum+i` simultaneously. It may give different results in different runs.

```
        program red_2
        use omp_lib
        implicit none
        integer:: sum=0,i
!$omp parallel do
        do i=1,1000
          sum=sum+i    ! parallel program may give wrong result
        end do
!$omp end parallel do
        print *, sum
        end program red_2
```

Program `red_3` uses the reduction clause that does not allow the expression `sum=sum+i` to be updated simultaneously by different threads. Naturally, this parallel program gives the correct result.

```
        program red_3
        use omp_lib
        implicit none
        integer:: sum=0,i
!$omp parallel do reduction(+:sum)
        do i=1,1000
          sum=sum+i
        end do
!$omp end parallel do
        print *, sum
        end program red_3
```

The clause reduction contains two items: an operator (+) and a variable (`sum`). Therefore, `sum=sum+i` will be calculated locally within each thread and finally will be merged with the global variable sum. The next two programs red_4 and red_5 use max and min intrinsic functions with reduction. For correctly using reduction, it is a must that finding maximum and minimum is independent of the order in which the locations are accessed, which is true in these cases.

```fortran
program red_4
use omp_lib
implicit none
integer:: mymax, i
integer, parameter::limit=10
integer, dimension(limit)::ia
real, dimension(limit):: a
call random_seed()
call random_number(a)
ia=a*1000.0
mymax=ia(1) !assume the first one is maximum - initialization
!$omp parallel do reduction(max:mymax)
     do i =2, limit
      mymax=max(mymax,ia(i))
     end do
!$omp end parallel do
     print *,'The array: ', ia
     print *, 'Maximum ', mymax
end program red_4


program red_5
use omp_lib
implicit none
integer:: mymin, i
integer, parameter::limit=10
integer, dimension(limit)::ia
real, dimension(limit):: a
call random_seed()
call random_number(a)  ! random number between 0 and 1
ia=a*1000.0            ! multiply by 1000 and convert to integer
mymin=ia(1)            ! assume the first one is minimum -
                       ! initialization
!$omp parallel do reduction(min:mymin)
     do i =2, limit
       mymin=min(mymin,ia(i))
     end do
!$omp end parallel do
     print *,'The array: ', ia
     print *, 'Minimum ', mymin
end program red_5
```

Finally, program red_6 shows how one can use two reductions with a do statement. This program adds 1 to 7 and also calculates factorial 7 using the do statement with a reduction clause.

```fortran
program red_6
use omp_lib
implicit none
```

```
          integer:: isum=0, iprod=1, i
!$omp parallel do reduction(+:isum) reduction(*:iprod)
          do i=1,7
            isum=isum+i
            iprod=iprod*i
          end do
!$omp end parallel do
          print *, "isum = ", isum, " iprod = ", iprod
          end program red_6
```

21.34 CRITICAL/END CRITICAL

The pair of instruction ensures that only a single thread accesses the enclosed code at any one time. While one thread is executing the code within the `critical/end critical` directive, if another thread reaches the critical section, it will wait until the current thread executing the critical section finishes its task. This ensures that only one thread can access the critical section at a time.

```
          program cri_1
          use omp_lib
          implicit none
          integer:: sum=0,i
!$omp parallel do
          do i=1,1000
!$omp critical
              sum=sum+i
!$omp end critical
          end do
!$omp end parallel do
          print *, sum
          end program cri_1
```

The critical section may have a name. All unlabeled sections form one critical block. Similarly, all critical sections having the same name form a single block. Critical sections having different names form separate blocks. For example, suppose there are two different independent blocks having different labels within a critical section. As the names are different, two threads may simultaneously enter into the two critical blocks. However, if the names of the two critical blocks are the same, as they are treated as a single block, when one thread enters into a critical block, no other thread can enter into the other block. It is thus recommended that it is better to use named critical blocks. The following program shows this feature. The max and min functions are not dependent on each other. Thus, two critical blocks have been created. When a thread executes, say, the findmax critical block, another thread safely enters into findmin critical block. If the name of the two blocks is made the same, say, find maxmin, once a thread uses the previous findmin critical block, now named as, say, maxmin, no other thread can enter any findmax (now named as maxmin) critical block.

```
          program cri_2
          use omp_lib
          implicit none
```

```
            integer:: mymax, mymin, i
            integer, parameter:: limit=10
            integer, dimension(limit):: ia
            real, dimension(limit)::a
            call random_seed()
            call random_number(a)        ! random number between 0 and 1
            ia=1000.0*a                   ! multiply by 1000 and
                                          ! convert to integer
            mymax=ia(1)                   ! set to the first location
            mymin=ia(1)                   ! set to first location
!$omp parallel do shared(mymax,mymin)    ! this is default
            do i=1, limit
!$omp critical (findmax)
              if(mymax < ia(i)) then ! mymax is less set mymax to the current ia
                mymax=ia(i)
              endif
!$omp end critical (findmax)
!$omp critical (findmin)
              if(mymin > ia(i)) then ! mymin is greater set mymin to the current ia
                mymin=ia(i)
              endif
!$omp end critical (findmin)
            end do
!$omp end parallel do
            print *, " Input data: "
            print *,ia
            print *, "Maximum: ",mymax
            print *, "Minimum: ",mymin
            end program cri_2
```

The output is as follows:

```
Input data:

        285     635     601     442     853     550
        811     799     554     292
Maximum:        853
Minimum:        285
```

21.35 LASTPRIVATE

The variable declared as private within a parallel region becomes undefined outside the parallel region. If the lastprivate clause is used, the original variable with the same name gets the last value of the private variable in the parallel region. The so-called last value is the value the variable should have obtained if the program is executed serially.

```
            program last_private
            use omp_lib
            implicit none
            integer::i
            integer::a
```

```
!$omp parallel do lastprivate(a)
        do i=1, 10    ! i is private variable by default
         a=i
         print *, 'Inside Parallel, a= ',a
        end do
!$omp end parallel do
        print *, 'Outside Parallel, a= ', a
        end program last_private
```

The output is as follows:

```
Inside Parallel, a=              1
Inside Parallel, a=              7
Inside Parallel, a=              8
Inside Parallel, a=              9
Inside Parallel, a=              4
Inside Parallel, a=              10
Inside Parallel, a=              5
Inside Parallel, a=              6
Inside Parallel, a=              2
Inside Parallel, a=              3
Outside Parallel,a=              10
```

In the preceding program, the last value of variable i should have been 10. Variable 'a' has this value outside the parallel region because of the lastprivate clause. The following program is another example of the lastprivate clause. If the program is executed serially, the final value of x would have been 40. The lastprivate clause ensures that the value of x declared before the parallel region is 40 (shown in bold).

```
        program last_private1
        use omp_lib
        implicit none
        integer::x
!$omp parallel sections lastprivate(x)
!$omp section
        x=10; print *, x, "First section"
!$omp section
        x=20; print *, x, "Second section"
!$omp section
        x=30; print *, x, "Third section"
!$omp section
        x=40; print *, x, "Fourth section"
!$omp end parallel sections
        print *, x, "Serial region"
        end program last_private1
```

The output is as follows:

```
10 First section
30 Third section
40 Fourth section
20 Second section
40 Serial region
```

21.36 ATOMIC

This directive is used to access a variable atomically within a parallel region. The system generates codes in such a way that only one thread can access the variable—other threads are prevented from having access to the variable until the atomic operation is over. This ensures no simultaneous access to a variable by different threads. The directive may be atomic read, atomic write, atomic update or atomic capture. The default is atomic update.

Atomic read: reads the shared variable atomically

Atomic write: writes the shared variable atomically

Atomic update: updates the variable atomically

Atomic capture: captures the original and final variable atomically

where:

atomic write statement:	`update_variable=expression`
atomic update statement:	`update_variable=update_variable operator expression`
	`update_variable=expression operator update_variable`
	`update_variable=intrinsic(update_variable, expression_list)`
	`update_variable=intrinsic(expression_list, update_variable)`
atomic capture statement:	`capture_variable=update_variable`

The `update_variable` and `capture_variable` are scalar variables of an intrinsic type. They cannot be an allocatable variable or a pointer. The allowed operators are `+`, `-`, `*`, `/`, `.and.`, `.or.`, `.eqv.`, `.neqv.`, `.or.` and `.xor.`. The allowed intrinsics are `max`, `min`, `iand`, `ior` and `ieor`.

Rules:

1. The expression cannot refer to the `update_variable`.
2. The comma-separated `expression_list` cannot refer to the `update_variable`.
3. The atomic directive without any clause is same as atomic update. This only acts on the statement following the atomic directive.
4. The expression associated with the atomic statement is not evaluated atomically.
5. The `capture_variable`, expression and `expression_list` are not permitted to access `update_variable` mentioned earlier.
6. Finally, the operator, intrinsic function and assignment must be intrinsic operator, function and assignment. They cannot be user-defined objects.

```
            program atomic_1
            use omp_lib
            integer:: isum=0, i
            integer, parameter:: siz=1000
!$omp parallel do
            do i=1, siz
!$omp atomic
                isum=isum+i
            end do
!$omp end parallel do
            print *, isum
            end program atomic_1

            program atomic_2
            use omp_lib
            implicit none
            integer, parameter:: siz=10
            integer, dimension(siz)::ia
            integer:: mymin, i
            real, dimension(siz)::a
            call random_seed()
            call random_number(a)
            ia=a*1000
            mymin=ia(1)
!$omp parallel do
            do i=2, siz
!$omp atomic
                mymin=min(mymin,ia(i))
            end do
            print *, ia; print *, mymin
            end program atomic_2
```

21.37 OMP BARRIER

The omp barrier creates a barrier such that no thread can proceed further until all threads reach omp barrier. We again consider the program given in Section 21.32. If omp barrier is placed after omp end master, no thread can proceed further until all threads including the master thread reach the barrier. Thus, in this case if 43.45 is typed in response to the read statement, all threads will print 43.45 because only after r has been updated within the master construct can all threads be permitted to proceed; r has been changed within the master construct and all print statements are executed after the updating is done by the read statement. Thus, this changed value will be reflected in all the outputs. The program logic is such that all threads must encounter the omp barrier construct; otherwise, a situation might crop up known as deadlock. There is an implied flash associated with the barrier construct to ensure memory consistency.

```
            program mast_2
            use omp_lib
            implicit none
            real:: r
            integer:: myid
            call omp_set_num_threads(3)
```

```
!$omp parallel private(myid)
        myid=omp_get_thread_num()
!$omp master
        read *, r
!$omp end master!
!$omp barrier
        select case (myid)
         case (0)
          print *, "Thread no ", myid, " r= ", r
         case (1)
          print *, "Thread no ", myid, " r= ", r
         case (2)
          print *, "Thread no ", myid, " r= ", r
        end select
!$omp end parallel
        end program mast_2
```

The output is as follows:

```
Thread No         0   r=     43.45000
Thread No         2   r=     43.45000
Thread No         1   r=     43.45000
```

21.38 THREADPRIVATE

The directive allows a named common block indicated by enclosing the name of the common block between two slashes (division symbol) or related variables (with save attribute) separated by comma to have private values within each thread. In other words, each thread maintains its own copy and the data of one thread is not visible to other threads.

```
        program thread_pvt1
        use omp_lib
        implicit none
        common /blk1/a,b
!$omp threadprivate(/blk1/)
        integer:: a,b
        integer::myid
        call omp_set_num_threads(4)
        print *, "Initialized through Blockdata: a,b =", a,b
        print *, "Printed through Subroutine Sub ........"
!$omp parallel private(myid)
        myid=omp_get_thread_num()
        call sub(myid)
!$omp end parallel
        call omp_set_num_threads(5)
        print *, "Printed through Subroutine Sub1 ........"
!$omp parallel private(myid)
        myid=omp_get_thread_num()
        call sub1(myid)
```

```
!$omp end parallel
        end program thread_pvt1
        subroutine sub(myid)
        implicit none            ! this subroutine prints the
                                 ! value initialized
        common/blk1/a,b          ! through blockdata
!$omp threadprivate(/blk1/)
        integer:: a,b
        integer::myid
        select case(myid)
        case(0)
          print *, "a= ",a," b= ",b, " thread no ",myid
          a=200                  ! private copies are modified
          b=300
        case(1)
          print *, "a= ",a," b= ",b, " thread no ",myid
          a=400                  ! private copies are modified
          b=500
        case(2)
          print *, "a= ",a," b= ",b, " thread no ",myid
          a=600                  ! private copies are modified
          b=700
        case(3)
          print *, "a= ",a," b= ",b, " thread no ",myid
          a=800                  ! private copies are modified
          b=900
        end select
        end subroutine sub
        subroutine sub1(myid)
        implicit none
        common/blk1/a,b
!$omp threadprivate(/blk1/)
        integer:: a,b
        integer::myid
        select case(myid)
        case(0)
          print *, "a= ",a," b= ",b, " thread no ",myid
        case(1)
          print *, "a= ",a," b= ",b, " thread no ",myid
        case(2)
          print *, "a= ",a," b= ",b, " thread no ",myid
        case(3)
          print *, "a= ",a," b= ",b, " thread no ",myid
        case(4)
          print *, "a= ",a," b= ",b, " thread no ",myid
        end select
        end subroutine sub1
        blockdata
        common/blk1/a,b
        integer a,b
!$omp threadprivate(/blk1/)   ! gfortran objects, ifort and nagfor accept
        data a,b/22,24/    ! initialize
        end blockdata
```

The output is as follows:

```
Initialized through Blockdata: a,b =  22   24
Printed through Subroutine Sub ........
a=      22  b=      24   thread no       0
a=      22  b=      24   thread no       1
a=      22  b=      24   thread no       2
a=      22  b=      24   thread no       3
Printed through Subroutine Sub1 ........
a=     200  b=     300   thread no       0
a=     800  b=     900   thread no       3
a=     400  b=     500   thread no       1
a=     600  b=     700   thread no       2
a=      22  b=      24   thread no       4
```

First, all private copies of the common blocks are initialized through blockdata. These are displayed through the subroutine sub. The private common blocks are modified within the subroutine sub, and subsequently the modified values are displayed through subroutine sub1. One important point may be noted. Before entering the second parallel region, there is call to the subroutine omp_set_num_threads with an argument 5. This means that one master and four slave threads will be created within the parallel region. The new thread having number equal to 4 will be initialized through blockdata. This is reflected in the output.

21.39 Rules for THREADPRIVATE

1. Blank common cannot be used with threadprivate.
2. !$omp threadprivate(/blk1/) or the name of appropriate block must be present in all subprograms.
3. Normal variables used with threadprivate must have save attribute.

21.40 COPYIN

The variables declared as threadprivate can be initialized by the current values of the master thread with this clause.

```
program thread_pvt_copyin
use omp_lib
implicit none
common/blk1/a,b
integer::a, b
integer::myid
```

```
!$omp threadprivate(/blk1/)
         call omp_set_num_threads(4)
         a=37; b=43
!$omp parallel private(myid) copyin(/blk1/)
         myid=omp_get_thread_num()
         select case(myid)
         case(0)
           print *, "a= ",a," b= ",b, " thread no ",myid
         case(1)
           print *, "a= ",a," b= ",b, " thread no ",myid
         case(2)
           print *, "a= ",a," b= ",b, " thread no ",myid
         case(3)
           print *, "a= ",a," b= ",b, " thread no ",myid
         end select
!$omp end parallel
         end program thread_pvt_copyin
```

The output is as follows:

```
a=      37  b=       43   thread no      3
a=      37  b=       43   thread no      1
a=      37  b=       43   thread no      0
a=      37  b=       43   thread no      2
```

The values of a and b in thread numbers 1, 2 and 3 are the same as the values of thread 0 (master thread) because of the copyin clause. If the copyin clause is removed from omp parallel, a and b of the master thread will not be copied to the private common blocks of the slaves. As they are initialized to zeros, the outputs will be as shown next:

```
a=      37  b=       43   thread no      0
a=       0  b=        0   thread no      1
a=       0  b=        0   thread no      3
a=       0  b=        0   thread no      2
```

If the variables are not in a common block, they must be declared as save variables.

```
         program thread_pvt_cop_save
         use omp_lib
         implicit none
         integer,save::a,b
         integer::myid
!$omp threadprivate(a,b)
         call omp_set_num_threads(4)
         a=37; b=43
!$omp parallel private(myid) copyin(a,b)
         myid=omp_get_thread_num()
         select case(myid)
          case(0)
           print *, "a= ",a," b= ",b, " thread no ",myid
          case(1)
           print *, "a= ",a," b= ",b, " thread no ",myid
          case(2)
           print *, "a= ",a," b= ",b, " thread no ",myid
```

```
        case(3)
         print *, "a= ",a," b= ",b, " thread no ",myid
        end select
!$omp end parallel
        end program thread_pvt_cop_save
```

21.41 ORDERED

This directive is used to conduct the iterative process within a do loop in an ordered way, that is, as if the iterations are performed serially. A particular thread has to wait if the previous iteration is not yet over. Consider the following expression: $f_2 = f_1 + f_0$, [$f(n+2) = f(n+1) + f(n)$], with $f_0 = 1$ and $f_1 = 1$. It is clear that in order to get f_2, f_1 and f_0 must be known. Therefore, the iterations are to be performed serially. For this purpose, omp do must have the directive ordered, and the block of code is to be surrounded by omp ordered and omp end ordered.

```
        program ord_1
        use omp_lib
        integer, dimension(0:10) :: f
        integer:: i
        f(0)=1; f(1)=1
!$omp parallel
!$omp do ordered
        do i=2, 10
!$omp ordered
           f(i)=f(i-1)+f(i-2)
!$omp end ordered
        end do
!$omp end do
!$omp end parallel
        print *, f
        end program ord_1
```

It may be verified that the outputs of this program are as follows:

```
1   1   2   3   5   8   13   21   34   55   89
```

21.42 COPYPRIVATE

If omp end contains a clause copyprivate(list-of-variables), the values of the variables are broadcast to all private variables mentioned in the copyprivate clause. For example, in the following program r is a private variable. It is read within the single block, and because of the presence of copyprivate with omp end single, this value of r is broadcast to the private r of threads 1 and 2.

```
        program cop_prv
        use omp_lib
        implicit none
```

```
        real:: r
        integer:: myid
        call omp_set_num_threads(3)
!$omp parallel private(myid, r)
        myid=omp_get_thread_num()
!$omp single
        read *, r ! input is 23.45
        print *, "Thread No ", myid,  " r inside single ",r
!$omp end single copyprivate(r)
        select case (myid)
         case (0)
          print *, "Thread No ", myid, " r= ", r
         case (1)
          print *, "Thread No ", myid, " r= ", r
         case (2)
          print *, "Thread No ", myid, " r= ", r
        end select
!$omp end parallel
        end program cop_prv
```

The output is as follows:

```
Thread No              0  r inside single     23.45000
Thread No              0  r=      23.45000
Thread No              2  r=      23.45000
Thread No              1  r=      23.45000
```

If the copyprivate clause is removed from omp end single, variable r inside threads 1 and 2 will remain undefined and therefore the corresponding print statement for r will display some garbage.

21.43 NOWAIT

It has been observed that in most cases discussed so far, there is an implied synchronization at the end of the parallel region. For example, for do unless all the threads reach the omp end do, the implied synchronization comes into play and threads reaching early to omp end do have to wait until all the threads reach there. To allow the thread that has reached the omp end parallel to continue without waiting for others, a nowait clause is used.

```
        integer:: i
!$omp parallel
!$omp do
        do i=1, 100
         .
        end do
!$omp end do nowait
!$omp end parallel
```

In this case, the implied synchronization is disabled and threads reaching omp end do will not wait for other threads to come to that point.

21.44 FLASH

During the execution of the parallel region, values of the variable may be held temporarily in the registers of the computer or buffers before they are actually written to their actual memory locations. For this reason, not all threads may see the same value of a particular variable. Flash is used to enforce memory consistency. Flash may contain a list of variables separated by a comma. In that case, the corresponding variables are affected. If the list is absent, all visible variables are flashed. The instruction is !$omp flash. This ensures that the latest copy of the variable is available in the memory.

21.45 Openmp LOCK

Lock is a mechanism by which a portion of the code is locked by a particular thread. This thread gets exclusive access to this portion of the code. All other threads, needing to access this portion of the code, have to wait to gain control of this code until the thread, which has the exclusive control over the code, releases the lock. Therefore, one thread at a time can access a portion of the code through this lock mechanism.

To use lock, it is necessary to initialize the lock through a subroutine omp_init_lock that takes one integer argument having kind=omp_lock_kind. This is defined in omp_lib.

```
integer(kind=omp_lock_kind):: lck
call omp_lock_init(lck)
```

Once initialized, now the lock can be applied by calling the subroutine omp_set_lock(lck). If the lock is set, only the same thread can release the lock when its job is over by calling the subroutine omp_unset_lock(lck) so that another thread, if necessary, may gain access to the lock. When the lock is no longer needed, a call to the subroutine omp_destroy_lock(lck) is used to destroy the lock. This call un-initializes the lock variable.

We illustrate the use of lock through our old program—add 1 to 1000 within parallel / end parallel.

```
program lock_1
use omp_lib
implicit none
integer, parameter::limit=1000
integer(kind=omp_lock_kind):: lck   ! lck is lock variable
integer:: sum=0, i
call omp_init_lock(lck)          ! initialize
```

```
!$omp parallel do
        do i=1, limit
          call omp_set_lock(lck)        ! lock applied
          sum=sum+i
          call omp_unset_lock(lck)      ! lock released
        end do
!$omp end parallel do
        print *, sum
        call omp_destroy_lock(lck)   ! lock destroyed, lck uninitialized
        end program lock_1
```

There is another method to apply lock. The function omp_test_lock takes one argument of kind=omp_lock_kind. It returns .true. if it is able to set the lock; otherwise it returns .false.. It does not wait whether it is able to set the lock or it is unable to set the lock, it just checks and continues. In the program lock_2, the do-while loop is repeatedly executed until it is able to apply lock (returns .true., .not. .true. is .false.).

```
        program lock_2
        use omp_lib
        implicit none
        integer, parameter::limit=1000
        integer(kind=omp_lock_kind):: lck   ! lck is lock variable
        integer:: sum=0, i
        call omp_init_lock(lck)              ! initialize
!$omp parallel do
        do i=1, limit
          do while (.not. omp_test_lock(lck)) ! wait till it can lock
          end do                              ! continues so long as
                                              ! it cannot lock
          sum=sum+i
          call omp_unset_lock(lck)            ! lock released
        end do
!$omp end parallel do
        print *, sum
        call omp_destroy_lock(lck)            ! lock destroyed
        end program lock_2
```

We give two more examples using a pointer variable. The examples use lock. The first one points to a target. In the second example, the pointer points to an unnamed location. The lock ensures that when the pointer accesses the shared variable, no other threads are allowed to access the variable pointed by the pointer. Both are shown next.

```
        program omp_pvt3
        use omp_lib
        integer, target::a
        integer, pointer::ap
        integer::id
        integer(kind=omp_lock_kind)::lck
        ap=>a
        call omp_set_num_threads(4)
        call omp_init_lock(lck)
!$omp parallel private(id)
        id=omp_get_thread_num()
        select case (id)
```

```
         case (0)
          call omp_set_lock(lck)
          ap=27
          print *, omp_get_thread_num(), ap
          call omp_unset_lock(lck)
         case(1)
          call omp_set_lock(lck)
          ap=28
          print *, omp_get_thread_num(), ap
          call omp_unset_lock(lck)
         case(2)
          call omp_set_lock(lck)
          ap=29
          print *, omp_get_thread_num(), ap
          call omp_unset_lock(lck)
         case(3)
          call omp_set_lock(lck)
          ap=30
          print *, omp_get_thread_num(), ap
          call omp_unset_lock(lck)
         end select
!$omp end parallel
          call omp_destroy_lock(lck)
          end program omp_pvt3
```

The output is as follows:

```
0        27
1        28
2        29
3        30
```

```
          program omp_pvt4
          use omp_lib
          integer, pointer::ap
          integer::id
          integer(kind=omp_lock_kind)::lck
          allocate(ap)                          ! pointing to unnamed
                                                ! location
          call omp_init_lock(lck)               ! initialize
          call omp_set_num_threads(5)
!$omp parallel private(id)
          id=omp_get_thread_num()
          select case (id)
          case (0)
            call omp_set_lock(lck)              !set lock
            ap=270
            print *, omp_get_thread_num(), ap
            call omp_unset_lock(lck)            ! clear lock
          case(1)
            call omp_set_lock(lck)
            ap=280
            print *, omp_get_thread_num(), ap
            call omp_unset_lock(lck)
```

```
      case(2)
        call omp_set_lock(lck)
        ap=290
        print *, omp_get_thread_num(), ap
        call omp_unset_lock(lck)
      case(3)
        call omp_set_lock(lck)
        ap=300
        print *, omp_get_thread_num(), ap
        call omp_unset_lock(lck)
      case(4)
        call omp_set_lock(lck)
        ap=310
        print *, omp_get_thread_num(), ap
        call omp_unset_lock(lck)
      end select
!$omp end parallel
      call omp_destroy_lock(lck)
      end program omp_pvt4
```

The output is as follows:

```
0        270
1        280
2        290
3        300
4        310
```

21.46 SCHEDULE

The clause associated with a parallelized do instruction controls how the iteration is distributed among the threads. Without the schedule clause, the system tries to distribute the load equally among the threads. For example, if the do loop index i varies between 1 and 12 and there are four threads within the parallel region, each thread is expected to perform 3 iterations—say, thread 0 will perform calculation for i= 1 to 3, thread 1 will perform calculation for i=4 to 6 and so on. If the number of cycles is say 10, one of the threads will perform fewer calculations.

This schedule clause takes two arguments: type and chunk_size. The first one, type, decides the way in which the distributions will take place. This can be static, dynamic, guided, runtime or auto. The second one is an optional argument called chunk (positive integer), which indicates the size of the work; auto does not have this chunk parameter. Program sch_1 is an omp do without any schedule clause. From the output it is clear that each of the four threads performs 3 iterations—thread 0 performs for i=1 to 3, thread 1 preforms for i=4 to 6, thread 2 performs for i=7 to 9 and thread 3 performs for i=10 to 12.

```
      program sch_1
      use omp_lib
      integer:: i, myid
      call omp_set_num_threads(4)
```

```
!$omp parallel private(myid)
        myid=omp_get_thread_num()
!$omp do
        do i=1,12
          print *, "Thread No ", myid," do index ", i
        end do
!$omp end do
!$omp end parallel
        end
```

The output is as follows:

```
Thread no        0  do index        1
Thread no        1  do index        4
Thread no        2  do index        7
Thread no        3  do index       10
Thread no        1  do index        5
Thread no        2  do index        8
Thread no        3  do index       11
Thread no        2  do index        9
Thread no        0  do index        2
Thread no        3  do index       12
Thread no        0  do index        3
Thread no        1  do index        6
```

21.47 STATIC SCHEDULE

The static parameter tries to distribute work evenly among the threads. The assignment of work is done at the beginning of the loop, and it remains fixed during execution. If chunk is specified, chunk number of iterations are grouped and the threads are assigned to these groups in a static round robin fashion in order of thread number. Now let us modify the line !$omp do to !$omp do schedule (static,4). In this case, 4 iterations are assigned to threads 0, 1 and 2, and thread number 3 remains idle (the loop in program sch_1 is from 1 to 12). If the instruction is !$omp do schedule (static,2), first each of the four threads will perform 2 iterations. The remaining four will be distributed between two other threads in a round robin fashion.

The output is as follows (static, 2):

```
Thread No        2  do index        5
Thread No        1  do index        3
Thread No        2  do index        6
Thread No        1  do index        4
Thread No        3  do index        7
Thread No        0  do index        1
Thread No        3  do index        8
Thread No        0  do index        2
Thread No        1  do index       11
Thread No        0  do index        9
Thread No        1  do index       12
Thread No        0  do index       10
```

TABLE 21.3

Effect of Chunk Size on the Distribution among Threads

Thread Number	0	1	2	3
No chunk	1-3	4-6	7-9	10-12
Chunk = 4	1-4	5-8	9-12	-
Chunk = 2	1-2, 9-10	3-4, 11-12	5-6	7-8
Chunk = 7	1-7	8-12	-	-

If the chunk is set to 7 instead of 4, thread 0 will handle iterations for i=1 to 7 and thread 1 will perform iterations for i= 8 to 12. Threads 2 and 3, though available for computation, will remain idle. The results are summarized in Table 21.3.

21.48 DYNAMIC SCHEDULE

When a dynamic schedule is specified, the total number of iterations is divided into chunks each of size chunk (the last one may be less than chunk size). If chunk is absent, it is assumed to be 1. After a thread finishes chunk number of iterations, it is given another chunk. This process continues until all chunks are exhausted. The last chunk may have less number of iterations than the chunk size. The output for !$omp do schedule(dynamic,2) is shown next. It is seen from the output that thread 3 has performed iterations for i=7, 8, 11, 12, whereas thread 0 has performed iterations for i=3 and 4.

The output is as follows:

```
Thread no        2   do index        1
Thread no        0   do index        3
Thread no        1   do index        5
Thread no        2   do index        2
Thread no        3   do index        7
Thread no        2   do index        9
Thread no        3   do index        8
Thread no        2   do index       10
Thread no        0   do index        4
Thread no        1   do index        6
Thread no        3   do index       11
Thread no        3   do index       12
```

The next example shows distributions of do loop among the threads when the schedule is (dynamic, 5).

```
        program sch_test
        use omp_lib
        implicit none
        integer::i, myid, a=0,b=0,c=0,d=0
        call omp_set_num_threads(4)     ! 4 threads for this run
!$omp parallel default(none) shared(a,b,c,d) private(myid)
!$omp barrier
```

```
!$omp do schedule(dynamic,5)
          do i=1,500
            myid=omp_get_thread_num()
            select case(myid)
              case(0)                 ! thread 0
                a=a+1
              case(1)                 ! thread 1
                b=b+1
              case(2)                 ! thread 2
                c=c+1
              case(3)                 ! thread 3
                d=d+1
            end select
          end do
!$omp end do
!$omp end parallel
    print *,"Thread 0, a= ", a
    print *,"Thread 1, b= ", b
    print *,"Thread 2, c= ", c
    print *,"Thread 3, d= ", d
    end program sch_test
```

An output for one run of the preceding program:

```
Thread 0, a=           305
Thread 1, b=            70
Thread 2, c=           110
Thread 3, d=            15
```

The output shows the number of iterations each thread has handled. It may be noted that these numbers will vary from one run to another.

21.49 GUIDED SCHEDULE

For guided schedule when no chunk size is specified, the default value 1 is taken. The chunk size parameter controls the size of the chunk. The thread executes one chunk of iteration, and after finishing, the operation asks for another chunk until all chunks are exhausted. When the chunk size is 1, the initial chunk is proportional to the number of iterations/number of threads. Subsequently, it is proportional to the number of iterations remaining/number of threads. For chunk size n, where n is greater than 1, the size of each chunk is calculated in a similar way with the restriction that is not less than n unless it is the last chunk. The preceding program with modification may be used to find out how the iterations have been distributed among the threads. We replace the line !$omp do schedule(dynamic,5) with !$omp do schedule(guided, 7). One typical result is shown next.

```
Thread 0, a=     288
Thread 1, b=     120
Thread 2, c=      76
Thread 3, d=      16
```

21.50 RUNTIME SCHEDULE

When a runtime schedule is selected, the environment variable OMP_SCHEDULE supplies the necessary parameter type (static, dynamic, guided, auto) and chunk size. Readers are advised to play with the parameter chunk and try to understand the output.

21.51 AUTO SCHEDULE

The compiler and the runtime system choose the most appropriate parameters to execute the parallel region.

21.52 Openmp Runtime Library Routines

There are 66 Openmp runtime library routines in Openmp version 5.0. In this section, only 14 essential runtime routines are discussed. Three such routines were introduced earlier. For the sake of completeness, they are also included in this section. To use these routines, there must be statement use omp_lib in the program.

omp_set_num_threads (integer): This subroutine sets the number of threads that the next parallel region will use. This is called outside the parallel region. The argument (constant or expression) is a positive integer greater than zero.

Example: call omp_set_num_threads (6)

The subsequent parallel region will use six threads.

omp_get_num_threads(): This function, if called within a parallel region, returns (integer) the number of threads available in the parallel region. If it is called in the serial region, it returns 1.

Example: print *, omp_get_num_threads ()

omp_get_thread_num(): Within the parallel region, threads are numbered between 0 and n-1, where n is the total number of threads in the team within the parallel region. The master thread gets a number 0. This function returns (integer) the thread number of the thread that has called the function. If it is called outside the parallel region, it returns 0, as outside the parallel region there is only one thread—the master thread.

Example: integer:: myid; myid=omp_get_thread_num ()

omp_get_max_threads(): The function returns (integer) the maximum value that will be returned when a call is made to omp_get_num_threads routine. This returns the same number both outside and inside the parallel regions.

Example: print *, omp_get_max_threads ()

omp_get_thread_limit(): This function returns (integer) the maximum number of threads available to a program.

Example: print *, omp_get_thread_limit ()

omp_get_num_procs(): This function returns (integer) the number of processors available to the calling program.

Example: `print *, omp_get_num_procs()`

omp_in_parallel(): This function returns `T (true)` if it called within a parallel region. If it returns `F (false)`, it is called within a serial region.

Example: `print *, omp_in_parallel()`

omp_set_dynamic(logical): This routine may be called with the argument `.true.` to adjust the number of available threads dynamically by the system within the parallel region. The system tries to manage the resource in the most efficient way. If it is necessary to have a fixed number of threads within the parallel region, this routine is called with the argument `.false.`, so the dynamic adjustment is switched off. The default is `.false.`.

Example: `call omp_set_dynamic(.true.)`

omp_get_dynamic(): This function returns `T (true)` or `F (false)` depending upon whether dynamic adjustment within the parallel region is allowed or not. The default is `.false.`.

Example: `print *, omp_get_dynamic()`

omp_set_nested(logical): The subroutine takes one logical as argument, `.true.` or `.false.`. If the argument is `.true.`, parallel nesting is allowed. If the argument is `.false.`, parallel nesting is disallowed.

Example: `call omp_set_nested(.true.)`

omp_get_nested(): This function returns (logical) the status of nested parallelism. If `T (true)`, nested parallelism is allowed; if `F (false)`, it is not allowed. The default is `.false.`.

Example: `print *, omp_get_nested()`

omp_set_max_active_levels(integer): This subroutine sets the number of nested levels allowed within parallel region. The default is compiler dependent.

Example: `call omp_set_max_active_levels(7) ! up to 7 nested level`

omp_get_max_active_levels(): This function returns the number of nested levels allowed within a parallel region. The default is compiler dependent.

Example: `print *, omp_get_max_active_levels()`

omp_get_level(): This function returns (integer) the current level number of the parallel region.

Example: `print *, omp_get_level() ! called within the parallel region`

omp_get_ancestor_thread_num(integer): This function takes an integer argument (level). It returns the thread number of the ancestor of the calling (current) thread at a given nested level.

For example, consider the following:

```
use omp_lib
call omp_set_nested(.true.)
call omp_set_num_threads(2)
!$omp parallel
print *, "Level-1",omp_get_thread_num()
```

```
!$omp parallel
print *, "Level-2",omp_get_thread_num()," Ancestor", &
  omp_get_ancestor_thread_num(1)
!$omp parallel
print *, "Level-3",omp_get_thread_num()," Ancestor ", &
  omp_get_ancestor_thread_num(2)
!$omp end parallel
!$omp end parallel
!$omp end parallel
end
```

The output is as follows:

```
Level-1        0
Level-1        1
Level-2        0  Ancestor        1
Level-2        1  Ancestor        1
Level-3        0  Ancestor        0
Level-3        1  Ancestor        0
Level-3        0  Ancestor        1
Level-3        1  Ancestor        1
Level-2        1  Ancestor        0
Level-2        0  Ancestor        0
Level-3        0  Ancestor        1
Level-3        1  Ancestor        1
Level-3        0  Ancestor        0
Level-3        1  Ancestor        0
```

21.53 Runtime Time Routines

There are two functions in this category: `omp_get_time` and `omp_wtick`. The function `omp_get_time` returns elapsed clock time in seconds. If two calls are made before and after a process, the difference is elapsed time in seconds to execute the code.

```
use omp_lib
real:: start,finish
integer::i, j, k
start=omp_get_wtime()
    .
finish=omp_get_wtime()
print *, finish-start   ! time taken to execute the code
end
```

The print statement will show the time (clock time) that has elapsed to execute the code. The other time function `omp_get_wtick` returns the number of seconds between successive clock ticks.

21.54 Environment Control

We have seen that we can control some of the parameters related to an Openmp construct in three different ways: (a) through an environment variable, (b) call to the runtime function and (c) using the clause associated with the construct. Consider a typical case. In these examples, the number of threads that will be available within a parallel region has been set by environment variable, call to the subroutine and using the clause associated with the construct as shown next:

```
SET OMP_NUM_THREADS=10
call omp_set_num_threads(8)
!$omp parallel num_threads(2)
```

The priority of the clause associated with !$omp parallel is more than the subroutine call, and the priority of the subroutine call is more than setting of the environment variable. In other words, the following relation holds for all the constructs whenever applicable.

  clause > subroutine call > setting of environment

In this case, the number of threads available within the parallel region is 2. If none of them is present, the default system value, which is system dependent, is assumed. However, if there is an if clause, it will have the highest priority.

```
         use omp_lib
         integer:: index=1
         call omp_set_num_threads(6)
!$omp parallel num_threads(3) if (index.eq.2)
         print *, " i am here"
!$omp end parallel
         end
```

In this case, the if returns false, so the omp parallel is ignored. The master thread moves serially ignoring the omp parallel directive.

21.55 Environment Variables

The environment variables are set from the command line of the operating system. Three important environment variables are shown in this section.

Linux (ubuntu):

```
<environmental variable>= <value>
export <environmental variable>
```

Windows

```
set <environmental variable>=<value>
```

omp_num_threads

```
set OMP_NUM_THREADS=10   (Windows)
OMP_NUM_THREADS=10 (Linux)
export OMP_NUM_THREADS
```

omp_nested

```
set OMP_NESTED TRUE   (Windows)
OMP_NESTED=true (Linux)
export OMP_NUM_THREADS
```

omp_dynamic

```
set OMP_DYNAMIC TRUE   (Windows)
OMP_DYNAMIC=true (Linux)
export OMP_DYNAMIC
```

omp_schedule

```
set OMP_SCHEDULE=STATIC/DYNAMIC/GUIDED/AUTO (Windows)
OMP_SCHEDULE= "STATIC,5"
export OMP_SCHEDULE
```

21.56 Programming Examples

1. Calculation of π using Monte Carlo method (algorithm has been discussed in Section 22.50):

```
      program pi_calc
      implicit integer(o)
      double precision:: x, y, sum=0.0d0, pi
      integer(kind=selected_int_kind(18)):: limit=80000000, i
      call omp_set_num_threads(4)
!$omp parallel
      call random_seed
!$omp do private(x,y) reduction(+:sum)
      do i=1, limit
       call random_number(x); call random_number(y)
       if (x*x+y*y <1.0d0) then
        sum=sum+1.0d0
       endif
      enddo
!$omp end do
!$omp end parallel
      pi=4.0d0*sum/limit
      print *, 'pi = ',pi
      end
```

2. Integration by trapezoidal rule:

```fortran
        program trap_1
        implicit none
        double precision::integral
        double precision, parameter::a=0.0d0, b=1.0d0 ! limits of integration
        integer, parameter::n=100
        double precision::h,x
        integer::i
        h=(b-a)/n
        integral=0.0d0
 !$omp parallel do private(x) reduction(+:integral)
        do i=1,n-1
         x=a+i*h; integral=integral+fun(x)
        end do
 !$omp end parallel do
        integral=(integral+(fun(a)+fun(b))/2.0d0)*h
        print *, "Result = ", integral
        contains
         double precision function fun(x)
         double precision::x
         fun=exp(x*x)
         end function fun
        end
```

3. Calculation of dot product:

```fortran
        program dot_p
        use omp_lib ! calculation of dot product
        implicit none
        integer :: i
        integer, parameter :: limit=20
        real, dimension(limit) :: a, b
        real:: res
        call omp_set_num_threads(4) ! 4 threads will be used
 !$omp parallel do
        do i = 1, limit ! i is private, a and b are shared by default
         a(i) = i * 5.0; b(i) = i * 10.0
        end do
 !$omp end parallel do
        res = 0.0
 !$omp parallel do reduction(+:res)
        do i = 1, limit
         res = res + (a(i) * b(i)) ! outer left and right brackets may
                                   ! be omitted
        end do
 !$omp end parallel do
        print *, 'The result is  = ',res
        end program dot_p
```

21.57 Final Word

One should be very careful while parallelizing serial code. It is always better to check the results of the serial code against the parallel code with same sets of data as far as practicable. The most important assumption usually made while parallelizing is that within the parallel region the calculation of some number by a particular thread should not depend on a value to be obtained from another thread. As the threads normally move asynchronously and the speed with which the threads move may be different, there should not be any dependency of one thread on another thread. A careful check of program logic will eliminate such errors. It must be remembered that the exact serial logic may not work in the parallel region. It may have to be modified to suit the needs of the parallel region.

22

Parallel Programming Using Message Passing Interface (MPI)

MPI is the abbreviation for message passing interface. It consists of a library of Fortran subroutines (also C functions) that the programmer calls explicitly to write parallel programs. A parallel computation, using MPI routines, is performed by a number of processors, each having its own local memory to execute the task assigned to it and communicate with each other through messages. A process cannot directly access the data of another process, and sharing of data is achieved only through messages. Usually processes run in different processors but may not always be running in different processors. The processors may also be heterogeneous. The source code is usually portable across the processors. The number of processors may be prescribed during the execution of the program. The basic principle of parallelizing is to divide the job among the processors so that a proper load balancing is maintained for efficient running of the program.

The serial program written with MPI routines is replicated in all processors, and the program is usually written in such a manner that each processor performs a specific task, and through message transfer, ultimately the final result is obtained using the partial results received from different processors.

22.1 MPI Module

This header file must be used in all MPI programs—use mpi. This file contains definitions, macros and function prototypes required for compiling MPI programs.

22.2 Compilation

The Intel ifort compiler and GCC gfortran compiler have been used to test all programs given in this chapter. To use the Intel compiler, the environment must be set using the files supplied by Intel.

For Windows: Two batch files supplied by Intel must be executed before using Intel compilers. The paths are shown in author's laptop. The paths may be different in other computers.

C:\"Program Files(x86)"\intelSWTools\parallel_studio_xe_2019\bin\psxevars.bat intel64

C:\"Program Files (x86)"\intelSWTools\compilers_and_libraries_2019\windows\mpi\intel64\bin\mpivars.bat

In fact, the first batch file is to be executed to set the environment before using the ifort compiler. Both batch files are to be executed before compiling and running programs containing calls to MPI routines. To compile and run an MPI program under Windows, the commands are as follows:

```
mpiifort a.f90 -o a [to compile a.for]
mpiexec -np 4 a [to execute the compiled MPI program with 4 processors].
```

For Linux (ifort): Again, here two files are to be executed in the current shell using the source Linux command to create the proper environment (in the author's laptop).

source /opt/intel/parallel_studio_xe_2019/bin/psxevars.sh

source /opt/intel/parallel_studio_xe_2019/compilers_and_libraries_2019/ linux/mpi/intel64 /bin/mpivars.sh

To compile and run MPI program under ifort (Linux), the commands are as follows:

```
mpiifort m.f90 -o m [to compile m.for]
mpiexec -np 4 m [to execute the compiled MPI program with 4 processors].
```

For Linux (gfortran): To compile and run MPI programs under gfortran, the instructions are as follows:

```
mpif90 r.f90 -o r [to compiler r.for]
mpirun -np 4 r [to execute the compiled MPI program with 4 processors]
```

22.3 Error Parameter of MPI Routines

Except for `mpi_wtime` and `mpi_wtick`, all MPI Fortran routines return an integer error value. This is the last argument of all Fortran subroutines. In this chapter, we always use this integer variable as `ierr`. It returns a named constant `mpi_success` if there is no error. It returns `mpi_err_other` if the subroutine returns an error. Normally, the job is aborted when there is an error. These named constants are defined in the module `mpi`.

22.4 MPI Version

MPI version and subversion can be obtained in two ways. The named integer constants `mpi_version` and `mpi_subversion` contain the version and subversion of `mpi` being used by the system. In addition, to make enquires about the version of MPI, the subroutine `mpi_get_version` may be called. The subroutine `mpi_get_version` takes 3 integers as its arguments. The first two arguments return the version and subversion, respectively.

```
program mpi_version_test
use mpi
```

```
integer:: ierr, ver, subver
print *, mpi_version, mpi_subversion
call mpi_get_version(ver, subver, ierr)
print *, ver, subver
end program mpi_version_test
```

The print statement displays 3 and 1, so the returned values from the subroutine mpi_get_version also indicate that the MPI version 3.1 is being used by the present system.

22.5 MPI_INIT

This subroutine must be called before using any other MPI routines. There are a few exceptions. For example, mpi_get_version can be called before calling mpi_init. This routine sets the environment of mpi. It returns an error code if it fails to initialize. The syntax is call mpi_init (ierr).

22.6 MPI_INITIALIZED

This routine may be called to find out whether mpi_init has been called or not. The routine takes two arguments: flag (logical) and ierr (integer). Flag returns true if mpi_init has been called; otherwise, it returns false.

```
    use mpi
    integer:: ierr
    logical:: flag
!   call mpi_init(ierr)
    call mpi_initialized(flag, ierr)
    print *, flag
    end
```

The output from this program is F (false). However, if the comment from call mpi_init is removed, it returns T (True). It may be noted that this routine can be called before calling mpi_init.

22.7 MPI_FINALIZE

This is the final MPI routine to be called at the end of all the MPI operations. This cleans up the MPI environment in a decent way. Once this is called, no other MPI routines can be called. This includes mpi_init. The syntax is call mpi_finalize(ierr). However, the routines like mpi_version can be called after calling mpi_finalize.

22.8 MPI Handles

Internally MPI maintains it own data structure that the programmer can access through an MPI handle. Many MPI routines use these handles as arguments both for input and output operations. In Fortran, the handles are usually integer constants or integer arrays (lower bound is 1). We frequently use the handle mpi_comm_world. This is an integer communicator consisting of all processors participating in a particular run.

22.9 About This Chapter

In this chapter, unless otherwise specified the following will be assumed for all programs or program segments: (a) The number of participating processors is four. (b) Most of the time the MPI handle is mpi_comm_world. (c) Count, record count and the like are non-zero positive integers. (d) The last parameter of Fortran subroutines is an integer ierr (there are two exceptions wtick and wtime) that returns the error code (success or failure of the subroutines).

22.10 Structure of a MPI Program

The structure of a typical MPI program is shown as follows:

```
use mpi
<declaration>
call mpi_init(ierr)
<statements>
call mpi_finalize(ierr)
end
```

All participating processors execute the same executable module. In other words, the code is replicated in all the processors.

```
use mpi
integer:: ierr
call mpi_init(ierr)
print *, 'I am using MPI'
call mpi_finalize(ierr)
end
```

If this program is executed with number of processors=4, all the processors will execute identical code and display the same message. The output will be as follows:

```
I am using MPI
I am using MPI
I am using MPI
I am using MPI
```

If the same code file is executed with np=10, the same character string will be displayed 10 times.

22.11 MPI_COMM_RANK

The system assigns a serial number (rank) to each of the participating processors. This starts at 0 and goes up to n-1, where n is the number of processors. Each processor may get its own rank by calling the subroutine mpi_comm_rank. The syntax is call mpi_comm_rank (mpi_comm_world, rank, ierr). The rank of the calling processor is returned through the integer variable rank.

```
use mpi
integer:: ierr, rank
call mpi_init(ierr)
call mpi_comm_rank(mpi_comm_world, rank, ierr)
print *, 'I am using MPI, my rank is: ', rank
call mpi_finalize(ierr)
end
```

The output is as follows:

```
I am using MPI, my rank is: 2
I am using MPI, my rank is: 0
I am using MPI, my rank is: 1
I am using MPI, my rank is: 3
```

This ordering of the outputs may change from one run to another, as it cannot be predicted how the different processors will move inside the MPI environment.

22.12 MPI_COMM_SIZE

This subroutine returns the size (integer) of the group associated with a communicator. In other words, it returns the total number of participating processors. The syntax is call mpi_comm_size (mpi_comm_world, size, ierr). In the following program, as the total number of participating processors is same for all processors, the value of size is same for all processors.

```
use mpi
integer:: ierr, rank, size
call mpi_init(ierr)
call mpi_comm_rank(mpi_comm_world, rank, ierr)
call mpi_comm_size(mpi_comm_world, size, ierr)
print *, 'I am using MPI, my rank is: ', rank, 'of total size ', size
call mpi_finalize(ierr)
end
```

The output is as follows:

```
I am using MPI, my rank is: 3 of total size 4
I am using MPI, my rank is: 2 of total size 4
I am using MPI, my rank is: 1 of total size 4
I am using MPI, my rank is: 0 of total size 4
```

22.13 Use of Rank in Controlling the Flow of the MPI Program

Until now, all processors execute the same set of instructions. Though the same code is replicated in all processors, normally all processors do not follow identical paths. One of the methods to alter the flow of program in different processors is through the variable rank (rank of the processors).

```
use mpi
integer:: ierr, rank, size
call mpi_init(ierr)
call mpi_comm_rank(mpi_comm_world, rank, ierr)
call mpi_comm_size(mpi_comm_world, size, ierr)
if (rank==0) then
 print *, 'Total size is', size ! only processor 0 executes print
endif
call mpi_finalize(ierr)
end
```

The output is as follows:

```
Total size is 4
```

Though all processors execute the same code, yet only processor 0 will execute the print statement, as the if statement returns true only for processor 0 (rank 0). The if statement returns false for processors 1, 2 and 3. Therefore, the print statement will be skipped for the processors 1, 2 and 3.

22.14 MPI_BARRIER

Normally, all processors participating in a computation do not move with the same speed. If it is necessary to synchronize their movement, this subroutine mpi_barrier is called. No process can cross the barrier unless all processors call this routine. Thus, if a processor reaches the barrier ahead of other processors, it has to wait until all other processors reach the barrier and call the mpi_barrier subroutine. One trivial application of mpi_barrier subroutine could be, say, one process reads a few variables from the keyboard. Unless the variables are read, other processes cannot proceed; they have to wait until the reading is over. The following program segment illustrates this point:

```
if (rank==0) then ! process 0 will read the data
 read *, a, b, c
end if
call mpi_barrier(mpi_comm_world, ierr) ! other processes will wait
                                        ! here
<next statement>
```

While processor 0 reads data from the keyboard, other processors will wait at the mpi_barrier, and after reading is over, all processes will be able to execute the instruction following the call mpi_barrier statement. This will ensure that the variables a, b, c get their values before other processes are allowed to proceed to execute the next statement.

22.15 Basic MPI Datatype in Fortran

Various intrinsic datatypes in Fortran are known to MPI by different names. These MPI names are used as arguments of MPI routines when it is necessary to specify type of the argument. Table 22.1 shows the correspondence between the MPI datatypes and the corresponding Fortran intrinsic types.

TABLE 22.1

Relation between MPI Datatype and Fortran Datatype

MPI Datatype	Fortran Datatype
mpi_integer	integer
mpi_real	real
mpi_double_precision	double precision
mpi_complex	complex
mpi_double_complex	double complex
mpi_character	character(1)
mpi_logical	logical
mpi_byte	none
mpi_packed	none

22.16 Point-to-Point Communication

In a point-to-point communication, one processor sends a message to another processor and the second processor receives the message. The message consists of two parts: an envelope and the message itself. The envelope consists of the source, destination, communicator and a tag to classify the message. The message body consists of the name of the buffer where the message is stored, type of the data and total number of items being sent. The receiver looks at the envelope, identifies the message and accepts the message. The whole process takes place within a communicator.

22.17 Communication Modes

There are four ways through which the message can be sent. They are synchronous sent, buffered sent, standard sent and ready sent.

Synchronous Sent

The transfer is considered successfully over when the receiver has received the message.

Buffered Sent

The message is transferred to a buffer. Unless there is an error, the transfer is considered over irrespective of the status of the receiver.

Standard Sent

The standard sent is considered over irrespective of the condition of the receiver. The message may wait in the communication network.

Ready Sent

A ready sent is somewhat similar to buffered sent. It completes immediately. If the matching receive has been posted, it is guaranteed to succeed. If no matching receive has yet been posted, the result of send-receive is not defined.

22.18 Message Sent and Received

The four `mpi` send modes are handled, respectively, by four subroutines: `mpi_ssend`, `mpi_bsend`, `mpi_send` and `mpi_rsend`. The subroutine `mpi_recv` is used to receive a message.

22.19 MPI_SEND and MPI_RECV

The subroutine `mpi_send` takes seven arguments. They are `buffer`, `count`, `datatype`, `destination`, `tag`, `communicator` and `ierr`. The `buffer` contains the message to be sent; `count` is the size of the message of type `datatype`; `destination`, as the name implies, is the `rank` of the `destination` processor; `tag` is an integer through which the receiver identifies the message; `communicator`, in our present case, is `mpi_comm_world`; and `ierr`, an integer, returns the success or failure of the subroutine (error code). The subroutine `mpi_recv` takes eight arguments: `buffer`, `count`, `datatype`, `source`, `tag`, `communicator`, `status` and `ierr`. The `buffer` is the place where the message will be stored; `count` is the size of the message of type `datatype`; `source` is the rank of the `source` processor; `tag` is an integer that must match with the `tag` of the `source` (Any source and Any tag will be discussed separately); `communicator`, in our present case, is `mpi_comm_world`; `status` is an integer array of size `mpi_status_size`; and `ierr`, an integer, returns the success or failure of the subroutine (error code).

```
program mp_send_recv
use mpi
implicit none
integer:: ierr, rank
integer, parameter::limit=10
integer:: status(mpi_status_size)
real, dimension(limit):: a
integer, dimension(limit)::x, y=0
call mpi_init(ierr)
call mpi_comm_rank(mpi_comm_world, rank, ierr)
if (rank==0) then
```

```
call random_seed()
call random_number(a) ! 10 random numbers for a
x=a*1000.0                ! converted to integer after multiplying by 1000
print *, 'x= '
print *, x
call mpi_send(x, limit, mpi_integer,1, 25,mpi_comm_world, ierr)
else if(rank==1) then
call mpi_recv(y, limit, mpi_integer,0,25,mpi_comm_world, status, ierr)
print *, 'y array of rank 1 received from x array of Rank 0 ='
print *,y ! y is same as x as it received data from processor 0
end if
call mpi_finalize(ierr)
end program mp_send_recv
```

The output is as follows:

```
x=
    364    598    715    838    846    28
    607    57    478    351
y array of rank 1 received from x array of Rank 0 =
    364    598    715    838    846    28
    607    57    478    351
```

In this program, random numbers are generated and integerized after multiplying by 1000. After the y array of processor 1 receives data from the x array of processor 0 through the subroutine mpi_send, y=x, and this is displayed by the print statement. The next example shows how a string of characters is transferred from one processor to another.

```
program send_recv_char
use mpi
implicit none
integer:: ierr, rank
integer:: status(mpi_status_size)
character(len=50)::msg, msg1
call mpi_init(ierr)
call mpi_comm_rank(mpi_comm_world, rank, ierr)
if (rank==0) then
 msg='Indian Association for the Cultivation of Science '
 print *, 'Message Transmitted: ', msg
 call mpi_send(msg, 50, mpi_character,1, 18, mpi_comm_world, ierr)
else if(rank==1) then
 call mpi_recv(msg1,50,mpi_character,0,18,mpi_comm_world, status, ierr)
 print *, 'Message Received: ', msg1
end if
call mpi_finalize(ierr)
end program send_recv_char
```

Both the subroutines mpi_send and mpi_recv do not return before the communication, they had initiated, has been completed.

The output is as follows:

```
Message Transmitted: Indian Association for the Cultivation of Science
Message Received: Indian Association for the Cultivation of Science
```

Send and receive can be combined and may be called through a single mpi call: mpi_sendrecv. This subroutine takes 13 arguments. They are send buffer, send count, send type, destination, send tag, receive buffer, receive count, receive type, source, receive tag, comm, status and ierr. This is illustrated with the following program. In this program, process 0 sends message to process 1 and receives one from process 2. Similarly, process 1 sends message to process 2 and receives one from process 0, and process 2 sends message to process 0 and receives one from process 1.

```
program send_recv_test
use mpi
integer:: rank, size, ierr, l1,l2
integer, parameter:: np=3 ! program will be executed with 3 processors
integer:: status(mpi_status_size)
real:: x, y
call mpi_init(ierr)
call mpi_comm_rank(mpi_comm_world, rank, ierr)
call mpi_comm_size(mpi_comm_world, size, ierr)
 if (np /=size) then
 print *, 'np ', np, 'is not equal to size ',size, '- error stop'
 error stop ! immediately terminate all the processes
 endif
x=(rank+5)*10.0 ! x of process 0, 1, 2 are 50, 60, 70, respectively
l1=mod(rank+1, size) ! l1 for rank 0, 1 and 2 are 1, 2, 0
l2=mod(rank+size-1, size) ! l2 for rank 0, 1, 2 are 2, 0, 1
print *, "Sending from ", rank, "to ", l1, "and receiving from ", l2
call mpi_sendrecv(x, 1, mpi_real, l1, 17, y, 1, mpi_real, &
 l2, 17, mpi_comm_world, status, ierr)
print *, rank, "send recv over"
call mpi_barrier(mpi_comm_world, ierr)
select case(rank)
 case(0)
  print *, "rank ", rank, y ! receives from process 2
 case(1)
  print *, "rank ", rank, y ! receives from process 0
 case(2)
  print *, "rank ", rank, y ! receives from process 1
end select
call mpi_finalize(ierr)
end program send_recv_test
```

The output is as follows:

```
Sending from  1  to  2 and receiving from    0
Sending from  0  to  1 and receiving from    2
    1 send recv over
Sending from  2  to  0 and receiving from    1
    2 send recv over
    0 send recv over
  rank    0 70.00000
  rank    2 60.00000
  rank    1 50.00000
```

The outputs show that process 0, 1 and 2 received messages from process 2, 0 and 1, respectively.

22.20 MPI_SSEND

This subroutine mpi_ssend is used for synchronous send. The subroutine mpi_send in the preceding program was used to send message in synchronous mode. Therefore, the result will be same if the subroutine mpi_send is replaced by mpi_ssend in this case.

22.21 MPI_BSEND

For this type of communication, the message is copied to a buffer and the subroutine mpi_bsend completes its task. The data is transmitted later. The size of the buffer is calculated using the subroutine mpi_pack_size having five arguments: count, datatype, comm, size and ierr, where count is an integer of type datatype, which specifies the size of the input; comm is mpi_comm_world; size is the returned value (bytes) as integer; and ierr is the error code. The size of the buffer should be greater than size value returned from the subroutine plus mpi_bsend_overhead defined in the module mpi. Subsequently, a call to subroutine mpi_buffer_attach is required to attach the user-supplied buffer for transmitting data. The subroutine mpi_buffer_attach takes three arguments: name of the buffer to be used for sending, buffer size obtained from the output of the subroutine mpi_pack_size plus mpi_bsend_overhead, third one being the usual ierr. Usually the existing buffer is detached when it is no longer required by calling the subroutine mpi_buffer_detach having the same arguments as mpi_buffer_attach. All these features are demonstrated in the following program:

```
program mp_bsend_recv
use mpi
implicit none
integer, parameter:: limit=24
integer:: ierr, rank, s1, sz
integer:: status(mpi_status_size)
real, dimension(limit):: a
integer, dimension(limit)::x, y=0
integer, allocatable, dimension(:)::z ! allocatable buffer
call mpi_init(ierr)
call mpi_comm_rank(mpi_comm_world, rank, ierr)
call mpi_pack_size(limit, mpi_integer, mpi_comm_world, s1,ierr)
sz=s1+mpi_bsend_overhead
allocate(z(sz/4 + 1)) ! allocated after calculating the size
if (rank==0) then ! one integer=4 bytes
 call random_seed()
 call random_number(a)
 x=a*1000.0
 print *, 'x= '
 print *, x
 call mpi_buffer_attach(z, sz, ierr) ! buffer attached
 call mpi_bsend(x, limit, mpi_integer,1,25,mpi_comm_world, ierr)
else if(rank==1) then
 call mpi_recv(y, limit, mpi_integer,0,25,mpi_comm_world, status, ierr)
 print *, 'y array of rank 1 received from x array of Rank 0 ='
```

```
   print *,y
   end if
   call mpi_buffer_detach(z, sz, ierr) ! buffer detached
   call mpi_finalize(ierr)
   deallocate(z) ! release the buffer
   end program mp_bsend_recv
```

The output is as follows:

```
x=
    554    302    100    827    508    801
    857     18    264    287    628    573
    742    936     71    491    195    702
    195    392    696    727    781    321
y array of rank 1 received from x array of Rank 0 =
    554    302    100    827    508    801
    857     18    264    287    628    573
    742    936     71    491    195    702
    195    392    696    727    781    321
```

22.22 MPI_RSEND

This type of send is similar to buffer send; it completes it task immediately. If the matching receive (recv) is available, the communication will always be successful. If no matching receive (recv) is posted, the result is undefined. In this chapter, we mainly use mpi_send and mpi_recv.

22.23 Deadlock

Deadlock is the situation when two or more processes cannot proceed because both of them depend on the other process to move first. In such cases, no messages could be sent and no messages could be received. Consider the following program:

```
program deadlock_1
use mpi
integer:: myrank, ierr, status(mpi_status_size), size
real, dimension(10):: x, y
call random_seed()
call mpi_init(ierr)
call mpi_comm_rank(mpi_comm_world, myrank, ierr)
call mpi_comm_rank(mpi_comm_world, size, ierr)
call random_number(x)
call random_number(y)
if (myrank==0) then
  call mpi_recv(y, 10, mpi_integer, 1, 12, mpi_comm_world, status, ierr)
  call mpi_send(x, 10, mpi_integer, 1, 18, mpi_comm_world, ierr)
else if (myrank==1) then
  call mpi_recv(y, 10, mpi_integer, 0, 18, mpi_comm_world, status, ierr)
```

```
    call mpi_send(x, 10, mpi_integer, 0, 12, mpi_comm_world, ierr)
  endif
  call mpi_finalize(ierr)
  print *, 'Job is over'
  end program deadlock_1
```

The processor 0 wants to receive message first from processor 1 and then send a message to processor 1. Again, processor 1 wants to receive message first from processor 0 that processor 0 cannot send unless it receives message from processor 0. This program will wait forever; it simply cannot move. This is a deadlock situation. This type of problem can be solved in many ways. One method is to rearrange the send and recv. In the program deadlock_2, the send and recv are interchanged when myrank is equal to 1 in the if statement. Now processor 0 will receive message from processor 1 and send message to processor 1, and processor 1 will be able to accept the message. After these operations, the program will normally terminate.

```
program deadlock_2
use mpi
integer:: myrank, ierr, status(mpi_status_size), size
real, dimension(10):: x, y
call random_seed()
call mpi_init(ierr)
call mpi_comm_rank(mpi_comm_world, myrank, ierr)
call mpi_comm_rank(mpi_comm_world, size, ierr)
call random_number(x)
call random_number(y)
if (myrank==0) then ! deadlock
 call mpi_recv(y, 10, mpi_integer, 1, 12, mpi_comm_world, status, ierr)
 call mpi_send(x, 10, mpi_integer, 1, 18, mpi_comm_world, ierr)
else if (myrank==1) then
 call mpi_send(x, 10, mpi_integer, 0, 12, mpi_comm_world, ierr)
 call mpi_recv(y, 10, mpi_integer, 0, 18, mpi_comm_world, status, ierr)
endif
call mpi_finalize(ierr)
print *, 'Job is over'
end program deadlock_2
```

22.24 Non-blocking Send and Receive

MPI provides another set of subroutines that does not block the calling process. These non-blocking send and recv need two calls to the subroutines for each communication. One initiates the communication, sometimes known as posting, and the second call finishes the communication. It may be noted that between these two calls, the program can execute other instructions, as it does not wait for completion like send and recv discussed in an earlier section. There are two different methods for finishing the communication. In the first method, a process can wait through a suitable subroutine call until the communication is over. The second method is to test repeatedly by a suitable subroutine call to check whether the data transfer is over or not. This type of non-blocking send and recv returns a handle called request (integer) that can be used to check the status of the communication.

```
program non_block_1
use mpi
integer:: myrank, ierr, status(mpi_status_size), size, request
real, dimension(10):: x, y
call random_seed()
call mpi_init(ierr)
call mpi_comm_rank(mpi_comm_world, myrank, ierr)
call mpi_comm_rank(mpi_comm_world, size, ierr)
call random_number(x); call random_number(y)
if (myrank==0) then
 call mpi_irecv(y, 10, mpi_integer, 1, 12, mpi_comm_world, request, ierr)
 call mpi_send(x, 10, mpi_integer, 1, 18, mpi_comm_world, ierr)
 call mpi_wait(request, status, ierr)
else if (myrank==1) then
 call mpi_irecv(y, 10, mpi_integer, 0, 18, mpi_comm_world, request, ierr)
 call mpi_send(x, 10, mpi_integer, 0, 12, mpi_comm_world, ierr)
 call mpi_wait(request, status, ierr)
endif
call mpi_finalize(ierr)
print *, 'Job is over'
end program non_block_1
```

In the program non_block_1, the processor 0 initiates irecv from processor 1 but does not wait. It then sends a message to processor 1, and after returning from the send routines, it waits until the recv finishes its work. Similarly, processor 1 initiates irecv from processor 0 but does not wait. It then sends a message to processor 0, and after returning from the send routine, it waits until the recv finishes its work. The next program uses mpi_test in place of mpi_wait.

```
program non_block_2
use mpi
integer:: myrank, ierr, status(mpi_status_size), size, request
real, dimension(10):: x, y
logical::flag, flag1
call random_seed()
call mpi_init(ierr)
call mpi_comm_rank(mpi_comm_world, myrank, ierr)
call mpi_comm_rank(mpi_comm_world, size, ierr)
call random_number(x)
if (myrank==0) then
 print *, 'Rank ', myrank; print *, x
endif
call random_number(y)
if (myrank==0) then
 call mpi_irecv(y, 10, mpi_integer, 1, 12, mpi_comm_world, request, ierr)
 call mpi_send(x, 10, mpi_integer, 1, 18, mpi_comm_world, ierr)
 flag=.false.
 do while (.not. flag) ! loops until flag becomes true
   call mpi_test(request, flag, status, ierr)
 end do
else if (myrank==1) then
 call mpi_irecv(y, 10, mpi_integer, 0, 18, mpi_comm_world, request, ierr)
 call mpi_send(x, 10, mpi_integer, 0, 12, mpi_comm_world, ierr)
 flag1=.false.
 do while(.not. flag1) ! loops until flag1 becomes true
```

```
    call mpi_test(request, flag1,status, ierr)
  enddo
  endif
  if (myrank==1) then
  print *, 'Rank ', myrank; print *, y
  endif
  call mpi_finalize(ierr)
  end program non_block_2
```

The output is as follows:

```
Rank   0
     0.8859678   0.7051181      0.9824182     0.9144161     0.8029735
     0.1468188   3.4957543E-02  0.3678806     0.5533310     0.8573631
Rank   1
     0.8859678   0.7051181      0.9824182     0.9144161     0.8029735
     0.1468188   3.4957543E-02  0.3678806     0.5533310     0.8573631
```

The subroutine mpi_test returns true when the isend or irecv finishes itsr work.
The logical variable flag or flag1 is set to .false. outside the do while loop. When
flag or flag1 becomes .true., the do while loop is exited. The x array of processor 0
is transferred to y array of processor 1.

22.25 Send Function-Naming Conventions in Blocking and Non-blocking Forms

There are eight subroutines in this category—four for blocking forms and four for non-blocking forms. The subroutines for non-blocking forms have 'i' prefixed to their blocking names (Table 22.2).

TABLE 22.2

Naming Conventions in Blocking and Non-blocking Forms

Send Mode	Blocking Function	Non-blocking Function
Standard	mpi_send	mpi_isend
Synchronous	mpi_ssend	mpi_issend
Ready	mpi_rsend	mpi_irsend
Buffered	mpi_bsend	mpi_ibsend

22.26 MPI_ANY_TAG and MPI_ANY_SOURCE

The recv can accept mpi_any_tag and mpi_any_source (wildcard) in place of a definite source and tag. The relevant program segment to reflect the changes is shown as follows:

```
if (myrank==0) then
  call random_seed(); call random_number(a)
  x=a*1000.0
```

```
    call random_number(a); y=a*1000.0
    print *, 'x= '; print *, x
    call mpi_send(x, limit, mpi_integer,1, 25,mpi_comm_world, ierr)
  else if(myrank==1) then
    call mpi_recv(y, limit, mpi_integer, mpi_any_source, &
     mpi_any_tag, mpi_comm_world, status, ierr)
    print *, 'y array of rank 1 received from x array of Rank 0 ='
    print *,y
  end if
```

22.27 REDUCTION

Reduction is an operation where the routine `mpi_reduce` collects data from each processor, reduces it to a single value and stores the result in one of the processors specified in the subroutine. We illustrate this procedure with the help of a few programs. The first program calculates the sum of the series 1 + 2 + 3 ... + 100. We divide the job among 4 participating processors, calculate the partial sums, add the partial sums and store the final result in the processor 0 (Figure 22.1).

```
program reduc_1
use mpi
implicit none ! reduction
integer:: ierr, rank, size, sum=0, i, np1, total
integer, parameter:: limit=100
call mpi_init(ierr)
call mpi_comm_rank(mpi_comm_world, rank, ierr)
call mpi_comm_size(mpi_comm_world, size, ierr)
np1=rank+1      ! np1 is 1, size=4
do i=np1, limit, size
  sum=sum+i
end do
call mpi_barrier(mpi_comm_world, ierr)
print *, 'Rank ', rank, 'Sum ', sum
call mpi_reduce(sum, total,1,mpi_integer, mpi_sum,0,mpi_comm_world, ierr)
if (rank==0) then
  print *, 'Sum of 1 + 2 + 3... +',limit, '= ', total
endif
call mpi_finalize(ierr)
end program reduc_1
```

The output is as follows:

```
Rank    1 Sum    1250
Rank    3 Sum    1300
Rank    2 Sum    1275
Rank    0 Sum    1225
Sum of 1 + 2+3 ... +100 = 5050
```

sum	1225	P0	total	5050
sum	1250	P1		
sum	1275	P2		
sum	1300	P3		

FIGURE 22.1
Before and after reduction (partial results of processors).

Let us first examine the `do` statement. As `np1` is `rank+1`, the value of `np1` in each processor is 1, 2, 3 and 4, respectively. The number of processors is 4, so the value of the variable `size` is 4. Therefore, the do loop in processor 0 is of the form do i=1, 100, 4; for processor 1, it is do i=2, 100, 4; for processor 3, it is do i=3, 100, 4; and for processor 4, it is do i=4, 100, 4. If we expand the do loop, we find that the do loop for processor 0 will add 1 + 5 + 9 + 13; processor 1 will add 2 + 6 + 10 + 14; processor 2 will add 3 + 7 + 11 + 15; and processor 3 will add 4 + 8 + 12 + 16. It is needless to point out that if partial sums are added, we get the total sum of the series 1 + 2 + 3 + 4 . . . +100. The mpi_reduce takes 8 arguments: (a) send buffer, where in this case the partial sum is stored; (b) receive buffer, where the total sum will be stored; (c) count, which is the number of items to be reduced; (d) datatype—in this case it is integer, that is, mpi_integer; (e) reduction_procedure—in this case it is mpi_sum; (f) rank of the receiving process—in this case it is 0; (g) communicator that is mpi_comm_world; and (h) ierr, like any other mpi subroutine. The fifth argument of the subroutine is the reduction procedure that may be some intrinsic arithmetic or a few library functions or user-defined operators. These are summarized in Table 22.3.

TABLE 22.3

Predefined Reduction Operations

Reduction Operations	Description
mpi_max	maximum
mpi_min	minimum
mpi_sum	sum
mpi_prod	product
mpi_land	logical and
mpi_band	bitwise and
mpi_lor	logical or
mpi_bor	bitwise or
mpi_lxor	logical exclusive or
mpi_bxor	bitwise exclusive or
mpi_maxloc	max value and location
mpi_minloc	min value and location

To use mpi_maxloc and mpi_minloc with mpi_reduce, three separate datatypes have been provided. These datatypes consist of a pair of intrinsic datatypes (Table 22.4).

TABLE 22.4

Special MPI Datatype for MPI_MAXLOC and MPI_MINLOC

Name	Description
mpi_2real	pair of reals
mpi_2double_precision	pair of double precision variable
mpi_2integer	pair of integers

The second example reduces an array for all the processors element by element.

```
program reduc_2
use mpi
implicit none
integer:: ierr, rank, size, i, np1, total
integer, parameter:: limit=5
real, dimension(limit)::a
```

```
integer, dimension(limit)::x, y
call mpi_init(ierr)
call mpi_comm_rank(mpi_comm_world, rank, ierr)
call mpi_comm_size(mpi_comm_world, size, ierr)
call random_seed()
call random_number(a)
x=1000.0*a
print *, 'Rank = ', rank, x
call mpi_barrier(mpi_comm_world, ierr)
call mpi_reduce(x, y,limit, mpi_integer, mpi_sum,0,mpi_comm_world, ierr)
if (rank==0) then
 print *,      'y =      ', y
endif
call mpi_finalize(ierr)
end program reduc_2
```

The output is as follows:

Rank	=	1	151	62	900	160	388
Rank	=	0	727	993	490	155	19
Rank	=	2	687	22	813	316	202
Rank	=	3	983	725	645	933	721
y	=		2548	1802	2848	1564	1330

In the second example, each of the processors generates 5 random numbers. The random numbers are then multiplied by 1000.0, integerized and stored in the array x. The x locations were reduced using the operator mpi_sum and stored in the array y. This essentially means that y(1) is the sum of all the four x(1)s of the four processors used in this computation. Similarly, y(2) is the sum of all the four x(2)s of the four processors used in this computation and so also the y(5). The reductions have been performed element-wise of x's corresponding to each processor. We now use the same program to calculate the minimum value of x's corresponding to each processor. The same program was used with mpi_min operator in place of mpi_sum. This time reduction is done element by element of x's. We replace the line of the preceding program call mpi_reduce (x, y,limit, mpi_integer, mpi_sum,0,mpi_comm_world, ierr) by

```
call mpi_reduce(x, y,limit, mpi_integer, mpi_min,0,mpi_comm_world, ierr).
```

The output is as follows:

Rank	=	1	403	498	4	413	903
Rank	=	0	275	132	427	25	52
Rank	=	3	475	419	830	726	531
Rank	=	2	643	755	85	954	199
y	=		275	132	4	25	52

The program reduc_4 reduces complex variables.

```
program reduc_4
use mpi
implicit none
integer:: ierr, rank, size, i, np1
integer, parameter:: limit=3 ! ifort compiler
complex, dimension(limit)::a, total=(0.0,0.0)
real, dimension(limit)::b, c
```

```
real::x=0.0, y=0.0
call mpi_init(ierr)
call mpi_comm_rank(mpi_comm_world, rank, ierr)
call mpi_comm_size(mpi_comm_world, size, ierr)
call random_seed()
call random_number(b)
call random_number(c)
a=cmplx(b, c)     ! complex a-array
print *, 'Rank = ',rank, a
call mpi_barrier(mpi_comm_world, ierr) ! wait
call mpi_reduce(a, total, limit, mpi_complex, mpi_sum,0, &
 mpi_comm_world, ierr)
if (rank==0) then
 print *, 'Using Reduction: ', total
endif
call mpi_finalize(ierr)
end program reduc_4
```

The output is as follows:

```
Rank = 1 (0.4939204,0.8224367) (0.1193415,0.9941474)(0.8722481,0.3784710)
Rank = 2 (0.5660955,0.1353348) (3.9733693E-02,0.6224477)(0.6981519,0.9876670)
Rank = 0 (0.9018703,0.5901801) (0.7135152,0.9574903)(0.2077041,0.3472721)
Rank = 3 (0.2700705,0.5192065) (0.3368205,0.1041190)(0.8658800,0.4720676)
Using Reduction: (2.231957,2.067158) (1.209411,2.678205)(2.643984,2.185478)
```

The result will vary from one run to another because each time different random number will be generated. The next program demonstrates the use of reduction with `mpi_maxloc`.

```
program reduc_5
use mpi
implicit none ! reduction
integer:: ierr, rank, size, i, np1, total
integer, parameter:: limit=5
real, dimension(limit)::a
integer, dimension(2,limit)::x, y
call mpi_init(ierr)
call mpi_comm_rank(mpi_comm_world, rank, ierr)
call mpi_comm_size(mpi_comm_world, size, ierr)
call random_seed()
call random_number(a)
do i=1, limit
 x(1,i)=1000.0*a(i)
 x(2,i)=rank
end do
print *,'Rank', rank, (x(1,i), i=1,limit)
call mpi_barrier(mpi_comm_world, ierr)
call mpi_reduce(x, y,limit, mpi_2integer, mpi_maxloc,0,mpi_comm_world, ierr)
if (rank==0) then
 print *, 'Maxima ',(y(1,i), i=1,limit)
 print *, 'rank ',(y(2,i), i=1,limit)
endif
```

```
     call mpi_finalize(ierr)
     end program reduc_5
```

The output is as follows:

Rank	0	774	502	245	943	58
Rank	2	467	0	355	464	223
Rank	3	361	437	878	531	757
Rank	1	880	65	723	876	524
Maxima		880	502	878	943	757
rank		1	0	3	0	3

To use the reduction operator `mpi_maxloc` or `mpi_minloc`, two dimensional arrays `x` and `y` are defined. The location `x(1, 1:limit)` is equated to `a(i)`, `i=1,10` and the location `x(2,1:limit)` is equated to corresponding ranks. The returned values are stored in `y(1,1:limit)` and `y(2, 1:limit)`; `y(1,1)`, `y(1,2)` ... contain the maximum among the first, second ... locations of the 'a' arrays for the processors, and `y(2,1)`, `y(2,2)` ... contain the corresponding ranks.

The reduction also allows the user to define his or her own operator for reduction. Consider the following program:

```
     program reduc_6
     use mpi
     implicit none
     integer:: ierr, rank, size, myop, len=5
     integer, dimension(5)::buf, kuf
     call mpi_init(ierr)
     call mpi_comm_rank(mpi_comm_world, rank, ierr)
     call mpi_comm_size(mpi_comm_world, size, ierr)
     select case(rank)
     case (0)
      buf=[1,2,3,4,5] ! four processors are assigned to some values
      print *,rank, buf
     case (1)
      buf=[11,12,13,14,15]
      print *,rank, buf
     case (2)
      buf=[21,22,23,24,25]
      print *,rank, buf
     case (3)
      buf=[31,32,33,34,35]
      print *,rank, buf
     end select
     call mpi_barrier(mpi_comm_world, ierr) !wait till processors initialized
     call mpi_op_create(myfun,.true., myop, ierr)
     call mpi_reduce(buf, kuf, len, mpi_integer, myop,0,mpi_comm_world, ierr)
     if (rank==0) then
      print *, kuf
     endif
     call mpi_op_free(myop, ierr)
     call mpi_finalize(ierr)
     contains
      subroutine myfun(myin, myout, len, dt)
      integer:: len, dt
      integer, dimension(len):: myin, myout
```

```
integer::i
do i=1,len
  myout(i)= myout(i)+myin(i)
enddo
end subroutine myfun
end program reduc_6
```

The subroutine `mpi_op_create` is necessary to create a user-defined global operator handle that can be used with `mpi_reduce` (also with `mpi_allreduce`, `mpi_reduce_scatter` and `mpi_scan`). It takes four parameters: (a) name of user-defined function that will do the actual operation; (b) commute—it is a logical, and if it is true, the expression used in the user-defined function is commutative; otherwise, it is false; (c) `operator handle`; and (d) usual error parameter of MPI routines. The user-defined subroutine has four arguments: `input`, `output`, `length` and `datatype`. The subroutine must have the expression `out(i)=out(i) op in(i)` for `i=1` to `len`. In the program `reduc_5`, the operator `mpi_sum` is replaced by a user-defined handle `myop`, which essentially calls the user-defined subroutine `myfun`. The subroutine `myfun` performs the same operations as program `reduc_2`; the only difference between the two programs is that program `reduc_2` uses the in-built op `mpi_sum`, whereas program `reduc_5` does the same calculation using the user-defined op `myop`. It is perhaps clear that this way the user can create his or her own operator other than one provided by `mpi`. The final program finds the global maximum from 4 processors. Each processor finds its own maximum in the usual way. Using `mpi_reduce`, the global maximum is found out and displayed.

```
program reduc_7
use mpi
implicit none
integer:: ierr, rank, size, i, gmax, mymax
integer, parameter:: limit=5
real, dimension(limit):: a
integer, dimension(limit)::x
call mpi_init(ierr)
call mpi_comm_rank(mpi_comm_world, rank, ierr)
call mpi_comm_size(mpi_comm_world, size, ierr)
call random_seed(); call random_number(a); x=1000.0*a
print *, 'Rank', rank, x
call mpi_barrier(mpi_comm_world, ierr)
do i=1, limit
  mymax=max(mymax, x(i))   ! max in each processor
end do
call mpi_reduce(mymax, gmax,1,mpi_integer, mpi_max,0,mpi_comm_world, ierr)
if (rank==0) then
  print *, 'Global Max Using reduce ', gmax
endif
call mpi_finalize(ierr)
end program reduc_7
```

The output is as follows:

```
Rank    3    394 779 762  720    199
Rank    2    711 468 193  519    596
Rank    0    355 33  597  438    952
Rank    1    180 164 332  260    554
Global Max Using reduce 952
```

22.28 MPI_SCAN

This routine does the reduction in a different way. The reduction is done elementwise, but there is one difference with the usual reduction process. The reduction for the `ith` processor is done using values of processors 0, 1, 2 . . . , i according to the operator given as the argument of scan (in the program scan_1, it is mpi_sum). The `jth` element of the received `ith` processor, in this case, is obtained by summing buf(j) for processor= 0 to i. Consider the fourth element of kuf (kuf is the received array) of the third processor. This is obtained by adding 4 + 14 + 24 + 34 that is 76 (Figure 22.2).

```
program scan_1
use mpi
implicit none
integer:: ierr, rank, size
integer, dimension(5)::buf, kuf
call mpi_init(ierr)
call mpi_comm_rank(mpi_comm_world, rank, ierr)
call mpi_comm_size(mpi_comm_world, size, ierr)
if(rank==0) then
 print *, 'Input to Scan.......'
endif
call mpi_barrier(mpi_comm_world, ierr)
select case(rank)
case (0)
 buf=[1,2,3,4,5] !four processors
 print *,'Rank ',rank, buf
case (1)
 buf=[11,12,13,14,15]
 print *,'Rank ',rank, buf
case (2)
 buf=[21,22,23,24,25]
 print *,'Rank ',rank, buf
case (3)
 buf=[31,32,33,34,35]
 print *,'Rank ',rank, buf
end select
call mpi_barrier(mpi_comm_world, ierr)
if(rank==0) then
print *, 'Output from Scan.....'
endif
call mpi_scan(buf, kuf,5,mpi_integer, mpi_sum, mpi_comm_world, ierr)
print *, 'Rank ', rank, kuf
call mpi_finalize(ierr)
end program scan_1
```

P0	1	2	3	4	5		1	2	3	4	5
P1	11	12	13	14	15		12	14	16	18	20
P2	21	22	23	24	25		33	36	39	42	45
P3	31	32	33	34	35		64	68	72	76	80

buf   kuf

FIGURE 22.2
Results of mpi_scan.

The output is as follows:

```
Input to Scan.......
Rank      0           1           2           3           4           5
Rank      3           31          32          33          34          35
Rank      2           21          22          23          24          25
Rank      1           11          12          13          14          15
```

```
Output from Scan.....
Rank      0         1         2         3         4         5
Rank      3        64        68        72        76        80
Rank      2        33        36        39        42        45
Rank      1        12        14        16        18        20
```

22.29 MPI_ALLREDUCE

This subroutine is same as mpi_reduce except that the results are returned to all the processors (Figure 22.3).

```
program allred_1
use mpi
implicit none
integer, parameter::limit=5
integer:: ierr, rank, size, myop
integer, dimension(limit)::buf, kuf
call mpi_init(ierr)
call mpi_comm_rank(mpi_comm_world, rank, ierr)
call mpi_comm_size(mpi_comm_world, size, ierr)
select case(rank)
case (0)
 buf=[1,2,3,4,5]
 print *, 'Rank ',rank
 print *,buf
case (1)
 buf=[11,12,13,14,15]
 print *, 'Rank ',rank
 print *,buf
case (2)
 buf=[21,22,23,24,25]
 print *, 'Rank ',rank
 print *,buf
case (3)
 buf=[31,32,33,34,35]
 print *, 'Rank ',rank
 print *,buf
end select
call mpi_barrier(mpi_comm_world, ierr)
if (rank==0)then
 print *, '----------------'
end if
kuf=0  ! element by element will be added and transferred to all
call mpi_allreduce(buf, kuf, limit, mpi_integer, mpi_sum, &
  mpi_comm_world, ierr)
select case(rank)
case (0)
 print *, 'Rank ',rank
 print *,kuf
case (1)
 print *, 'Rank ',rank
```

```
      print *,kuf
case (2)
print *, 'Rank ',rank
 print *,kuf
 case (3)
 print *, 'Rank ',rank
 print *,kuf
 end select
 call mpi_finalize(ierr)
 end program allred_1
```

P0	1	2	3	4	5		64	68	72	76	80
P1	11	12	13	14	15		64	68	72	76	80
P2	21	22	23	24	25		64	68	72	76	80
P3	31	32	33	34	35		64	68	72	76	80
			buf						kuf		

FIGURE 22.3
Results of mpi_allreduce.

The output is as follows:

```
Rank                 0
        1           2          3          4          5
Rank                 3
       31          32         33         34         35
Rank                 1
       11          12         13         14         15
Rank                 2
       21          22         23         24         25
----------------
Rank                 0
       64          68         72         76         80
Rank                 3
       64          68         72         76         80
Rank                 1
       64          68         72         76         80
Rank                 2
       64          68         72         76         80
```

The arguments of mpi_allreduce are input buffer, output buffer, count, reduction procedure, communicator and ierr. Final outputs of the array kuf are same for
all the processors. The next example adds 1 to 100 using 4 processors. In the next example,
the partial sums are assembled into sum and the reduced result is sent to all the processors:

```
program allred_2
use mpi
implicit none
integer:: ierr, rank, size, sum=0, i, np1, total
integer, parameter:: limit=100
call mpi_init(ierr)
call mpi_comm_rank(mpi_comm_world, rank, ierr)
call mpi_comm_size(mpi_comm_world, size, ierr)
np1=rank+1
do i=np1, limit, size
 sum=sum+i
end do
call mpi_allreduce(sum, total, 1, mpi_integer, mpi_sum, mpi_comm_world, ierr)
select case(rank)
 case (0)
 print *, 'Rank ', rank, 'Sum of 1 + 2+3... +',limit, '= ', total
 case (1)
 print *, 'Rank ', rank, 'Sum of 1 + 2+3... +',limit, '= ', total
```

```
case (2)
print *, 'Rank ', rank, 'Sum of 1 + 2+3... +',limit, '= ', total
case (3)
print *, 'Rank ', rank, 'Sum of 1 + 2+3... +',limit, '= ', total
end select
call mpi_finalize(ierr)
end program allred_2
```

The output is as follows:

```
Rank       2 Sum of 1 + 2 + 3... +       100 =       5050
Rank       3 Sum of 1 + 2 + 3... +       100 =       5050
Rank       0 Sum of 1 + 2 + 3... +       100 =       5050
Rank       1 Sum of 1 + 2 + 3... +       100 =       5050
```

22.30 MPI_REDUCE_SCATTER_BLOCK

This subroutine also reduces the arrays of each processor element by element according to op, which is mpi_sum in this case. At the end of the operation, it distributes the result depending on the value of the variable recount. It tries to divide the results evenly among the processors. For example, in this case, the recount is 3 and the number of elements is 10. The routines send the first 3 results to processor 0, next 3 to processor 1 and next 3 to processor 2. The last one is sent to processor 3. If the recount is set to 5, first 5 are sent to processor 0 and the rest are sent to processor 1. Processors 2 and 3 do not receive any data. The output of the program with recount equal to 3 is displayed next (Figures 22.4 and 22.5).

P0	1	2	3	4	5	6	7	8	9	10
P1	11	12	13	14	15	16	17	18	19	20
P2	21	22	23	24	25	26	27	28	29	30
P3	31	32	33	34	35	36	37	38	39	40

P0	64	68	72							
P1	76	80	84							
P2	88	92	96							
P3	100	0	0							

FIGURE 22.4
reduce_scatter_block contents of buf.

FIGURE 22.5
reduce_scatter_block contents of kuf.

```
program red_scat_1
use mpi
implicit none
integer:: ierr, rank, size, k
integer, dimension(10)::buf, kuf
integer:: reccount=3
call mpi_init(ierr)
call mpi_comm_rank(mpi_comm_world, rank, ierr)
call mpi_comm_size(mpi_comm_world, size, ierr)
select case(rank)
case (0)
 buf=[1,2,3,4,5,6,7,8,9,10]
 print *,'Rank', rank, buf
case (1)
 buf=[11,12,13,14,15,16,17,18,19,20]
 print *,'Rank', rank, buf
case (2)
```

```
      buf=[21,22,23,24,25,26,27,28,29,30]
      print *,'Rank', rank, buf
     case (3)
      buf=[31,32,33,34,35,36,37,38,39,40]
      print *,'Rank', rank, buf
     end select
     call mpi_barrier(mpi_comm_world, ierr)
     kuf=0
     call mpi_reduce_scatter_block(buf, kuf, reccount, mpi_integer, &
      mpi_sum, mpi_comm_world, ierr)
     select case(rank)
     case (0)
      print *, 'Rank',rank,(kuf(k),k=1,reccount)
     case (1)
      print *, 'Rank',rank,(kuf(k),k=1,reccount)
     case (2)
      print *, 'Rank',rank,(kuf(k),k=1,reccount)
     case (3)
      print *, 'Rank',rank,(kuf(k),k=1,reccount)
     end select
    call mpi_finalize(ierr)
    end program red_scat_1
```

The output is as follows:

Rank	0	1	2	3	4	5	6	7	8	9	10
Rank	2	21	22	23	24	25	26	27	28	29	30
Rank	3	31	32	33	34	35	36	37	38	39	40
Rank	1	11	12	13	14	15	16	17	18	19	20
Rank		3		100		0		0			
Rank		2		88		92		96			
Rank		0		64		68		72			
Rank		1		76		80		84			

22.31 MPI_REDUCE_SCATTER

This routine also reduces the arrays of the processors element by element. However, the result is distributed according to the array recv. In the program red_scat_2, the array recv is defined as [3, 4, 2, 1]. This dictates that processor 0 will receive the first 3 outputs, processor 1 will get the next 4 results, processor 2 will receive the next 2 results and the last result will go to processor 3 (Figure 22.6).

```
program red_scat_2
use mpi
implicit none
integer:: ierr, rank, size, k
integer, dimension(10)::buf, kuf
integer, dimension(4):: recv=[3,4,2,1]
call mpi_init(ierr)
call mpi_comm_rank(mpi_comm_world, rank, ierr)
call mpi_comm_size(mpi_comm_world, size, ierr)
```

P0	64	68	72	
P1	76	80	84	88
P2	92	96		
P3	100			

FIGURE 22.6
kuf after mpi_reduce_scatter.

```
    select case(rank)
    case (0)
     buf=[1,2,3,4,5,6,7,8,9,10]
     print *,'Rank', rank, buf
    case (1)
     buf=[11,12,13,14,15,16,17,18,19,20]
     print *,'Rank', rank, buf
    case (2)
     buf=[21,22,23,24,25,26,27,28,29,30]
     print *,'Rank', rank, buf
    case (3)
     buf=[31,32,33,34,35,36,37,38,39,40]
     print *,'Rank', rank, buf
    end select
    call mpi_barrier(mpi_comm_world, ierr)
    if(rank==0) then
     print *, '------'
    endif
    kuf=0
    call mpi_reduce_scatter(buf, kuf, recv,&
      mpi_integer, mpi_sum, mpi_comm_world, ierr)
    select case(rank)
    case (0)
     print *, 'Rank',rank, (kuf(k),k=1,recv(1))
    case (1)
     print *, 'Rank',rank, (kuf(k),k=1,recv(2))
    case (2)
     print *, 'Rank',rank, (kuf(k),k=1,recv(3))
    case (3)
     print *, 'Rank',rank, (kuf(k),k=1,recv(4))
    end select
   call mpi_finalize(ierr)
   end program red_scat_2
```

The output is as follows:

```
Rank    1    11   12   13   14   15   16   17   18   19   20
Rank    3    31   32   33   34   35   36   37   38   39   40
Rank    2    21   22   23   24   25   26   27   28   29   30
Rank    0    1    2    3    4    5    6    7    8    9    10
------
Rank    2    92   96
Rank    3    100
Rank    0    64   68   72
Rank    1    76   80   84   88
```

22.32 MPI_BROADCAST

This subroutine broadcasts a message from one of the processors to all the processors belonging to the same group. The arguments are the message, size of the message, type of the message, rank of the broadcasting processor, comm handle and error code. At the end of the successful completion of the subroutine, the buffer of the broadcasting processor will be copied to the corresponding buffer of all the participating processors.

```fortran
program bcast_1
use mpi
implicit none
integer:: ierr, rank
character(len=50)::msg='XXX'
call mpi_init(ierr)
call mpi_comm_rank(mpi_comm_world, rank, ierr)
if (rank==0) then
 msg='IACS, Kolkata'
 print *, 'Message Transmitted from Rank 0: ', msg
end if
call mpi_bcast(msg, 13, mpi_character,0, mpi_comm_world, ierr)
select case(rank)
case(0)
 print *, 'After Broadcast Rank ', rank, msg
case(1)
 print *, 'After Broadcast Rank ', rank, msg
case(2)
 print *, 'After Broadcast Rank ', rank, msg
case(3)
 print *, 'After Broadcast Rank ', rank, msg
end select
call mpi_finalize(ierr)
end program bcast_1
```

The output is as follows:

```
Message Transmitted from Rank 0: IACS, Kolkata
After Broadcast Rank 0 IACS, Kolkata
After Broadcast Rank 2 IACS, Kolkata
After Broadcast Rank 3 IACS, Kolkata
After Broadcast Rank 1 IACS, Kolkata
```

22.33 MPI_GATHER

All processors, including the root processor, send their send buffer to the receive buffer. These are stored in the receive buffer in order of rank. The arguments are send buffer, send count, send type, receive buffer, receive count, receive type, root, comm handle and ierr. In the following example, the partial sums of each processor are sent to the nbuf array, and subsequently these partial sums are added to get the total sum (Table 22.5 and Figure 22.7).

```fortran
program gather_1
use mpi
implicit none
integer:: ierr, rank, size, sum=0, i, np1, total=0
integer, parameter:: limit=100
integer, parameter:: tsz=10
```

```
integer, dimension(tsz):: nbuf ! 10 processors
call mpi_init(ierr)
call mpi_comm_rank(mpi_comm_world, rank, ierr)
call mpi_comm_size(mpi_comm_world, size, ierr)
if (rank==0) then
 if (size.ne. tsz) then
  print *, 'This program will be '&
   'tested for', tsz,'processors.... STOP'
 stop
 endif
endif
np1=rank+1
do i=np1, limit, size
 sum=sum+i
end do
call mpi_gather(sum,1,mpi_integer, nbuf,1,mpi_integer, 0,&
 mpi_comm_world, ierr)
if (rank==0) then
 do i=1, size
 total=total+ nbuf(i)
 end do
 print *, '1+2+3+...',limit, &
  total
endif
call mpi_finalize(ierr)
end program gather_1
```

TABLE 22.5

Partial Values of Sum

P0	sum	460
P1	sum	470
P2	sum	480
P3	sum	490
P4	sum	500
P5	sum	510
P6	sum	520
P7	sum	530
P8	sum	540
P9	sum	550

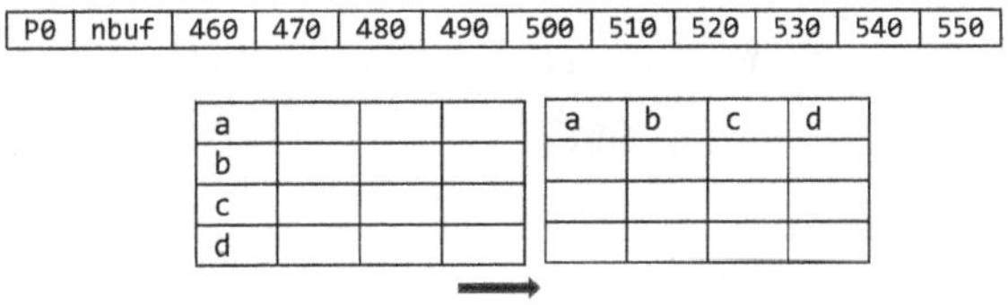

FIGURE 22.7
nbuf after gather operation.

The output is as follows:

```
1+2+3... 100 5050
```

22.34 MPI_ALLGATHER

This subroutine is almost same as mpi_gather except that the results are stored in all processors. In the calling sequence, the root is not required.

```
program all_gather_1
use mpi
implicit none
integer:: ierr, rank, size, sum=0, i, np1, total=0, j
integer, parameter:: limit=100
integer, parameter:: tsz=10
integer, dimension(tsz):: nbuf ! maximum 10 processors
call mpi_init(ierr)
call mpi_comm_rank(mpi_comm_world, rank, ierr)
call mpi_comm_size(mpi_comm_world, size, ierr)
if (rank==0) then
 if (size.ne. tsz) then
   print *, 'This program will be tested for', tsz,'processors... STOP'
 stop
 endif
```

```
    endif
    np1=rank+1
    do i=np1, limit, size
     sum=sum+i
    end do
    call mpi_allgather(sum,1,mpi_integer, nbuf,1,mpi_integer,&
     mpi_comm_world, ierr)
    do j=0, size-1
    if (rank == j) then
     total=0
     do i=1, size
      total=total+ nbuf(i)
     end do
    print *, 'Obtained from rank', j, 'buffer,1+2+3+...',limit, total
    endif
    end do
    call mpi_finalize(ierr)
    end program all_gather_1
```

The output is as follows:

```
Obtained from rank    6 buffer,1+2+3+...    100    5050
Obtained from rank    0 buffer,1+2+3+...    100    5050
Obtained from rank    7 buffer,1+2+3+...    100    5050
Obtained from rank    2 buffer,1+2+3+...    100    5050
Obtained from rank    5 buffer,1+2+3+...    100    5050
Obtained from rank    4 buffer,1+2+3+...    100    5050
Obtained from rank    3 buffer,1+2+3+...    100    5050
Obtained from rank    1 buffer,1+2+3+...    100    5050
Obtained from rank    8 buffer,1+2+3+...    100    5050
Obtained from rank    9 buffer,1+2+3+...    100    5050
```

22.35 MPI_SCATTER

This subroutine is used to distribute data from a particular processor to all the processors in the group. The calling sequence is send buffer, send count, send type, receive buffer, receive count, receive type, rank of the sending process, comm and ierr (Figure 22.8).

```
    program scatter_1
    use mpi
    implicit none
    integer:: ierr, rank, size, sum=0, &
     i, np1, recb
    integer, parameter:: limit=10
    integer, parameter:: tsz=10
    real, dimension(tsz)::kbuf
    real, dimension(2*tsz):: tbuf
    call mpi_init(ierr)
    call mpi_comm_rank(mpi_comm_world, rank, ierr)
    call mpi_comm_size(mpi_comm_world, size, ierr)
```

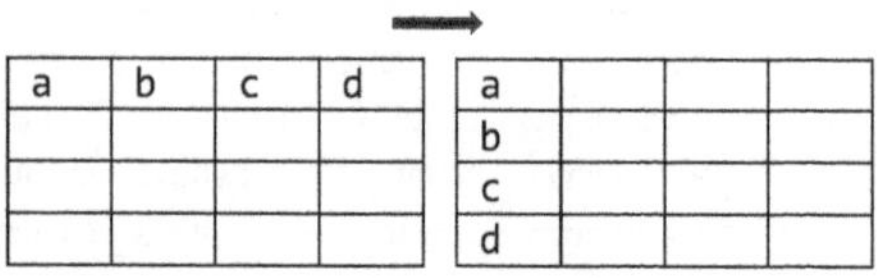

FIGURE 22.8
Scatter.

```
    if (rank==0) then
     if (size.ne. tsz) then
       print *, 'This program will be tested for ', tsz, &
       'processors.... STOP'
     stop
     endif
    endif
    if (rank==0) then
     call random_seed()
     call random_number(tbuf)
     print *, 'Input Random Numbers:'
     print *, tbuf
     print *
     print *, 'Scattered Numbers:'
    endif
    call mpi_scatter(tbuf,2,mpi_real, kbuf,2,mpi_real, 0,&
      mpi_comm_world, ierr)
    do i=0, size-1
     if (rank==i) then
       print *,'Rank ',rank, kbuf(1), kbuf(2)
     end if
    end do
    call mpi_finalize(ierr)
    end program scatter_1
```

The output is as follows:

```
Input Random Numbers:
0.4503087    0.7345567    4.9310908E-02   0.3960744    0.1108879
0.7151722    0.6935076    0.9092112       0.1196077    0.5351772
0.9144143    0.6792647    0.2402318       0.4082699    0.6213635
0.5636747    0.9890307    0.6365536       0.7559081    0.8984796


Scattered Numbers:
Rank    0    0.4503087       0.7345567
Rank    1    4.9310908E-02   0.3960744
Rank    8    0.9890307       0.6365536
Rank    3    0.6935076       0.9092112
Rank    2    0.1108879       0.7151722
Rank    4    0.1196077       0.5351772
Rank    7    0.6213635       0.5636747
Rank    6    0.2402318       0.4082699
Rank    5    0.9144143       0.6792647
Rank    9    0.7559081       0.8984796
```

22.36 MPI_SCATTERV

This subroutine offers more control than scatter. It allows sending of variable amounts of data to each process by having an additional integer array, say, scount. The elements of the array are the amount of data sent to different processors. For example, in the program scatterv_1, 2 real numbers are sent to processor 0 and 3 real numbers are sent to processor 1. Moreover, the source of these numbers is specified by having another argument (integer array) disp, which gives the displacement of the sent numbers from the input buffer by the content of disp. This means that the data transfer to processor 0 would start from tbuf(2) [displacement 1], and as scount(1) is 2, two real numbers will be transferred. Similarly, processor 1 will receive 3 real numbers [scount(2)=3] starting from tbuf(3).

```fortran
program scatterv_1
use mpi
implicit none
integer:: ierr, rank, size, sum=0, i, np1, recb, j
integer, parameter:: limit=30
integer, parameter:: tsz=10
integer, dimension(10)::scount=[2,3,1,4,1,2,3,1,4,2]
integer, dimension(10)::disp=[1,2,3,4,5,6,7,8,9,10]
real, dimension(tsz)::kbuf !initialize
real, dimension(2*tsz):: tbuf
call mpi_init(ierr)
call mpi_comm_rank(mpi_comm_world, rank, ierr)
call mpi_comm_size(mpi_comm_world, size, ierr)
if (rank==0) then
 if (size.ne. tsz) then
 print *, 'This program will be tested for', tsz,&
'processors only.... STOP'
 stop
 endif
endif
if (rank==0) then
 call random_seed()
 call random_number(tbuf)
 print *, 'Input Random Numbers:'
 print *, tbuf
 print *
endif
call mpi_scatterv(tbuf, scount, disp, mpi_real, kbuf,30,mpi_real, 0,&
  mpi_comm_world, ierr)
do i=0, size-1
 if (rank==i) then
  print *,'Rank ',rank, (kbuf(j), j=1,scount(i+1))
 end if
end do
call mpi_finalize(ierr)
end program scatterv_1
```

The output is as follows:

```
Input Random Numders:
    0.3416618      0.8728436      0.9098957      0.1684904      0.4226614
    0.6095756      0.9302325      0.9395322      0.4069776      0.9602857
    0.3967422      0.6538010      0.6165169      0.3158682      8.3719306E-02
    0.8421853      0.9888359      0.7295285      0.1102109      0.2529527
Rank 0 0.8728436 0.9098957
Rank 1 0.9098957 0.1684904 0.4226614
Rank 2 0.1684904
```

```
Rank 3 0.4226614 0.6095756 0.9302325 0.9395322
Rank 5 0.9302325 0.9395322
Rank 4 0.6095756
Rank 6 0.9395322 0.4069776 0.9602857
Rank 7 0.4069776
Rank 8 0.9602857 0.3967422 0.6538010 0.6165169
Rank 9 0.3967422 0.6538010
```

22.37 MPI_ALLTOALL

In this case, each processor sends data to each receiver. The jth block of the ith processor of the sender becomes ith block of the jth processor of the receiver. The arguments are sendbuf, sendcount, datatype, recvbuf, recvcount, datatype, comm and ierr (Figure 22.9).

```
program alltoall_1
use mpi
implicit none
integer:: ierr, rank, size
integer, dimension(4)::buf, kuf
call mpi_init(ierr)
call mpi_comm_rank(mpi_comm_world, rank, ierr)
call mpi_comm_size(mpi_comm_world, size, ierr)
if(rank==0) then
 print *, 'Input to Alltoall.......'
endif
call mpi_barrier(mpi_comm_world, ierr)
select case(rank)
case (0)
 buf=[1,2,3,4] ! four processors are assigned to some values
 print *,'Rank ',rank, buf
case (1)
 buf=[11,12,13,14]
 print *,'Rank ',rank, buf
case (2)
 buf=[21,22,23,24]
 print *,'Rank ',rank, buf
case (3)
 buf=[31,32,33,34]
 print *,'Rank ',rank, buf
end select
call mpi_barrier(mpi_comm_world, ierr)
if(rank=0) then
 print *, 'Output from Alltoall.....'
endif
call mpi_alltoall(buf,1,mpi_integer, kuf,1,mpi_integer, &
    mpi_comm_world, ierr)
print *, 'Rank ', rank, kuf
call mpi_finalize(ierr)
end program alltoall_1
```

P0	1	2	3	4
P1	11	12	13	14
P2	21	22	23	24
P3	31	32	33	34

P0	1	11	21	31
P1	2	12	22	32
P2	3	13	23	33
P3	4	14	24	34

FIGURE 22.9
Altoall.

The output is as follows:

```
Input    to   Alltoall.......
   Rank   0   1   2   3   4
   Rank   1   11  12  13  14
   Rank   3   31  32  33  34
   Rank   2   21  22  23  24

Output from Alltoall.....
   Rank   3   4   14  24  34
   Rank   2   3   13  23  33
   Rank   1   2   12  22  32
   Rank   0   1   11  21  31
```

Here send buffer is buf and send count is 1. The buf of processor 0 will send 1 integer (mpi_integer) to the receive buffer kuf (1) of all the processors. So kuf (1) of processor 0 will get 1 from buf (1) of processor 0, kuf (1) of processor 1 will receive 2 from buf (2) of processor 0, kuf (1) of processor 2 will get 3 from buf (3) of processor 0 and kuf (1) of processor 3 will receive 4 from buf (4) of processor 0. Similarly, kuf (2) of processors 0, 1, 2 and 3 will receive 11, 12, 13 and 14 from buf (1), buf (2), buf (3) and buf (4) of processor 1, respectively. The other outputs may be explained in a similar way.

22.38 Derived Data Types

MPI allows creation of user-defined data structure based on sequences of basic primitive data types. These user-defined data types are contiguous data even if the original data is non-contiguous. This is explained in subsequent sections.

22.39 MPI_TYPE_CONTIGUOUS

This routine replicates data into contiguous locations. The arguments are count (integer), oldtype and newtype. The oldtype and newtype (both integers) are old and new handles. The newtype datatype is created by concatenating count number of old datatypes. In the program deri_1, a new datatype mytype has been created concatenating 3 integers (mpi_integer). To use mytype for communication, it is to be committed through the subroutine mpi_type_commit. It is necessary to call this routine for all user-derived datatypes. This subroutine takes two arguments: mytype (the newly created datatype) and ierr. When the datatype is no longer required, it is set free using the subroutine mpi_free. This routine also takes two arguments: mytype (the newly created datatype) and ierr.

```
      program deri_1
      use mpi
      implicit none
      common/blk1/a, b, c !   contiguous
```

```
integer:: a, b, c
integer::mytype, ierr, rank, size
integer::stat (mpi_status_size)
call mpi_init(ierr)
call mpi_comm_rank(mpi_comm_world, rank, ierr)
call mpi_comm_size(mpi_comm_world, size, ierr)
call mpi_type_contiguous(3, mpi_integer, mytype, ierr)
call mpi_type_commit(mytype, ierr) ! new datatype; 3 integers
if (rank==0) then
 a=10; b=20; c=30
 print *, 'Input'
 print *, 'a, b, c ', a, b, c
endif
call mpi_barrier(mpi_comm_world, ierr)
if (rank==0) then
 call mpi_send(a, 1, mytype, 1, 17, mpi_comm_world, ierr)
 print *, 'Sent to processor 1, a, b, c ', a, b, c
else if (rank==1) then
 call mpi_recv(a, 1, mytype, 0, 17, mpi_comm_world, stat, ierr)
 print *, 'Received, Rank ',rank, 'a, b, c ', a, b, c
end if
call mpi_finalize(ierr)
end program deri_1
```

The output is as follows:

```
Input
 a, b, c                         10    20    30
 Sent to processor   1,   a, b, c    10    20    30
 Received, Rank      1    a, b, c    10    20    30
```

The next example transfers columns 2 to 5 of x (2-D array of size 5 × 5) of processor 0 to processors 1 to 4.

```
program deri_2
use mpi
implicit none
integer, parameter:: leng=5
integer, dimension(leng, leng)::x ! executed with 5 processors
integer, dimension(leng)::y
integer::mytype, ierr, myrank, size, i, t
integer::stat(mpi_status_size)
call mpi_init(ierr)
call mpi_comm_rank(mpi_comm_world, myrank, ierr)
call mpi_comm_size(mpi_comm_world, size, ierr)
t=20
if(myrank==0) then    ! executed with 5 processors
 x=reshape([[(i, i=1,25)]], [leng, leng], order= [1,2]) ! array initialized
 print *, 'x='
 print *, x
endif
call mpi_type_contiguous(leng, mpi_integer, mytype, ierr)
call mpi_type_commit(mytype, ierr)
if (myrank==0) then
 do i=1, size-1
```

```
  call mpi_send(x(1,i+1),1,mytype, mod(i, size),t, mpi_comm_world, ierr)
  enddo
else
  call mpi_recv(y, leng, mpi_integer,0,t, mpi_comm_world, stat, ierr)
  print *, 'Rank ',myrank,'y= ',y
endif
call mpi_type_free(mytype, ierr)
call mpi_finalize(ierr)
end program deri_2
```

The output is as follows:

```
x=
      1         2         3         4         5         6
      7         8         9        10        11        12
     13        14        15        16        17        18
     19        20        21        22        23        24
     25
 Rank   1    y=        6         7         8         9        10
 Rank   2    y=       11        12        13        14        15
 Rank   3    y=       16        17        18        19        20
 Rank   4    y=       21        22        23        24        25
```

The result will be same if `call mpi_recv` is replaced by `call mpi_recv` `(y, 1, mytype, 0, t, mpi_comm_world, stat, ierr)`.

22.40 MPI_TYPE_VECTOR

This is similar to `mpi_type_contiguous`, but it allows having gap (strides) between the blocks. The arguments are number of blocks, number of elements in each block, stride, old datatype and new datatype. The stride is the spacing between the beginnings of each block in terms of the number of elements.

```
program deri_3
use mpi
implicit none ! executed with np=4
integer, parameter:: leng=5
integer, dimension(leng, leng)::x
integer, dimension(6)::y
integer::mytype, ierr, myrank, size, i, t=20
integer::stat(mpi_status_size)
call mpi_init(ierr)
call mpi_comm_rank(mpi_comm_world, myrank, ierr)
call mpi_comm_size(mpi_comm_world, size, ierr)
if (myrank==0) then
 x=reshape([(i, i=1,25)], [leng, leng], order= [1,2])
 print *, 'x='; print *, x
endif
call mpi_type_vector(3,2,4,mpi_integer, mytype, ierr) ! no of blocks =3
call mpi_type_commit(mytype, ierr) ! blocksize=2, stride=4
if (myrank==0) then
```

```
   do i=1, size-1
     call mpi_send(x(1,i+1),1,mytype, mod(i, size),t, mpi_comm_world, ierr)
   enddo
 else
   call mpi_recv(y, 6, mpi_integer, 0, t, mpi_comm_world, stat, ierr)
   print *, 'Rank ',myrank,'y= ',y
 endif
 call mpi_type_free(mytype, ierr)
 call mpi_finalize(ierr)
 end program deri_3
```

The output is as follows:

```
x=
      1          2  3       4       5       6
      7          8  9      10      11      12
     13         14 15      16      17      18
     19         20 21      22      23      24
     25
Rank   1   y=   6    7     10      11      14      15
Rank   3   y=   16   17    20      21      24      25
Rank   2   y=   11   12    15      16      19      20
```

TABLE 22.6

(see text)

1 (x_{11})	6 (x_{12})	11(x_{13})	16(x_{14})	21(x_{15})
2 (x_{21})	7 (x_{22})	12(x_{23})	17(x_{24})	22(x_{25})
3 (x_{31})	8 (x_{32})	13(x_{33})	18(x_{34})	23(x_{35})
4 (x_{41})	9 (x_{42})	14(x_{43})	19(x_{44})	24(x_{45})
5 (x_{51})	10(x_{52})	15(x_{53})	20(x_{54})	25(x_{55})

An explanation of the output is perhaps necessary. It is known that a Fortran array is stored column-wise. Thus, the elements of array x are shown in Table 22.6.

The first 3 arguments of call type_vector are 3, 2 and 4. There will be 3 blocks; each will have 2 elements, and they are separated by 4 elements. We now consider the call mpi_send statement. The first argument is x(1, i+1). When i=1, this is x(1,2) that is the 6th element of the array x. As the count is 2, the first 2 elements of the send buffer are 6 and 7. The stride is 4, so the next two elements are 10 and 11 (count 4 from x_{12}). Similarly, the next two elements are 14 and 15 (count 4 from x_{52}). The other outputs can be explained in a similar way.

22.41 MPI_TYPE_CREATE_HVECTOR

This is almost same as mpi_type_vector. The only difference is that the stride is expressed in bytes. In the preceding program, if the statement call mpi_type_vector(3,2,4,mpi_integer, mytype, ierr) is changed to call mpi_type_create_hvector(3,2,16,mpi_integer, mytype, ierr), identical results will be obtained. Note that for hvector, in this case, the stride is 4 * 4 as integers take 4 bytes.

22.42 MPI_TYPE_INDEXED

For this subroutine, an additional item, an integer array `disp` of input datatype, is required. The calling sequence is `count`, array of block lengths, array of displacements, old datatype, new datatype and `ierr`.

```
program deri_4
use mpi
implicit none
integer, parameter:: leng=5
integer, dimension(leng, leng)::x
integer, dimension(9)::y
integer::mytype, ierr, myrank, size, i, t=22, tb
integer, dimension(3):: bl
integer, dimension(3):: disp
integer::stat(mpi_status_size)
call mpi_init(ierr)
call mpi_comm_rank(mpi_comm_world, myrank, ierr)
call mpi_comm_size(mpi_comm_world, size, ierr)
bl(1)=3; bl(2)=2; bl(3)=4          ! block length [3,2,4]
tb=bl(1)+bl(2)+bl(3)               ! total 9, dimension of y
disp(1)=2; disp(2)=7; disp(3)=13   ! displacement [2, 7, 13]
if (myrank==0) then
 x=reshape([(i, i=1,25)], [leng, leng])
 print *, 'x='
 print *, x
endif
call mpi_type_indexed(3,bl, disp, mpi_integer, mytype, ierr)
call mpi_type_commit(mytype, ierr)
if (myrank==0) then
 do i=1, size-1
   call mpi_send(x(i,1),1,mytype, mod(i, size),t, mpi_comm_world, ierr)
 enddo
else
 call mpi_recv(y, tb, mpi_integer, 0, t, mpi_comm_world, stat, ierr)
 print *, 'Rank ',myrank,'y= ',(y(i), i=1,tb)
endif
call mpi_type_free(mytype, ierr)
call mpi_finalize(ierr)
end program deri_4
```

The output is as follows:

```
X=
        1       2       3       4       5       6
        7       8       9      10      11      12
       13      14      15      16      17      18
       19      20      21      22      23      24
       25
Rank    1   y=  3   4   5       8       9      14      15      16      17
Rank    3   y=  5   6   7      10      11      16      17      18      19
Rank    2   y=  4   5   6       9      10      15      16      17      18
```

An explanation is probably required to understand the output. We discuss the output associated with rank 1 (processor 1). The number of blocks is 3; block lengths are 3, 2 and 4; displacements are 2, 7 and 13. The x array contains integers 1 to 25, stored column-wise. Therefore, the first block will contain 3, 4, 5 [displacement 2 and block length 3], the second block will contain 8 and 9 [displacement 7 and block length 2] and the third block will contain 14, 15, 16 and 17 [displacement 13 and block length 4]. The whole chunk of data 3, 4, 5, 8, 9, 14, 15, 16, 17 is sent to processor 1 by processor 0. The outputs associated with ranks 2 and 3 may be explained in a similar way.

22.43 MPI_TYPE_CREATE_HINDEXED

This routine is almost same as the mpi_type_indexed. Two changes are to be made to the preceding program to use this routine: (a) the array disp should be of mpi_address_kind like integer (kind=mpi_address_kind), dimension(3):: disp (b) and the displacement should be in bytes. Since the integer takes 4 bytes, the displacements calculated in the previous example should be multiplied by 4. Thus, in this case, disp(1)=2 * 4; disp(2)=7 * 4; disp(3)=13 * 4. Output will be same as the previous example when mpi_type_create_hindexed is called with appropriate parameters.

22.44 MPI_TYPE_CREATE_INDEXED_BLOCK

The calling sequence of the subprogram is length of the array of displacement, block count, array of displacements, old type, new type and ierr.

```
program deri_5
use mpi
implicit none
integer, parameter:: leng=5
integer, dimension(leng, leng)::x
integer, dimension(9)::y
integer::mytype, ierr, myrank, size, i, t, tb
integer, dimension(3):: disp
integer::stat(mpi_status_size)
call mpi_init(ierr)
call mpi_comm_rank(mpi_comm_world, myrank, ierr)
call mpi_comm_size(mpi_comm_world, size, ierr)
t=22
disp(1)=2; disp(2)=8; disp(3)=12
if (myrank==0) then
 x=reshape([(i, i=1,25)], [leng, leng])
 print *, 'x='
 print *, x
endif
call mpi_type_create_indexed_block(3,3,disp, mpi_integer, mytype, ierr)
```

```
  call mpi_type_commit(mytype, ierr)
  if (myrank==0) then
   do i=1, size-1
    call mpi_send(x(1,i),1,mytype, mod(i, size),t, mpi_comm_world, ierr)
   enddo
  else
   call mpi_recv(y, 9, mpi_integer, 0, t, mpi_comm_world, stat, ierr)
   print *, 'Rank ',myrank,'y= ', (y(i), i=1,9)
  endif
  call mpi_type_free(mytype, ierr)
  call mpi_finalize(ierr)
  end program deri_5
```

The output i as follows:

x=

	1	2	3	4	5	6			
	7	8	9	10	11	12			
	13	14	15	16	17	18			
	19	20	21	22	23	24			
	25								
Rank	1	y=	3	4	5	9	10	11	13	14	15
Rank	2	y=	8	9	10	14	15	16	18	19	20
Rank	3	y=	13	14	15	19	20	21	23	24	25

Let us try to explain the output corresponding to rank 2 (processor 2). The first argument of mpi_send is x(1,i), which is x(1,2) when i=2. This is the starting location for this data transfer. The number of blocks is 3, and the block size is 3. The displacement will be counted from x(1,2), which is the 6th location of the array, as the array is stored column-wise. Displacement 2 from this location and with the block count 3 means 8, 9, 10. In a similar manner displacement, 8 and 12 with the block size 3 means 14, 15, 16, 18, 19 and 20. Thus, the subroutine mpi_send will send 8, 9, 10, 14, 15, 16, 18, 19, 20 to processor 1. Following the identical logic, the other outputs of ranks 1 and 3 can be explained.

22.45 MPI_TYPE_CREATE_HINDEXED_BLOCK

This routine is almost same as the previous one. There are two differences: the disp array has a kind=mpi_address_kind and the displacement is expressed in bytes. As an integer takes 4 bytes, disp(1), disp(2) and disp(3) of the previous example are to be multiplied by 4. The outputs are same as the previous example when mpi_type_create_hindexed_block is used.

22.46 MPI_TYPE_CREATE_STRUCT

This subroutine is the most general type constructor. The calling sequence of this subroutine is number of blocks, integer array containing block length, integer array containing displacements, integer array containing types, new datatype handle and ierr. We first define a module containing 6 variables of intrinsic type and define a derived type variable mydata. The array dtype describes the type of variables within the module sequentially. All elements are of block length 1. The addresses of the variables within the module from mydata are calculated using the routine mpi_get_address and stored in the array addr. This array is subsequently used to calculate the distance of each variable within the module from mydata. These are stored in the array dist. The routine mpi_create_struct takes 6 arguments: number of blocks, array of block length, array of displacements, array of types, new datatype (handle) and ierr. Next, the routine mpi_commit is called.

```fortran
module mymodule
  type mytype
    integer::a
    real:: b
    double precision::c
    complex::d
    logical::e
    character:: f
  end type mytype
end module mymodule
program deri_6
use mpi
use mymodule
implicit none
type(mytype):: mydata
integer:: mpitype
integer:: ierr, rank, size
integer::status(mpi_status_size)
integer, dimension(6):: blength
integer(kind=mpi_address_kind), dimension(6):: dist
integer, dimension(7):: addr
integer, dimension(6):: dtype
call mpi_init(ierr)
call mpi_comm_rank(mpi_comm_world, rank, ierr)
call mpi_comm_size(mpi_comm_world, size, ierr)
dtype(1)=mpi_integer ! define the datatypes
dtype(2)=mpi_real
dtype(3)=mpi_double_precision
dtype(4)=mpi_complex
dtype(5)=mpi_logical
dtype(6)=mpi_character
blength(1)=1 ! block length is 1 same for all the variables
blength(2)=1
blength(3)=1
blength(4)=1
blength(5)=1
blength(6)=1
call mpi_get_address(mydata, addr(1), ierr) !address
call mpi_get_address(mydata%a, addr(2), ierr)
call mpi_get_address(mydata%b, addr(3), ierr)
call mpi_get_address(mydata%c, addr(4), ierr)
call mpi_get_address(mydata%d, addr(5), ierr)
call mpi_get_address(mydata%e, addr(6), ierr)
call mpi_get_address(mydata%f, addr(7), ierr)
dist(1)=addr(2)-addr(1) ! distance from mydata
```

```
      dist(2)=addr(3)-addr(1)
      dist(3)=addr(4)-addr(1)
      dist(4)=addr(5)-addr(1)
      dist(5)=addr(6)-addr(1)
      dist(6)=addr(7)-addr(1)
      call mpi_type_create_struct(6,blength, dist, dtype, mpitype, ierr)
      call mpi_type_commit(mpitype, ierr) ! committed to use as mpitype
       if(rank==0) then
        mydata%a=10
        mydata%b=3.14
        mydata%c=12.123456789d0
        mydata%d=cmplx(2.7,5.9)
        mydata%e=.true.
        mydata%f='S'
        print *, 'Input Data to Rank ', rank
        print *, mydata%a, mydata%b, mydata%c, mydata%d, mydata%e, mydata%f
       endif
      call mpi_barrier(mpi_comm_world, ierr)
      call mpi_bcast(mydata,1,mpitype,0,mpi_comm_world, ierr)! processor=2
      if (rank==1) then
       print *, 'Data Transferred through Broadcast to rank ', rank
       print *, mydata%a, mydata%b, mydata%c, mydata%d, mydata%e, mydata%f
      endif
      call mpi_finalize(ierr)
      end program deri_6
```

The output is as follows:

```
Input Data to Rank 0
    10   3.140000    12.1234567890000    (2.700000,5.900000) T  S
Data Transferred through Broadcast to rank 1
    10   3.140000    12.1234567890000    (2.700000,5.900000) T  S
```

22.47 MPI_PACK and MPI_UNPACK

The mpi_pack subroutine is used to pack data before sending to a contiguous buffer. The process, which receives this packed data, unpacks the data from the buffer using the mpi_unpack subroutine. The argument to mpi_pack routines are input buffer, number of data item, datatype, output buffer, output buffer size in bytes, current position in buffer in bytes, comm and ierr. The position (say, pos) is set to 0 at the beginning, and the routine increments the variable automatically so that the next variable after packing is added to the end of the previous variable in the buffer. The calling sequence of unpack is input buffer, size of the input buffer in bytes, current position in bytes, output buffer, number of items to be unpacked, datatype, comm and ierr. The program pack_unpack first packs and then unpacks five variables from the buffer ch having size equal to 100 characters. The variable pos is set to 0 before unpacking, and its increment is taken care of by the mpi_unpack routine depending on the type of the variable being unpacked.

```
      program pack_unpack
      use mpi
      implicit none
```

```
real::a, b
integer::c, ierr, size, pos, rank
double precision::d
complex::e
character, dimension(100)::ch
call mpi_init(ierr)
call mpi_comm_rank(mpi_comm_world, rank, ierr)
call mpi_comm_size(mpi_comm_world, size, ierr)
if (rank==0) then
 a=2.453
 b=10.1725
 c=100
 d=2.732d4
 e=cmplx(7.12, 8.57)
 pos=0 ! starts with zero
 call mpi_pack(a, 1, mpi_real, ch, 100, pos, mpi_comm_world, ierr)
 call mpi_pack(b, 1, mpi_real, ch, 100, pos, mpi_comm_world, ierr)
 call mpi_pack(c, 1, mpi_integer, ch, 100, pos, mpi_comm_world, ierr)
 call mpi_pack(d,1,mpi_double_precision, ch,100,pos, mpi_comm_world, ierr)
 call mpi_pack(e,1,mpi_complex, ch,100,pos, mpi_comm_world, ierr)
endif
call mpi_bcast(ch, 100, mpi_packed, 0, mpi_comm_world, ierr)
call mpi_barrier(mpi_comm_world, ierr)
if (rank==1) then
 pos=0 ! again starts with zero before unpacking
 call mpi_unpack(ch, 100, pos, a, 1, mpi_real, mpi_comm_world, ierr)
 call mpi_unpack(ch, 100, pos, b, 1, mpi_real, mpi_comm_world, ierr)
 call mpi_unpack(ch, 100, pos, c, 1, mpi_integer, mpi_comm_world, ierr)
 call mpi_unpack(ch,100,pos, d,1,mpi_double_precision,&
  mpi_comm_world, ierr)
 call mpi_unpack(ch, 100, pos, e, 1, mpi_complex, mpi_comm_world, ierr)
 print *, a, b, c, d, e
endif
call mpi_finalize(ierr)
end program pack_unpack
```

The output is as follows:

```
2.453000    10.17250    100    27320.0000000000    (7.120000,8.570000)
```

22.48 MPI_COMM_SPLIT

This subroutine is used to partition a group, say, mpi_comm_world, into disjoint subgroups according to the second parameter, color. The calling sequence of this subroutine is comm, color (integer), key (integer), new communicator and ierr. The argument color controls the assignment of the subsets. All processes having the same color constitute a single group. The rank within the subgroup is determined by the third parameter, key, of the subroutine. In case of a tie, the original rank of the processes determines the rank within the subgroup. A process may supply its color as mpi_undefined, and in that case, new communicator is returned as mpi_comm_null. The call to mpi_comm_split creates new communicators. The number of such new communicators depends on the number of different colors. All the new communicators have same name.

```fortran
program split_1
use mpi
implicit none
integer:: ierr, myrank, size, myrow, errorcode ! ifort compiler
integer:: rowcolumn, rowrank, rowsize, buf, kuf
integer, parameter::np=16
call mpi_init(ierr)
call mpi_comm_rank(mpi_comm_world, myrank, ierr)
call mpi_comm_size(mpi_comm_world, size, ierr)
if(np.ne. size) then
 if(myrank==0) then
   print *, 'The program will be executed with', np,' processors'
   call mpi_abort(mpi_comm_world, errorcode, ierr)
 endif
else
 myrow=myrank/4 ! integer division, myrow is 0 for myrank=0,1,2,3
 buf=myrank ! there will be 4 colors.
 call mpi_comm_split(mpi_comm_world, myrow, myrank, rowcolumn, ierr)
 call mpi_comm_rank(rowcolumn, rowrank, ierr)! rank within new communicator
 call mpi_comm_size(rowcolumn, rowsize, ierr)! size of each new communicator
 call mpi_allreduce(buf, kuf, 1, mpi_integer, mpi_sum, rowcolumn, ierr)
 print *, 'rowrank ', rowrank, 'rowsize ', rowsize, kuf
 call mpi_comm_free(rowcolumn, ierr) ! allreduce treats 4 groups separately
 call mpi_finalize(ierr)
endif
end program split_1
```

The output is as follows:

```
rowrank    2    rowsize    4    22
rowrank    0    rowsize    4    22
rowrank    1    rowsize    4    22
rowrank    3    rowsize    4    22
rowrank    0    rowsize    4    38
rowrank    0    rowsize    4    54
rowrank    2    rowsize    4    54
rowrank    3    rowsize    4    54
rowrank    3    rowsize    4    38
rowrank    2    rowsize    4    38
rowrank    0    rowsize    4     6
rowrank    1    rowsize    4    38
rowrank    2    rowsize    4     6
rowrank    3    rowsize    4     6
rowrank    1    rowsize    4    54
rowrank    1    rowsize    4     6
```

Readers are advised to trace the program and verify the output.

22.49 Timing Routines

There are two functions in this category. They are `mpi_wtime` and `mpi_wtick`. Neither of these functions requires any argument. The function `mpi_wtime` returns a double precision number, indicating the wall clock time as if timing starts and finishes with a clock.

```
double precision:: d1, d2, elapsed_time
d1=mpi_wtime()
<some calculations>
d2=mpi_wtime()
elasp_time=d2-d1 ! difference; so actual clock does not matter
```

The other function `mpi_wtick` also returns a double precision number. It is the number of seconds between two successive wall clock ticks.

22.50 Programming Examples

Following are a few programming examples for demonstration. In the first program, two people are playing table tennis. The first person sends the ping-pong ball to the second person; the second person receives the ball and sends it back to the first person. The program stops after 10 such exchanges.

```
program ping_pong
use mpi
integer:: myrank, mypartner
integer:: size, ierr
integer, parameter:: exchange=10
integer:: count=0
integer:: status(mpi_status_size)
call mpi_init(ierr)
call mpi_comm_rank(mpi_comm_world, myrank, ierr)
call mpi_comm_size(mpi_comm_world, size, ierr)
if (size.ne. 2) then
 print *, 'Size should be 2, stop'
 stop
endif
mypartner= mod((myrank+1),2)
do while (count < exchange)
 if (myrank == mod(count,2)) then
  count=count+1
  call mpi_send(count,1,mpi_integer, mypartner,25,mpi_comm_world, ierr)
  print *, myrank, 'sent count to ', mypartner, 'Count ', count
 else
  call mpi_recv(count,1,mpi_integer, mypartner,25, &
   mpi_comm_world, status, ierr)
   print *, myrank, 'received count from ', mypartner, 'Count ', count
```

```
      endif
      enddo
      call mpi_finalize(ierr)
      end
```

The output is as follows:

```
0 sent count to 1 Count 1
1 received count from 0 Count 1
1 sent count to 0 Count 2
1 received count from 0 Count 3
1 sent count to 0 Count 4
1 received count from 0 Count 5
1 sent count to 0 Count 6
1 received count from 0 Count 7
1 sent count to 0 Count 8
1 received count from 0 Count 9
1 sent count to 0 Count 10
0 received count from 1 Count 2
0 sent count to 1 Count 3
0 received count from 1 Count 4
0 sent count to 1 Count 5
0 received count from 1 Count 6
0 sent count to 1 Count 7
0 received count from 1 Count 8
0 sent count to 1 Count 9
0 received count from 1 Count 10
```

In the second program, 10 children are standing on the edge of a circle. The first child throws a ring to the second child who accepts the ring and throws the ring to the third child. The process continues until the first child gets back the ring. This is simulated by using mpi_send and mpi_recv and transferring an integer from processor 0 to processor 1 and then processor 1 to processor 2 and so on. The program stops when processor 0 gets back the integer from processor 9.

```
      program ring
      use mpi
      implicit none
      integer:: rnk
      integer:: sz, ierr, dat
      integer:: status(mpi_status_size)
      call mpi_init(ierr)
      call mpi_comm_rank(mpi_comm_world, rnk, ierr)
      call mpi_comm_size(mpi_comm_world, sz, ierr)
      if (rnk.ne. 0) then
       call mpi_recv(dat,1,mpi_integer, rnk-1,25,mpi_comm_world, status, ierr)
       print *, 'Processor',rnk,'received data from ',rnk-1,'data = ', dat
      else
       dat=100
      endif
      call mpi_send(dat,1,mpi_integer, mod(rnk+1,sz),25,mpi_comm_world, ierr)
      if(rnk==0) then
       call mpi_recv(dat,1,mpi_integer, sz-1,25,mpi_comm_world, status, ierr)
       print *, 'Processor',rnk,'received data from ',sz-1,'data = ',dat
```

```
     endif
     call mpi_finalize(ierr)
     end
```

The output is as follows:

```
Process 1 received data from process   0 data =    100
Process 2 received data from process   1 data =    100
Process 3 received data from process   2 data =    100
Process 4 received data from process   3 data =    100
Process 5 received data from process   4 data =    100
Process 6 received data from process   5 data =    100
Process 7 received data from process   6 data =    100
Process 8 received data from process   7 data =    100
Process 9 received data from process   8 data =    100
Process 0 received data from process   9 data =    100
```

The third program calculates the number of prime numbers between 1 and n.

```
     program prime_no
     use mpi
     implicit none
     integer:: myrank, size, ierr, n, total, i, j, primes
     integer:: istart=1, iend=16384, factor=2!total prime no between 1 and n
     integer:: status(mpi_status_size)
     call mpi_init(ierr)
     call mpi_comm_rank(mpi_comm_world, myrank, ierr) ! rank of processors
     call mpi_comm_size(mpi_comm_world, size, ierr) ! total no processors
     n=istart
     if (myrank==0) then
      print *, 'n primes'
     endif
     do while (n<=iend) ! calculate up to iend
      call mpi_bcast(n,1,mpi_integer,0,mpi_comm_world, ierr)
     !   current value of n to all processors
      call prime_calc(n, myrank, size, total)
     ! subroutine calculates prime no for each processor
      call mpi_reduce(total, primes, 1, mpi_integer, mpi_sum, 0, &
       mpi_comm_world, ierr) ! reduced
      if (myrank==0) then
       print *, n, primes
      endif
      n=n*factor ! end point changed
     end do
     call mpi_finalize(ierr)
     contains
     subroutine prime_calc(n, myrank, size, total)
     implicit none
     integer:: n, myrank, size, total
     integer:: prime
     total=0
     do i = 2+myrank, n, size ! job is distributed among the processors
      prime=1  ! due to presence of myrank and size in the do statement
      do j = 2, i-1
        if (mod(i, j)==0) then ! reminder zero, not prime
```

```
      prime=0 !exit from do. no need to check further
      exit
    endif
  end do ! normal exit means it is a prime so add 1 to total
  total=total+prime
end do
end subroutine prime_calc
end program prime_no
```

The output is as follows:

n	primes
1	0
2	1
4	2
8	4
16	6
32	11
64	18
128	31
256	54
512	97
1024	172
2048	309
4096	564
8192	1028
16384	1900

The fourth program integrates a function using the trapezoidal rule. The job is distributed over 4 processors, and finally the partial result is reduced to get the final result.

```
program trap
use mpi
implicit none
integer:: rank, size, np, ierr
integer::n=8000, i
double precision:: h, sum=0.0d0,x, a=3.0d0, b=5.0d0, result
h=(b-a)/dble(n)   ! integrate x² dx, limits 3 to 5
call mpi_init(ierr)
call mpi_comm_rank(mpi_comm_world, rank, ierr)
call mpi_comm_size(mpi_comm_world, size, ierr)
sum=0.0d0
np=rank+1
x=a+np*h
do i=np+1, n, size
 sum=sum+f(x)
 x=x+(size)*h
end do
call mpi_barrier(mpi_comm_world, ierr) ! wait
call mpi_reduce(sum, result, 1, mpi_double_precision,&
 mpi_sum, 0, mpi_comm_world, ierr)
if(rank==0) then
 result=result+0.5d0*(f(a)+f(b))
 result=result*h
 print *, 'Result of Integration: ',result
```

```
      endif
      call mpi_finalize(ierr)
      contains
        double precision function f(x)
        double precision::x
       f=x*x
        return
        end function f
      end program trap
```

The output is as follows:

```
Result of Integration: 32.6666666874981
```

The final program calculates the value of π using the Monte Carlo method. Random numbers between 0 and 1 are generated, and the numbers are checked whether they are inside a circle having a diameter equal to 1 using the relation $(x^2+y^2 <= 1)$. If x^2+y^2 is less than 1, the point is within the circle. If the relation is false, the point is outside the circle. The value of π is the 4 * (number of points inside the circle/total number of points).

```
      program pi_calc
      use mpi
      implicit none
      integer:: ierr, i, myrank, size
      integer, parameter:: sz=selected_int_kind(18)
      integer(kind=sz), parameter:: limit=100000 !random number per processor
      real:: count=0.0, total=0.0, pi
      real, dimension(limit)::x, y
      call mpi_init(ierr)
      call mpi_comm_rank(mpi_comm_world, myrank, ierr)
      call mpi_comm_size(mpi_comm_world, size, ierr)
      call random_seed(); call random_number(x); call random_number(y)
      do i=1, limit
        if(x(i)**2+y(i)**2 <= 1.0) then
        count=count+1
        endif
      enddo
!     print *, count
      call mpi_reduce(count, total,1,mpi_real, mpi_sum,0,mpi_comm_world, ierr)
      call mpi_barrier(mpi_comm_world, ierr)
      if (myrank==0) then
        pi=4.0*total/(size*limit)
        print *, 'The value of π', pi
      endif
      call mpi_finalize(ierr)
      end program pi_calc
```

The output is as follows:

```
The value of π 3.140980
```

The reader may run this program to see the effect of total number of random numbers on the final value of π.

22.51 Final Word

The MPI library contains more than 300 routines; only 10% of them have been discussed in this chapter. It is hoped that the readers will be able to write useful parallel programs to solve their numerical problems using the routines discussed in this chapter. The MPI manual contains detailed description of all the MPI routines. The manual is available on the internet. Interested readers may consult the manual and use the MPI functions not covered in this book.

In this book, three methods were discussed for writing parallel programs. The decision to select a particular method depends on the problem. Any particular problem (e.g., integration by trapezoidal rule) can be programmed with any of the three methods discussed in this book: Coarray, OpenMP and MPI. However, the ease of writing the program, debugging it and its speed of getting to a result will be different. For a small program that may only be used a few times, these methods may not be worth the effort (apart from as a learning exercise), but if you are writing code to fit into a climate model needing a lot of computation on large datasets, speed is of the essence. A deep analysis of the problem to be solved will be needed to decide which method or mix of methods to use, with advice from others who have worked on difficult problems with the methods.

Finally, in parallel computing, increasing the number of processors without any limit does not necessarily increase the speed of computation. The theoretical limit can be assessed by using Amdahl's prescription as reevaluated by Gustafson and Barsis.

Appendix A: ASCII Character Set

Dec	Octal	Hex	Symbol	Dec	Octal	Hex	Symbol	Dec	Octal	Hex	Symbol
0	000	00	^@	46	056	2e	.	92	134	5c	\
1	001	01	^A	47	057	2f	/	93	135	5d	]
2	002	02	^B	48	060	30	0	94	136	5e	^
3	003	03	^C	49	061	31	1	95	137	5f	_
4	004	04	^D	50	062	32	2	96	140	60	`
5	005	05	^E	51	063	33	3	97	141	61	a
6	006	06	^F	52	064	34	4	98	142	62	b
7	007	07	^G	53	065	35	5	99	143	63	c
8	010	08	^H	54	066	36	6	100	144	64	d
9	011	09	^I	55	067	37	7	101	145	65	e
10	012	0a	^J	56	070	38	8	102	146	66	f
11	013	0b	^K	57	071	39	9	103	147	67	g
12	014	0c	^L	58	072	3a	:	104	150	68	h
13	015	0d	^M	59	073	3b	;	105	151	69	i
14	016	0e	^N	60	074	3c	<	106	152	6a	j
15	017	0f	^O	61	075	3d	=	107	153	6b	k
16	020	10	^P	62	076	3e	>	108	154	6c	l
17	021	11	^Q	63	077	3f	?	109	155	6d	m
18	022	12	^R	64	100	40	@	110	156	6e	n
19	023	13	^S	65	101	41	A	111	157	6f	o
20	024	14	^T	66	102	42	B	112	160	70	p
21	025	15	^U	67	103	43	C	113	161	71	q
22	026	16	^V	68	104	44	D	114	162	72	r
23	027	17	^W	69	105	45	E	115	163	73	s
24	030	18	^X	70	106	46	F	116	164	74	t
25	031	19	^Y	71	107	47	G	117	165	75	u
26	032	1a	^Z	72	110	48	H	118	166	76	v
27	033	1b	^[	73	111	49	I	119	167	77	w
28	034	1c	^\	74	112	4a	J	120	170	78	x
29	035	1d	^]	75	113	4b	K	121	171	79	y
30	036	1e	^^	76	114	4c	L	122	172	7a	z
31	037	1f	^_	77	115	4d	M	123	173	7b	{
32	040	20	space	78	116	4e	N	124	174	7c	|
33	041	21	!	79	117	4f	O	125	175	7d	}
34	042	22	"	80	120	50	P	126	176	7e	~
35	043	23	#	81	121	51	Q	127	177	7f	DEL
36	044	24	$	82	122	52	R				
37	045	25	%	83	123	53	S				
38	046	26	&	84	124	54	T				
39	047	27	,	85	125	55	U				
40	050	28	(	86	126	56	V				
41	051	29	)	87	127	57	W				
42	052	2a	*	88	130	58	X				
43	053	2b	+	89	131	59	Y				
44	054	2c	'	90	132	5a	Z				
45	055	2d	–	91	133	5b	[				

Appendix B: Order and Execution Sequence

Program, Function, Subroutine, Module, Submodule or Block data statement		
Use statement		
Import statement		
Format and Entry statements	Implicit none	
	Parameter statements	Implicit statements
	Parameter and Data statements	Derived-type definitions, interface blocks, type declarations, enumeration definitions, procedure declarations, specification statements, and statement function statements
	Data statements	Executable constructs
Contains statements		
Internal or module subprograms		
End statement		

Vertical lines indicate that the statements may be interspaced; horizontal lines indicate that the statements cannot be interspaced.

Appendix C: Library Functions

ABS	(A)	E	Absolute value
ACHAR	(I, [.KIND])	E	Character from ASCII code value
ACOS	(X)	E	Arccosine (inverse cosine) function
ACOSH	(X)	E	Inverse hyperbolic cosine function
ADJUSTL	(STRING)	E	Left-adjusted string value
ADJUSTR	(STRING)	E	Right-adjusted string value
AIMAG	(Z)	E	Imaginary part of a complex number.
AINT	(A [, KIND])	E	Truncation toward 0 to a whole number
ALL	(MASK) or (MASK, DIM)	T	Array reduced by .AND. operator
ALLOCATED	(ARRAY) or (SCALAR) variable	I	Allocation status of allocatable
ANINT	(A [, KIND])	E	Nearest whole number
ANY	(MASK) or (MASK, DIM)	T	Array reduced by .OR. operator
ASIN	(X)	E	Arcsine (inverse sine) function
ASINH	(X)	E	Inverse hyperbolic sine function
ASSOCIATED	(POINTER [, TARGET])	I	Pointer association status inquiry
ATAN	(X) or (Y, X)	E	Arctangent (inverse tangent) function
ATAN2	(Y, X)	E	Arctangent (inverse tangent) function
ATANH	(X)	E	Inverse hyperbolic tangent function
ATOMIC_ADD	(ATOM, VALUE [, STAT])	A	Atomic addition
ATOMIC_AND	(ATOM, VALUE [, STAT])	A	Atomic bitwise AND
ATOMIC_CAS	(ATOM, OLD, COMPARE, VALUE [, STAT])	A	Atomic compare and swap
ATOMIC_DEFINE	(ATOM, VALUE [, STAT])	A	Define a variable atomically
ATOMIC_FETCH_ADD	(ATOM, VALUE, OLD[, STAT])	A	Atomic fetch and add
ATOMIC_FETCH_AND	(ATOM, VALUE, OLD[, STAT])	A	Atomic fetch and bitwise AND
ATOMIC_FETCH_OR	(ATOM, VALUE, OLD[, STAT])	A	Atomic fetch and bitwise OR
ATOMIC_FETCH_XOR	(ATOM, VALUE, OLD[, STAT])	A	Atomic fetch and bitwise exclusive OR
ATOMIC_OR	(ATOM, VALUE [, STAT])	A	Atomic bitwise OR
ATOMIC_REF	(VALUE, ATOM [, STAT])	A	Reference a variable atomically
ATOMIC_XOR	(ATOM, VALUE [, STAT])	A	Atomic bitwise exclusive OR
BESSEL_J0	(X)	E	Bessel function of the first kind, order 0
BESSEL_J1	(X)	E	Bessel function of the first kind, order 1
BESSEL_JN	(N, X)	E	Bessel function of the first kind, order N
BESSEL_JN	(N1, N2, X)	T	Bessel functions of the first kind
BESSEL_Y0	(X)	E	Bessel function of the second kind, order 0

(Continued)

BESSEL_Y1	(X)	E	Bessel function of the second kind, order 1
BESSEL_YN	(N, X)	E	Bessel function of the second kind, order N
BESSEL_YN	(N1, N2, X)	T	Bessel functions of the second kind
BGE	(I, J)	E	Bitwise greater than or equal to
BGT	(I, J)	E	Bitwise greater than
BIT_SIZE	(I)	I	Number of bits in integer model 16.3
BLE	(I, J)	E	Bitwise less than or equal to
BLT	(I, J)	E	Bitwise less than
BTEST	(I, POS)	E	Test single bit in an integer
CEILING	(A [, KIND])	E	Least integer greater than or equal to A
CHAR	(I [, KIND])	E	Character from code value
CMPLX	(X [, KIND]) or (X [, Y, KIND])	E	Conversion to complex type
CO_BROADCAST	(A, SOURCE_IMAGE [, STAT, ERRMSG])	C	Broadcast value to images
CO_MAX	(A [, RESULT_IMAGE, STAT, ERRMSG])	C	Compute maximum value across images
CO_MIN	(A [, RESULT_IMAGE, STAT, ERRMSG])	C	Compute minimum value across images
CO_REDUCE	(A, OPERATION [, RESULT_IMAGE, STAT, ERRMSG])	C	Generalized reduction across images
CO_SUM	(A [, RESULT_IMAGE, STAT, ERRMSG])	C	Compute sum across images
COMMAND_ARGU-MENT_COUNT	()	T	Number of command arguments
CONJG	(Z)	E	Conjugate of a complex number
COS	(X)	E	Cosine function
COSH	(X)	E	Hyperbolic cosine function
COSHAPE	(COARRAY [, KIND])	I	Sizes of codimensions of a coarray
COUNT	(MASK [, DIM, KIND])	T	Logical array reduced by counting true values
CPU_TIME	(TIME)	S	Processor time used
CSHIFT	(ARRAY, SHIFT [, DIM])	T	Circular shift of an array
DATE_AND_TIME	([DATE, TIME, ZONE, VALUES])	S	Date and time
DBLE	(A)	E	Conversion to double precision real
DIGITS	(X)	I	Significant digits in numeric model
DIM	(X, Y)	E	Maximum of X ? Y and zero
DOT_PRODUCT	(VECTOR_A, VECTOR_B)	T	Dot product of two vectors
DPROD	(X, Y)	E	Double precision real product
DSHIFTL	(I, J, SHIFT)	E	Combined left shift
DSHIFTR	(I, J, SHIFT)	E	Combined right shift
EOSHIFT	(ARRAY, SHIFT [, BOUNDARY, DIM])	T	End-off shift of the elements of an array
EPSILON	(X)	I	Model number that is small compared to 1
ERF	(X)	E	Error function

(Continued)

ERFC	(X)	E	Complementary error function
ERFC_SCALED	(X)	E	Scaled complementary error function
EVENT_QUERY	(EVENT, COUNT [, STAT])	S	Query event count
EXECUTE_COM- MAND_LINE	(COMMAND [, WAIT, EXITSTAT, CMDSTAT, CMDMSG])	S	Execute a command line
EXP	(X)	E	Exponential function
EXPONENT	(X)	E	Exponent of floating-point number
EXTENDS_TYPE_OF	(A, MOLD)	I	Dynamic type extension inquiry
FAILED_IMAGES	([TEAM, KIND])	T	Indices of failed images
FINDLOC	(ARRAY, VALUE, DIM [, MASK, KIND, BACK]) or (ARRAY, VALUE [, MASK, KIND, BACK])	T	Location(s) of a specified value
FLOOR	(A [, KIND])	E	Greatest integer less than or equal to A
FRACTION	(X)	E	Fractional part of number
GAMMA	(X)	E	Gamma function
GET_COMMAND	([COMMAND, LENGTH, STATUS, ERRMSG])	S	Get program invocation command
GET_COMMAND_- ARGUMENT	(NUMBER [, VALUE, LENGTH, STATUS, ERRMSG])	S	Get program invocation argument
GET_ENVIRON- MENT_VARIABLE	(NAME [, VALUE, LENGTH, STATUS, TRIM_NAME, ERRMSG])	S	Get environment variable
GET_TEAM	([LEVEL])	T	Team
HUGE	(X)	I	Largest model number
HYPOT	(X, Y)	E	Euclidean distance function
IACHAR	(C [, KIND])	E	ASCII code value for character
IALL	(ARRAY, DIM [, MASK]) or (ARRAY [, MASK])	T	Array reduced by IAND function
IAND	(I, J)	E	Bitwise AND
IANY	(ARRAY, DIM [, MASK]) or (ARRAY [, MASK])	T	Array reduced by IOR function
IBCLR	(I, POS)	E	I with bit POS replaced by 0
IBITS	(I, POS, LEN)	E	Specified sequence of bits
IBSET	(I, POS)	E	I with bit POS replaced by 1
ICHAR	(C [, KIND])	E	Code value for character
IEOR	(I, J)	E	Bitwise exclusive OR
IMAGE_INDEX	(COARRAY, SUB) or (COARRAY, SUB, TEAM) or (COARRAY, SUB, TEAM_NUMBER)	I	Image index from cosubscripts
IMAGE_STATUS	(IMAGE [, TEAM])	T	Image execution state
INDEX	(STRING, SUBSTRING [, BACK, KIND])	E	Character string search
INT	(A [, KIND])	E	Conversion to integer type
IOR	(I, J)	E	Bitwise inclusive OR

(Continued)

IPARITY	(ARRAY, DIM [, MASK]) or (ARRAY [, MASK])	T	Array reduced by IEOR function
ISHFT	(I, SHIFT)	E	Logical shift
ISHFTC	(I, SHIFT [, SIZE])	E	Circular shift of the rightmost bits
IS_CONTIGUOUS	(ARRAY)	I	Array contiguity test (8.5.7)
IS_IOSTAT_END	(I)	E	IOSTAT value test for end of file
IS_IOSTAT_EOR	(I)	E	IOSTAT value test for end of record
KIND	(X)	I	Value of the kind type parameter of X
LBOUND	(ARRAY [, DIM, KIND])	I	Lower bound(s)
LCOBOUND	(COARRAY [, DIM, KIND])	I	Lower cobound(s) of a coarray
LEADZ	(I)	E	Number of leading 0 bits
LEN	(STRING [, KIND])	I	Length of a character entity
LEN_TRIM	(STRING [, KIND])	E	Length without trailing blanks
LGE	(STRING_A, STRING_B)	E	ASCII greater than or equal to
LGT	(STRING_A, STRING_B)	E	ASCII greater than
LLE	(STRING_A, STRING_B)	E	ASCII less than or equal to
LLT	(STRING_A, STRING_B)	E	ASCII less than
LOG	(X)	E	Natural logarithm
LOG_GAMMA	(X)	E	Logarithm of the absolute value of the gamma function
LOG10	(X)	E	Common logarithm
LOGICAL	(L [, KIND])	E	Conversion between kinds of logical
MASKL	(I [, KIND])	E	Left-justified mask
MASKR	(I [, KIND])	E	Right-justified mask
MATMUL	(MATRIX_A, MATRIX_B)	T	Matrix multiplication
MAX	(A1, A2 [, A3,...])	E	Maximum value
MAXEXPONENT	(X)	I	Maximum exponent of a real model
MAXLOC	(ARRAY, DIM [, MASK, KIND, BACK]) or (ARRAY[, MASK, KIND, BACK])	T	Location(s) of maximum value
MAXVAL	(ARRAY, DIM [, MASK]) or (ARRAY [, MASK])	T	Maximum value(s) of array
MERGE	(TSOURCE, FSOURCE, MASK)	E	Expression value selection
MERGE_BITS	(I, J, MASK)	E	Merge of bits under mask
MIN	(A1, A2 [, A3,...])	E	Minimum value
MINEXPONENT	(X)	I	Minimum exponent of a real model
MINLOC	(ARRAY, DIM [, MASK, KIND, BACK]) or (ARRAY[, MASK, KIND, BACK])	T	Location(s) of minimum value
MINVAL	(ARRAY, DIM [, MASK]) or (ARRAY [, MASK])	T	Minimum value(s) of array
MOD	(A, P)	E	Remainder function
MODULO	(A, P)	E	Modulo function
MOVE_ALLOC	(FROM, TO [, STAT, ERRMSG])	PS	Move an allocation
MVBITS	(FROM, FROMPOS, LEN, TO, TOPOS)	ES	Copy a sequence of bits
NEAREST	(X, S)	E	Adjacent machine number
NEW_LINE	(A)	I	Newline character

(Continued)

NINT	(A [, KIND])	E	Nearest integer
NORM2	(X) or (X, DIM)	T	L2 norm of an array
NOT	(I)	E	Bitwise complement
NULL	([MOLD])	T	Disassociated pointer or unallocated entry
NUM_IMAGES	() or (TEAM) or (TEAM_ NUMBER)	T	Number of images
OUT_OF_RANGE	(X, MOLD [, ROUND])	E	Whether a value cannot be converted safely
PACK	(ARRAY, MASK [, VECTOR])	T	Array packed into a vector
PARITY	(MASK) or (MASK, DIM)	T	Array reduced by .NEQV. operator
POPCNT	(I)	E	Number of 1 bits
POPPAR	(I)	E	Parity expressed as 0 or 1
PRECISION	(X)	I	Decimal precision of a real model
PRESENT	(A)	I	Presence of optional argument
PRODUCT	(ARRAY, DIM [, MASK]) or (ARRAY [, MASK])	T	Array reduced by multiplication
RADIX	(X)	I	Base of a numeric model
RANDOM_INIT	(REPEATABLE, IMAGE_ DISTINCT)	S	Initialize the pseudorandom number generator
RANDOM_NUMBER	(HARVEST)	S	Generate pseudorandom number(s)
RANDOM_SEED	([SIZE, PUT, GET])	S	Restart or query the pseudorandom number generator
RANGE	(X)	I	Decimal exponent range of a numeric model (16.4)
RANK	(A)	I	Rank of a data object
REAL	(A [, KIND])	E	Conversion to real type
REDUCE	(ARRAY, OPERATION, DIM [, MASK, IDENTITY, ORDERED]) or (ARRAY, OPERATION [, MASK, IDENTITY, ORDERED])	T	General reduction of array
REPEAT	(STRING, NCOPIES)	T	Repetitive string concatenation
RESHAPE	(SOURCE, SHAPE [, PAD, ORDER])	T	Arbitrary shape array construction
RRSPACING	(X)	E	Reciprocal of relative spacing of model
SAME_TYPE_AS	(A, B)	I	Dynamic type equality test
SCALE	(X, I)	E	Real number scaled by radix power
SCAN	(STRING, SET [, BACK, KIND])	E	Character set membership search
SELECTED_CHAR_ KIND	(NAME)	T	Character kind selection
SELECTED_INT_ KIND	(R)	T	Integer kind selection
SELECTED_REAL_KIND	([P, R, RADIX])	T	Real kind selection
SET_EXPONENT	(X, I)	E	Real value with specified exponent
SHAPE	(SOURCE [, KIND])	I	Shape of an array or a scalar
SHIFTA	(I, SHIFT)	E	Right shift with fill
SHIFTL	(I, SHIFT)	E	Left shift
SHIFTR	(I, SHIFT)	E	Right shift
SIGN	(A, B)	E	Magnitude of A with the sign of B
SIN	(X)	E	Sine function

(Continued)

SINH	(X)	E	Hyperbolic sine function
SIZE	(ARRAY [, DIM, KIND])	I	Size of an array or one extent
SPACING	(X)	E	Spacing of model numbers
SPREAD	(SOURCE, DIM, NCOPIES)	T	Value replicated in a new dimension
SQRT	(X)	E	Square root
STOPPED_IMAGES	([TEAM, KIND])	T	Indices of stopped images
STORAGE_SIZE	(A [, KIND])	I	Storage size in bits
SUM	(ARRAY, DIM [, MASK]) or (ARRAY [, MASK])	T	Array reduced by addition
SYSTEM_CLOCK	([COUNT, COUNT_RATE, COUNT_MAX])	S	Query system clock
TAN	(X)	E	Tangent function
TANH	(X)	E	Hyperbolic tangent function
TEAM_NUMBER	([TEAM])	T	Team number
THIS_IMAGE	([TEAM])	T	Index of the invoking image
THIS_IMAGE	(COARRAY [, TEAM]) or (COARRAY, DIM [,	T	Cosubscript(s) for this image
TINY	(X)	I	Smallest positive model number
TRAILZ	(I)	E	Number of trailing zero bits
TRANSFER	(SOURCE, MOLD [, SIZE])	T	Transfer physical representation
TRANSPOSE	(MATRIX)	T	Transpose of an array of rank 2
TRIM	(STRING)	T	String without trailing blanks
UBOUND	(ARRAY [, DIM, KIND])	I	Upper bound(s)
UCOBOUND	(COARRAY [, DIM, KIND])	I	Upper cobound(s) of a coarray
UNPACK	(VECTOR, MASK, FIELD)	T	Vector unpacked into an array

`IEEE_ARITHMETIC` Module Procedure Summary

Procedure	Arguments	Class	Description
IEEE_CLASS	(X)	E	Classify number
IEEE_COPY_SIGN	(X, Y)	E	Copy sign
IEEE_FMA	(A, B, C)	E	Fused multiply–add operation
IEEE_GET_ROUNDING_MODE	(ROUND_VALUE[, RADIX])	S	Get rounding mode
IEEE_GET_UNDERFLOW_MODE	(GRADUAl)	S	Get underflow mode
IEEE_IN	(A, ROUND [, KIND])	E	Conversion to integer type
IEEE_IS_FINITE	(X)	E	Whether a value is finite
IEEE_IS_NAN	(X)	E	Whether a value is an IEEE NaN
IEEE_IS_NEGATIVE	(X)	E	Whether a value is negative
IEEE_IS_NORMAL	(X)	E	Whether a value is a normal
IEEE_LOGB	(X)	E	Exponent
IEEE_MAX_NUM	(X, Y)	E	Maximum numeric value
IEEE_MAX_NUM_MAG	(X, Y)	E	Maximum magnitude numeric value
IEEE_MIN_NUM	(X, Y)	E	Minimum numeric value
IEEE_MIN_NUM_MAG	(X, Y)	E	Minimum magnitude numeric value
IEEE_NEXT_AFTER	(X, Y)	E	Adjacent machine number
IEEE_NEXT_DOWN	(X)	E	Adjacent lower machine number

(Continued)

Procedure	Arguments	Class	Description
IEEE_NEXT_UP	(X)	E	Adjacent higher machine number
IEEE_QUIET_EQ	(A, B)	E	Quiet compares equal to
IEEE_QUIET_GE	(A, B)	E	Quiet compares greater than or equal to
IEEE_QUIET_GT	(A, B)	E	Quiet compares greater than
IEEE_QUIET_LE	(A, B)	E	Quiet compares less than or equal to
IEEE_QUIET_LT	(A, B)	E	Quiet compares less than or equal to
IEEE_QUIET_NE	(A, B)	E	Quiet compares not equal to
IEEE_REAL	(A [, KIND])	E	Conversion to real type
IEEE_REM	(X, Y)	E	Exact remainder
IEEE_RINT	(X)	E	Round to integer
IEEE_SCALB	(X, I)	E	$X \times 2^I$
IEEE_SELECTED_REAL_KIND	([P, R, RADIX])	T	IEEE kind type parameter value
IEEE_SET_ROUNDING_MODE	(ROUND_VALUE [, RADIX])	S	Set rounding mode

`IEEE_ARITHMETIC` Module Procedure Summary

Procedure	Arguments	Class	Description
IEEE_SET_ UNDERFLOW_ -MODE	(GRADUAL)	S	Set underflow mode
IEEE_SIGNALING_EQ	(A, B)	E	Signaling compares equal to
IEEE_SIGNALING_GE	(A, B)	E	Signaling compares greater than or equal to
IEEE_SIGNALING_GT	(A, B)	E	Signaling compares greater than
IEEE_SIGNALING_LE	(A, B)	E	Signaling compares less than or equal to
IEEE_SIGNALING_LT	(A, B)	E	Signaling compares less than
IEEE_SIGNALING_NE	(A, B)	E	Signaling compares not equal to
IEEE_SIGNBIT	(X)	E	Test sign bit
IEEE_SUPPORT_ DATATYPE	([X])	I	Query IEEE arithmetic support
IEEE_SUPPORT_ DENORMAL	([X])	I	Query subnormal number support
IEEE_SUPPORT_DIVIDE	([X])	I	Query IEEE division support
IEEE_SUPPORT_INF	([X])	I	Query IEEE infinity support
IEEE_SUPPORT_IO	([X])	I	Query IEEE formatting support
IEEE_SUPPORT_NAN	([X])	I	Query IEEE NaN support
IEEE_SUPPORT_ ROUNDING	(ROUND_VALUE[, X])	T	Query IEEE rounding support
IEEE_SUPPORT_SQRT	([X])	I	Query IEEE square root support
IEEE_SUPPORT_ SUBNORMAL	([X])	I	Query subnormal number support
IEEE_SUPPORT_ STANDARD	([X])	I	Query IEEE standard support
IEEE_SUPPORT_UNDER- FLOW_CONTROL	([X])	I	Query underflow control support
IEEE_UNORDERED	(X, Y)	E	Whether two values are unordered
IEEE_VALUE	(X, CLASS)	E	Return number in a class

`IEEE_EXCEPTIONS` Module Procedure Summary

Procedure	Arguments	Class	Description
IEEE_GET_FLAG	(FLAG, FLAG_VALUE)	ES	Get an exception flag
IEEE_GET_HALTING_MODE	(FLAG, HALTING)	ES	Get a halting mode
IEEE_GET_MODES	(MODES)	S	Get floating-point modes
IEEE_GET_STATUS	(STATUS_VALUE)	S	Get floating-point status
IEEE_SET_FLAG	(FLAG, FLAG_VALUE)	PS	Set an exception flag
IEEE_SET_HALTING_MODE	(FLAG, HALTING)	PS	Set a halting mode
IEEE_SET_MODES	(MODES)	S	Set floating-point modes
IEEE_SET_STATUS	(STATUS_VALUE)	S	Restore floating-point status
IEEE_SUPPORT_FLAG	(FLAG [, X])	T	Query exception support
IEEE_SUPPORT_HALTING	(FLAG)	T	Query halting mode support

In the intrinsic module `IEEE_ARITHMETIC`, the elemental functions listed are provided for all real X and Y.

Appendix D: Priority of Operators

Category of Operation	Operators	Precedence
Extension	Defined unary operators	High
Numeric	**	
Numeric	/ or *	
Numeric	Unary + or −	
Numeric	Binary + or −	
Character	//	
Relational	.eq., .ne., .gt., .ge., .lt., .le., ==, /=, >, >=, <, <=	
Logical	.not.	
Logical	.and.	
Logical	.or.	
Logical	.eqv., .neqv.	
Extension	Defined binary operator	Low

Appendix E: Statements Allowed in Scoping Units

Type of Scoping Unit	Mail Program	Module or Submodule	Block Data	External Subprogram	Module Subprogram	Internal Subprogram	Interface Body
USE statement	Y	Y	Y	Y	Y	Y	Y
IMPORT statement	N	N	N	N	N	N	Y
ENTRY statement	N	N	N	Y	Y	N	N
FORMAT statement	Y	N	N	Y	Y	Y	N
Misc Declaration (1)	Y	Y	Y	Y	Y	Y	Y
DATA statement	Y	Y	Y	Y	Y	Y	N
Derived Type definition	Y	Y	Y	Y	Y	Y	Y
Interface block	Y	Y	N	Y	Y	Y	Y
Executable statement	Y	N	N	Y	Y	Y	N
CONTAINS statement	Y	Y	N	Y	Y	N	N
Statement Function statement	Y	N	N	Y	Y	Y	N

(1) PARAMETER, IMPLICIT, Type Declaration, Enumeration Definition, Procedure Declaration and Specification statements.

(2) Module subprogram is not included in the scoping unit of a module.

(3) Y stands for yes and N stands for no.

Appendix F: Obsolescent Features

Several features of Fortran language have been declared as obsolescent. Although compilers still support these features, there is a possibility that these will be de-implemented from the future release of the Fortran compiler. The new Fortran program should not use these statements.

1. Arithmetic IF. Block IF is a better choice.
2. Sharing the terminal statement of nested DO or using the terminal statement other than END DO or CONTINUE.
3. Alternate RETURN. This makes the program much unstructured. This can be replaced by proper use of SELECT CASE.
4. Computed GO TO. SELECT CASE is a better choice.
5. Statement function. Internal subprogram is a better choice.
6. DATA statements within the executable. This is rather confusing. It is better to keep this statement with other specification statements at the beginning of the program unit.
7. Assumed length character function.
8. Fixed form source. Now free form is preferred.
9. CHARACTER * form of character declaration. This can be replaced by CHARACTER(LEN=..).
10. ENTRY statement. This also makes the program much unstructured. This can be avoided by using the module judiciously.
11. Statement function.
12. COMMON and Block data declaration.
13. Equivalence statement.

Appendix G: Deleted Features

The following features of old Fortran have been declared as obsolete or deleted from the present standard:

1. Real and double precision DO index variable
2. Branching to an END IF from outside the block IF
3. PAUSE statement
4. The ASSIGN statement, assigned GO TO and assigned format specifier
5. The edit descriptor H
6. Vertical format control—interpretation of the first character of the output line as a carriage control character

Appendix H: Arithmetic Operations with Complex Numbers

Arithmetic operations (binary) involving two complex numbers x1 = a1 + i b1 and x2 = a2 + i b2 are defined as follows:

```
x1 + x2 = (a1 + a2) + i (b1 + b2)
x1 - x2 = (a1 - a2) + i (b1 - b2)
x1*x2 = (a1.a2 - b1.b2) + i (a1.b2 + b1.a2)
x1/x2 = (a1.a2 + b1.b2)/(a2**2 + b2**2) + i [(b1.a2-a1.b2)/(a2**2 + b2**2)]
x1**x2 = exp((a2 + ib2)*cmplx((log(sqrt(a1**2 + b1**2))),atan2(b1,a1)))
```

where `atan2` is a library function that evaluates the inverse tangent of (b/a).

Appendix I: Mapping of 15 Dimensioned Array into Single Dimension

For a 15-dimensional array of dimensions (I, J, K, L, M, N, O, P, Q, R, S, T, U, V, W), the (i, j, k, l, m, n, o, p, q, r, s, t, u, v, w) th location is mapped onto a single dimension as follows:

```
i+(j-1)*I+(k-1)*I*J+(l-1)*I*J*K+(m-1)*I*J*K*L+(n-1)*I*J*K*L*M+(o-1)*I*J*K
    *L*M*N+
        (p-1)*I*J*K*L*M*N*O+(q-1)*I*J*K*L*M*N*O*P+(r-1)*I*J*K*L*M*N*O*P*Q+
        (s-1)*I*J*K*L*M*N*O*P*Q*R+(t-1)*I*J*K*L*M*N*O*P*Q*R*S+
        (u-1)*I*J*K*L*M*N*O*P*Q*R*S*T+(v-1)*I*J*K*L*M*N*O*P*Q*R*S*T*U+
        (w-1)*I*J*K*L*M*N*O*P*Q*R*S*T*U*V
=i+(I*((j-1)+J*((k-1)+K*((l-1)+L*((m-1)+M*((n-1)+N*((o-1)+O*((p-1)+P*((q-1)+
        Q*((r-1)+R*((s-1)+S*((t-1)+T*((u-1)+U*((v-1)+V*((w-1)))))))))))))))
```

References

ISO/IEC JTC 1/SC 22/WG5/N2137. 6 July 2017.

Varying Length Character String in Fortran—ISO/IEC 1539-2:1994(E).

OpenMP Application Programming Interface, Version 5.0 rev 2, November 2017.

Message-Passing Interface Standard 2018, Draft Specification, Message Passing Interface Forum, November 2018.

Index